# Beginning Algebra

## eText Reference

**Kirk Trigsted**

University of Idaho

**Kevin Bodden**

Lewis & Clark Community College

**Randy Gallaher**

Lewis & Clark Community College

**PEARSON**

Boston   Columbus   Indianapolis   New York   San Francisco   Upper Saddle River
Amsterdam   Cape Town   Dubai   London   Madrid   Milan   Munich   Paris   Montreal   Toronto
Delhi   Mexico   São Paulo   Sydney   Hong Kong   Seoul   Singapore   Taipei   Tokyo

**Editorial Director, Mathematics:** Christine Hoag
**Editor in Chief:** Paul Murphy
**Acquisitions Editor:** Dawn Giovanniello
**Editorial Assistant:** Chelsea Pingree
**Design Manager:** Andrea Nix
**Art Director:** Heather Scott
**Senior Production Supervisor:** Ron Hampton
**Math Media Producer:** Ceci Fleming
**Project Manager, MathXL:** Eileen Moore
**Marketing Manager:** Michelle Renda
**Marketing Assistant:** Alica Frankel
**Market Development Manager:** Dona Kenly
**Manufacturing Manager:** Carol Melville

To my wife, Wendy, and our children, Benjamin, Emily, Gabrielle, and Isabelle.

**—Kirk Trigsted**

To my wife, Angie, and my children, Shawn, Payton, and Logan, for their love and support; and to my parents for their encouragement and consultation.

**—Kevin Bodden**

To my wife, Karen, and my children, Ethan, Ben, and Annie, for their inspiration, encouragement, and patience.

**—Randy Gallaher**

**PEARSON**

**ISBN-13:** 978-0-321-73850-9
**ISBN-10:**    0-321-73850-0

# Contents

# Preface

## Introduction

This *Beginning Algebra* eText Reference contains the pages of Kirk Trigsted, Kevin Bodden, and Randy Gallaher's *Beginning Algebra* eText in a portable, spiral-bound format. The structure of the *Beginning Algebra* eText Reference helps students organize their notes by providing them with space to summarize the videos and animations. Students can also use it to review the eText's material anytime, anywhere.

## A Note to Students

This textbook was created for you! Unlike a traditional text, we wanted to create content that gives you, the reader, the ability to be an active participant in your learning. The eText was specifically designed to be read online. The eText pages have large, readable fonts and were designed without the need to scroll. Throughout the material, we have carefully placed thousands of hyperlinks to definitions, previous chapters, interactive videos, animations, and other important content. Many of the videos and animations allow you to actively participate and interact. Take some time to "click around" and get comfortable with the navigation of the eText and explore its many features.

Before you attempt each homework assignment, read the appropriate section(s) of the eText. At the beginning of each section (starting in Chapter R), you will encounter a feature called Things to Know. This feature includes all the prerequisite objectives from previous sections that you will need to successfully learn the material presented in the new section. If you do not have a basic understanding of these objectives, click on the desired hyperlinks and rework through the objectives, taking advantage of any videos or animations.

An additional feature of the eText is the inclusion of an audio icon (◀))). By clicking on this icon, you can listen to the text as you read. While you read through the pages of the eText, use the margins of this *Beginning Algebra* eText Reference to take notes, summarize key points, and list helpful tips and reminders. An additional option, if made available by your instructor, is to use the *Guided Notebook* to guide your note taking as you work through the eText. The *Guided Notebook* provides more specific direction on how to proceed through the material while providing more space for note taking.

Try testing yourself by working through the corresponding You Try It exercises. Remember, you learn math by doing math! The more time you spend working through the videos, animations, and exercises, the more you will understand. If your instructor assigns homework in MyMathLab or MathXL, rework the exercises until you get them right. Be sure to go back and read the eText at anytime while you are working on your homework. This text caters to your educational needs. We hope you enjoy the experience!

## A Note to Instructors

Today's students have grown up in a technological world where everything is "clickable." We have taught with MyMathLab for many years and have experienced firsthand how fewer and fewer students have been referring to their traditional textbooks, opting instead to use electronic resources. As the use of technology plays an ever increasing role in how we are teaching our students, it seems only natural to have a textbook that mirrors the way our students are learning. We are excited to have written a text that was conceived from the ground up to be used as an online interactive tool. Unlike traditional printed textbooks, this eText was specifically designed for students to read online while working in MyMathLab. Therefore, we wrote this text entirely from an online perspective, with MyMathLab and its existing functionality specifically in mind. Every hyperlink, video, and animation has been strategically integrated within the context of each page to engage the student and maximize his or her learning experience. All of the interactive media was designed so students can actively participate while they learn math. Many of the interactive videos and animations require student interaction, giving specific feedback for incorrect responses.

We are proponents of students learning terms and definitions. Therefore, we have created hyperlinks throughout the text to the definitions of important mathematical terms. We have also inserted a tremendous amount of "just-in-time review" throughout the text by creating links to prerequisite topics. Students have the ability to reference these review materials with just a click of the mouse. You will see that the exercise sets are concise and nonrepetitive. Since MyMathLab will be used as the main assessment engine, there is no need for a repetitive exercise set in the hardcopy version of the text. Every exercise is available for you to assign within MyMathLab or MathXL. For the first time, instructors can assess reading assignments. We have created five conceptual reading assessment questions for every section of the eText, giving the student specific feedback for both correct and incorrect responses. Each feedback message directs them back to the appropriate location for review. Our hope is that every student that uses the eText will have a positive learning experience.

## Acknowledgments

First of all, we want to thank our wives for their loving support, encouragement, and sacrifices. Thanks to our children for being patient when we needed to work and for reminding us when we needed to take breaks. We could not have completed this project without generous understanding and forgiveness from each of you.

Writing this textbook has been one of the most difficult and rewarding experiences of our lives. We truly could not have accomplished our goal without the support and contributions of so many wonderful, talented people. These extraordinary talents deserve every accolade we can give them. From Pearson, we would like to thank Dawn Giovanniello, our editor and friend, for believing in us; Greg Tobin, Chris Hoag, and Paul Murphy for their continued support; Heather Scott and Andrea Nix for their design brilliance; Dona Kenly, Michelle Renda, and Tracy Rabinowitz for their marketing expertise; Ceci Fleming and Ruth Berry for taking care of all the details; Eileen Moore, Rebecca Williams, and the rest of the MathXL development team for an amazing job; Brian Morris and the art team for their great eye for detail; Trish O'kane of PreMedia Global; and Ron Hampton, our production manager, for all his support.

Along the way, we had the help of so many people from around the world and from many different walks of life. The contributions of Susan Knights and Elaine Page cannot be measured. Their detailed reviews and numerous suggestions have improved this text considerably. Our "Friday conversations" were invaluable throughout the entire writing process. Thank you for everything! We appreciate everyone else who made this book a reality: Alice Champlin and Anthony T. J. Kilian at Magnitude Entertainment for creating quality videos and animations; Pamela Trim for reading the entire eText and catching the little things; Andrew Roberts for creating the answers; and Donna Densmore for her work on the *Guided Notebook*.

The following list attempts to recognize all of the reviewers. Please accept our deepest apologies if we have inadvertently omitted anyone. We have benefited greatly from you honest feedback, constructive criticism, and thoughtful suggestions. Each of you has helped us create a resource we truly believe will be effective in helping students learn mathematics. We are deeply grateful and we genuinely thank you all from the bottom of our hearts.

–Kevin Bodden
Randy Gallaher

Gregory Bloxom, Pensacola Junior College

Mary Ann Teel, University of North Texas

Mike May, SJ, Saint Louis University

Jeanette Shotwell, Central Texas College

Sharon Jackson, Brookhaven College

Tom Blackburn, Northeastern Illinois University

Disa Beaty, Rose State College

Charlotte Ellen Bell, Richland College

David Bell, Florida Community College at Jacksonville

Dr. Marsha Lake, Brevard Community College (FL)–Titusville Campus

Stacey Moore, Wallace State Community College–Hanceville, AL

Linda Parrish, Brevard Community College

Natalie Rivera, Estrella Mountain Community College

Denise Nunley, Glendale Community College

Arunas Dagys, Saint Xavier University

Karen Egedy, Baton Rouge Community College

Evelyn Porter, Utah Valley University

Mary Jo Anhalt, Bakersfield College, Delano Center

Debby Casson, University of New Mexico

Shawna Haider, Salt Lake Community College

Carla K. Kulinsky, Salt Lake Community College

Kenneth Takvorian, Mount Wachusett Community College

Phil Veer, Johnson County Community College

Amadou Hama, Kennedy-King College, Chicago

Al Hemenway, Los Angeles Mission College

Tom Pulver, Waubonsee Community College

Thomas Hartfield, Gainesville State College

Shirley Brown, Weatherford College

Rita LaChance, University of Maine–Augusta

Dr. Said Ngobi, Victor Valley College

Dr. Mary Wagner-Krankel, St. Mary's University

Gail Brooks, McLennan Community College

Daniela Kojouharov, Tarrant County College, SE

Amberlee Bosse, University of Phoenix

Mark Chapman, Baker College

Jeremy Coffelt, Blinn College

Donna Densmore, Bossier Parish Community College

Elizabeth Howell, North Central Texas College

Joyce Lindstrom, St. Charles Community College

Frank Marfai

Lisa Sheppard, Lorain County Community College

# CHAPTER R
# Review

# R.1 Operations on Fractions

## OBJECTIVES

**1** Write Fractions in Equivalent Forms

**2** Write Fractions in Simplest Form

**3** Multiply and Divide Fractions

**4** Add and Subtract Fractions

### OBJECTIVE 1    WRITE FRACTIONS IN EQUIVALENT FORMS

A **fraction** is a number of the form $\frac{a}{b}$, where $a$ and $b$ are **whole numbers** and $b \neq 0$. $a$ is called the **numerator**, and $b$ is called the **denominator**. Consider the fraction $\frac{7}{8}$ (see **figure**). The numerator 7 is smaller in value than the denominator 8. Such a fraction is called a **proper fraction** and represents a quantity less than 1.

A fraction with a numerator that is larger than or equal to the denominator, such as $\frac{11}{8}$ or $\frac{8}{8}$, is called an **improper fraction** and represents a quantity of 1 or larger.

A **mixed number** is a number of the form $a\frac{b}{c}$, where $a, b,$ and $c$ are nonzero whole numbers, and $b < c$. $a\frac{b}{c}$ means $a + \frac{b}{c}$, but the $+$ symbol is not written. $a$ is the **whole-number part** and $\frac{b}{c}$ is the **fraction part** of the mixed number.

An improper fraction can always be written as a mixed number or whole number, and vice versa. For example, $\frac{11}{8} = 1\frac{3}{8}$ (see **figure**) and $\frac{8}{8} = 1$ (see **figure**). Launch this **popup box** to review how to convert between mixed numbers and improper fractions.

### Example 1  Converting Between Improper Fractions and Mixed Numbers

*My video summary*  Complete the chart.

|  | Mixed Number or Whole Number | Improper Fraction |
|---|---|---|
| a. | $6\dfrac{7}{8}$ | |
| b. | | $\dfrac{137}{9}$ |
| c. | $9\dfrac{7}{16}$ | |
| d. | | $\dfrac{78}{6}$ |

**Solution**  Try to complete the chart on your own. View the **answers**, or watch this **video** for fully worked solutions.

**You Try It**   Work through this You Try It problem.

Work Exercise 1 in this textbook or in the *MyMathLab* Study Plan.

Two or more fractions are **equivalent fractions** if they represent the same value. For example, $\dfrac{1}{2} = \dfrac{2}{4} = \dfrac{4}{8} = \dfrac{8}{16}$ (see **figure**). Multiplying or dividing both the **numerator** and **denominator** of a fraction by the same nonzero number results in an equivalent fraction.

---

**Property of Equivalent Fractions**

If $a, b$, and $c$ are numbers, then

$$\frac{a}{b} = \frac{a \cdot c}{b \cdot c} \quad \text{and} \quad \frac{a}{b} = \frac{a \div c}{b \div c}$$

as long as $b$ and $c$ are not equal to 0.

---

### Example 2  Finding Equivalent Fractions

*My video summary*  Find the **equivalent fraction** with the given **numerator** or **denominator**.

a. $\dfrac{9}{10}$; denominator 60         b. $\dfrac{24}{56}$; numerator 3

**Solutions**

a. To obtain a denominator of 60, we must multiply the current denominator 10 by 6. This means we must also multiply the current numerator 9 by 6:

$$\frac{9}{10} = \frac{9 \cdot 6}{10 \cdot 6} = \frac{54}{60}$$

b. Try to find this equivalent fraction on your own. View the **answer**, or watch this **video** for fully worked solutions to both parts.

**You Try It**   Work through this **You Try It** problem.

Work Exercises 2–6 in this textbook or in the *MyMathLab* Study Plan.

## OBJECTIVE 2   WRITE FRACTIONS IN SIMPLEST FORM

 *My video summary*

Before proceeding, you may want to watch this **video** for a review of **prime factorization**.

A fraction is in **simplest form**, or **lowest terms**, when the **numerator** and **denominator** have no **factors** in common other than 1. A **simplified fraction** is in simplest form.

The fraction $\dfrac{12}{30}$ is not in simplest form because $12 = 2 \cdot 2 \cdot 3$ and $30 = 2 \cdot 3 \cdot 5$ have the **common factors** 2 and 3. To simplify a fraction, we can divide both the numerator and denominator by each common factor. We show this process by placing a slash mark through each common factor and replacing it with a 1, the result of dividing the common factor by itself. For example,

$$\frac{12}{30} = \frac{2 \cdot 2 \cdot 3}{2 \cdot 3 \cdot 5} = \frac{\overset{1}{\cancel{2}} \cdot 2 \cdot \overset{1}{\cancel{3}}}{\underset{1}{\cancel{2}} \cdot \underset{1}{\cancel{3}} \cdot 5} = \frac{1 \cdot 2 \cdot 1}{1 \cdot 1 \cdot 5} = \frac{2}{5}.$$

---

**Simplifying Fractions**

If $a$, $b$, and $c$ are numbers, then $\dfrac{a \cdot c}{b \cdot c} = \dfrac{a \cdot \overset{1}{\cancel{c}}}{b \cdot \underset{1}{\cancel{c}}} = \dfrac{a \cdot 1}{b \cdot 1} = \dfrac{a}{b}$ for $b \neq 0$ and $c \neq 0$.

---

**CAUTION** Literally replacing each divided-out factor with a 1 can lead to unnecessary clutter. It is, therefore, all right to leave the 1's unwritten. If every factor in the numerator divides out, however, be sure to include a 1 in the numerator of the simplified fraction.

### Example 3  Simplifying Fractions

Write each fraction in **simplest form**.

a. $\dfrac{12}{14}$     b. $\dfrac{28}{45}$     c. $\dfrac{7}{21}$     d. $\dfrac{36}{54}$

### Solutions

a. Factor the numerator and denominator:   $\dfrac{12}{14} = \dfrac{2 \cdot 2 \cdot 3}{2 \cdot 7}$

Divide out the common factor 2:   $= \dfrac{\cancel{2} \cdot 2 \cdot 3}{\cancel{2} \cdot 7}$

Multiply the remaining factors:   $= \dfrac{2 \cdot 3}{7} = \dfrac{6}{7}$

The fraction $\dfrac{12}{14}$ simplifies to $\dfrac{6}{7}$.

**b.** Factor the numerator and denominator: $\dfrac{28}{45} = \dfrac{2 \cdot 2 \cdot 7}{3 \cdot 3 \cdot 5}$

Because there are no common **prime** factors in the numerator and

denominator, the fraction $\dfrac{28}{45}$ is already written in simplest form.

*My video summary*   **c.–d.** Try to simplify these fractions on your own. View the **answers**, or watch this **video** for fully worked solutions.

**You Try It**   **Work through this You Try It problem.**

**Work Exercises 7–12 in this textbook or in the *MyMathLab* Study Plan.**

OBJECTIVE 3   MULTIPLY AND DIVIDE FRACTIONS

---

**Multiplying Fractions**

If $a, b, c,$ and $d$ are numbers and $b$ and $d$ are not equal to 0, then $\dfrac{a}{b} \cdot \dfrac{c}{d} = \dfrac{a \cdot c}{b \cdot d}$.

---

### Example 4  Multiplying Fractions

Multiply and **simplify**.

**a.** $\dfrac{5}{6} \cdot \dfrac{9}{25}$   **b.** $\dfrac{4}{45} \cdot \dfrac{15}{8}$   **c.** $18 \cdot \dfrac{5}{9}$   **d.** $\dfrac{5}{14} \cdot 6\dfrac{3}{10}$

### Solutions

**a.**   Write the original problem:   $\dfrac{5}{6} \cdot \dfrac{9}{25}$

Multiply numerators; multiply denominators:   $= \dfrac{5 \cdot 9}{6 \cdot 25}$

Find the **prime factorization** of each factor:   $= \dfrac{5 \cdot 3 \cdot 3}{2 \cdot 3 \cdot 5 \cdot 5}$

Divide out the common factors:   $= \dfrac{\cancel{5} \cdot \cancel{3} \cdot 3}{2 \cdot \cancel{3} \cdot \cancel{5} \cdot 5}$

$= \dfrac{3}{2 \cdot 5}$

Multiply the remaining factors:   $= \dfrac{3}{10}$

**b.**   Write the original problem:   $\dfrac{4}{45} \cdot \dfrac{15}{8}$

Multiply numerators; multiply denominators:   $= \dfrac{4 \cdot 15}{45 \cdot 8}$

Find the **prime factorization** of each factor:   $= \dfrac{2 \cdot 2 \cdot 3 \cdot 5}{3 \cdot 3 \cdot 5 \cdot 2 \cdot 2 \cdot 2}$

Finish the rest of this problem on your own. See the **complete solution** and check your answer.

*My video summary*   **c.–d.** Try working these problems on your own. See the **answers**, or watch this **video** for detailed solutions.

**Note:** When **multiplying fractions**, if a number in a numerator has a **factor** in common with a number in a denominator, then this **common factor** can be divided out of both numbers first before multiplying the fractions. This generally results in fewer shown steps. See these **alternate solutions** for Examples 4a and 4b.

**You Try It**    Work through this You Try It problem.

**Work Exercises 13–18 in this textbook or in the *MyMathLab* Study Plan.**

Two nonzero numbers are **reciprocals**, or **multiplicative inverses**, if their **product** is 1. For example, the reciprocal of $\dfrac{5}{8}$ is $\dfrac{8}{5}$ because $\dfrac{5}{8} \cdot \dfrac{8}{5} = 1$. Recall that we can find the reciprocal of a nonzero fraction by switching its **numerator** and **denominator**.

To divide two fractions, **multiply** the first fraction by the **reciprocal** of the second fraction.

---

**Dividing Fractions**

If $a, b, c,$ and $d$ are numbers and $b, c,$ and $d$ are not equal to 0, then

$$\frac{a}{b} \div \frac{c}{d} = \frac{a}{b} \cdot \frac{d}{c} = \frac{a \cdot d}{b \cdot c}.$$

---

## Example 5   Dividing Fractions

*My video summary*    ▣   Divide and **simplify**.

**a.** $\dfrac{2}{5} \div \dfrac{7}{3}$      **b.** $\dfrac{8}{10} \div 6$      **c.** $2\dfrac{3}{5} \div 3\dfrac{5}{7}$

**Solutions**

**a.**

| | |
|---|---|
| Write the original problem: | $\dfrac{2}{5} \div \dfrac{7}{3}$ |
| Change to multiplication by the reciprocal: | $= \dfrac{2}{5} \cdot \dfrac{3}{7}$   ←   The reciprocal of $\dfrac{7}{3}$ is $\dfrac{3}{7}$. |
| Multiply numerators; multiply denominators: | $= \dfrac{2 \cdot 3}{5 \cdot 7}$ |
| No factors divide out; multiply factors: | $= \dfrac{6}{35}$ |

**b.**

| | |
|---|---|
| Write the original problem: | $\dfrac{8}{10} \div 6$ |
| Change to multiplication by the reciprocal: | $= \dfrac{8}{10} \cdot \dfrac{1}{6}$   ←   The reciprocal of 6 is $\dfrac{1}{6}$. |

Try to finish this problem on your own, then check your **answer**. Watch the **video** for complete solutions to parts a–c.

**c.** Try working this problem on your own. View the **answer**, or watch the **video** for complete solutions to parts a–c.

When dividing fractions, never divide out common factors before the problem has been changed to multiplication.

**Incorrect**             **Correct**

 **You Try It**    Work through this You Try It problem.

**Work Exercises 19–24 in this textbook or in the *MyMathLab* Study Plan.**

## OBJECTIVE 4   ADD AND SUBTRACT FRACTIONS

Two fractions are **like fractions** if they have the same **denominator**, called a **common denominator**.

---

**Adding and Subtracting Like Fractions**

If $a, b,$ and $c$ are numbers and $c \neq 0$, then

$$\frac{a}{c} + \frac{b}{c} = \frac{a+b}{c} \quad \text{and} \quad \frac{a}{c} - \frac{b}{c} = \frac{a-b}{c}.$$

---

### Example 6   Adding and Subtracting Like Fractions

Add or subtract. **Simplify** if necessary.

a. $\dfrac{2}{15} + \dfrac{11}{15}$             b. $\dfrac{9}{10} - \dfrac{5}{10}$

### Solutions

a. The two like fractions share the common denominator 15. We add the two **numerators** 2 and 11 and place the **sum** over 15:

$$\frac{2}{15} + \frac{11}{15} = \frac{\overbrace{2+11}^{\text{Add numerators}}}{\underset{\uparrow}{15}} = \frac{13}{15}$$

Common denominator

Because 13 and 15 have no **common factors**, the answer is in **simplest form**.

b. $\dfrac{9}{10} - \dfrac{5}{10} = \dfrac{\overbrace{9-5}^{\text{Subtract numerators}}}{10} = \dfrac{4}{10} = \dfrac{\overbrace{2\cdot 2}^{\text{Factor}}}{5\cdot 2} = \dfrac{2\cdot \cancel{2}}{5\cdot \cancel{2}} = \dfrac{2}{5}$

Common denominator

 **You Try It**    Work through this You Try It problem.

**Work Exercises 25–26 in this textbook or in the *MyMathLab* Study Plan.**

To add or subtract fractions with different denominators, we must first write each fraction as an **equivalent fraction** with a **common denominator**. We typically use the **least common denominator (LCD)** which is of the smallest number that is **divisible** by all the denominators in the group. Launch this **popup box** to review steps for finding the LCD.

---

**Steps for Adding or Subtracting Fractions**

**Step 1.** If necessary, find the LCD of the fractions.

**Step 2.** Write each fraction as an equivalent fraction with the LCD as its denominator.

**Step 3.** Add or subtract the numerators of the **like fractions** and write the result over the LCD.

**Step 4.** Simplify if necessary.

---

### Example 7  Adding and Subtracting Fractions

*My interactive video summary*

 Add or subtract. **Simplify** if necessary.

a. $\dfrac{1}{12} + \dfrac{4}{9}$    b. $\dfrac{11}{30} - \dfrac{7}{24}$    c. $1\dfrac{5}{6} + 4\dfrac{7}{9}$

### Solutions

a. Launch this **popup box** to see that the **LCD** is 36.

$$\text{Begin with the original problem:} \quad \frac{1}{12} + \frac{4}{9}$$

$$\text{Write each fraction with a denominator of LCD} = 36: \quad = \frac{1 \cdot 3}{12 \cdot 3} + \frac{4 \cdot 4}{9 \cdot 4}$$

$$= \frac{3}{36} + \frac{16}{36}$$

$$\text{Add numerators } 3 + 16 = 19; \text{ write result over } 36: \quad = \frac{19}{36}$$

b. Launch this **popup box** to see that the LCD is 120. Try to finish the problem on your own, then check your **answer**. Watch the **interactive video** to see a complete solution.

c. Try working this problem on your own, then check your **answer**. Watch the **interactive video** to see a complete solution.

**You Try It**    Work through this You Try It problem.

**Work Exercises 27–32 in this textbook or in the *MyMathLab*  Study Plan.**

# R.1 Exercises

**You Try It**

**1.** Complete the chart by providing the equivalent mixed number, whole number, or improper fraction.

|  | Mixed Number or Whole Number | Improper Fraction |
|---|---|---|
| **a.** | $7\dfrac{5}{9}$ |  |
| **b.** |  | $\dfrac{245}{8}$ |
| **c.** | $8\dfrac{5}{12}$ |  |
| **d.** |  | $\dfrac{234}{9}$ |

In Exercises 2–6, find the equivalent fraction with the given numerator or denominator.

**You Try It**

**2.** $\dfrac{4}{5}$; denominator 15        **3.** $\dfrac{8}{48}$; denominator 12        **4.** $\dfrac{7}{15}$; numerator 105

**5.** $\dfrac{35}{140}$; numerator 7        **6.** 13; denominator 6

In Exercises 7–12, write each fraction in simplest form.

**You Try It**

**7.** $\dfrac{6}{8}$        **8.** $\dfrac{24}{54}$        **9.** $\dfrac{16}{45}$

**10.** $\dfrac{15}{60}$        **11.** $\dfrac{88}{120}$        **12.** $\dfrac{105}{140}$

In Exercises 13–18, multiply and simplify.

**You Try It**

**13.** $\dfrac{3}{7}\cdot\dfrac{5}{8}$        **14.** $\dfrac{5}{9}\cdot\dfrac{6}{10}$        **15.** $\dfrac{9}{10}\cdot\dfrac{35}{6}$

**16.** $\dfrac{75}{125}\cdot\dfrac{5}{36}$        **17.** $32\cdot\dfrac{5}{8}$        **18.** $2\dfrac{1}{4}\cdot3\dfrac{1}{5}$

In Exercises 19–24, divide and simplify.

**You Try It**

**19.** $\dfrac{3}{8}\div\dfrac{2}{5}$        **20.** $\dfrac{7}{18}\div\dfrac{2}{3}$        **21.** $\dfrac{18}{28}\div\dfrac{10}{16}$

**22.** $\dfrac{20}{81}\div8$        **23.** $36\div\dfrac{9}{4}$        **24.** $4\dfrac{1}{8}\div2\dfrac{3}{4}$

In Exercises 25–32, add or subtract. Simplify if necessary.

**You Try It**

**25.** $\dfrac{7}{32}+\dfrac{11}{32}$        **26.** $\dfrac{23}{30}-\dfrac{7}{30}$        **27.** $\dfrac{5}{6}+\dfrac{2}{15}$

**28.** $\dfrac{4}{9}+\dfrac{7}{12}$        **29.** $\dfrac{17}{18}-\dfrac{4}{9}$        **30.** $\dfrac{11}{30}-\dfrac{7}{24}$

**You Try It**

**31.** $3\dfrac{5}{9}+1\dfrac{7}{12}$        **32.** $7\dfrac{5}{6}-2\dfrac{14}{15}$

# R.2 Decimals and Percents

## THINGS TO KNOW

Before working through this section, be sure you are familiar with the following concepts:

|  |  | VIDEO | ANIMATION | INTERACTIVE |
|--|--|-------|-----------|-------------|

**You Try It**

1. Write Fractions in Simplest Form
   (Section R.1, **Objective 2**)

**You Try It**

2. Multiply and Divide Fractions
   (Section R.1, **Objective 3**)

## OBJECTIVES

**1** Add and Subtract Decimals

**2** Multiply and Divide Decimals

**3** Convert Between Decimals and Fractions

**4** Convert Between Percents and Decimals or Fractions

---

OBJECTIVE 1   ADD AND SUBTRACT DECIMALS

As with **fractions**, a **decimal number**, or simply a **decimal**, can be used to describe a part of a whole. The position of each **digit** in a decimal determines its **place value** as shown in the place-value chart below.

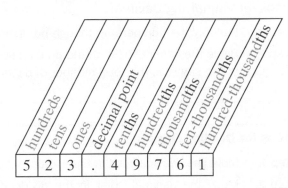

**Place Value Chart**

We add and subtract **decimals** much like we add and subtract **whole numbers**.

---

**Steps for Adding or Subtracting Decimals**

**Step 1.**   Write the numbers vertically so that the **decimal points** line up.

**Step 2.**   Add or subtract the corresponding place values, working from right to left. Carry or borrow as needed.

**Step 3.**   Write the decimal point in the answer so that it lines up with the decimal points in the problem.

---

**Note:** If one number has more decimal places than the other, we can fill in the "missing" place values with 0's for placeholders.

## Example 1  Adding and Subtracting Decimals

 *My video summary*   Add or subtract.

**a.** $16.93 + 4.695$

**b.** $53.72 - 6.589$

**c.** $143 + 8.67$

**d.** $7.2 - 0.036$

**Solutions**  We follow the **steps for adding or subtracting decimals**.

```
        1 1 1   ← Carrying
      16.930  ← 0 place holder
a.   + 4.695
      21.625
         ↑
   Align decimal points
```

```
      4  13  6  11  10  ← Borrowing
      5  3 . 7  2  0  ← 0 place holder
b.   -  6 . 5  8  9
      4 7 . 1  3  1
             ↑
      Align decimal points
```

**c.–d.** Try working these problems on your own, then check your **answers**. Watch the **video** for complete solutions to all four problems.

**You Try It**  Work through this You Try It problem.

**Work Exercises 1–6 in this textbook or in the** *MyMathLab* **Study Plan.**

**OBJECTIVE 2  MULTIPLY AND DIVIDE DECIMALS**

Multiplying and dividing **decimals** are similar to multiplying and dividing **whole numbers**, but we must be careful to write the **decimal point** in the proper position.

---

**Steps for Multiplying Decimals**

**Step 1.**  Multiply the numbers as though no decimal points were present.

**Step 2.**  Place the decimal point in the **product** so that the number of **digits** to its right is the same as the total number of digits to the right of the decimal points in the **factors**.

---

**Steps for Dividing Decimals**

**Step 1.**  Move the decimal point in the **divisor** to the right until it is behind the last digit.

**Step 2.**  Move the decimal point in the **dividend** to the right the same number of places as the decimal point was moved in Step 1. Add zeros as placeholders if necessary.

**Step 3.**  Divide. Place the decimal point in the **quotient** directly above the location of the moved decimal point in the dividend.

---

**Note:** When dividing decimals, the quotient will be either a **terminating** or a **repeating** **decimal**.

## Example 2  Multiplying and Dividing Decimals

Multiply or divide.

**a.** $2.5 \cdot 3.7$

**b.** $8.925 \div 7.5$

**c.** $1.49 \cdot 0.04$

**d.** $7.5 \div 0.33$

## Solutions

a. We follow the **steps for multiplying decimals**. The **factors** have a total of two **digits** behind the decimal points, so the **product** will have two digits behind its decimal point.

$$
\begin{array}{r}
2.5 \leftarrow 1 \text{ decimal place} \\
\times\ 3.7 \leftarrow 1 \text{ decimal place} \\
\hline
175 \\
750 \\
\hline
9.25 \leftarrow 2 \text{ decimal places}
\end{array}
$$

b. We follow the **steps for dividing decimals**. The **divisor** has one **digit** behind the decimal point. So, we move the decimal points in both the divisor and **dividend** one place to the right.

$$
7.5\overline{)8.925} \qquad \Rightarrow \qquad
\begin{array}{r}
1.19 \\
75.\overline{)89.25} \\
\underline{75}\phantom{.00} \\
142\phantom{0} \\
\underline{75}\phantom{0} \\
675 \\
\underline{675} \\
0 \leftarrow \text{Terminates}
\end{array}
$$

The quotient is 1.19.

 ▣ **c.–d.** Try working these problems on your own. View the **answers**, or watch this **video** for detailed solutions.

**You Try It**   Work through this You Try It problem.

Work Exercises 7–14 in this textbook or in the *MyMathLab* Study Plan.

---

OBJECTIVE 3   CONVERT BETWEEN DECIMALS AND FRACTIONS

Notice the **decimal** 0.7 and the **fraction** $\dfrac{7}{10}$ are both read *seven tenths*. Reading decimals naturally enables us to write them as fractions. Launch this **popup box** to review complete steps.

### Example 3  Writing Decimals as Fractions

*My video summary*   ▣ Write each decimal as a fraction. **Simplify** if necessary.

a.  0.23          b.  0.125          c.  8.6

### Solutions

a.  0.23 is read *twenty-three hundredths*. In fraction form, we write $\dfrac{23}{100}$.

b.–c.  Try working these problems on your own. View the **answers**, or watch this **video** for complete solutions to all three parts.

**You Try It**   Work through this You Try It problem.

**Work Exercises 15–18 in this textbook or in the _MyMathLab_ Study Plan.**

To write a fraction as a decimal, divide the **numerator** by the **denominator**. The **quotient** will either be a **terminating** or **repeating decimal**.

### Example 4  Writing Fractions as Decimals

*My video summary*    Write each fraction as a decimal.

a. $\dfrac{9}{11}$    b. $\dfrac{7}{20}$    c. $\dfrac{19}{8}$

**Solutions**  We divide the **numerator** by the **denominator**. Note that **decimal points** are understood to be behind the last digit of each whole number. We can place zeros behind the decimal point to continue dividing.

a. $\dfrac{9}{11}$   $\Rightarrow$

$$
\begin{array}{r}
0.8181... \\
11\overline{)9.0000} \\
88 \\
\overline{\phantom{0}20} \\
11 \\
\overline{\phantom{0}90} \\
88 \\
\overline{\phantom{0}20} \\
11 \\
\overline{\phantom{00}9}
\end{array}
$$

$\left. \right\}$ Repeating pattern

The pattern "81" keeps **repeating**, so $\dfrac{9}{11} = 0.\overline{81}$.

**b.–c.** Try working these problems on your own, then check your **answers**. Watch this **video** to view detailed solutions for all three parts.

**You Try It**   Work through this You Try It problem.

**Work Exercises 19–22 in this textbook or in the _MyMathLab_ Study Plan.**

CAUTION When writing a repeating decimal, use as few digits as possible to show the repeating pattern.

| Correct | Incorrect | Incorrect |
|---------|-----------|-----------|
| $0.\overline{81}$ | $0.\overline{8181}$ | $0.8181$ |

OBJECTIVE 4   CONVERT BETWEEN PERCENTS AND DECIMALS OR FRACTIONS

Recall that the word **percent** means "per hundred" or "out of 100." A percent is denoted by the **percent symbol** %. Multiplying or dividing by 100 % is equivalent to multiplying or dividing by 1. So, a value does not change when it is multiplied or divided by 100 %. We can use this fact to convert between percents and **decimals** or **fractions**.

---

### Writing a Percent as a Decimal or Fraction

To write a percent as a decimal or fraction, divide the percent by 100% and write the answer in either decimal or fraction form.

**Note:** This is the same as dropping the percent symbol (%) and either moving the decimal point two places to the left or writing the number over 100 and simplifying the fraction.

---

### Writing a Decimal or Fraction as a Percent

To write a decimal or fraction as a percent, multiply it by 100%.

**Note:** This is the same as either moving the decimal point two places to the right or multiplying the fraction by 100, and including the percent symbol (%).

---

**CAUTION** Percents that are less than 100% will have decimal forms that are less than 1 and fraction forms that are **proper fractions**. Percents that are larger than 100% will have decimal forms that are larger than 1 and fraction forms that are **improper fractions** (or mixed numbers).

### Example 5  Converting Between Percents and Decimals or Fractions

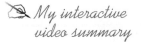

 Complete the chart with appropriate **percents**, **decimals**, and **fractions**.

|     | Percent | Decimal | Fraction |
| --- | --- | --- | --- |
| a. | 28% | | |
| b. | | | $\dfrac{17}{40}$ |
| c. | | 0.875 | |
| d. | | | $\dfrac{13}{12}$ |

### Solution

a. To convert 28% to a decimal or fraction, divide it by 100%, and write the answer in decimal or fraction form accordingly.

$$28\% = \frac{\overset{\text{Divide by 100\%}}{28.\%}}{100\%} = \frac{28.\cancel{\%}}{100\cancel{\%}} = 0.28$$

Two places to the left

$$28\% = \frac{\overset{\text{Divide by 100\%}}{28\%}}{100\%} = \frac{28\cancel{\%}}{100\cancel{\%}} = \frac{28}{100} = \frac{2\cdot2\cdot7}{2\cdot2\cdot5\cdot5} = \frac{\cancel{2}\cdot\cancel{2}\cdot7}{\cancel{2}\cdot\cancel{2}\cdot5\cdot5} = \frac{7}{25}$$

**b.** To convert $\dfrac{17}{40}$ to a percent, multiply by 100%.

$$\frac{17}{40} = \frac{17}{40} \cdot 100\% = \left( \frac{17}{\underset{2}{\cancel{40}}} \cdot \frac{\overset{5}{\cancel{100}}}{1} \right)\% = \frac{85}{2}\% = 42\frac{1}{2}\% \text{ or } 42.5\%$$

To convert $\dfrac{17}{40}$ to a decimal, we divide the 17 by 40.

$$\frac{17}{40} \quad \Rightarrow \quad \begin{array}{r} 0.425 \\ 40\overline{)17.000} \\ \underline{16\ 0} \\ 1\ 00 \\ \underline{80} \\ 200 \\ \underline{200} \\ 0 \end{array} \leftarrow \text{Terminates}$$

**c.–d.** Try to complete the chart on your own. View the **answers**, or watch this **interactive video** for fully worked solutions to all four parts.

**You Try It**    Work through this You Try It problem.

**Work Exercises 23–28 in this textbook or in the *MyMathLab* Study Plan.**

# R.2 Exercises

In Exercises 1–6, add or subtract.

**You Try It**

**1.** $82.52 + 71.69$     **2.** $24.39 + 3.184$      **3.** $91.42 - 8.653$

**4.** $96.385 - 73.64$   **5.** $19.4 + 1.66$      **6.** $245 - 8.63$

In Exercises 7-14, multiply or divide. Write any repeating decimals with as few digits as possible.

**You Try It**

**7.** $4.2 \cdot 3.9$        **8.** $7.2 \cdot 0.16$         **9.** $0.132 \cdot 0.6$         **10.** $1.57 \cdot 0.26$

**11.** $45.24 \div 5.2$     **12.** $11.5 \div 1.38$      **13.** $0.726 \div 0.33$      **14.** $3.7 \div 1.1$

In Exercises 15–18, write each decimal as a fraction. Simplify if necessary.

**You Try It**

**15.** $0.57$          **16.** $0.625$          **17.** $0.072$          **18.** $9.6$

In Exercises 19–22, write each fraction as a decimal. Write any repeating decimals with as few digits as possible.

**You Try It**

**19.** $\dfrac{3}{8}$          **20.** $\dfrac{4}{15}$          **21.** $\dfrac{37}{16}$          **22.** $\dfrac{8}{27}$

**You Try It**

In Exercises 23–28, convert between percents, decimals, and fractions to complete the chart. Write any repeating decimals with as few digits as possible.

|      | Percent | Decimal | Fraction |
|------|---------|---------|----------|
| **23.** | 7.5% | | |
| **24.** | | | $\dfrac{13}{125}$ |
| **25.** | | 0.36 | |
| **26.** | 140% | | |
| **27.** | | | $\dfrac{16}{15}$ |
| **28.** | | 0.072 | |

CHAPTER ONE

# Real Numbers and Algebraic Expressions

## 1.1  The Real Number System

### THINGS TO KNOW

Before working through this section, be sure you are familiar with the following concepts:

|  | VIDEO | ANIMATION | INTERACTIVE |
|--|-------|-----------|-------------|

You Try It

1.  Write Fractions in Simplest Form (Section R.1, **Objective 2**)

You Try It

2.  Convert Between Percents and Decimals or Fractions (Section R.2, **Objective 4**)

### OBJECTIVES

1  Classify Real Numbers

2  Plot Real Numbers on a Number Line

3  Find the Opposite of a Real Number

4  Find the Absolute Value of a Real Number

5  Use Inequality Symbols to Order Real Numbers

6  Translate Word Statements Involving Inequalities

### OBJECTIVE 1  CLASSIFY REAL NUMBERS

A **set** is a collection of objects. Each object in a set is called an **element** or a **member** of the set. Typically, capital letters name sets, and braces { } group the list of elements in a set. For example, the days of the week form a set. If we name this set *D*, we write:

$$D = \{\text{Sunday, Monday, Tuesday, Wednesday, Thursday, Friday, Saturday}\}.$$

If *S* represents the set of single-digit numbers, then

$$S = \{0, 1, 2, 3, 4, 5, 6, 7, 8, 9\}.$$

A set that contains no elements is called an **empty set** or **null set**. We represent this set with a pair of **empty braces** { } or the **null symbol** ∅. For example, let *A* represent the set of all days of the week that begin with the letter B. Because there are no such days, we write $A = \{\ \}$ or $A = \varnothing$.

 The symbol {∅} is not the empty set. It is the set containing the element ∅. Be sure to use the symbols { } or ∅ to represent the empty set, not {∅}.

Both *D* and *S* are **finite sets**, meaning that each **set** has a fixed number of **elements** (7 days, 10 single-digit numbers). When sets are **infinite**, they have an unlimited number of elements. For example, the set of positive **even numbers** is written as

$$E = \{2, 4, 6, 8, \ldots\}.$$

The three dots, . . . , are called an *ellipsis* and indicate the given pattern continues. We encounter a variety of numbers every day. A plane might carry 120 passengers, and a netbook may cost $349.99. Your friend may pay $\frac{1}{2}$ the rent for your apartment.

The numbers 120, 349.99, and $\frac{1}{2}$ are examples of *real numbers*. Let's look at the smaller sets of numbers that make up the set of real numbers.

When counting books on a shelf, we count them as 1, 2, 3, 4, and so on. These values form the set of *natural numbers*, or *counting numbers*, represented by the symbol $\mathbb{N}$.

---

**Definition    Natural Numbers**

A **natural number**, or **counting number**, is an element of the set $\mathbb{N} = \{1, 2, 3, 4, 5, \ldots\}$.

---

Adding the element 0 to the set of natural numbers gives us the set of *whole numbers*, represented by the symbol $\mathbb{W}$.

---

**Definition    Whole Numbers**

A **whole number** is an element of the set $\mathbb{W} = \{0, 1, 2, 3, 4, 5, \ldots\}$.

---

 Notice that every natural number is also a whole number because all natural numbers are included within the set of whole numbers.

Adding the negative numbers, $-1, -2, -3, -4, \ldots$ to the set of whole numbers forms the set of *integers*, represented by $\mathbb{Z}$.

---

**Definition    Integers**

An **integer** is an **element** of the set $\mathbb{Z} = \{\ldots, -4, -3, -2, -1, 0, 1, 2, 3, 4, \ldots\}$.

---

Notice that every whole number is also an integer. What does this tell us about **natural numbers**? View the **answer**.

Integers are still "whole" in the sense that they are not **fractions**. Adding fractions to the set of integers gives us the set of *rational numbers*, represented by $\mathbb{Q}$.

A **rational number** can be expressed as the **quotient** of two **integers** $p$ and $q$, written as $\frac{p}{q}$, when $q \neq 0$. $\frac{p}{q}$ is called the **fraction form** of the rational number such that $p$ is the **numerator** and $q$ is the **denominator**. Examples of rational numbers include $\frac{3}{4}, -\frac{7}{8}$, and $\frac{13}{11}$.

---

**Definition   Rational Numbers**

A **rational number** is a number that can be written as a fraction, $\frac{p}{q}$, where $p$ and $q$ are integers and $q \neq 0$.

---

**Rational numbers** can be written in **decimal form**. When a **fraction** is written as a **decimal**, the result is either a **terminating decimal**, such as $\frac{3}{4} = 0.75$ and $-\frac{7}{8} = -0.875$, or a **repeating decimal**, such as $\frac{5}{6} = 0.8333\ldots = 0.8\overline{3}$ and $\frac{13}{11} = 1.181818\ldots = 1.\overline{18}$.

Every **integer** is also a **rational number** because we can write the integer with a denominator of 1. For example, $5 = \frac{5}{1}, 0 = \frac{0}{1}$, and $-8 = \frac{-8}{1}$. What does this tell us about **whole numbers** and **natural numbers**? View the **answer**.

Not all numbers can be written as the **quotient** of two **integers**. Such numbers belong to the set of *irrational numbers*.

---

**Definition   Irrational Numbers**

An **irrational number** is a number that cannot be written as a fraction, $\frac{p}{q}$, where $p$ and $q$ are integers. In its decimal form, an irrational number does not repeat or terminate.

---

**Square roots** and other numbers expressed with **radicals** are often **irrational numbers**. For example, $\sqrt{3} = 1.7320508\ldots$ and $\sqrt{2} = 1.41421356\ldots$ have decimal forms that continue indefinitely without a repeating pattern, so each is an irrational number. The number $\pi = 3.14159265\ldots$ is another example of an irrational number.

 Not all numbers expressed with radicals are irrational numbers. For example, $\sqrt{9} = 3$ and $\sqrt{225} = 15$ are rational numbers. In fact, $\sqrt{9} = 3$ and $\sqrt{225} = 15$ are also natural numbers, whole numbers, and integers.

Combining the set of **rational numbers** with the set of irrational numbers forms the set of *real numbers*, represented by $\mathbb{R}$.

---

**Definition   Real Numbers**

A **real number** is any number that is either rational or irrational.

---

Figure 1 shows the following relationships involving the set of real numbers:

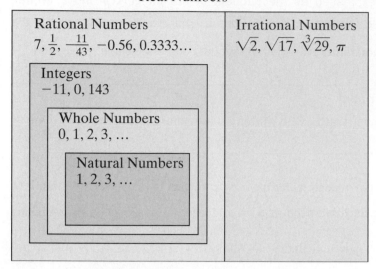

**Figure 1** The set of real numbers

- Every real number is either rational or irrational.
- All natural numbers are whole numbers.
- All whole numbers are integers.
- All integers are rational numbers.

### Example 1  Classifying Real Numbers

Classify each **real number** as a **natural number, whole number, integer, rational number,** and/or **irrational number.** Each number may belong to more than one **set.**

a.  8

b.  −4.8

c.  $\sqrt{10}$

d.  −7

e.  $-\dfrac{4}{7}$

f.  $\sqrt{25}$

g.  0

h.  $3.\overline{45}$

### Solutions

a.  Because 8 is one of the natural (or counting) numbers, we can use **Figure 1** to classify it as a natural number, a whole number, an integer, and a rational number.

b.  Because −4.8 is **negative,** it cannot be a natural number or a whole number. Also, because it contains a fractional portion (.8), it cannot be an integer. Notice that −4.8 is a **terminating decimal.** Therefore, we can classify −4.8 as a rational number.

c.  $\sqrt{10} = 3.1622776\ldots$ is not a terminating or **repeating decimal,** so it cannot be a rational number. Therefore, $\sqrt{10}$ is an irrational number.

 *My video summary*    **d–h.** Try classifying these numbers on your own. Check your **answers,** or watch this **video** for the complete solutions.

Before classifying a **real number,** check to see if it can be **simplified.** For example, the rational number $\dfrac{6}{2}$ can be simplified as $\dfrac{6}{2} = 3$, so it is also an **integer,** a **whole number,** and a **natural number.**

**You Try It**   Work through this You Try It problem.

Work Exercises 1–8 in this textbook or in the *MyMathLab* ® Study Plan.

OBJECTIVE 2   PLOT REAL NUMBERS ON A NUMBER LINE

The **real number line**, also known simply as the number line, is a **graph** that represents the **set** of all **real numbers**. Every point on the number line corresponds to exactly one real number, and every real number corresponds to exactly one point on the number line. The point is called the **graph** of that real number, and the real number corresponding to a point is called the **coordinate** of that point.

The point 0 is called the **origin** of the number line. Numbers located to the left of the origin are **negative numbers**, and numbers located to the right of the origin are **positive numbers**. The number 0 is neither negative nor positive. Figure 2 shows us that any two **consecutive integers** are located one unit apart. The word *consecutive* means that the integers follow in order, one after another, without interruption.

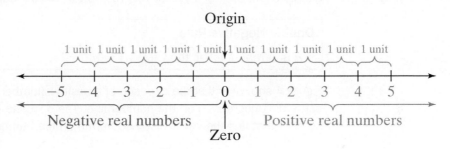

**Figure 2**  The real number line

We **plot**, or **graph**, a real number by placing a solid circle (●) at its location on the number line.

**Example 2  Plotting Real Numbers on the Number Line**

Plot the following **set** of numbers on the **number line**.

$$\left\{-3, -\frac{3}{2}, 1, 2.25\right\}$$

**Solution**  We plot each number by placing a solid circle at its location on the number line. See Figure 3.

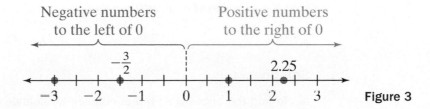

Figure 3

Notice that the sign of each number indicates whether the number lies on the left or right side of 0. Because $-\frac{3}{2} = -1\frac{1}{2}$, its point lies midway between $-2$ and $-1$. However, because $2.25 = 2\frac{1}{4}$, its point lies one-fourth of the way between 2 and 3.

**You Try It**   **Work through this You Try It problem.**

**Work Exercises 9 and 10 in this textbook or in the *MyMathLab*® Study Plan.**

## OBJECTIVE 3   FIND THE OPPOSITE OF A REAL NUMBER

For every **real number** other than 0 on the **number line**, there is a corresponding number located the same distance away from 0 on the number line but in the opposite direction. These pairs of numbers are called *opposites*.

> **Definition   Opposites**
>
> Two numbers are **opposites**, if they are located the same distance away from 0 on the number line but lie on opposite sides of 0.

We use a negative sign $(-)$ to represent "the opposite of." So, $-8$ reads as "the opposite of eight," and $-(-8)$ reads as "the opposite of negative eight." The opposite of negative eight equals eight, so, $-(-8) = 8$. This rule is true for all real numbers.

> **Double-Negative Rule**
>
> If $a$ is a real number, then $-(-a) = a$.

Figure 4 shows us that the opposite of a **positive number** is the **negative number** that lies the same distance from 0 on the number line. The opposite of a negative number is the positive number that lies the same distance from 0 on the number line. Zero is its own opposite.

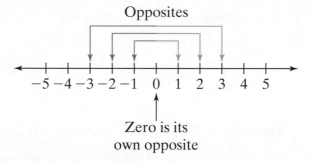

**Figure 4**

> **Finding the Opposite of a Real Number**
>
> To find the opposite of a **real number**, change its **sign**.

### Example 3  Finding Opposites of Real Numbers

Find the **opposite** of each real number.

a.  22          b.  $-\dfrac{4}{5}$          c.  6.4          d.  0

### Solutions

a.  To find the opposite of a real number, we change its sign. Because 22 is **positive**, the opposite of 22 is $-22$.

b.  Because $-\dfrac{4}{5}$ is **negative**, the opposite of $-\dfrac{4}{5}$ is $\dfrac{4}{5}$.

c.  The opposite of 6.4 is $-6.4$.

d.  Because 0 is its own opposite, the opposite of 0 is 0.

**You Try It**   **Work through this You Try It problem.**

**Work Exercises 11–15 in this textbook or in the *MyMathLab*® Study Plan.**

OBJECTIVE 4   FIND THE ABSOLUTE VALUE OF A REAL NUMBER

As defined earlier in this section and illustrated in **Figure 2**, the distance from 0 to 1 on a **real number line** is the **unit distance** for the number line. The concept of unit distance is related to the concept of *absolute value*.

---

**Definition   Absolute Value**

The **absolute value** of a **real number** $a$, written as $|a|$, is the distance from 0 to $a$ on the number line.

---

The symbol $|4|$ reads as "the absolute value of four," but $|-4|$ reads as "the absolute value of negative four." Because both 4 and $-4$ are 4 units from 0 on the number line, both absolute values equal 4. So, $|4| = 4$ and $|-4| = 4$. See Figure 5.

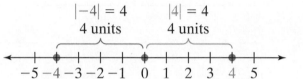

$$|-4| = 4 \qquad |4| = 4$$
$$\text{4 units} \qquad \text{4 units}$$

**Figure 5**

 Because distance cannot be negative, the absolute value of a real number will always be **non-negative**.

What does Figure 5 tell us about the absolute value of opposites? See the **answer**.

**Example 4  Finding Absolute Values**

Find each absolute value.

a. $|3|$        b. $|-5|$        c. $|-1.5|$        d. $\left|-\dfrac{7}{2}\right|$        e. $|0|$

**Solutions**  See Figure 6 for an illustration of parts a and b.

a. $|3| = 3$
   The absolute value of 3 is 3 because 3 lies 3 units away from 0 on the **number line**.
b. $|-5| = 5$
   The absolute value of $-5$ is 5 because $-5$ lies 5 units away from 0 on the number line.

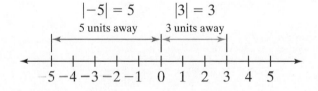

$$|-5| = 5 \qquad |3| = 3$$
$$\text{5 units away} \qquad \text{3 units away}$$

**Figure 6**

*My video summary*   **c–e.**  Try to find each absolute value on your own. Check your **answers**, or watch this video for the complete solutions.

**You Try It**   Work through this You Try It problem.

Work Exercises **16–19** in this textbook or in the *MyMathLab*® Study Plan.

## OBJECTIVE 5   USE INEQUALITY SYMBOLS TO ORDER REAL NUMBERS

To find the **order** of two **real numbers** $a$ and $b$ means to determine if the first number $a$ is smaller than, larger than, or equal to the second number $b$. We can find the *order* of any two real numbers by comparing their locations on the **number line**. Because numbers on the number line increase as we read from left to right, as we move further to the right, the larger the number.

---

**Order of Real Numbers**

1. If $a$ is located to the left of $b$ on the real number line, as shown in Figure 7(a), then $a$ **is less than** $b$, written as $a < b$.
2. If $a$ is located to the right of $b$, as shown in Figure 7(b), then $a$ **is greater than** $b$, written as $a > b$.
3. If $a$ and $b$ are located at the same position on the number line, as shown in Figure 7(c), then $a$ **is equal to** $b$, written as $a = b$.

---

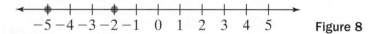

(a) $a < b$      (b) $a > b$      (c) $a = b$      **Figure 7**

The symbol for equality is the **equal sign**, $=$. Symbols for less than, $<$, and greater than, $>$, are called **inequality symbols**. We use these symbols to order real numbers. For example, because $-5$ is located to the left of $-2$ on the **number line**, as shown in Figure 8, we say that $-5$ *is less than* $-2$ and write $-5 < -2$. We say that $-2$ *is greater than* $-5$ and write $-2 > -5$.

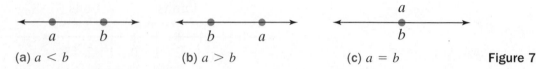

**Figure 8**

When using the symbols $<$ or $>$ to **compare** or order numbers, the symbol should point to the smaller of the two numbers, as shown in Figure 9.

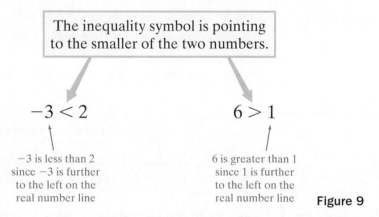

The inequality symbol is pointing to the smaller of the two numbers.

$-3 < 2$

$-3$ is less than 2 since $-3$ is further to the left on the real number line

$6 > 1$

6 is greater than 1 since 1 is further to the left on the real number line

**Figure 9**

Figure 10 shows that $0.5$ and $\dfrac{1}{2}$ share the same position on the number line. So, we say that $0.5$ *equals* $\dfrac{1}{2}$ and write $0.5 = \dfrac{1}{2}$.

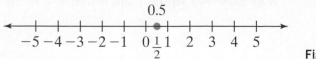

**Figure 10**

**Example 5** Ordering Real Numbers Using Equality and Inequality Symbols

Fill in the blank with the correct symbol, $<, >$, or $=$, to make a true statement.

**a.** 0 ____ 3     **b.** $-3.7$ ____ $-1.5$     **c.** $-\dfrac{5}{4}$ ____ $-1.25$     **d.** $\dfrac{4}{5}$ ____ $\dfrac{5}{9}$

**Solutions**

**a.** View 0 **and** 3 **plotted** on the **number line.** Because 0 is to the left of 3 on the number line, 0 *is less than* 3. So, we write $0 < 3$.

**b.** View $-3.7$ and $-1.5$ **plotted** on the number line. Because $-3.7$ is to the left of $-1.5$ on the number line, $-3.7$ *is less than* $-1.5$. So, we write $-3.7 < -1.5$.

**c.** View $-\dfrac{5}{4}$ and $-1.25$ **plotted** on the number line. The two numbers share the same position, so they are equal, or $-\dfrac{5}{4} = -1.25$.

**d.** To compare $\dfrac{4}{5}$ and $\dfrac{5}{9}$ more easily, first we convert both **fractions to decimals:** $\dfrac{4}{5} = 0.8$ and $\dfrac{5}{9} = 0.\overline{5}$. On the number line, 0.8 is to the right of $0.\overline{5}$, so $\dfrac{4}{5}$ *is greater than* $\dfrac{5}{9}$, or $\dfrac{4}{5} > \dfrac{5}{9}$. View the **numbers plotted** on the number line.

**You Try It**    **Work through this You Try It problem.**

**Work Exercises 20–28 in this textbook or in the *MyMathLab*®** Study Plan.

In addition to $<$ and $>$, three other **inequality symbols** can be used to **compare** real numbers: *less than or equal to* ($\leq$), *greater than or equal to* ($\geq$), and *not equal to* ($\neq$). Notice that $\leq$ and $\geq$ allow for equality as part of the comparison. For this reason, inequalities of the forms $a \leq b$ and $a \geq b$ are **non-strict inequalities.** Similarly, inequalities of the forms $a < b, a > b$, and $a \neq b$ are **strict inequalities** because they do not allow for equality.

For non-strict inequalities, we need to remember that the inequality contains two parts ($<$ or $>$, and equality). In order for the inequality to be true, only one of the two parts must be a true statement.

$$a \leq b \begin{cases} \text{True if } a < b & (4 \leq 8 \text{ is true because } 4 < 8 \text{ is true}) \\ \text{True if } a = b & \left(3.5 \leq \dfrac{7}{2} \text{ is true because } 3.5 = \dfrac{7}{2} \text{ is true}\right) \end{cases}$$

$$a \geq b \begin{cases} \text{True if } a > b & (9 \geq 6 \text{ is true because } 9 > 6 \text{ is true}) \\ \text{True if } a = b & \left(2.5 \geq \dfrac{5}{2} \text{ is true because } 2.5 = \dfrac{5}{2} \text{ is true}\right) \end{cases}$$

**Example 6** Using Inequality Symbols

Determine if each statement is true or false.

**a.** $\dfrac{7}{10} \leq 0.7$     **b.** $-8 \geq 4$     **c.** $-2 \geq -4$

**d.** $\dfrac{7}{3} \leq 1.\overline{3}$     **e.** $-\dfrac{9}{4} \neq -2.75$

**Solutions**

**a.** In order for the statement to be true, we need either $\frac{7}{10} < 0.7$ or $\frac{7}{10} = 0.7$ to be true.

The comparison may be easier if we first convert the **fraction to a decimal**. In decimal form, $\frac{7}{10} = 0.7$ So, the statement $\frac{7}{10} \le 0.7$ is true because $\frac{7}{10} = 0.7$.

**b.** In order for the statement to be true, we need either $-8 > 4$ or $-8 = 4$ to be true. Because $-8$ lies to the left of 4 on the **real number line**, we have $-8 < 4$. So, the statement $-8 \ge 4$ is false.

 *My video summary*    **c–e.** Try to answer these problems on your own. Check your **answers** or watch this **video** for the complete solution.

**You Try It**   Work through this You Try It problem.

**Work Exercises 29–33** in this textbook or in the *MyMathLab*® Study Plan.

## OBJECTIVE 6   TRANSLATE WORD STATEMENTS INVOLVING INEQUALITIES

When solving application problems, understanding the language of mathematics is important in order to translate word statements into **mathematical expressions**. When working with inequalities, it will be important to use key words to distinguish between the various **inequality symbols**. There are many key words or phrases that describe inequality symbols. Figure 11 lists some examples.

| Inequality | Key Words | Word Statement | Mathematical Expression |
|:---:|---|---|---|
| $<$ | is less than, is fewer than | Six is fewer than eight | $6 < 8$ |
| $>$ | is more than, is greater than | 10 is more than two | $10 > 2$ |
| $\le$ | is less than or equal to, at most, no more than | Seven is less than or equal to nine; Four is no more than five | $7 \le 9$; $4 \le 5$ |
| $\ge$ | is greater than or equal to, at least, no less than | 15 is greater than or equal to 13; Eight is no less than three | $15 \ge 13$; $8 \ge 3$ |
| $\ne$ | not equal to, different than | 12 is not equal to 15 | $12 \ne 15$ |

**Figure 11**
Key words for inequalities

### Example 7  Translating Word Statements Involving Inequalities

Write a mathematical expression for each word phrase.

**a.** Fourteen is greater than ten

**b.** Twenty-four is no more than thirty

**c.** Nine is not equal to eighteen

**Solutions** Try translating the statements on your own, then check your **answers**.

**You Try It**   Work through this You Try It problem.

**Work Exercises 34–38** in this textbook or in the *MyMathLab*® Study Plan.

### Example 8  Translating Word Statements Involving Inequalities

Use **real numbers** and write an inequality that represents the given comparison.

a. Orchid Island Gourmet Orange Juice sells for $6, which is more than Florida's Natural Premium Orange Juice that sells for $3.

b. In 2008 there were 4983 identity thefts reported in Colorado, which is different than the 4433 reported in Missouri during the same year. (*Source: census.gov/compendia/*)

c. On May 20, 2010, the Dow Jones Industrial Average closed at 10,068.01 which was less than on May 19, 2010, when it closed at 10,444.37. (*Source: dowjonesclose.com/*)

### Solutions

a. The key words "is more than" indicate we should use the **inequality symbol** '>'. We represent the cost of the Orchid Island juice with the **integer** 6 and the Florida's Natural juice with the integer 3.

| cost of Orchid Island juice | is more than | cost of Florida Natural juice |
|:---:|:---:|:---:|
| 6 | > | 3 |

*My video summary*   **b–c.** Try translating these statements on your own. Check your **answers**, or watch this video for the solutions.

**You Try It**  Work through this You Try It problem.

Work Exercises 39–42 in this textbook or in the *MyMathLab*® Study Plan.

## 1.1 Exercises

In Exercises 1–8, classify each real number as a natural number, whole number, integer, rational number, and/or irrational number. Each number may belong to more than one set.

**You Try It**   1. $-15$

2. $9.5$

3. $-\sqrt{11}$

4. $\dfrac{2}{5}$

5. $-\dfrac{24}{3}$

6. $18$

7. $4.\overline{35}$

8. $\sqrt{36}$

In Exercises 9 and 10, plot each set of numbers on a number line.

9. $\left\{-3, 2.5, -\dfrac{7}{4}, 4\right\}$

10. $\left\{\dfrac{3}{4}, -4.\overline{2}, 0, 5, -3\right\}$

**You Try It**

In Exercises 11–15, find the opposite of each real number.

**11.** $-17$

**12.** $29$

**13.** $12.45$

**14.** $-\dfrac{7}{12}$

**15.** $4.\overline{27}$

**You Try It**

In Exercises 16–19, find each absolute value.

**16.** $|32|$

**17.** $|-12.5|$

**18.** $|-9|$

**19.** $\left|\dfrac{15}{28}\right|$

**You Try It**

In Exercises 20–28, fill in the blank with the correct symbol, $<$, $>$, or $=$, to make a true statement.

**You Try It**

**20.** $-6.9$ ____ $-3.2$

**21.** $5$ ____ $0$

**22.** $-\dfrac{1}{2}$ ____ $\dfrac{1}{3}$

**23.** $\dfrac{9}{4}$ ____ $2.25$

**24.** $15$ ____ $30$

**25.** $-3\dfrac{3}{4}$ ____ $-\dfrac{15}{4}$

**26.** $1.4$ ____ $-\dfrac{1}{5}$

**27.** $\dfrac{7}{3}$ ____ $\dfrac{9}{4}$

**28.** $-4.35$ ____ $1\dfrac{1}{3}$

In Exercises 29–33, determine if each statement is true or false.

**You Try It**

**29.** $-3 \le -7$

**30.** $\dfrac{22}{5} \le 4.6$

**31.** $-2 \ne 2$

**32.** $8.4 \ge 8.40$

**33.** $6.40 \ge 6\dfrac{4}{5}$

In Exercises 34–38, write a mathematical expression for each word phrase.

**You Try It**

**34.** Seventeen is greater than twelve

**35.** Forty-five is less than or equal to fifty

**36.** Eight is at least three

**37.** Four is different than five

**38.** Thirty is greater than or equal to fifteen

In Exercises 39–42, use real numbers and write an inequality that compares the given numbers.

**You Try It**

**39.** In September 2009, 54% of teens reported texting friends on a daily basis. This is greater than the 38% who reported daily texting in February 2008. (*Source: pewinternet.org/*)

**40.** The average price per gallon for unleaded gas in the U.S. on May 17, 2010 was $2.864, which was different than the average price of $2.309 on May 17, 2009. (*Source: eia.doe.gov*)

**41.** The average high temperature for Phoenix, AZ in January is 66°F, which is less than the average high temperature of 70°F for February. (*Source: phoenix.about.com/*)

**42.** Between 1990 and 2000, the population of Philadelphia, PA decreased by 68,000 while the population of Chicago, IL increased by 112,000. (*Source: infoplease.com/*)

# 1.2 Adding and Subtracting Real Numbers

## THINGS TO KNOW

Before working through this section, be sure you are familiar with the following concepts:

| | VIDEO | ANIMATION | INTERACTIVE |
|---|---|---|---|

**You Try It**

**1.** Write Fractions in Simplest Form (Section R.1, **Objective 2**)

**2.** Add and Subtract Fractions (Section R.1, **Objective 4**)
**You Try It**

**You Try It**
3. Add and Subtract Decimals
   (Section R.2, **Objective 1**)

**You Try It**
4. Find the Absolute Value of a Real Number
   (Section 1.1, **Objective 4**)

## OBJECTIVES

**1** Add Two Real Numbers with the Same Sign

**2** Add Two Real Numbers with Different Signs

**3** Subtract Real Numbers

**4** Translate Word Statements Involving Addition or Subtraction

**5** Solve Applications Involving Addition or Subtraction of Real Numbers

OBJECTIVE 1    ADD TWO REAL NUMBERS WITH THE SAME SIGN

The result of adding two **real numbers** is called the **sum** of the numbers. The numbers being added are called **terms**, or **addends**. How we add terms depends on their **signs** (**positive** or **negative**). First, let's add two real numbers with the same sign.

> **Adding Two Real Numbers with the Same Sign**
>
> If two real numbers being added have the same sign (both positive or both negative), then we add as follows:
>
> **Step 1** Find the absolute value of each term.
>
> **Step 2.** Add the two results from Step 1.
>
> **Step 3.** Use the common sign as the sign of the **sum**.

**Example 1  Adding Two Real Numbers with the Same Sign**

Add.

**a.** $2 + 5$          **b.** $-4 + (-3)$

**Solutions**

**a.** First, we find the absolute values, $|2| = 2$ and $|5| = 5$. Then we add the results, $2 + 5 = 7$. Because the common sign is positive, the sum will be positive. Thus, $2 + 5 = 7$.

**b.** First, we find the absolute values, $|-4| = 4$ and $|-3| = 3$. Then, we add the results, $4 + 3 = 7$. Because the common sign is negative, the sum will be negative. So, $-4 + (-3) = -7$.

**You Try It**    Work through this You Try It problem.

**Work Exercises 1–4 in this textbook or in the *MyMathLab*®  Study Plan.**

To visualize addition, we can use a **number line** to add **real numbers**. View this **popup** to see Example 1a worked using a number line. View this **popup** to see Example 1b. The number line method is more involved than using absolute values, so we generally use the absolute value approach.

Another way to think about adding real numbers is to consider money gained or lost. View an **example**.

### Example 2  Adding Two Real Numbers with the Same Sign

Add.

a.  $-3.65 + (-7.45)$    b.  $\dfrac{4}{5} + \dfrac{13}{5}$    c.  $-\dfrac{3}{5} + \left(-\dfrac{7}{2}\right)$    d.  $-3\dfrac{1}{3} + \left(-5\dfrac{1}{4}\right)$

### Solutions

a.  We find the absolute values, $|-3.65| = 3.65$ and $|-7.45| = 7.45$. Both numbers are negative, so the sum will be negative. Because $3.65 + 7.45 = 11.10$, we get $-3.65 + (-7.45) = -11.10$.

b.  We find the **absolute values**, $\left|\dfrac{4}{5}\right| = \dfrac{4}{5}$ and $\left|\dfrac{13}{5}\right| = \dfrac{13}{5}$. Both numbers are **positive**, so the **sum** will be positive.

$$\dfrac{4}{5} + \dfrac{13}{5} = \dfrac{\overbrace{4 + 13}^{\text{Add the numerators}}}{\underbrace{5}_{\text{Keep the common denominator}}} = \dfrac{17}{5}$$

$$\underbrace{\phantom{\dfrac{4}{5} + \dfrac{13}{5}}}_{\text{Common denominator}}$$

The sum is $\dfrac{17}{5}$.

*My video summary*    c.  Because both numbers are **negative**, we add the absolute values and make the sum negative.

$$\left|-\dfrac{3}{5}\right| = \dfrac{3}{5} \quad \text{and} \quad \left|-\dfrac{7}{2}\right| = \dfrac{7}{2}$$

To add $\dfrac{3}{5}$ and $\dfrac{7}{2}$, we need a **common denominator**.

$$\underbrace{\dfrac{3}{5} + \dfrac{7}{2}}_{\text{LCD} = 10} = \dfrac{3}{5} \cdot \dfrac{2}{2} + \dfrac{7}{2} \cdot \dfrac{5}{5} = \dfrac{6}{10} + \dfrac{35}{10}$$

Finish adding the numbers on your own. Check your **answer**, or watch this **video** for the complete solution.

*My video summary*  d.  Try adding the numbers on your own. Check your **answer**, or watch this **video** for the complete solution.

**You Try It**    Work through this **You Try It** problem.

Work Exercises 5–10 in this textbook or in the *MyMathLab* ® Study Plan.

## OBJECTIVE 2  ADD TWO REAL NUMBERS WITH DIFFERENT SIGNS

Sometimes we must add two numbers with different **signs**. For example, if the value of one share of Apple Inc. stock increased \$8 one day and decreased \$3 the next, then the share increased in value by \$5 over the two days, or $8 + (-3) = 5$.

When adding two **real numbers** with different signs, we can use **absolute value**.

---

**Adding Two Real Numbers with Different Signs**

If two real numbers being added have different signs (one **positive** and one **negative**), then we add as follows:

**Step 1.** Find the absolute value of each **term**.

**Step 2.** Subtract the smaller result from the larger result from Step 1.

**Step 3.** Use the sign of the term with the larger absolute value as the sign of the **sum**.

---

### Example 3 Adding Two Real Numbers with Different Signs

Add.

**a.** $7 + (-4)$   **b.** $-6 + 4$   **c.** $6 + (-6)$

**Solutions**

**a.** First, we find the absolute value of each **term**.

$$|7| = 7 \quad \text{and} \quad |-4| = 4$$

The terms have different signs so we subtract the smaller result from the larger result.

$$7 - 4 = 3$$

The larger absolute value result belongs to the positive number, so the sum will be positive.

$$7 + (-4) = 3$$

**b.** Find the **absolute value** of each number.

$$|-6| = 6 \quad \text{and} \quad |4| = 4$$

Subtract the smaller result from the larger result.

$$6 - 4 = 2$$

Because the larger absolute value result belongs to the **negative** number, the **sum** will be negative.

$$-6 + 4 = -2$$

**c.** Find the absolute values.

$$|6| = 6 \quad \text{and} \quad |-6| = 6$$

The numbers have the same absolute value, so the **difference** is 0.

$$6 - 6 = 0$$

Therefore,

$$6 + (-6) = 0.$$

**You Try It**  Work through this You Try It problem.

**Work Exercises 11–16 in this textbook or in the *MyMathLab*® Study Plan.**

We can use a **number line** to visualize the addition of real numbers with different signs. View this **popup** to see Example 3a worked using a number line. View this **popup** to see Example 3b.

In Example 3c, the two numbers are **opposites** with a sum of 0. This result shows why we refer to opposites as **additive inverses**.

---

**Adding a Real Number and Its Opposite**

The sum of a real number $a$ and its opposite, or **additive inverse**, $-a$, is 0.

$$a + (-a) = 0$$

---

We now summarize our procedures for adding two **real numbers**.

---

**Adding Two Real Numbers**

**Step 1.** If the **signs** of the two **terms** are the same (both **positive** or both **negative**), add their **absolute values** and use the common sign as the sign of the **sum**.

**Step 2.** If the signs of the two terms are different (one positive and one negative), subtract the smaller absolute value from the larger absolute value and use the sign of the term with the larger absolute value as the sign of the sum.

---

**Example 4  Adding Two Real Numbers**

Add.

**a.** $-12 + (-9)$    **b.** $7 + (-18)$    **c.** $\dfrac{4}{3} + \dfrac{5}{6}$    **d.** $-5.7 + 12.3$

**Solutions**

**a.** Because both numbers are negative, we add the absolute values and make the sum negative.

$$\left|-12\right| = 12 \quad \text{and} \quad \left|-9\right| = 9$$

$12 + 9 = 21$, so $-12 + (-9) = -21$.

**b.** Find the absolute value of each number.

$$\left|7\right| = 7 \quad \text{and} \quad \left|-18\right| = 18$$

Subtract the smaller result from the larger result.

$$18 - 7 = 11$$

Because the larger absolute value result belongs to the negative number, the sum will be negative.

$$7 + (-18) = -11$$

*My video summary*  **c–d.** Try to work these problems on your own. Start by finding the absolute value of each number. Add or subtract the absolute values as required, then determine the sign of the result. Check your **answers**, or watch this **video** for the complete solutions.

**You Try It**   Work through this **You Try It** problem.

Work Exercises 17–20 in this textbook or in the ***MyMathLab*** ® Study Plan.

## OBJECTIVE 3   SUBTRACT REAL NUMBERS

The next **arithmetic operation** is **subtraction**. The result of subtracting two **real numbers** is called the **difference** of the numbers. The number being subtracted is called the **subtrahend**, and the number being subtracted from is called the **minuend**.

$$\underset{\underset{23}{\downarrow}}{\text{minuend}} \; - \; \underset{\underset{7}{\downarrow}}{\text{subtrahend}} \; = \; \underset{\underset{16}{\downarrow}}{\text{difference}}$$

Using the concept of **opposite**, we can define subtraction in terms of **addition**. We subtract a number by adding its opposite.

---

**Subtracting Two Real Numbers**

If $a$ and $b$ represent two real numbers, then

$$a - b = a + (-b).$$

---

### Example 5   Subtracting Two Real Numbers

*My interactive video summary*

📷 Subtract.

a. $15 - 6$   b. $9 - 17$   c. $-4.5 - 3.2$   d. $\dfrac{7}{3} - \dfrac{2}{5}$

e. $-4.9 - (-2.5)$   f. $7\dfrac{3}{4} - \left(-2\dfrac{1}{5}\right)$   g. $4 - (-4)$

**Solutions**  For each **subtraction**, we add the opposite.

a. Add the opposite of 6:  $15 - 6 = 15 + (-6)$
   We now choose the appropriate **addition rule**. The **signs** are different. $|15| = 15$ and $|-6| = 6$. Because 15 has the larger **absolute value**, this **sum** is **positive**. We subtract 6 from 15 to find the result, $15 - 6 = 9$. The **difference** is 9.

b. Add the **opposite** of 17:  $9 - 17 = 9 + (-17)$
   The **signs** are different. $|9| = 9$ and $|-17| = 17$. Because $-17$ has the larger **absolute value**, this **sum** is **negative**. We subtract 9 from 17 and make the result negative. $17 - 9 = 8$, so $9 + (-17) = -8$. The **difference** is $-8$.

c. Add the opposite of 3.2:  $-4.5 - 3.2 = -4.5 + (-3.2)$
   Both $-4.5$ and $-3.2$ are negative, so the sum is negative. $|-4.5| = 4.5$ and $|-3.2| = 3.2$. Adding gives $4.5 + 3.2 = 7.7$, so $-4.5 + (-3.2) = -7.7$. The difference is $-7.7$.

d. Add the opposite of $\dfrac{2}{5}$:  $\dfrac{7}{3} - \dfrac{2}{5} = \dfrac{7}{3} + \left(-\dfrac{2}{5}\right)$

   The signs are different. $\left|\dfrac{7}{3}\right| = \dfrac{7}{3}$ and $\left|-\dfrac{2}{5}\right| = \dfrac{2}{5}$. The sum is positive because $\dfrac{7}{3}$

   has the larger absolute value. We subtract $\dfrac{2}{5}$ from $\dfrac{7}{3}$ to find the result.

$$\underset{\text{LCD} = 15}{\underbrace{\dfrac{7}{3} - \dfrac{2}{5}}} = \dfrac{7}{3} \cdot \dfrac{5}{5} - \dfrac{2}{5} \cdot \dfrac{3}{3} = \dfrac{35}{15} - \dfrac{6}{15} = \underset{\substack{\text{Combine the}\\\text{numerators}}}{\dfrac{35 - 6}{15}} = \dfrac{29}{15}$$

   The difference is $\dfrac{29}{15}$.

**e–g.** Try to work these problems on your own. Check your **answers**, or watch this **interactive video** for the complete solutions.

**You Try It**     Work through this **You Try It** problem.

**Work Exercises 21–30 in this textbook or in the *MyMathLab*®  Study Plan.**

OBJECTIVE 4    TRANSLATE WORD STATEMENTS INVOLVING ADDITION OR SUBTRACTION

When solving application problems, we use the language of mathematics to translate word statements into **mathematical expressions**. There are many key words and phrases that indicate **addition**. Figure 12 shows some examples.

| Key Word | Word Phrase | Mathematical Expression |
|---|---|---|
| Sum | The *sum* of $-2$ and 5 | $-2 + 5$ |
| Increased by | 4 *increased by* 7 | $4 + 7$ |
| Added to | 3 *added to* 8 | $8 + 3$ |
| More than | 6 *more than* $-4$ | $-4 + 6$ |

**Figure 12**   Key words meaning addition

## Example 6   Translating Word Statements Involving Addition

Write a mathematical expression for each word phrase.

**a.** Five more than $-8$

**b.** 8.4 increased by 0.17

**c.** The sum of $-4$ and $-10$

**d.** 15 added to $-30$

## Solutions

**a.** The key words "more than" mean addition. What is being added? 5 and $-8$.

$$\overset{\text{Addition}}{\underbrace{\underset{\substack{\text{The two quantities} \\ \text{being added}}}{\text{Five more than } -8}}}$$

So, "five more than $-8$" is written as $-8 + 5$.

**b.** The key words "increased by" mean **addition**. What is being added? 0.17 is being added to 8.4.

$$\overset{\text{Addition}}{\underbrace{\underset{\substack{\text{The two quantities} \\ \text{being added}}}{8.4 \text{ increased by } 0.17}}}$$

So, "8.4 increased by 0.17" translates to $8.4 + 0.17$.

c. The keyword "**sum**" indicates addition. $-4$ and $-10$ are being added. The key word "and" separates the two quantities and can be translated to a $+$ sign.

$$\underset{\substack{\text{The two quantities}\\\text{being added}}}{\text{The } \overbrace{\text{sum of}}^{\text{Addition}} \underline{-4} \text{ and } \underline{-10}}$$

So, "the sum of $-4$ and $-10$" is written as $-4 + (-10)$.

d. The key words are "added to." The numbers $-30$ and $15$ are being added.

$$\underset{\substack{\text{The two quantities}\\\text{being added}}}{\underline{15} \overbrace{\text{added to}}^{\text{Addition}} \underline{-30}}$$

So, "15 added to $-30$" translates to $-30 + 15$.

**You Try It**   **Work through this You Try It problem.**

**Work Exercises 31–36 in this textbook or in the *MyMathLab*®  Study Plan.**

As with **addition**, many keywords or phrases indicate **subtraction**. Figure 13 shows some examples.

| Key Word | Word Phrase | Mathematical Expression |
|---|---|---|
| Difference | The *difference* of 9 and 7 | $9 - 7$ |
| Decreased by | 18 *decreased by* 4 | $18 - 4$ |
| Subtracted from | 6 *subtracted from* 3 | $3 - 6$ |
| Less than | 10 *less than* 25 | $25 - 10$ |

**Figure 13**  Key words for subtraction

 Unlike addition, the order of numbers in subtraction matters, so we have to pay close attention to wording. For example, $10 - 2$ (the difference of 10 and 2) is not the same as $2 - 10$. (the difference of 2 and 10)

### Example 7  Translating Word Statements Involving Subtraction

Write a mathematical expression for each word phrase.

a. Fifteen subtracted from 22

b. The **difference** of 7 and 12

c. Eight decreased by 11

d. 20 less than the difference of 4 and 9

e. The **sum** of 8 and 13, decreased by 5

**Solutions**

a. The key words "subtracted from" mean **subtraction**. What is being subtracted? 15 is being subtracted *from* 22.

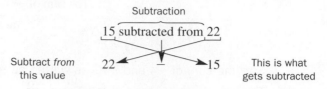

So, "fifteen subtracted from 22" is written as $22 - 15$.

b. "**Difference**" means subtraction. 12 is being subtracted from 7. Note in **Figure 13** that with a difference, the second number is subtracted from the first.

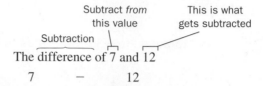

"The difference of 7 and 12" translates to $7 - 12$.

c. "Decreased by" indicates **subtraction**, and the numbers 8 and 11 are being subtracted. Because 8 is being decreased, we write this value first.

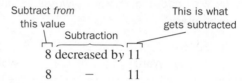

"Eight decreased by 11" translates to $8 - 11$.

*My video summary*  d. The key words "less than" and "difference" suggest two subtractions. Try to complete this translation on your own. Check your **answer**, or watch this **video** for a complete solution.

*My video summary* e. This problem involves both **addition** and subtraction. Try to do this translation on your own. Check your **answer**, or watch this **video** for a complete solution.

**You Try It**   Work through this You Try It problem.

Work Exercises 37–42 in this textbook or in the *MyMathLab*® Study Plan.

OBJECTIVE 5   SOLVE APPLICATIONS INVOLVING ADDITION OR SUBTRACTION OF REAL NUMBERS

Addition and subtraction can be used to solve a variety of application problems that involve finding a **sum** or **difference**. Looking for key words is important because they will tell us which **operation** to perform and on which quantities.

**Example 8  Solving Applications Involving Addition or Subtraction**

On May 20, 2010 the Dow Jones Industrial Average lost 376.36 points. The next day it gained 125.38 points. What was the total result for the two days? (*Source: Yahoo! Finance*)

**Solution** The loss of 376.36 points can be expressed as $-376.36$ while the gain of 125.38 points can be expressed as $+125.38$. To find the total result, we **add**.

$$\underbrace{-376.36}_{\text{May 20}} + \underbrace{125.38}_{\text{May 21}}$$

The numbers have different signs so we find the **absolute value** of each number and subtract the smaller result from the larger.

$$|-376.36| = 376.36 \quad \text{and} \quad |125.38| = 125.38$$

$$376.36 - 125.38 = 250.98$$

Because the larger absolute value result belongs to the **negative** number, the sum will be negative.

$$-376.36 + 125.38 = -250.98$$

The total result for the two days was a loss of 250.98 points.

### Example 9  Solving Applications Involving Addition or Subtraction

*My video summary*  The record high temperature in Alaska was $100°F$ recorded in 1915 at Fort Yukon. The record low was $-80°F$ recorded in 1971 at Prospect Creek Camp. What is the difference between these record high and low temperatures? (*Source: National Climatic Data Center*)

**Solution** Try to solve this problem on your own. Check your **answer**, or watch this **video** for a complete solution.

**You Try It**  Work through this You Try It problem.

Work Exercises 43–46 in this textbook or in the **MyMathLab**® Study Plan.

## 1.2  Exercises

In Exercises 1–20, add the real numbers.

**You Try It**

**1.** $7 + 8$

**2.** $-2 + (-3)$

**3.** $-23 + (-48)$

**4.** $29 + 85$

**You Try It**

**5.** $-2.78 + (-1.25)$

**6.** $15.06 + 0.78$

**7.** $-\dfrac{4}{3} + \left(-\dfrac{7}{3}\right)$

**8.** $\dfrac{3}{4} + \dfrac{7}{8}$

**You Try It**

**9.** $-\dfrac{2}{3} + \left(-\dfrac{12}{5}\right)$

**10.** $\left(-2\dfrac{1}{3}\right) + \left(-3\dfrac{1}{2}\right)$

**11.** $6 + (-2)$

**12.** $(-5) + 8$

**You Try It**

**13.** $-3 + 1$

**14.** $7 + (-9)$

**15.** $-12 + 12$

**16.** $(-2.79) + 4.21$

**You Try It**

**17.** $\left(-\dfrac{13}{4}\right) + \left(\dfrac{13}{4}\right)$

**18.** $\left(-\dfrac{2}{7}\right) + \left(\dfrac{7}{2}\right)$

**19.** $\left(-\dfrac{7}{3}\right) + \left(2\dfrac{1}{3}\right)$

**20.** $(3.87) + (-10.4)$

In Exercises 21–30, subtract the real numbers.

**You Try It**

**21.** $15 - 9$

**22.** $8 - 11$

**23.** $\dfrac{7}{3} - \dfrac{3}{7}$

**24.** $-3.1 - 5.9$

**25.** $-7 - (-3)$      **26.** $-1\frac{2}{3} - 3\frac{1}{4}$      **27.** $-\frac{2}{9} - \frac{5}{18}$      **28.** $2.1 - 1.2$

**29.** $\frac{5}{8} - \left(-\frac{3}{4}\right)$      **30.** $-2.1 - (-3.2)$

In Exercises 31–36, write a mathematical expression for each word phrase.

**You Try It**

**31.** 5 added to $-7$               **32.** $-3$ increased by 12

**33.** 8.1, added to 2.1 more than 3.7      **34.** The sum of $-\frac{3}{5}$ and $\frac{5}{3}$

**35.** Twelve more than $-15$          **36.** $2\frac{3}{4}$ increased by $\frac{1}{2}$

In Exercises 37–42, write a mathematical expression for each word phrase.

**You Try It**

**37.** The difference of 12 and $-13$      **38.** $-18$ decreased by 15

**39.** 5 more than the difference of 2 and 7      **40.** 8 subtracted from 5, decreased by 3

**41.** Two-thirds subtracted from $\frac{5}{9}$      **42.** 34 less than the sum of $-3$ and 19

In Exercises 43–46, solve the given application problem.

**You Try It**

**43.** On May 18, 2010 the Dow Jones Industrial Average lost 114.88 points. The next day it lost 66.58 points. What was the total result for the two days? Express this as a real number. (*Source: Yahoo! Finance*)

**44.** For the 2008 tax year, the IRS received 131,543,000 individual tax returns. Of those, 90,639,000 were filed electronically. How many individual tax returns were sent by mail for 2008? (*Source: irs.gov*)

**45.** The lowest point in the Atlantic Ocean is the Milwaukee Deep in the Puerto Rico Trench, at an elevation of $-8605$ m. The lowest point in the Pacific Ocean is the Challenger Deep in the Mariana Trench, at an elevation of $-10,924$ m. What is the difference in elevation between the Milwaukee Deep and the Challenger Deep? (*Source: www.worldatlas.com*)

**46.** The record high temperature in Montana was 47.2°C recorded in 1937 at Medicine Lake. The record low was $-56.7$°C recorded in 1954 at Rogers Pass. What is the difference between these record high and low temperatures? (*Source: National Climatic Data Center*)

# 1.3   **Multiplying and Dividing Real Numbers**

## THINGS TO KNOW

Before working through this section, be sure you are familiar with the following concepts:

VIDEO      ANIMATION      INTERACTIVE

**You Try It**

**1.** Write Fractions in Simplest Form (Section R.1, **Objective 2**)     

**You Try It**

**2.** Multiply and Divide Fractions (Section R.1, **Objective 3**)

**You Try It**

3. Multiply and Divide Decimals
   (Section R.2, **Objective 2**)

**You Try It**

4. Find the Absolute Value of a Real Number
   (Section 1.1, **Objective 4**)

## OBJECTIVES

**1** Multiply Real Numbers

**2** Divide Real Numbers

**3** Translate Word Statements Involving Multiplication or Division

**4** Solve Applications Involving Multiplication or Division

---

OBJECTIVE 1   **MULTIPLY REAL NUMBERS**

Multiplication is simply repeated **addition**. For example, $4(6)$ means that 6 is added four times, and $5(-6)$ means that $-6$ is added five times. Figure 14 displays this process visually.

$$4(6) = 6 + 6 + 6 + 6 = 24$$
$$5(-6) = (-6) + (-6) + (-6) + (-6) + (-6) = -30 \quad \textbf{Figure 14}$$

The result of multiplying two **real numbers** is called the **product** of the numbers. The numbers being multiplied are called **factors**. The **sign** of a product depends on the signs of the factors.

The following gives a summary for multiplying two real numbers.

---

**Multiplying Two Real Numbers**

Multiply the **absolute values** of the two **factors** to get the absolute value of the **product**. Determine the sign of the product using the following rules:

**1.** If the signs of the two factors are the same (both **positive** or both **negative**), then the product is positive.

**2.** If the signs of the two factors are different (one positive and one negative), then the product is negative.

**3.** The product of a real number and 0 is equal to 0 (**Multiplication Property of Zero**).

View this **illustration** of how the factor signs affect the sign of the product.

---

**Example 1  Multiplying Two Real Numbers**

Multiply.

**a.** $5 \cdot 13$      **b.** $6(-7)$      **c.** $0 \times 15$      **d.** $(-4)(-12)$

**Solutions**

**a.** $|5| = 5$ and $|13| = 13$. The **absolute value** of the **product** is $5 \cdot 13 = 65$. Since 5 and 13 have the same sign, the product is positive. So, $5 \cdot 13 = 65$.

**b.** $|6| = 6$ and $|-7| = 7$. The absolute value of the product is $6 \cdot 7 = 42$. One **factor** is positive and the other is negative, so the product is negative. So, $6(-7) = -42$.

**c.** One of the factors is 0, so the product is 0.

**d.** $|-4| = 4$ and $|-12| = 12$. The absolute value of the product is $4 \cdot 12 = 48$. Since $-4$ and $-12$ have the same sign, the product is positive. So, $(-4)(-12) = 48$.

**You Try It**  Work through this You Try It problem.

Work Exercises 1–8 in this textbook or in the *MyMathLab*® Study Plan.

You may wish to review multiplying fractions in **Section R.1** before working through Example 2.

### Example 2  Multiplying Two Real Numbers

Multiply.

**a.** $\left(-\dfrac{3}{4}\right)\left(-\dfrac{7}{9}\right)$    **b.** $5 \cdot \dfrac{3}{10}$    **c.** $\dfrac{3}{8} \times 0$    **d.** $\left(-\dfrac{2}{3}\right)\left(\dfrac{6}{14}\right)$

### Solutions

**a.** $\left|-\dfrac{3}{4}\right| = \dfrac{3}{4}$ and $\left|-\dfrac{7}{9}\right| = \dfrac{7}{9}$. The absolute value of the product is

$$\underbrace{\dfrac{3}{4} \cdot \dfrac{7}{9} = \dfrac{3^1}{4} \cdot \dfrac{7}{9^3}}_{\text{Simplify}} = \dfrac{1 \cdot 7}{4 \cdot 3} = \dfrac{7}{12}.$$

Because $-\dfrac{3}{4}$ and $-\dfrac{7}{9}$ have the same sign, the **product** is **positive**. So,

$$\left(-\dfrac{3}{4}\right)\left(-\dfrac{7}{9}\right) = \dfrac{7}{12}.$$

*My video summary*  **b–d.** Try to finish these problems on your own. Remember to write **fractions** in **lowest terms**. Check your **answers**, or watch this **video** for the complete solutions.

**You Try It**  Work through this You Try It problem.

Work Exercises 9–14 in this textbook or in the *MyMathLab*® Study Plan.

### Example 3  Multiplying Two Real Numbers

Multiply.

**a.** $(1.4)(-3.5)$    **b.** $(8.32)(0)$    **c.** $-\dfrac{3}{5} \times 6\dfrac{1}{3}$    **d.** $(4)(5.8)$

### Solutions

**a.** The **factors** have different signs, so the product is **negative**.
Since $(1.4)(3.5) = 4.9$ (View the **details**), the product is $(1.4)(-3.5) = -4.9$.

**b.** Since one of the factors is 0, the product is 0. So, $(8.32)(0) = 0$.

*My video summary*  **c–d.** Try to work these problems on your own. Remember to write **mixed numbers** as **improper fractions** before multiplying, and write fractions in lowest terms. Check your **answers**, or watch this **video** for the complete solutions.

**You Try It**  Work through this You Try It problem.

Work Exercises 15–20 in this textbook or in the *MyMathLab*® Study Plan.

## OBJECTIVE 2  DIVIDE REAL NUMBERS

Our next **arithmetic operation** is **division**. The result of dividing two **real numbers** is called the **quotient** of the numbers. The number being divided is called the **dividend**, and we divide by a number called the **divisor**.

$$\underset{20}{\overset{\text{dividend}}{\downarrow}} \quad \underset{\div}{\overset{\text{divisor}}{\downarrow}} \quad \underset{4}{} \quad \underset{=}{} \quad \underset{5}{\overset{\text{quotient}}{\downarrow}} \qquad \text{dividend} \rightarrow \frac{20}{4} = 5 \leftarrow \text{quotient} \quad \text{divisor} \rightarrow$$

In Section 1.2, we saw that **subtraction** is defined in terms of **addition**. We subtract a number by adding its **opposite**, or **additive inverse**.

$$a - b = a + (-b)$$

Similarly, we can define division in terms of **multiplication**. Consider the following:

$$10 \div 5 = \frac{10}{5} = \frac{10 \cdot 1}{1 \cdot 5} = \frac{10}{1} \cdot \frac{1}{5} = 10 \cdot \frac{1}{5}$$

The number $\frac{1}{5}$ is called the **reciprocal**, or **multiplicative inverse**, of 5. Two numbers are reciprocals if their **product** is 1. So, 5 and $\frac{1}{5}$ are reciprocals of each other because $5 \cdot \frac{1}{5} = 1$.

---

**Definition   Reciprocals (or Multiplicative Inverses)**

Two numbers are **reciprocals**, or **multiplicative inverses**, if their product is 1.

---

 The number 0 does not have a reciprocal. Find out **why**.

Every nonzero **real number** $b$ has a **reciprocal**, given by $\frac{1}{b}$. We can find the reciprocal of a **rational number** by inverting or "flipping" the **fraction**. View some **examples**.

 The reciprocal of a number will have the same sign as the number. For example, the reciprocal of $\frac{3}{4}$ is $\frac{4}{3}$, and the reciprocal of $-\frac{2}{3}$ is $-\frac{3}{2}$.

To divide by a number, we multiply by its reciprocal.

---

**Definition   Division of Two Real Numbers**

If $a$ and $b$ represent two **real numbers** and $b \neq 0$, then

$$a \div b = \frac{a}{b} = a \cdot \frac{1}{b}.$$

---

### Example 4  Dividing Two Real Numbers

Divide.

**a.** $\dfrac{-60}{4}$      **b.** $\dfrac{3}{25} \div \left(-\dfrac{9}{20}\right)$

### Solutions

**a.** $\dfrac{-60}{4} = \underbrace{-60 \cdot \dfrac{1}{4}}_{\substack{\text{Change to} \\ \text{multiplication} \\ \text{by the reciprocal}}} = -15 \leftarrow$ the reciprocal of 4 is $\frac{1}{4}$

**b.** $\dfrac{3}{25} \div \left(-\dfrac{9}{20}\right) = \underbrace{\dfrac{3}{25} \cdot \left(-\dfrac{20}{9}\right)}_{\substack{\text{Change to} \\ \text{multiplication} \\ \text{by the reciprocal}}}$

$\left|\dfrac{3}{25}\right| = \dfrac{3}{25}$ and $\left|-\dfrac{20}{9}\right| = \dfrac{20}{9}$, so the **absolute value** of the **product** is

$$\dfrac{3}{25} \cdot \dfrac{20}{9} = \underbrace{\dfrac{3^1}{25^5} \cdot \dfrac{20^4}{9^3}}_{\text{Simplify}} = \dfrac{1 \cdot 4}{5 \cdot 3} = \dfrac{4}{15}.$$

Since $\dfrac{3}{25}$ and $-\dfrac{20}{9}$ have different **signs**, the product is **negative**.

$$\dfrac{3}{25} \div \left(-\dfrac{9}{20}\right) = -\dfrac{4}{15}$$

We can use other techniques to **divide** real numbers in special cases. For example, dividing two **integers** is the same process as simplifying a **fraction** to **lowest terms**. Also, when dividing **decimals**, we can use **long division** and then determine the appropriate sign for the **quotient**.

 Because division is defined in terms of **multiplication**, the rules for finding the sign of a quotient are the same as the rules for finding the sign of a product.

---

**Dividing Two Real Numbers Using Absolute Value**

Divide the **absolute values** of the two **real numbers** to find the absolute value of the **quotient**. Determine the **sign** of the quotient by the following rules:

1. If the signs of the two numbers are the same (both **positive** or both **negative**), then their quotient is positive.

2. If the signs of the two numbers are different (one positive and one negative), then their quotient is negative.

3. The quotient of 0 and any non-zero real number is 0. So, $\dfrac{0}{b} = 0$ for $b \neq 0$. **(See why.)**

4. Division by zero, or $\dfrac{a}{0}$, is undefined.

---

## Example 5  Dividing Two Real Numbers

Divide.

**a.** $(-8) \div (-36)$      **b.** $\dfrac{0}{5}$      **c.** $\dfrac{12}{35} \div \left(-\dfrac{27}{14}\right)$      **d.** $15 \div \left(\dfrac{3}{4}\right)$

### Solution

**a.** The two numbers are the same sign so the quotient is positive. Dividing the **absolute values**, we get

$$\dfrac{8}{36} = \dfrac{2 \cdot 4}{9 \cdot 4} = \dfrac{2 \cdot \cancel{4}}{9 \cdot \cancel{4}} = \dfrac{2}{9}$$

The quotient is $\dfrac{2}{9}$.

b. Since the quotient of zero and a non-zero real number is $0$, $\dfrac{0}{5} = 0$.

*My video summary*  c. The two numbers have different **signs** so we divide the **absolute values** of the numbers and make the **quotient** negative. To divide, we multiply the first **fraction** by the **reciprocal** of the second fraction.

$$\frac{12}{35} \div \frac{27}{14} = \frac{12}{35} \cdot \frac{14}{27}$$

Try to finish this problem on your own. Check your **answer**, or watch this **video** for the complete solution.

*My video summary*  d. Try to work this problem on your own. Check your **answer**, or watch this **video** for the complete solution.

**You Try It**   Work through this You Try It problem.

Work Exercises 21–29 in this textbook or in the **MyMathLab** Study Plan.

**Example 6  Dividing Two Real Numbers**

Divide.

a. $\dfrac{48.6}{-3}$    b. $-7\dfrac{2}{5} \div (-3)$    c. $\dfrac{-59.4}{4.5}$    d. $6\dfrac{5}{8} \div 2\dfrac{1}{4}$

**Solutions**

a. The two numbers have different **signs** so we divide the **absolute values** of the numbers and make the quotient negative. (View the details.)

$$\begin{array}{r} 16.2 \\ 3\overline{)48.6} \end{array}$$

The quotient is $-16.2$.

b. The two numbers have the same **sign** (both negative) so the **quotient** will be positive. We change the **mixed number** to an **improper fraction** and then divide the **absolute values** of the two numbers.

$$7\frac{2}{5} \div 3 = \frac{37}{5} \div 3 = \underbrace{\frac{37}{5} \cdot \frac{1}{3}}_{\substack{\text{Change to} \\ \text{multiplication} \\ \text{by the} \\ \text{reciprocal}}} = \frac{37}{15} = 2\frac{7}{15}$$

So, $-7\dfrac{2}{5} \div (-3) = 2\dfrac{7}{15}$.

*My video summary*  c-d. Try to complete these problems on your own. Check your **answers**, or watch this **video** for the complete solutions.

**You Try It**   Work through this You Try It problem.

Work Exercises 30–34 in this textbook or in the **MyMathLab** Study Plan.

CAUTION  If a quotient is **negative**, it can be written in one of three ways: $-\dfrac{a}{b} = \dfrac{-a}{b} = \dfrac{a}{-b}$.

OBJECTIVE 3   TRANSLATE WORD STATEMENTS INVOLVING MULTIPLICATION OR DIVISION

We have already seen many key words and phrases that indicate **addition** or **subtraction**. Figure 15 shows some examples of key words and phrases for **multiplication**.

| Key Word | Word Phrase | Mathematical Expression |
|---|---|---|
| Product | The *product* of 7 and 8 | $7(8)$<br>[or $7 \times 8, 7 \cdot 8, 7*8$] |
| Times | $-7$ *times* $-9$ | $(-7)(-9)$ |
| Of (with fractions) | One-third *of* 27 | $\frac{1}{3}(27)$ |
| Of (with percents) | 15% *of* 200 | $0.15(200)$ |
| Twice | *Twice* 7 | $2(7)$ |

**Figure 15** Key words meaning multiplication

Just as there are many keywords to indicate multiplication, Figure 15 illustrates different **mathematical symbols** that can be used to represent multiplication. The most common symbols are the cross, $\times$, and the dot, $\cdot$. **Grouping symbols** such as parentheses can also be used to indicate multiplication. (Note: The asterisk, *, is also used to display multiplication, primarily on calculator view screens. View an **example**.)

### Example 7  Translating Word Statements Involving Multiplication

Write a **mathematical expression** for each word phrase.

**a.** The product of 3 and $-6$

**b.** 30% of 50

**c.** Three times the sum of 10 and 4

**d.** Three-fourths of 20, increased by 7

**e.** The difference of 2 and the product of 8 and 15

**f.** 3 increased by 15, times 4

### Solutions

**a.** The key word "**product**" indicates **multiplication**. What is being multiplied? 3 and $-6$. Notice that the key word "and" separates the quantities that are being multiplied.

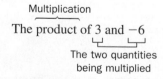

"The product of 3 and $-6$" is written as $3(-6)$.

**b.** The key word "of" indicates **multiplication** because it follows a percent.

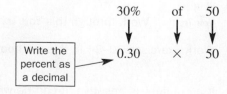

So, "30% of 50" is written as $0.30 \times 50$ or $0.30(50)$.

c. "Times" means multiplication. Three and the **sum** are being multiplied. "Sum" indicates **addition**, and the numbers 10 and 4 are being added.

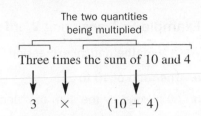

The sum is treated as a single quantity, so we put it within **grouping symbols.** "Three times the sum of 10 and 4" translates to $3 \times (10 + 4)$ or $3(10 + 4)$.

d. The comma separates the phrase into two parts: "Three-fourths of 20" and "increased by 7." The key word "of" indicates **multiplication,** and the numbers $\frac{3}{4}$ and 20 are being multiplied. We stop at the comma when determining the second **factor.** The phrase "increased by" indicates **addition.** We add 7 to everything before the comma: "three-fourths of 20."

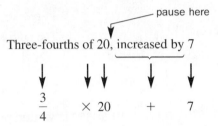

"Three-fourths of 20, increased by 7" translates to $\left( \frac{3}{4} \times 20 \right) + 7$ or $\frac{3}{4}(20) + 7$.

📹 *My video summary*    🎬  e. Try to translate this phrase on your own. Check your **answer,** or watch this **video** for the complete solution.

📹 *My video summary*    🎬  f. Try to translate this phrase on your own. Remember that the comma is used to separate phrases into parts. Check your **answer,** or watch this **video** for the complete solution.

**You Try It**    Work through this You Try It problem.

**Work Exercises 35–41 in this textbook or in the *MyMathLab*® Study Plan.**

How would the result in Example 7d change if the phrase did not contain a comma? **Find out.**

As with **multiplication,** there are many key words and phrases that indicate **division.** Figure 16 shows some examples.

| Key Word | Word Phrase | Mathematical Expression |
|----------|-------------|--------------------------|
| Quotient | The *quotient* of 12 and 6 | $\frac{12}{6}$ [or $12 \div 6$, $12/6$] |
| Divided by | 24 *divided by* $-3$ | $24 \div (-3)$ |
| Per | 3 tutors *per* 50 students | $\frac{3 \text{ tutors}}{50 \text{ students}}$ |
| Ratio | The *ratio* of 4 to 9 | $\frac{4}{9}$ |

**Figure 16**
Key words meaning division

**1.3** Multiplying and Dividing Real Numbers    **1-29**

When translating word statements involving division, typically the **numerator** is the first number or expression given and the **denominator** is the second number or expression.

### Example 8  Translating Word Statements Involving Division

Write a **mathematical expression** for each word phrase.

**a.** The ratio of 10 to 35

**b.** 60 divided by the sum of 3 and 7

**c.** The quotient of 20 and 4

**d.** The difference of 12 and 7, divided by the difference of 8 and $-3$

### Solutions

**a.** The key word "ratio" indicates **division**. What is being divided? The first number, 10, is being divided by the second number, 35. Remember that the order of the given numbers is important.

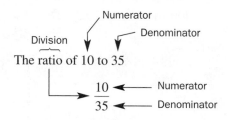

"The ratio of 10 to 35" is written as $\dfrac{10}{35}$.

**b.** "Divided by" means **division**. 60 is being divided by the **sum**, and "sum" means addition. What is being added? 3 and 7 are being added.

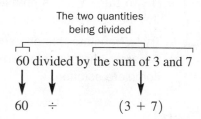

"60 divided by the sum of 3 and 7" translates to $60 \div (3 + 7)$ or $\dfrac{60}{3 + 7}$.

Note that the sum is grouped as a single quantity either by using a fraction bar or **grouping symbols**.

*My video summary*  **c–d.** Try to translate these phrases on your own. Check your **answers**, or watch this **video** for the complete solutions.

**You Try It**   Work through this You Try It problem.

**Work Exercises 42–48 in this textbook or in the *MyMathLab*®** **Study Plan.**

How would the result in Example 8d have changed if the phrase did not contain a comma? **Find out.**

OBJECTIVE 4    SOLVE APPLICATIONS INVOLVING MULTIPLICATION OR DIVISION

As with **addition** and **subtraction**, **multiplication** and **division** occur frequently in application problems.

### Example 9  Solving Applications Involving Multiplication or Division

For their 8 home games in the 2009 regular season, the Denver Broncos had a total attendance of 600,928. Assuming equal attendance at all home games, how many people attended each home game? (*Source: espn.go.com*)

**Solution**  The number attending each home game can be found by dividing the total attendance number by the number of home games.

$$\frac{600,928}{8} = 75,116$$

There were 75,116 people in attendance at each home game (See the **work**).

### Example 10  Solving Applications Involving Multiplication or Division

The amount of acid in a solution can be found by multiplying the volume of solution by the percent of acid in the solution (written in decimal form). How much acid is in 20 liters of a 3% solution?

**Solution**  We find the amount of acid by multiplying the solution volume by the percent of the solution. In decimal form, 3% is written as 0.03.

$$20(0.03) = 0.6$$

The solution contains 0.6 liter of acid.

**You Try It**    Work through this You Try It problem.

**Work Exercises 49–54 in this textbook or in the *MyMathLab*®  Study Plan.**

## 1.3  Exercises

In Exercises 1–20, multiply the real numbers.

**You Try It**

**1.** $(3)(-2)$

**2.** $4 \times 5$

**3.** $(-6)(-1)$

**4.** $9 \cdot 17$

**5.** $14(-5)$

**6.** $27 \cdot 0$

**7.** $(-18)(-52)$

**8.** $(-7)(-7)$

**You Try It**

**9.** $\left(-\frac{3}{2}\right)\left(-\frac{8}{9}\right)$

**10.** $\frac{3}{5}\left(-\frac{2}{7}\right)$

**11.** $0 \cdot \frac{4}{25}$

**12.** $(-7)\left(\frac{5}{14}\right)$

**13.** $\frac{4}{3} \times 8$

**14.** $\frac{7}{12} \cdot \frac{12}{5}$

**15.** $(5)(-2.6)$

**16.** $(-3.5)(4.2)$

**17.** $(-4.59)(0)$

**18.** $(-2.3)(-10.5)$

**19.** $8 \times 2\frac{1}{4}$

**20.** $\left(-3\frac{2}{9}\right)(6)$

**You Try It**

In Exercises 21–34, divide the real numbers.

**21.** $\dfrac{42}{-6}$

**22.** $160 \div (-20)$

**23.** $\dfrac{-64}{-4}$

**24.** $\dfrac{0}{-8}$

**25.** $-\dfrac{5}{8} \div \left(-\dfrac{3}{2}\right)$

**26.** $-15 \div \dfrac{5}{2}$

**27.** $\dfrac{1}{3} \div \dfrac{1}{4}$

**28.** $0 \div \left(-\dfrac{2}{5}\right)$

**29.** $\dfrac{\frac{5}{12}}{\frac{3}{10}}$

**30.** $\dfrac{-5}{0.4}$

**31.** $\dfrac{-0.6}{-4}$

**32.** $\dfrac{47.25}{-6.3}$

**33.** $-13 \div 4\dfrac{1}{3}$

**34.** $\left(3\dfrac{1}{3}\right) \div \left(-1\dfrac{1}{2}\right)$

In Exercises 35–48, write a mathematical expression for each word phrase.

**35.** 22% of 18

**36.** The product of 5 and −20

**37.** Two-thirds of −20

**38.** 3 more than twice 0.37

**39.** 19 decreased by 12, times $\dfrac{7}{8}$

**40.** 19, decreased by 12 times $\dfrac{7}{8}$

**41.** Eighteen more than the product of 0.7 and the difference of 20 and 4

**42.** The quotient of 24 and −7

**43.** 5 chaperones per 60 students

**44.** The ratio of 7 to 12

**45.** 3.76 divided by 5

**46.** 26% of 20, divided by 8

**47.** 30 increased by 12, divided by $\dfrac{7}{8}$

**48.** The difference of 5 and 9, divided by the difference of 10 and 4

In Exercises 49–54, solve each application by using multiplication or division.

**49.** In 2010, first-class postage for a standard postcard was $0.28. At this price, how much would it cost to mail 9 postcards? (*Source: usps.com*)

**50.** A serving of chicken rice soup contains 1.5 grams of fat. If a can of soup contains 2.5 servings, how many total grams of fat are in the can?

**51.** For 80 home games in the 2009 season, the Chicago Cubs had a total attendance of about 3,168,800. Assuming equal attendance at all home games, how many people attended each home game? (*Source: espn.go.com*)

**52.** A farmer decides to retire and distribute his 357 acres of land equally among his 5 children. How many acres will each child receive?

**53.** How much acid is in 40 liters of a 30% solution?

**54.** How much acid is in 7.5 gallons of a 7% solution?

# 1.4 Exponents and Order of Operations

## THINGS TO KNOW

Before working through this section, be sure you are familiar
with the following concepts:

|  |  | VIDEO | ANIMATION | INTERACTIVE |
|--|--|-------|-----------|-------------|

 **You Try It**   **1.** Add Two Real Numbers with Different Signs
(Section 1.2, **Objective 2**)

**You Try It**   **2.** Subtract Real Numbers
(Section 1.2, **Objective 3**)

**You Try It**   **3.** Multiply Real Numbers
(Section 1.3, **Objective 1**)

**You Try It**   **4.** Divide Real Numbers
(Section 1.3, **Objective 2**)

## OBJECTIVES

**1** Evaluate Exponential Expressions

**2** Use the Order of Operations to Evaluate Numeric Expressions

----

OBJECTIVE 1   EVALUATE EXPONENTIAL EXPRESSIONS

In **Section 1.3**, we learned that multiplication is repeated addition. Similarly, we can represent repeated multiplication by using *exponents*.

Consider the **product** $2 \cdot 2 \cdot 2 \cdot 2 \cdot 2$. The **factor** 2 is repeated 5 times. We can write this product as the *exponential expression* $2^5$. The number 5 is called the **exponent**, or **power**, and indicates that the factor 2 is repeated 5 times. The number 2 is called the **base** and indicates the factor that is being repeated.

$$\underbrace{2 \cdot 2 \cdot 2 \cdot 2 \cdot 2}_{5 \text{ factors of 2}} = 2^{\underset{\uparrow}{5}} \leftarrow \text{exponent}$$
$$\text{base}$$

We read $2^5$ as "2 raised to the fifth power."

----

**Definition   Exponential Expression**

If $a$ is a **real number** and $n$ is a **natural number**, then the **exponential expression** $a^n$ represents the product of $n$ factors of $a$.

$$a^n = \underbrace{a \cdot a \cdot a \cdot \ldots \cdot a}_{n \text{ factors of } a}$$

$a$ is the **base**, and $n$ is the **exponent** or **power**. We read $a^n$ as "$a$ raised to the $n$th power" or "$a$ to the $n$th."

----

When we raise a number to the first **power**, the exponent is 1. Any real number raised to the first power is equal to itself. For example, $2^1 = 2, 3^1 = 3$, and so on. If no exponent is written, it is assumed to be an **exponent of 1**.

We **evaluate** numeric **exponential expressions** by multiplying the **factors**. For example, $2^5 = 2 \cdot 2 \cdot 2 \cdot 2 \cdot 2 = 32$.

### Example 1 Evaluate Exponential Expressions

Evaluate each exponential expression.

**a.** $4^3$        **b.** $\left(\dfrac{2}{3}\right)^4$        **c.** $(0.3)^2$

### Solutions

**a.** $4^3$ means to multiply 3 factors of 4, so $4^3 = 4 \cdot 4 \cdot 4 = 64$.

**b.** $\left(\dfrac{2}{3}\right)^4$ means to multiply 4 factors of $\dfrac{2}{3}$, so $\left(\dfrac{2}{3}\right)^4 = \dfrac{2}{3} \cdot \dfrac{2}{3} \cdot \dfrac{2}{3} \cdot \dfrac{2}{3} = \dfrac{16}{81}$.

**c.** $(0.3)^2 = (0.3)(0.3) = 0.09$

**You Try It**   Work through this You Try It problem.

Work Exercises 1–5 in this textbook or in the *MyMathLab*® Study Plan.

When a negative **sign** is involved, we must see if the negative sign is part of the **base**. For example, consider $(-3)^2$. The parentheses around $-3$ indicate that the negative sign is part of the base.

$$\underset{\text{Base is } -3}{\underbrace{(-3)}}{}^{\overset{\text{Exponent}}{}2} = \overset{\overset{\text{Two factors}}{\text{of } -3}}{\overbrace{(-3)(-3)}} = 9.$$

In the expression $-3^2$, the negative sign is not part of the base. Instead, we must find the **opposite** of $3^2$.

$$-3^{\overset{\text{Exponent}}{}2} = \overset{\overset{\text{Two factors}}{\text{of } 3}}{-\overbrace{(3 \cdot 3)}} = -9.$$

Base is 3      Take the opposite

### Example 2 Evaluate Exponential Expressions

Evaluate each exponential expression.

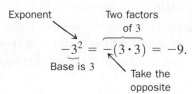

*My video summary*   **a.** $(-5)^3$        **b.** $-4^2$        **c.** $(-2)^4$

### Solutions

**a.** The parentheses around the $-5$ indicate that the base is $-5$, and the negative sign is part of the base. So, $(-5)^3 = (-5)(-5)(-5) = -125$.

**b.** Because there are no parentheses around the $-4$, the negative **sign** is not part of the base. Instead, we find the **opposite** of $4^2$. So, $-4^2 = -(4 \cdot 4) = -16$.

**c.** Try to evaluate this expression on your own. Check your **answer**, or watch this **video** for the complete solutions to all three parts.

**You Try It**   Work through this You Try It problem.

Work Exercises 6–12 in this textbook or in the *MyMathLab*® Study Plan.

When the base of an exponential expression is negative, how do we use the **exponent** to determine the sign of the result? **Find out.**

OBJECTIVE 2   USE THE ORDER OF OPERATIONS TO EVALUATE NUMERIC EXPRESSIONS

The **numeric expression** $2 + 3 \cdot 4$ contains two **operations**: addition and multiplication. Depending on which operation is performed first, a different answer will result.

Add first, then multiply:   $2 + 3 \cdot 4 = \quad 5 \cdot 4 = 20$

Different Results

Multiply first, then add:   $2 + 3 \cdot 4 = 2 + 12 = 14$

For this reason, mathematicians have agreed on a specific **order of operations**.

---

**Order of Operations**

1. **Parentheses (or other grouping symbols)**   Evaluate operations within parentheses (or other **grouping symbols**) first, starting with the innermost set and working out.

2. **Exponents**   Work from left to right and evaluate any **exponential expressions** as they occur.

3. **Multiplication and Division**   Work from left to right and perform any **multiplication** or **division** operations as they occur.

4. **Addition and Subtraction**   Work from left to right and perform any **addition** or **subtraction** operations as they occur.

---

Because we perform multiplication before addition, the correct evaluation for our numeric expression above is $2 + 3 \cdot 4 = 14$, not 20.

View this **tip** on remembering the order of operations.

### Example 3  Using Order of Operations to Evaluate Numeric Expressions

**Simplify** each expression.

a.  $10 - 4^2$                    b.  $12 \div 4 + 8$

**Solutions**  For each expression, we follow the **order of operations**.

a.  The two **operations** are subtraction and exponents. Exponents have priority over subtraction, so we evaluate the exponent first.

Begin with the original expression:   $10 - 4^2$

Evaluate $4^2$:   $= 10 - 16$

Subtract:   $- -6 \leftarrow 10 - 16 = 10 + (-16)$

b.  The two operations are division and addition. Because division has priority over addition, we divide first.

Begin with the original expression:   $12 \div 4 + 8$

Divide $12 \div 4$:   $= 3 + 8$

Add:   $= 11$

**You Try It**   **Work through this You Try It problem.**

**Work Exercises 13–20 in this textbook or in the *MyMathLab*® Study Plan.**

In the **order of operations**, multiplication and division are given the same priority and are performed from left to right as *either* occurs. It is incorrect to do all the multiplication first and then all the division. View this **illustration**.

Similarly, addition and subtraction are given equal priority and are performed from left to right as *either* occurs. It is incorrect to do all the addition first and then all the subtraction. View this **illustration**.

### Example 4  Using Order of Operations to Evaluate Numeric Expressions

Simplify each expression.

 **a.** $15 - 3 + 6 - 8 + 7$          **b.** $3 \cdot 15 \div 5 \cdot 6 \div 2$

**Solutions** For each expression, follow the order of operations. Try to do these on your own. Check your **answers**, or watch the **video** for complete solutions.

### Example 5  Using Order of Operations to Evaluate Numeric Expressions

Simplify each expression.

**a.** $5 + (4 - 2)^2 - 3^2$          **b.** $[5 - 9]^2 + 12 \div 4$

**Solutions** For each expression, we follow the order of operations.

**a.** Begin with the original expression:     $5 + (4 - 2)^2 - 3^2$

      Evaluate $4 - 2$ inside parentheses:     $= 5 + (2)^2 - 3^2$

           Evaluate $2^2$ and $3^2$:     $= 5 + 4 - 9$

                 Add $5 + 4$:     $= 9 - 9$

               Subtract $9 - 9$:     $= 0$

**b.** Begin with the original expression:     $[5 - 9]^2 + 12 \div 4$

      Evaluate $5 - 9$ inside brackets:     $= [-4]^2 + 12 \div 4$

              Evaluate $[-4]^2$:     $= 16 + 12 \div 4$

            Divide $12 \div 4$:     $= 16 + 3$

               Add $16 + 3$:     $= 19$

**You Try It**   Work through this You Try It problem.

Work Exercises 21–27 in this textbook or in the ***MyMathLab*** Study Plan.

Because **grouping symbols** have the highest priority in **order of operations**, we can use them to control the order in which computations are completed. This can be especially helpful if using a calculator to do computations.

### Example 6  Using Order of Operations to Evaluate Numeric Expressions

 Simplify each expression.

**a.** $(-5 + 8) \cdot 3$       **b.** $(10 - 4)^2$       **c.** $12 \div (4 + 8)$

**Solutions** Try to do these problems on your own. Check your **answers**, or watch the **video** for complete solutions. Compare your results in parts b and c to the results from **Example 3**. Notice how the parentheses affect which operation is performed first and the final result.

**You Try It**   Work through this You Try It problem.

Work Exercises 28–30 in this textbook or in the ***MyMathLab*** Study Plan.

Parentheses, ( ), brackets, [ ], and braces, { }, are the most common **grouping symbols**. Fraction bars, ——, however, and **absolute value** symbols, | |, are also treated as grouping symbols. A fraction bar separates the expression into two parts: the **numerator** and **denominator**. The numerator is grouped together and the denominator is grouped together. View this **illustration**.

Grouping the numerator together and grouping the denominator together is particularly important when using a calculator to **evaluate** an expression. A calculator will strictly follow order of operations and cannot guess what you meant to enter. View this **calculator example**.

We must **simplify expressions** separately in the numerator and denominator of a **fraction** before dividing. Similarly, we simplify expressions within absolute value symbols before evaluating the absolute value.

$$3^2 - 5 \cdot 2 + |3^2 - 8 \div 4|$$

Absolute value is treated as a grouping symbol so we simplify inside the absolute value first

Once the expression within the absolute value is simplified, the absolute value can be evaluated with the same priority as **exponents**. For example, in the expression $-5 \cdot |-6|$, we find the absolute value first and then perform the multiplication.

$$\text{Original expression:} \quad -5 \cdot |-6|$$
$$\text{Evaluate } |-6|: \quad = -5 \cdot 6$$
$$\text{Multiply } -5 \cdot 6: \quad = -30$$

## Example 7 Using Order of Operations with Special Grouping Symbols

**Simplify** each expression.

a. $\dfrac{-2(3) + 6^2}{(-4)^2 - 1}$

b. $|7^2 - 5(3)| \div 2 + 8$

### Solutions

a. The **fraction bar** is a **grouping symbol**. We begin by simplifying the numerator and denominator separately before dividing.

$$\text{Begin with the original expression:} \quad \dfrac{-2(3) + 6^2}{(-4)^2 - 1}$$

$$\text{Evaluate the exponents:} \quad = \dfrac{-2(3) + 36}{16 - 1} \quad \begin{array}{l} \leftarrow 6^2 = 36 \\ \leftarrow (-4)^2 = 16 \end{array}$$

$$\text{Multiply in the numerator:} \quad = \dfrac{-6 + 36}{16 - 1}$$

$$\text{Add and subtract:} \quad = \dfrac{30}{15} \quad \begin{array}{l} \leftarrow -6 + 36 = 30 \\ \leftarrow 16 - 1 = 15 \end{array}$$

$$\text{Divide:} \quad = 2$$

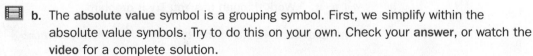 b. The **absolute value** symbol is a grouping symbol. First, we simplify within the absolute value symbols. Try to do this on your own. Check your **answer**, or watch the **video** for a complete solution.

**You Try It** Work through this You Try It problem.

**Work Exercises 31–36** in this textbook or in the *MyMathLab* Study Plan.

Some **numeric expressions** contain **nested grouping symbols** (grouping symbols within grouping symbols). In such situations, we begin with the innermost set of grouping symbols and work our way outward. To avoid confusion, we can use different types of grouping symbols if more than one set is required.

### Example 8  Using Order of Operations with Nested Grouping Symbols

Simplify each expression.

**a.** $[2^3 - 3(5 - 7)^2] \div 6 - 9$

**b.** $\dfrac{|-5^2 + 2^3| - 10}{4^2 - 6 \cdot 5}$

### Solutions

**a.** Begin with the original expression: $\quad [2^3 - 3(\underbrace{5 - 7})^2] \div 6 - 9$

Innermost grouping 1st

$\qquad$ Evaluate $(5 - 7)$: $\quad = [2^3 - 3(-2)^2] \div 6 - 9$

Outermost grouping 2nd

$\qquad$ Evaluate exponents $2^3$ and $(-2)^2$: $\quad = [8 - 3(4)] \div 6 - 9$

$\qquad$ Multiply $3(4)$: $\quad = [8 - 12] \div 6 - 9$

$\qquad$ Subtract $8 - 12$: $\quad = (-4) \div 6 - 9$

$\qquad$ Divide $-4 \div 6$: $\quad = -\dfrac{2}{3} - 9 \; \leftarrow \; \boxed{\dfrac{-4}{6} = \dfrac{-2 \cdot 2}{3 \cdot 2} = \dfrac{-2}{3} = -\dfrac{2}{3}}$

$\qquad$ Subtract $-\dfrac{2}{3} - 9$: $\quad = -\dfrac{29}{3} \; \leftarrow \; \boxed{-\frac{2}{3} - 9 = -\frac{2}{3} - \frac{27}{3} = -\frac{2}{3} + \left(-\frac{27}{3}\right) = -\frac{29}{3}}$

*My video summary*  **b.** The **absolute value** bars act as grouping symbols. First, we simplify within the absolute value symbols. Try to do this problem on your own, then check your **answer**. Watch this **video** for a complete solution.

**You Try It**  Work through this You Try It problem.

Work Exercises 37–40 in this textbook or in the *MyMathLab*® Study Plan.

### Example 9  Using Order of Operations to Evaluate Numeric Expressions

*My interactive video summary*  Simplify each expression.

**a.** $\dfrac{3}{10} \cdot \dfrac{5}{2} - \dfrac{1}{2}$ $\qquad$ **b.** $36 \div \dfrac{8}{3^2 - 5} + (-2)^3$ $\qquad$ **c.** $\dfrac{\left|\dfrac{1}{3} - \dfrac{3}{5}\right|}{4} \div \dfrac{1}{2} - 1$

**Solutions**  For each expression, follow the **order of operations**. Check your **answers**, or watch the **interactive video** for the complete solutions.

**You Try It**  Work through this You Try It problem.

Work Exercises 41–48 in this textbook or in the *MyMathLab*® Study Plan.

# 1.4 Exercises

In Exercises 1–12, evaluate each exponential expression.

**You Try It**

**1.** $2^8$    **2.** $(0.2)^5$    **3.** $\left(\dfrac{3}{4}\right)^4$    **4.** $\left(\dfrac{3}{2}\right)^4$    **5.** $1^9$    **6.** $(-3)^4$

**You Try It**

**7.** $-3^4$    **8.** $-\left(\dfrac{3}{5}\right)^3$    **9.** $-\left(-\dfrac{3}{5}\right)^3$    **10.** $-(0.1)^4$    **11.** $-(-0.1)^4$    **12.** $0^3$

In Exercises 13–48, simplify each expression using the order of operations.

**You Try It**

**13.** $3\cdot 2 + 4$    **14.** $3 + 2\cdot 4$    **15.** $3 - 2^2$    **16.** $6 \div 3 - 1$

**17.** $\dfrac{1}{2} \div \dfrac{1}{3} + \dfrac{1}{4}$    **18.** $-6 + 3 - 1$    **19.** $6 \div 2\cdot\dfrac{1}{3}$    **20.** $(3 + 2)\cdot 4$

**You Try It**

**21.** $20 - 4\cdot 9 + 12$    **22.** $70 - 42 \div 7 + 9$    **23.** $8 \div 4 + 3^2$    **24.** $2(-3 + 5)^2 - (8 - 5)^2$

**25.** $(2\cdot 3)^2 + 2\cdot 3^2$    **26.** $2(9 - 5)^3$    **27.** $7\cdot 2 + 3^2$    **28.** $2^2 - 5\cdot 2 + (2^2 - 5)\cdot 2$

**You Try It**

**29.** $6 \div (3 - 1) + 6 \div 3 - 1$    **30.** $(40 \div 2)^2 - 40 \div 2^2$

**31.** $\dfrac{2 + 5\cdot 3}{2\cdot 5 + 3}$    **32.** $|-2|^3 - |3|^2$

**You Try It**

**33.** $\dfrac{|6 - 11| + 4}{4^2 - 2(5)}$    **34.** $\dfrac{5^2 - 3^2}{5 - 3}$    **35.** $\dfrac{3(15)-5}{(-2)^2 + 3}$    **36.** $\dfrac{2 - 7\cdot 1}{3\cdot 2 - 1} \div \dfrac{2\cdot 7 + 1}{3 + 2\cdot 1}$

**You Try It**

**37.** $[5 + 3(10 - 7)^2]$    **38.** $\dfrac{(3 + 2^3)^2 - 11}{7(4 - 5)-3}$

**39.** $|1 - 3|^3\cdot\left[6\cdot\dfrac{2}{3} + 4\left(1 + \dfrac{3}{4}\right)\right]$    **40.** $\left\{\left[(-2)^3 + \dfrac{2}{3}\cdot 6\right] \div 12\right\}^5$

**You Try It**

**41.** $12 \div 4 \div 3$    **42.** $24 \div (6\cdot 2)$    **43.** $12 \div 4\cdot 3$    **44.** $24 \div (6 \div 2)$

**45.** $1 + 2\cdot 3 - 4 \div 5$    **46.** $7 - 2(3 - 1)$    **47.** $3^2 + 2(3 - 5)$    **48.** $\dfrac{\left(\dfrac{3}{2} + 1\right) \div \dfrac{1}{2}}{2^6 - 8\cdot 4}$

# 1.5 Variables and Properties of Real Numbers

## THINGS TO KNOW

Before working through this section, be sure you are familiar with the following concepts:

VIDEO    ANIMATION    INTERACTIVE

**You Try It**

**1.** Find the Opposite of a Real Number (Section 1.1, **Objective 3**)

**You Try It**

**2.** Add Two Real Numbers with Different Signs (Section 1.2, **Objective 2**)

**You Try It**

**3.** Subtract Real Numbers (Section 1.2, **Objective 3**)

**You Try It**

**4.** Multiply Real Numbers (Section 1.3, **Objective 1**)

**You Try It**

**5.** Use the Order of Operations to Evaluate Numeric Expressions (Section 1.4, **Objective 2**)

## OBJECTIVES

**1** Evaluate Algebraic Expressions

**2** Use the Commutative and Associative Properties

**3** Use the Distributive Property

**4** Use the Identity and Inverse Properties

---

OBJECTIVE 1    EVALUATE ALGEBRAIC EXPRESSIONS

So far, we have focused only on **numeric expressions**. Now we expand our discussion to include *algebraic expressions*.

A **variable** is a symbol (usually a letter) that represents a changing value. For example, suppose the letter $x$ represents the number of iTunes downloads purchased by a user. This might be 20 for one user and 2000 for another. A **constant** has a value that never changes. For example, 3 is a constant because the value of 3 is always the same.

An **algebraic expression** is a variable or a combination of variables, **constants**, **operations**, and **grouping symbols**. Examples of algebraic expressions include

$$\underset{\text{a number}}{x} \quad, \quad \underset{\substack{\text{five times} \\ \text{a number}}}{5z} \quad, \quad \underset{\substack{\text{the square} \\ \text{of a number}}}{x^2} \quad, \quad \underset{\substack{\text{a number} \\ \text{increased by 7}}}{w + 7} \quad, \quad \underset{\substack{\text{three less than} \\ \text{five times a number}}}{5a - 3} \quad, \text{ and } \quad \underset{\substack{\text{twice the sum of} \\ \text{two numbers}}}{2(x + y)} \quad.$$

If a constant appears next to a variable, then we multiply the number and the variable. For example, $5z = 5 \cdot z$ and $5a - 3 = 5 \cdot a - 3$.

**Algebraic expressions** can describe quantities in a general way. For example, if the common price for an iPhone app is $0.99 (*Source:* pinchmedia.com), then the algebraic expression

$0.99x$ gives us the cost for purchasing $x$ apps at this price. To determine the cost for a specific number of apps, we replace the variable with a number and then **simplify** the resulting **numeric expression**. This process is called **evaluating** the expression for a given value of the variable.

For example, to find the cost for purchasing 6 apps, we substitute 6 for $x$ and simplify the resulting numeric expression, or **evaluate** the algebraic expression $0.99x$ for $x = 6$:

$$0.99x = 0.99(6) = 5.94$$

The value of the algebraic expression $0.99x$ is 5.94 when $x = 6$, so 6 apps cost $5.94. When the value of the **variable** changes, the value of the algebraic expression will change as well.

---

**Evaluate Algebraic Expressions**

To **evaluate an algebraic expression**, substitute the given values for the variables and simplify the resulting numeric expression using the **order of operations**.

---

### Example 1  Evaluating Algebraic Expressions

Evaluate each **algebraic expression** for the given values of the variables. Remember to follow **order of operations** when simplifying.

**a.**  $3x + 7$ for $x = 5$

**b.**  $a^2 - 3$ for $a = -4$

### Solutions

**a.**  Begin with the original algebraic expression:  $3x + 7$

Substitute 5 for $x$:  $= 3(5) + 7$

Multiply:  $= 15 + 7$

Add:  $= 22$

**b.**  Begin with the original algebraic expression:  $a^2 - 3$

Substitute $-4$ for $a$:  $= (-4)^2 - 3$

Evaluate the exponent:  $= 16 - 3$

Subtract:  $= 13$

**You Try It**   Work through this You Try It problem.

**Work Exercises 1–5 in this textbook or in the *MyMathLab*®  Study Plan.**

When substituting a given value for a variable, write the value within parentheses to help avoid mathematical mistakes.

In Example 1, notice that we put parentheses around given values for a variable during the **substitution** process. View this **popup** to see how the problem would be different if we had not used parentheses in part (b).

### Example 2  Evaluating Algebraic Expressions

Evaluate each **algebraic expression** for the given values of the **variables**. Remember to follow **order of operations** when simplifying.

**a.**  $5x^2 + 9$ for $x = 7$

**b.**  $-2(m + 3) - 5$ for $m = 8$

## Solutions

**a.** Begin with the original algebraic expression:  $5x^2 + 9$

Substitute 7 for $x$: $= 5(7)^2 + 9$

Evaluate the exponent: $= 5(49) + 9$

Multiply: $= 245 + 9$

Add: $= 254$

*My video summary*  **b.** Begin with the original algebraic expression:  $-2(m + 3) - 5$

Substitute 8 for $m$: $= -2((8) + 3) - 5$

Try to finish **simplifying** on your own. Check your **answer**, or watch this **video** for the complete solution.

**You Try It**  Work through this You Try It problem.

**Work Exercises 6–10 in this textbook or in the *MyMathLab*®** Study Plan.

### Example 3  Evaluating Algebraic Expressions

Evaluate each **algebraic expression** for the given values of the **variables**.

**a.** $\dfrac{x^2 + 6}{5x - 2}$ for $x = 2$

**b.** $|3y - 4| + 7y - 1$ for $y = -3$

## Solutions

**a.** Begin with the original algebraic expression: $\dfrac{x^2 + 6}{5x - 2}$

Substitute 2 for $x$: $= \dfrac{(2)^2 + 6}{5(2) - 2}$

Evaluate the exponent: $= \dfrac{4 + 6}{5(2) - 2}$

Multiply: $= \dfrac{4 + 6}{10 - 2}$

Add and subtract: $= \dfrac{10}{8}$

Simplify: $= \dfrac{5}{4}$

*My video summary*  **b.** Try evaluating this expression on your own using the **order of operations**. Check your **answer**, or watch this **video** for the complete solution.

**You Try It**  Work through this You Try It problem.

**Work Exercises 11–16 in this textbook or in the *MyMathLab*®** Study Plan.

Given values for the **variables**, we can evaluate **algebraic expressions** involving more than one variable.

### Example 4  Evaluating Algebraic Expressions

Evaluate each algebraic expression for the given values of the variables.

**a.** $12a + 7b$ for $a = -4$ and $b = 12$

**b.** $x^2 - 2xy + 3y^2$ for $x = 3$ and $y = -1$

**Solutions**

a. Begin with the original expression:  $12a + 7b$

Substitute $-4$ for $a$ and $12$ for $b$:  $= 12(-4) + 7(12)$

Multiply:  $= -48 + 84$

Add:  $= 36$

 *My video summary*   b. Try to **simplify** on your own by following the **order of operations**. Check your **answer**, or watch this **video** for the complete solutions.

**You Try It**  Work through this You Try It problem.

**Work Exercises 17–24 in this textbook or in the *MyMathLab*® Study Plan.**

OBJECTIVE 2   USE THE COMMUTATIVE AND ASSOCIATIVE PROPERTIES

In this chapter, we have reviewed how to simplify **numeric expressions** involving **addition**, **subtraction**, **multiplication**, and **division**. To help with this process, let's review and name several properties of **real numbers**.

The **commutative property of addition** states that the *order* of the **terms** in addition does not affect the **sum**. For example, $4 + 7 = 11$ and $7 + 4 = 11$, so $4 + 7 = 7 + 4$.

> **Commutative Property of Addition**
>
> If $a$ and $b$ are real numbers, then $a + b = b + a$.

The **commutative property of multiplication** states that the order of the **factors** does not affect the **product**. For example, $8(-4) = -32$ and $(-4)(8) = -32$, so $8(-4) = (-4)(8)$.

> **Commutative Property of Multiplication**
>
> If $a$ and $b$ are real numbers, then $a \cdot b = b \cdot a$.

**CAUTION** Subtraction and division **do not** have commutative properties. For example, $9 - 2 = 7$ but $2 - 9 = -7$. Similarly, $6 \div 3 = 2$ but $3 \div 6 = 0.5$.

### Example 5  Using the Commutative Properties

Use the given property to rewrite each statement. Do not simplify.

a. **Commutative property of multiplication:**  $-2(6) = $ _____

b. **Commutative property of addition:**   $5.03 + 9.2 = $ _____

**Solutions**

a. Using the commutative property of multiplication, we can change the order of the factors: $-2(6) = 6(-2)$.

b. The commutative property of addition allows us to change the order of the **terms**. $5.03 + 9.2 = 9.2 + 5.03$.

**You Try It**  Work through this You Try It problem.

**Work Exercises 25–30 in this textbook or in the *MyMathLab*® Study Plan.**

The **associative property of addition** states that regrouping terms does not affect the sum. For example, $(7 + 2) + 5 = 9 + 5 = 14$ and $7 + (2 + 5) = 7 + 7 = 14$, so $(7 + 2) + 5 = 7 + (2 + 5)$.

---

**Associative Property of Addition**

If $a$, $b$, and $c$ are real numbers, then $(a + b) + c = a + (b + c)$.

---

The **associative property of multiplication** states that regrouping **factors** does not affect the product. For example, $(-10 \cdot 4) \cdot 6 = (-40) \cdot 6 = -240$ and $-10 \cdot (4 \cdot 6) = -10 \cdot (24) = -240$, so $(-10 \cdot 4) \cdot 6 = -10 \cdot (4 \cdot 6)$.

---

**Associative Property of Multiplication**

If $a$, $b$, and $c$ are real numbers, then $(a \cdot b) \cdot c = a \cdot (b \cdot c)$.

---

 Subtraction and division **do not** have associative properties. For example, $(7 - 4) - 2 = 3 - 2 = 1$ but $7 - (4 - 2) = 7 - 2 = 5$. Similarly, $(80 \div 8) \div 2 = 10 \div 2 = 5$ but $80 \div (8 \div 2) = 80 \div 4 = 20$.

### Example 6  Using the Associative Properties

Use the given property to rewrite each statement. Do not simplify.

a. **Associative property of addition:** $\left(\dfrac{2}{3} + \dfrac{1}{6}\right) + \dfrac{5}{6} = $ _____

b. **Associative property of multiplication:** $5 \cdot (2 \cdot 13) = $ _____

### Solutions

a. Using the associative property of addition, we can change the grouping of the **terms**:
$$\left(\frac{2}{3} + \frac{1}{6}\right) + \frac{5}{6} = \frac{2}{3} + \left(\frac{1}{6} + \frac{5}{6}\right).$$

b. The associative property of multiplication allows us to change the grouping of the factors:
$$5 \cdot (2 \cdot 13) = (5 \cdot 2) \cdot 13.$$

**You Try It**    Work through this You Try It problem.

**Work Exercises 31–38 in this textbook or in the *MyMathLab*® Study Plan.**

The **commutative** and **associative** properties will be helpful when we later simplify expressions and solve equations.

### Example 7  Using the Commutative and Associative Properties

Use the commutative and associative properties to simplify each expression.

a. $(3 + x) + 7$

b. $(8y)\left(\dfrac{1}{2}\right)$

## Solutions

a.  Commutative property of addition: $(3 + x) + 7 = \overset{\overset{\text{Changed order}}{\text{of the terms}}}{(x + 3)} + 7$

Associative property of addition: $= \overset{\overset{\text{Changed grouping}}{\frown}}{x + (3 + 7)}$

Add 3 and 7: $= x + \overset{\overset{3 + 7 = 10}{\frown}}{10}$

b.  Commutative property of multiplication: $(8y)\left(\dfrac{1}{2}\right) = \overset{\overset{\text{Changed order}}{\text{of the terms}}}{\dfrac{1}{2}(8y)}$

Associative property of multiplication: $= \overset{\overset{\text{Changed grouping}}{\frown}}{\left(\dfrac{1}{2} \cdot 8\right)y}$

Multiply $\dfrac{1}{2}$ and 8: $= \overset{\overset{\frac{1}{2} \cdot 8 = 4}{\frown}}{4y}$

**You Try It**    **Work through this You Try It problem.**

**Work Exercises 39–42 in this textbook or in the *MyMathLab*®  Study Plan.**

## OBJECTIVE 3   USE THE DISTRIBUTIVE PROPERTY

Another important property is the **distributive property**, which states that **multiplication** *distributes* over **addition** (or **subtraction**). This means that multiplying a number by a sum is **equivalent** to first multiplying each term by the number and then summing the results. For example, compare $5(3 + 4) = 5(7) = 35$ with $5 \cdot 3 + 5 \cdot 4 = 15 + 20 = 35$. Because both expressions simplify to 35, they are equivalent: $5(3 + 4) = 5 \cdot 3 + 5 \cdot 4$.

---

**Distributive Property**

If $a$, $b$, and $c$ are **real numbers**, then $a(b + c) = ab + ac$.

---

Because multiplication is **commutative**, we can also write the distributive property as

$$(b + c)a = ba + ca.$$

In **Section 1.2**, we defined subtraction by using addition. So the distributive property also applies to subtraction.

$$a(b - c) = ab - ac$$

The **distributive property** extends to sums (or differences) involving more than two **terms**.

$$a(b + c + d) = ab + ac + ad$$

See **why this works.**

The distributive property will be helpful because it allows us to remove **grouping symbols** so we can combine terms, then regroup in a simpler form.

### Example 8  Using the Distributive Property

Use the **distributive property** to remove parentheses and write the **product** as a **sum**. Simplify if possible.

**a.** $9(x + 2)$          **b.** $(7x - 5) \cdot 3$

### Solutions

**a.**
Apply the distributive property:    $9(x + 2) = 9 \cdot x + 9 \cdot 2$

Find each product:    $= 9x + 18$

**b.**
Apply the **distributive property:**    $(7x - 5) \cdot 3 = 7x \cdot 3 - (5) \cdot 3$

Commutative property of multiplication:    $= 3 \cdot 7x - (5) \cdot 3$

Find each product:    $= 21x - 15$

**You Try It**  Work through this You Try It problem.

**Work Exercises 43–50 in this textbook or in the *MyMathLab*®️ Study Plan.**

In **Section 1.1**, we used a negative **sign** to indicate the **opposite** of a real number. We can also find a number's opposite if we multiply the number by $-1$. For example, the opposite of 8 is $-1 \cdot 8 = -8$, and the opposite of $-3$ is $-1(-3) = -(-3) = 3$.

The **expression** $-1 \cdot (3z - 4)$ could also be written as $-(3z - 4)$. This simply means we want to find the opposite of the expression inside the parentheses. The result $-3z + 4$ is found by writing the opposite of each **term** within the parentheses. This leads us to the following rule.

---

**Opposite of an Expression**

If a negative sign appears in front of an expression within parentheses, remove the parentheses and then change the sign of each term in the expression. The result is the **opposite of the expression**.

---

### Example 9  Using the Distributive Property

Use the **distributive property** to remove parentheses and write the **product** as a **sum**. Simplify if possible.

**a.** $2(4y + 3z - 5)$      **b.** $-6(3y - 8)$      **c.** $-(2a - 7b + 8)$

### Solutions

**a.**
Begin with the original expression:    $2(4y + 3z - 5)$

Apply the distributive property:    $= 2 \cdot 4y + 2 \cdot 3z - 2 \cdot 5$

Associative property of multiplication:    $= (2 \cdot 4)y + (2 \cdot 3)z - (2 \cdot 5)$

Find each product:    $= 8y + 6z - 10$

*My video summary*    **b–c.** Try to work these problems on your own. Check your **answers**, or watch this **video** for the complete solutions.

**You Try It**  Work through this You Try It problem.

**Work Exercises 51–60 in this textbook or in the *MyMathLab*®️ Study Plan.**

### Example 10 Using the Distributive Property

Use the **distributive property** to write each sum as a product.

*My video summary*  **a.** $9 \cdot x + 9 \cdot 4$ **b.** $4x + 4y$

**Solutions** Remove the common **factor** from each **term**, then write the remaining expression in parentheses. Work through the following, or watch the **video** for the solutions.

**a.** Begin with the original expression: $9 \cdot x + 9 \cdot 4$

Remove the common factor: $= (9 \cdot x + 9 \cdot 4)$

Remaining expression in parentheses: $= 9(x + 4)$

**b.** Begin with the original expression: $4x + 4y$

Remove the common factor: $= (4x + 4y)$

Remaining expression in parentheses: $= 4(x + y)$

**You Try It** Work through this You Try It problem.

Work Exercises 61–64 in this textbook or in the **MyMathLab**® Study Plan.

## OBJECTIVE 4    USE THE IDENTITY AND INVERSE PROPERTIES

We complete our discussion of the properties of **real numbers** with the *identity* and *inverse properties*.

If 0 is added to any real number, then the **sum** is the original number. For example, $4 + 0 = 4$ and $0 + 7 = 7$. This is called the **additive identity**.

---

**Identity Property of Addition**

If $a$ is a real number, then $a + 0 = a$ and $0 + a = a$.

---

Similarly, if 1 is multiplied by a real number, then the **product** is the original number. For example, $5 \cdot 1 = 5$ and $1 \cdot 9 = 9$. This is called the **multiplicative identity**.

---

**Identity Property of Multiplication**

If $a$ is a real number, then $a \cdot 1 = a$ and $1 \cdot a = a$.

---

The **opposite** of a real number is also called the **additive inverse** of the number because the sum of a number and its inverse is zero. For example, $-6$ is the opposite of 6, and $-6$ is the additive inverse of 6 because $6 + (-6) = 0$.

---

**Inverse Property of Addition**

If $a$ is a real number, then there is a **unique** real number $-a$ such that

$$a + (-a) = -a + a = 0.$$

---

Similarly, the **reciprocal** of a **real number** is called the **multiplicative inverse** of the number because the product of a number and its reciprocal is 1. For example, the reciprocal, or multiplicative inverse, of $\frac{5}{3}$ is $\frac{3}{5}$ because $\frac{5}{3} \cdot \frac{3}{5} = 1$.

---

**Inverse Property of Multiplication**

If $a$ is a nonzero real number, then there is a unique real number $\frac{1}{a}$ such that

$$a \cdot \frac{1}{a} = \frac{1}{a} \cdot a = 1.$$

---

### Example 11 Using the Identity and Inverse Properties

Identify the property of real numbers illustrated in each statement.

**a.** $-4 \cdot 1 = -4$    **b.** $(-5 + 5) + x = 0 + x$

**c.** $0 + y = y$    **d.** $\frac{1}{2} \cdot 2x = x$

**Solutions** Try to identify each property on your own. Check your **answers**.

**You Try It**   Work through this You Try It problem.

**Work Exercises 65–68 in this textbook or in the *MyMathLab*® Study Plan.**

# 1.5 Exercises

In Exercises 1–16, evaluate each algebraic expression for the given value of the variable.

**You Try It**

**1.** $2x + 6$ for $x = 7$

**2.** $8 - 5x$ for $x = -3$

**3.** $4w^2$ for $w = -6$

**4.** $\frac{3}{5}x - 7$ for $x = 10$

**5.** $\frac{3}{4}x - 9$ for $x = -\frac{5}{3}$

**6.** $\frac{2.5y + 3.8}{4}$ for $y = 1.4$

**You Try It**

**7.** $3a^2 - 10$ for $a = 4$

**8.** $-2(n + 4) + 9$ for $n = \frac{1}{2}$

**9.** $\frac{4b + 3}{5}$ for $b = -2$

**10.** $4|5 - x| + 1$ for $x = -8$

**11.** $-z^2 + 3z$ for $z = -1$

**12.** $5 - 4y + 2y^2$ for $y = 3$

**You Try It**

**13.** $\frac{2x + 6}{x - 5}$ for $x = 10$

**14.** $\frac{7 - 3x}{x^2 + 1}$ for $x = -2$

**15.** $|2a - 3| + a - 8$ for $a = -5$

**16.** $\frac{2(c + 3)}{c - 3}$ for $c = \frac{2}{3}$

In Exercises 17–24, evaluate each algebraic expression for the given values of the variables.

**17.** $2x - 3y$ for $x = -4$ and $y = 5$

**18.** $20a + 9b$ for $a = -\frac{1}{4}$ and $b = \frac{2}{3}$

**19.** $m^2 - 4mn - 12n^2$ for $m = 1$ and $n = -3$

**20.** $a^2 - b^2$ for $a = -3$ and $b = -5$

**21.** $|3x^2 - y|$ for $x = 3$ and $y = -2$        **22.** $mx + b$ for $m = \dfrac{1}{5}, x = 15$, and $b = -4$

**23.** $b^2 - 4ac$ for $a = 1, b = -3$, and $c = -6$

**24.** $\dfrac{y_2 - y_1}{x_2 - x_1}$ for $x_1 = 9, x_2 = -3, y_1 = -4$, and $y_2 = 2$

In Exercises 25–30, use a commutative property to complete each statement. Do not simplify.

**You Try It**

**25.** $5 + 9 = $ _____      **26.** $-6 \cdot y = $ _____      **27.** $11 + x = $ _____

**28.** $\dfrac{1}{5}x = $ _____      **29.** $8.3(-3) = $ _____      **30.** $m + (-8) = $ _____

In Exercises 31–38, use an associative property to complete each statement. Do not simplify.

**You Try It**

**31.** $(3 + 6) + 15 = $ _____        **32.** $-3(4 \cdot 17) = $ _____

**33.** $(7y) \cdot 4 = $ _____        **34.** $\dfrac{3}{4} + \left(x + \dfrac{5}{9}\right) = $ _____

**35.** $(1.45 + a) + b = $ _____        **36.** $0.05(20x) = $ _____

**37.** $-6 + (6 + c) = $ _____        **38.** $\dfrac{3}{5}\left(\dfrac{5}{3}w\right) = $ _____

In Exercises 39–42, use the commutative and associative properties to simplify each expression.

**You Try It**

**39.** $-2 + (7 + x)$     **40.** $-4(2m)$     **41.** $(3z) \cdot 8$     **42.** $\left(\dfrac{1}{3} + y\right) + \dfrac{2}{5}$

In Exercises 43–60, use the distributive property to remove parentheses and write the product as a sum. Simplify if possible.

**You Try It**

**43.** $3(a + b)$     **44.** $10(x - 4)$     **45.** $4(2y - 5)$     **46.** $3.2(5x + 2)$

**47.** $(8x + 3) \cdot 4$     **48.** $-1 \cdot (2z + 7)$     **49.** $(3x + 2y)(-4)$     **50.** $\dfrac{2}{3}(3x + 9)$

**You Try It**

**51.** $-(7 - n)$     **52.** $0.5(4x - 1)$     **53.** $(3x - 16) \cdot \dfrac{1}{4}$     **54.** $-4\left(\dfrac{3}{2}m - \dfrac{1}{10}\right)$

**55.** $0.2(1.5 + 3.4x)$     **56.** $-(3a - 2b - 4)$     **57.** $12(x + 4 - 5y)$     **58.** $-2(4w - 9z)$

**59.** $3(5x + y - 7)$     **60.** $(4r + s + 5)(-0.4)$

In Exercises 61–64, use the distributive property to write each sum as a product.

**You Try It**

**61.** $4 \cdot x + 4 \cdot 1$     **62.** $10a + 10b$     **63.** $\dfrac{1}{2} \cdot y - \dfrac{1}{2} \cdot 6$     **64.** $(-5)m + (-5)(2)$

In Exercises 65–68, identify the property of real numbers illustrated in each statement.

**You Try It**

**65.** $20 \cdot 1 = 20$     **66.** $x + 14 + (-14) = x$     **67.** $\left(\dfrac{1}{3} \cdot 3\right)x = 1x$     **68.** $0 + 9 = 9$

# 1.6 Simplifying Algebraic Expressions

## THINGS TO KNOW

Before working through this section, be sure you are familiar with the following concepts:

|  |  | VIDEO | ANIMATION | INTERACTIVE |

**You Try It**

1. Translate Word Statements Involving Addition or Subtraction (Section 1.2, **Objective 4**) 🎞️

**You Try It**

2. Translate Word Statements Involving Multiplication or Division (Section 1.3, **Objective 3**) 🎞️

**You Try It**

3. Use the Order of Operations to Evaluate Numeric Expressions (Section 1.4, **Objective 2**) 🎞️  ▶️

## OBJECTIVES

**1** Identify Terms, Coefficients, and Like Terms of an Algebraic Expression

**2** Simplify Algebraic Expressions

**3** Write Word Statements as Algebraic Expressions

**4** Solve Applied Problems Involving Algebraic Expressions

---

OBJECTIVE 1   IDENTIFY TERMS, COEFFICIENTS, AND LIKE TERMS OF AN ALGEBRAIC EXPRESSION

In **Section 1.2**, we learned that numbers being added are called *terms*. Similarly, the **terms** of an **algebraic expression** are the quantities being added. For example, in the expression $9x^2 + 6x + 4$, the terms are $9x^2, 6x$, and 4. Because **subtraction** is defined by adding the opposite, subtracted quantities are "negative" terms. In the expression $7x - 3y$, the terms are $7x$ and $-3y$ because $7x - 3y = \underbrace{7x}_{\text{term}} + \underbrace{(-3y)}_{\text{term}}$.

> **Definition   Term**
>
> The **terms** of an algebraic expression are the quantities being added. Terms containing variable **factors** are called **variable terms**, while terms without variables are called **constant terms**.

The numeric factor of a **term** is called the **coefficient** of the term. Every term in an algebraic expression has a coefficient. The expression $9x^2 + 6x + 4$ has three terms, so it has 3 coefficients. Similarly, the expression $7x - 3y$ has two terms, so it has 2 coefficients.

| Expression | Coefficients |
|:---:|:---:|
| $9x^2 + 6x + 4$ | 9, 6, 4 |
| $7x - 3y$ | 7, −3 |

> **Definition   Coefficient (or Numerical Coefficient)**
>
> A **coefficient**, or **numerical coefficient**, is the numeric factor of a term.

Remember the coefficient of a term includes the sign. For example, $4x - 8y = 4x + (-8y)$, so the coefficients are 4 and $-8$.

### Example 1  Identifying Terms and Coefficients

Determine the number of **terms** in each expression and list the **coefficients** for each term.

**a.** $3x^2 + 7x - 3$    **b.** $4x^3 - \dfrac{3}{2}x^2 + x - 1$    **c.** $3x^2 - 2.3x + x - \dfrac{3}{4}$

### Solution

**a.** $3x^2 + 7x - 3 = \underbrace{3x^2}_{\text{term 1}} + \underbrace{7x}_{\text{term 2}} + \underbrace{(-3)}_{\text{term 3}}$

The expression has three terms, so there are three coefficients.

| Term | Coefficient |
|------|-------------|
| $3x^2$ | 3 |
| $7x$ | 7 |
| $-3$ | $-3$ |

**b.** $4x^3 - \dfrac{3}{2}x^2 + x - 1 = \underbrace{4x^3}_{\text{term 1}} + \underbrace{\left(-\dfrac{3}{2}x^2\right)}_{\text{term 2}} + \underbrace{(1x)}_{\text{term 3}} + \underbrace{(-1)}_{\text{term 4}}$

The expression has four **terms**, so there are four **coefficients**.

| Term | Coefficient |
|------|-------------|
| $4x^3$ | 4 |
| $-\dfrac{3}{2}x^2$ | $-\dfrac{3}{2}$ |
| $x$ | 1 |
| $-1$ | $-1$ |

  **c.** Try this problem on your own. Check your **answer**, or watch this **video** for the complete solution.

**You Try It**   Work through this You Try It problem.

Work Exercises 1–6 in this textbook or in the *MyMathLab*® Study Plan.

Two terms are **like terms** if their variable factors are exactly the same. For example, the terms $5x^2y$ and $4x^2y$ are **like terms** because they have the same variable factor, $x^2y$. The terms $6a^2b$ and $9ab^2$ are not like terms because the variable factors are not exactly the same. All **constants** are like terms.

**Example 2  Identifying Like Terms**

Identify the like terms in each **algebraic expression**.

a.  $5x^2 + 3x - 6 + 4x^2 - 7x + 10$

b.  $3.5a^2 + 2.1ab + 6.9b^2 - ab + 8a^2$

**Solutions**

a.  $5x^2$ and $4x^2$ are like terms because they have the same variable factor, $x^2$. Similarly, $3x$ and $-7x$ are like terms with the same variable factor of $x$. Lastly, $-6$ and $10$ are like terms because they are both constants.

 *My video summary*  ▣ **b.** Compare the variable factors to identify like terms. Check your **answer**, or watch this **video** for the complete solution.

**You Try It**   **Work through this You Try It problem.**

**Work Exercises 7–10 in this textbook or in the *MyMathLab*® Study Plan.**

OBJECTIVE 2   SIMPLIFY ALGEBRAIC EXPRESSIONS

When an algebraic expression contains **like terms**, we **simplify the expression** by combining the like terms. To **combine like terms**, we use the **distributive property** in reverse.

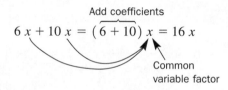

We can also use the **commutative** and **associative** properties of addition to reorder terms in an **algebraic expression** so like terms are grouped, or collected, together.

**Example 3  Combining Like Terms**

 *My interactive video summary*  ▣ Simplify each algebraic expression by combining like terms.

a.  $5x - 2x$          b.  $6x^2 - 12x - 3x^2 + 4x$    c.  $3z - 2z^2 + 7z^2$

d.  $6x^2 + 2x + 4x + 3$    e.  $-3x + 5 - y + x - 8$

**Solutions**

a.  $5x$ and $-2x$ are like terms because they have the same **variable factor** $x$. To combine like terms, we reverse the distributive property.

$$\begin{aligned} \text{Begin with the original expression:} \quad & 5x - 2x \\ \text{Reverse the distributive property:} \quad & = (5 - 2)x \\ \text{Subtract 5 and 2:} \quad & = 3x \end{aligned}$$

b.  $6x^2$ and $-3x^2$ are **like terms**, as are $-12x$ and $4x$.

$$\begin{aligned} \text{Begin with the original expression:} \quad & 6x^2 - 12x - 3x^2 + 4x \\ \text{Rearrange to collect like terms:} \quad & = 6x^2 - 3x^2 - 12x + 4x \\ \text{Reverse the distributive property:} \quad & = (6 - 3)x^2 + (-12 + 4)x \end{aligned}$$

Add and subtract: $= 3x^2 + (-8)x$
$= 3x^2 - 8x$

**c–e.** Try to **simplify** the remaining expressions on your own. View the **answers**, or watch this **interactive video** for the complete solutions to all five parts.

**You Try It**  Work through this **You Try It** problem.

Work Exercises 11–23 in this textbook or in the ***MyMathLab***® Study Plan.

Before we **combine like terms**, often we must first remove **grouping symbols** using the distributive property. In general, an algebraic expression is **simplified** if all grouping symbols are removed and like terms have been combined.

---

**Simplifying an Algebraic Expression**

**1.** Remove grouping symbols using the distributive property.

**2.** Combine like terms by using the distributive property in reverse.

---

### Example 4  Simplifying Algebraic Expressions

*My interactive video summary*

Simplify each algebraic expression.

**a.** $3(x - 4) + 2$    **b.** $8(x + 6) + 7x$

**c.** $5(x - 6) - 3(x - 7)$    **d.** $2(5z + 1) - (3z - 2)$

### Solutions

**a.** Use the **distributive property** to remove the parentheses, then **combine like terms**.

Begin with the original expression:  $3(x - 4) + 2$
Distribute the 3:  $= 3x - 3(4) + 2$
Multiply $3 \cdot 4$:  $= 3x - 12 + 2$
Combine like terms:  $= 3x - 10 \leftarrow -12 + 2 = -10$

**b.** Begin with the original expression:  $8(x + 6) + 7x$
Distribute the 8:  $= 8x + 8(6) + 7x$
Multiply $8 \cdot 6$:  $= 8x + 48 + 7x$
Combine like terms:  $= 15x + 48 \leftarrow 8x + 7x = 15x$

**c–d.** Try to simplify these expressions on your own. Each contains two sets of grouping symbols. Use the distributive property to remove both sets, then collect and combine like terms. Check your **answers**, or watch this **interactive video** for the complete solutions to all four parts.

**You Try It**  Work through this **You Try It** problem.

Work Exercises 24–32 in this textbook or in the ***MyMathLab***® Study Plan.

## OBJECTIVE 3    WRITE WORD STATEMENTS AS ALGEBRAIC EXPRESSIONS

In algebra, problem solving often involves writing word statements as **algebraic expressions**. When translating word statements, look for key words or phrases that translate into **arithmetic operations**. Table 2 provides a review of some phrases and their corresponding operations.

| Addition | Subtraction | Multiplication | Division |
|---|---|---|---|
| add | subtract | multiply | divide |
| plus | minus | times | divided by |
| sum | difference | product | quotient |
| increased by | decreased by | of | per |
| more than | less | double | ratio |
| total | less than | triple | |
| | | twice | |

**Table 2**

In Sections 1.2 and 1.3, we translated word statements into **numeric expressions**. Now we can use **variables** to represent any unknown values within verbal descriptions. Consider the following:

$$\underbrace{\text{The sum of a number}}_{\substack{\text{Unknown} \\ \text{value}}} \text{ and } \underbrace{10}_{\substack{\text{Known} \\ \text{value}}}$$

When translating this word statement, we use the variable $x$ to represent the unknown value.

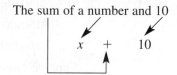

## Example 5  Writing Word Statements as Algebraic Expressions

  Write each word statement as an **algebraic expression**. Use $x$ to represent the unknown number.

a. Twenty decreased by a number

b. The product of sixteen and a number

c. Five more than twice a number

d. Three-fourths of the square of a number

e. The quotient of 12 and a number, increased by the number

f. The sum of a number and 4, divided by the difference of the number and 9.

## Solutions

a. Looking at **Table 2**, the phrase "decreased by" indicates **subtraction**. What is being subtracted? 20 is decreased by an unknown number $x$, so $x$ is being subtracted from 20. "Twenty decreased by a number" is written as $20 - x$.

**b.** The word "product" means **multiplication**. 16 and an unknown number $x$ are being multiplied. "The product of sixteen and a number" translates to $16x$.

**c.** "More than" indicates **addition**. 5 and twice $x$ are being added, which means 5 and $2x$. "Five more than twice a number $x$" is written as $5 + 2x$.

**d–f.** Try to finish these problems on your own. View the **answers**, or watch this **video** for the complete solutions.

**You Try It**    Work through this You Try It problem.

Work Exercises 33–38 in this textbook or in the *MyMathLab*® Study Plan.

### Example 6  Writing Word Statements as Algebraic Expressions

**a.** The longest side of a triangle is four units longer than five times the length of the shortest side. Express the length of the longest side in terms of the shortest side, $a$.

**b.** Michelle invests $d$ dollars in one account and $6750 less than this amount in the second account. Express the amount she invests in the second account in terms of $d$.

**c.** The state of Texas has 10 fewer institutes of higher education than twice the number in Virginia. If we let $n =$ the number of institutes in Virginia, express the number in Texas, in terms of $n$. (*Source: Statistical Abstract, 2010*)

### Solutions

**a.** The key word '*times*' means **multiplication**, while the key words 'longer than' imply **addition**. Letting $a =$  the length of the shortest side, we get

$$\underbrace{\text{four units}}_{4} \quad \underbrace{\text{longer than}}_{+} \quad \underbrace{\text{five}}_{5} \quad \underbrace{\text{times}}_{\cdot} \quad \underbrace{\text{the length of the shortest side}}_{a}$$

The length of the longest side is $4 + 5a$, or $5a + 4$, units.

**b.** The key words 'less than' mean **subtraction**. What is being subtracted? 6750 is being subtracted *from* the amount in the first account.

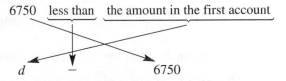

The amount invested in the second account is $d - 6750$ dollars.

 **c.** Try to work this problem on your own. View the **answer**, or watch this **video** for the complete solution.

**You Try It**    Work through this You Try It problem.

Work Exercises 39–42 in this textbook or in the *MyMathLab*® Study Plan.

OBJECTIVE 4    SOLVE APPLIED PROBLEMS INVOLVING ALGEBRAIC EXPRESSIONS

**Simplifying** algebraic expressions can be helpful when solving applied problems.

### Example 7 Solving Applied Problems Involving Algebraic Expressions

The perimeter of a rectangle is the sum of the lengths of the sides of the rectangle. Use the following rectangle to answer the questions.

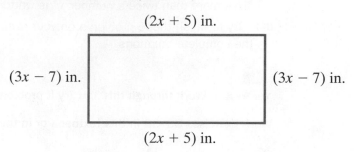

$(2x + 5)$ in.

$(3x - 7)$ in.          $(3x - 7)$ in.

$(2x + 5)$ in.

a. Write a **simplified algebraic expression** that represents the perimeter of the rectangle.

b. Use your result from part (a) to find the perimeter if $x = 7$.

### Solutions

a. To find the perimeter, we add the lengths of the four sides.

Add the lengths of the four sides: $(2x + 5) + (3x - 7) + (2x + 5) + (3x - 7)$

Remove grouping symbols: $= 2x + 5 + 3x - 7 + 2x + 5 + 3x - 7$

Rearrange to collect like terms: $= \underline{2x + 3x + 2x + 3x} + \underline{5 - 7 + 5 - 7}$

Combine like terms: $= 10x + (-4)$

Simplify: $= 10x - 4$

The perimeter of the rectangle is $(10x - 4)$ in.

b. Evaluate the expression $10x - 4$ for $x = 7$.

Begin with the original algebraic expression:  $10x - 4$

Substitute 7 for $x$:  $= 10(7) - 4$

Multiply:  $= 70 - 4$

Subtract:  $= 66$

If $x = 7$, the perimeter of the rectangle would be 66 inches.

### Example 8 Solving Applied Problems Involving Algebraic Expressions

Based on data from the National Fire Protection Association, the number of residential property fires, in thousands, is given by $111x^2 - 1366x + 6959$, where $x =$ the number of years after 2000. The number of vehicle fires, in thousands, is given by $-52.5x + 775$. (*Source: nfpa.org*)

a. Write a **simplified algebraic expression** for the difference between the number of residential property fires and the number of vehicle fires.

b. Use your result from part (a) to estimate the difference in 2010.

*My video summary*    **Solutions** Try to work this problem on your own. Check your **answer**, or watch this **video** for the complete solution.

**You Try It**   Work through this You Try It problem.

Work Exercises 43–46 in this textbook or in the ***MyMathLab***® Study Plan.

# 1.6 Exercises

In Exercises 1–6, determine the number of terms in the expression and then list the coefficients for each term.

**You Try It**

**1.** $3x - 5$

**2.** $5x^2 - 6x + 4$

**3.** $-\dfrac{3}{4}x^2 - 9$

**4.** $w^3 - 5w^2 + w - 10$

**5.** $a^4 - \dfrac{1}{2}a^2 + 8$

**6.** $3x^2 - x - \dfrac{4}{3}$

In Exercises 7–10, identify the like terms in each algebraic expression.

**You Try It**

**7.** $2a - b + 3 + 2b - 4a$

**8.** $x^2 - 4x + 2y + 3x^2 - y$

**9.** $6.3m - 4.5n^2 + 3.1mn + 2.2n^2 - 8m$

**10.** $5m - 3 + 6n + 4 - m$

In Exercises 11–32, simplify each algebraic expression by collecting and combining like terms.

**You Try It**

**11.** $-6z + 13z$

**12.** $8y + \dfrac{5}{3}y$

**13.** $5x^2 - 3x + 7x - 8$

**14.** $ab - 4.3ab + 7.8ab - 4.1$

**15.** $7x + 4x^2 + 6x^2$

**16.** $14 - \dfrac{1}{2}y + \dfrac{2}{3}y^2 - \dfrac{1}{4}y^2$

**17.** $8a - 2b + 14a$

**18.** $6m - 20 - 12m + 50$

**19.** $4.7m + 6.9 - 4.7m + 5.4$

**20.** $4x + y - 3 + x^2 - 6y + 7$

**21.** $-2c^2 + 0.7c^2 - 0.05c^2$

**22.** $\dfrac{x}{3} + \dfrac{3}{4} + \dfrac{x}{6} - \dfrac{5}{12}$

**23.** $3x^2 - 5x + 6 - 7x^2 + 4x - 8$

**24.** $6(x + 3) - 4$

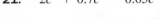

**You Try It**

**25.** $-(4x - 6) + 9$

**26.** $5(2t - 11) - 4t$

**27.** $-3(1 - 5x) + 4x - 5$

**28.** $\dfrac{4}{3}\left(9x + \dfrac{15}{28}\right) - x + \dfrac{3}{7}$

**29.** $-5(w + 3) + 2(6w - 1)$

**30.** $3.5(y^2 + 4y - 9) + 1.8y^2 + 6.3$

**31.** $4(x^2 - 3x + 5) + 6(3x^2 + 5x - 2)$

**32.** $3(y^2 - 5) - 7(1 + 4y - 3y^2)$

In Exercises 33–38, write each word phrase as an algebraic expression. Use $x$ to represent the unknown value.

**You Try It**

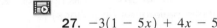

**33.** The sum of a number and eight

**34.** Twice the difference of a number and seventeen

**35.** The quotient of ten and a number, increased by two

**36.** A number decreased by twenty-five

**37.** One more than three-fifths of a number

**38.** The product of a number and three, minus the product of nine and the number

**You Try It**

**39.** A board is cut into two pieces. The longer piece is 18 cm longer than three times the shorter piece. If $s$ = the length of the shorter piece, express the length of the longer piece in terms of $s$.

**40.** Stephanie has $d$ dollars to invest and splits the money between two accounts. If she invests $3000 in one account, write an algebraic expression in terms of $d$ for the amount she invests in the second account.

**41.** China is building a network of global navigation satellites to be completed in 2020. The total number of satellites in the network will be one less than three times the number expected to be operational in 2012. Let $n$ = the number of expected satellites in the network in 2012. Express the number of satellites in the completed network in terms of $n$. (*Source:* www.gpsdaily.com)

**42.** In 2009, the percentage of 12 year-olds who owned a cell phone was 5 points higher than three times the percentage in 2004. If $p$ is the percentage in 2004, write an algebraic expression in terms of $p$ for the percentage in 2009.

**You Try It**

**43.** The perimeter of a triangle is the sum of the lengths of the sides of the triangle. Use the following triangle to answer the questions.

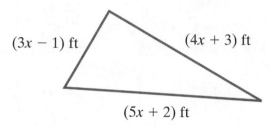

$(3x - 1)$ ft      $(4x + 3)$ ft

$(5x + 2)$ ft

    **a.** Write a simplified algebraic expression that represents the perimeter of the rectangle.

    **b.** Use your result from part (a) to find the perimeter if $x = 5$.

**44.** A wire with total length $(5x + 4.6)$ meters is cut into two pieces such that one piece measures $(3x - 2.5)$ meters.

    **a.** Write a simplified algebraic expression that represents the length of the second piece.

    **b.** Use your result from part (a) to find the length of the second piece if $x = 4.5$.

**45.** Based on data from the U.S. Department of Agriculture, the average annual per capita consumption of non-alcoholic beverages other than tap water (in gallons) is given by $0.44x + 150$, where $x$ = the number of years after 2000. The average annual per capita consumption of alcoholic beverages (in gallons) is given by $0.05x + 25$. (*Source: census.gov/compendia/statab/2010*)

    **a.** Write a simplified algebraic expression that represents the total average per capita beverage consumption.

    **b.** Use your result from part (a) to estimate the total average per capita beverage consumption in 2012.

**46.** To build a bookcase, Brennan needs two side boards and four shelf boards. Side boards have length $(7x - 2)$ inches and shelf boards have length $(3x + 6)$ inches.

    **a.** Write a simplified algebraic expression that represents the total length of wood that Brennan would need to build a bookcase.

    **b.** Use your result from part (a) to find the total length if $x = 6$.

CHAPTER TWO

# Linear Equations and Inequalities in One Variable

## CHAPTER TWO CONTENTS

## 2.1 The Addition and Multiplication Properties of Equality

**THINGS TO KNOW**

Before working through this section, be sure you are familiar with the following concepts:

VIDEO    ANIMATION    INTERACTIVE

**You Try It**  1. Use the Order of Operations to Evaluate Numeric Expressions (Section 1.4, **Objective 2**)

**You Try It**  2. Evaluate Algebraic Expressions (Section 1.5, **Objective 1**)

**You Try It**  3. Simplify Algebraic Expressions (Section 1.6, **Objective 2**)

**OBJECTIVES**

1 Identify Linear Equations in One Variable

2 Determine If a Given Value Is a Solution to an Equation

3 Solve Linear Equations Using the Addition Property of Equality

4 Solve Linear Equations Using the Multiplication Property of Equality

5 Solve Linear Equations Using Both Properties of Equality

## OBJECTIVE 1   IDENTIFY LINEAR EQUATIONS IN ONE VARIABLE

An **equation** is a statement that two quantities are equal. For example, the numeric equation

$$\underbrace{12 + 4}_{\text{Quantity}} = \underbrace{16}_{\text{Quantity}}$$

means that the quantities $12 + 4$ and $16$ are equal. An equation that contains one or more **variables** is called an **algebraic equation**. An algebraic equation indicates that two **algebraic expressions** are equal. Algebraic equations contain an equality symbol $(=)$, whereas algebraic expressions do not.

For example, $5x - 3$ and $4x + 6$ are both algebraic *expressions*, but

$$\underbrace{5x + 3}_{\substack{\text{Algebraic} \\ \text{expression}}} \quad \underbrace{=}_{\substack{\text{Equality} \\ \text{symbol}}} \quad \underbrace{4x + 6}_{\substack{\text{Algebraic} \\ \text{expression}}}$$

*My interactive video summary*

🎬 is an algebraic *equation*. Work through this **interactive video** to practice distinguishing between expressions and equations.

An **equation in one variable** contains a single variable. Each of the following is an example of an equation in one variable:

$$\underbrace{2x + 3 = -7,}_{\text{One variable} \rightarrow x} \quad \underbrace{a^2 - 3a + 8 = 0,}_{\text{One variable} \rightarrow a} \quad \text{and} \quad \underbrace{z - \frac{2}{5} = \frac{3}{4}z}_{\text{One variable} \rightarrow z}$$

**Note:** Even if the same variable appears multiple times, the equation is still considered to be "in one variable."

The equations $2x + 3 = -7$ and $z - \dfrac{2}{5} = \dfrac{3}{4}z$ are both examples of *linear equations in one variable*.

---

**Definition   Linear Equation in One Variable**

A **linear equation in one variable** is an equation that can be written in the form $ax + b = c$, where $a$, $b$, and $c$ are real numbers and $a \neq 0$.

---

Linear equations are also **first-degree equations** because the **exponent** on the variable is **understood to be** 1. Until we learn how to simplify equations, this is how we will identify linear equations. If an equation contains a variable raised to an exponent other than 1, then the equation is *nonlinear*. View this **popup** to see other characteristics of **nonlinear equations**.

*My interactive video summary*

🎬 Watch this **interactive video** to practice distinguishing between linear and nonlinear equations. It is important to understand that an equation does not have to be in the form $ax + b = c$ to be considered linear. Rather, the equation must be able *to be written in the form* $ax + b = c$.

### Example 1  Identifying Linear Equations in One Variable

Determine if each is a linear equation in one variable. If not, state why.

a. $4x + 3 - 2x$

b. $4x + 2 = 3x - 1$

c. $x^2 + 3x = 5$

d. $2x + 3y = 6$

## Solutions

a. $4x + 3 - 2x$ is not a **linear equation in one variable** because it is not an **equation**.

b. $4x + 2 = 3x - 1$ is a linear equation in one variable.

c. $x^2 + 3x = 5$ is an equation in one variable, but it is nonlinear because $x^2$ has an **exponent** other than 1.

d. $2x + 3y = 6$ is not a linear equation in one variable because it has more than one **variable**.

**You Try It**    Work through this You Try It problem.

**Work Exercises 1–6 in this textbook or in the *MyMathLab*®** Study Plan.

OBJECTIVE 2    DETERMINE IF A GIVEN VALUE IS A SOLUTION TO AN EQUATION

We can check the truth of an **equation** by simplifying both sides of the equation and comparing the results. Consider the following two equations:

$$-3 + (-2)^2 = -7 + 1 \qquad |10 - 26| + 4 = 3^2 + 11$$

Simplify both sides of each equation and compare the results to determine if the statement is true. View this **popup** to see the steps. Simplifying the first equation results in a false statement, so the equation is not true. Simplifying the second equation results in a true statement, so the equation is true.

When we *solve* an **algebraic equation**, we look for all values for the **variable(s)** that make the equation true. Values that result in a true statement are called *solutions* of the equation.

---

**Definition**    **Solve**

To **solve** an equation means to find its **solution set**, or the set of all values that make the equation true.

---

**Definition**    **Solution**

A **solution** is a value that, when substituted for a variable, makes the equation true.

---

 Remember that we *simplify* expressions (as in **Section 1.6**) but we *solve* equations.

To determine if a given value is a **solution** to an equation, we **substitute** the value for the variable, **simplify** both sides of the equation, and compare the results to see if the statement is true. For example, consider the equation $5x + 2 = x - 6$. You can **view the checks** to see if either 3 or $-2$ is a solution.

 When substituting values for a variable, it is best to use parentheses around the substituted value to avoid mistakes in performing operations.

### Example 2  Determining If a Given Value Is a Solution to an Equation

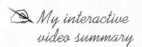

Determine if the given value is a solution to the equation.

a. $-3x + 5 = 8 - 4x$;  $x = 3$

b. $2(y + 1) + 8 = 7 - y$;  $y = -3$

c. $|a - 6| - 1 = 9 + a^2$;  $a = -2$

d. $\dfrac{3}{5}w - \dfrac{1}{2} = -\dfrac{3}{10}w$;  $w = \dfrac{5}{9}$

**Solutions** First, we substitute the given value for the **variable** and simplify. If the resulting statement is true, then the value is a solution to the equation.

**a.** Begin with the original equation: $\quad -3x + 5 = 8 - 4x$

Substitute 3 for $x$: $\quad -3(3) + 5 \overset{?}{=} 8 - 4(3)$

Simplify: $\quad -9 + 5 \overset{?}{=} 8 - 12$

Compare: $\quad -4 = -4 \quad$ True

The final statement is true, so $x = 3$ is a solution to the equation.

**b.** Begin with the original equation: $\quad 2(y + 1) + 8 = 7 - y$

Substitute $-3$ for $y$: $\quad 2((-3) + 1) + 8 \overset{?}{=} 7 - (-3)$

Simplify: $\quad 2(-2) + 8 \overset{?}{=} 7 + 3$

$-4 + 8 \overset{?}{=} 7 + 3$

Compare: $\quad 4 = 10 \quad$ False

The final statement is false, so $y = -3$ is not a solution to the equation.

**c.–d.** Try completing parts c and d on your own, then view the **answers**. Watch this **interactive video** to see the complete solutions to all four parts.

**You Try It**  **Work through this You Try It problem.**

**Work Exercises 7–12 in this textbook or in the *MyMathLab*® Study Plan.**

OBJECTIVE 3  SOLVE LINEAR EQUATIONS USING THE ADDITION PROPERTY OF EQUALITY

All the solutions to an equation form the **solution set** of the equation. Two or more equations with the same solution set are called **equivalent equations**. When solving a **linear equation in one variable**, we look for simpler equivalent equations until we find one that ends with an **isolated variable** of the form:

$$variable = value \quad \text{or} \quad value = variable.$$

To find simpler equivalent equations, we use the *properties of equality* to add, subtract, multiply, and/or divide both sides of an equation by the same quantity without changing its solution set.

Consider the true statement, $15 = 15$. Figure 1a shows that we can think of an equation as a balance scale in which both sides are equal. Notice that if we add 5 to the left side of the equation, the two sides are no longer equal (Figure 1b), and the equality statement is no longer true. To maintain equality (balance), we also need to add 5 to the right side of the equation (Figure 1c).

**Figure 1**

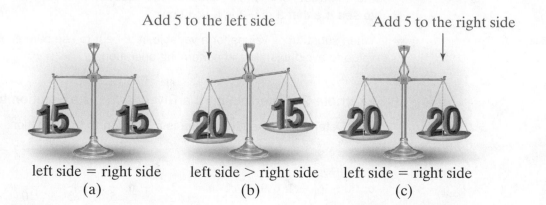

Add 5 to the left side          Add 5 to the right side

left side = right side     left side > right side     left side = right side
(a)                        (b)                        (c)

**Figure 1** illustrates the addition property of equality.

---

**Addition Property of Equality**

Let $a$, $b$, and $c$ be real numbers or **algebraic expressions**. Then,

$$a = b \quad \text{and} \quad a + c = b + c$$

are equivalent equations.

---

Using the addition property of equality, we can add the same value to both sides of an **equation** without changing its **solution set**. Because **subtraction** is defined in terms of addition, this property also holds true for subtraction:

$$a = b \quad \text{and} \quad a - c = b - c \quad \text{are equivalent equations.}$$

## Example 3  Solving Linear Equations Using the Addition Property of Equality

Solve.

**a.** $x - 5 = 3$

**b.** $y + \dfrac{2}{3} = \dfrac{1}{5}$

## Solutions

**a.** To solve the equation, we want to get the variable $x$ on one side of the equation.

Begin with the original equation: $\quad x - 5 = 3$

Add 5 to both sides: $\quad x \underbrace{- 5 + 5}_{0} = 3 + 5$

Simplify: $\quad x = 8$

To check, we **substitute** 8 for $x$ in the original equation to see if a true statement results. View the **check**.

**b.** To **solve** the equation, we want to get the variable $y$ by itself on one side of the equation.

Begin with the original equation: $\quad y + \dfrac{2}{3} = \dfrac{1}{5}$

Subtract $\dfrac{2}{3}$ from both sides: $\quad y + \underbrace{\dfrac{2}{3} - \dfrac{2}{3}}_{0} = \dfrac{1}{5} - \dfrac{2}{3}$

Simplify: $\quad y = \dfrac{1}{5} - \dfrac{2}{3}$

Write each fraction with a denominator of LCD = 15: $\quad y = \dfrac{1 \cdot 3}{5 \cdot 3} - \dfrac{2 \cdot 5}{3 \cdot 5}$

$$y = \dfrac{3}{15} - \dfrac{10}{15}$$

Subtract: $\quad y = -\dfrac{7}{15}$

Check this result by **substituting** $-\dfrac{7}{15}$ for $y$ in the original equation. View the **check**.

**You Try It**   Work through this **You Try It** problem.

**Work Exercises 13–16 in this textbook or in the *MyMathLab*® Study Plan.**

The **addition property of equality** can be applied to **variable terms** as well as **constants**. Example 4 illustrates this idea.

### Example 4 Solving Linear Equations Using the Addition Property

Solve.

**a.** $3z - 4.5 = 4z$          **b.** $12 - 7x = -6x$

**Solutions**

**a.** To solve the equation we want to get the variable $z$ by itself on one side of the equation.

$$\text{Begin with the original equation:} \qquad 3z - 4.5 = 4z$$

$$\text{Subtract } 3z \text{ from both sides:} \qquad \underbrace{3z - 3z}_{0} - 4.5 = 4z - 3z$$

$$\text{Simplify:} \qquad\qquad -4.5 = z$$

The solution set is $\{-4.5\}$. View the **check**.

**b.** To solve the equation we want to get the variable $x$ by itself on one side of the equation. Add $7x$ to both sides and **simplify**. View the **solution and check**.

**You Try It**    Work through this **You Try It** problem.

Work Exercises 17–20 in this textbook or in the *MyMathLab*® Study Plan.

We can also apply the **addition property of equality** to both **constants** and **variable terms** within the same equation.

### Example 5 Solving Linear Equations Using the Addition Property

Solve.

**a.** $5x - 3 = 6x + 2$          **b.** $-5w + 27 = 13 - 4w$

**Solutions**  Try solving these two equations on your own. Apply the addition property of equality twice within each equation: once for the constant and once for the variable term. View the **answers**, or watch this **video** for complete solutions.

**You Try It**    Work through this **You Try It** problem.

Work Exercises 21–25 in this textbook or in the *MyMathLab*® Study Plan.

OBJECTIVE 4    SOLVE LINEAR EQUATIONS USING THE MULTIPLICATION PROPERTY OF EQUALITY

Not all equations can be solved using the **addition property of equality**. For example, the equation $\dfrac{x}{5} = -3$ cannot be solved by adding or subtracting the same value from both sides. Instead, we apply the **multiplication property of equality**.

---

**Multiplication Property of Equality**

Let $a$, $b$, and $c$ be real numbers or **algebraic expressions** with $c \neq 0$. Then,

$$a = b \quad \text{and} \quad a \cdot c = b \cdot c$$

are equivalent equations.

---

Using the multiplication property of equality, we can multiply both sides of an **equation** by the same **nonzero** value without changing the **solution set** of the equation. Because **division** is defined in terms of multiplication, this property also holds true for division:

$$a = b \quad \text{and} \quad \frac{a}{c} = \frac{b}{c} \quad \text{are equivalent equations.}$$

 Multiplying or dividing both sides of an equation by zero is not allowed. Do you see **why?**

### Example 6 Solving Linear Equations Using the Multiplication Property of Equality

Solve.

**a.** $\dfrac{x}{5} = -3$ 　　　　　　　**b.** $-4x = -32$

### Solutions

**a.** We want to **isolate** the variable $x$ on one side of the equation. While we have the variable *term* by itself, we do not have the *variable* by itself. To do this, we need to multiply both sides of the equation by 5.

Begin with the original equation: 　$\dfrac{x}{5} = -3$

Multipy both sides by 5: 　$\underbrace{5 \cdot \dfrac{x}{5}}_{5 \cdot \frac{1}{5} = 1} = 5 \cdot (-3)$

Simplify: 　$x = -15$

The solution set is $\{-15\}$ View the check.

**b.** Again we have the **variable term** by itself but not the variable. To isolate the variable, we divide both sides of the equation by $-4$.

Begin with the original equation: 　$-4x = -32$

Divide both sides by $-4$: 　$\dfrac{-4x}{-4} = \dfrac{-32}{-4} \quad \leftarrow \left( \dfrac{-4}{-4} = 1 \right)$

Simplify: 　$x = 8$

The solution set is $\{8\}$. View the **check**.

**You Try It**　Work through this You Try It problem.

Work Exercises 26–29 in this textbook or in the *MyMathLab* Study Plan.

### Example 7 Solving Linear Equations Using the Multiplication Property of Equality

*My video summary*　Solve.

**a.** $\dfrac{4}{3}x = 52$ 　　　　　**b.** $2.2x = 6.93$

**Solutions** Try solving these equations on your own. Apply the **multiplication property of equality** to isolate the variable. View the **answers**, or watch this **video** for complete solutions.

**You Try It**　Work through this You Try It problem.

Work Exercises 30–36 in this textbook or in the *MyMathLab* Study Plan.

### OBJECTIVE 5   SOLVE LINEAR EQUATIONS USING BOTH PROPERTIES OF EQUALITY

To solve some equations, we must use both the addition and multiplication **properties of equality**.

First, we use the addition property to get the **variable terms** on one side of the equation and the **constants** on the other. Then we use the multiplication property to isolate the variable.

### Example 8  Solving Linear Equations in One Variable

*My video summary*   Use the properties of equality to solve each equation.

a. $\dfrac{2}{5}x - 4 = 4$         b. $7n = 3n - 8$

**Solutions**  Work through each problem, or watch this **video** to see the complete solutions to both parts.

a.  Begin with the original equation:  $\dfrac{2}{5}x - 4 = 4$

Add 4 to both sides:  $\dfrac{2}{5}x - 4 + 4 = 4 + 4$  ← Addition property of equality

Simplify:  $\dfrac{2}{5}x = 8$

Multiply both sides by $\dfrac{5}{2}$:  $\dfrac{5}{2} \cdot \dfrac{2}{5}x = \dfrac{5}{2} \cdot 8$  ← Multiplication property of equality

Simplify:  $x = 20$

The solution set is $\{20\}$. **View the check** of this **solution** in the original equation.

b.  Begin with the original equation:  $7n = 3n - 8$

Subtract $3n$ from both sides:  $7n - 3n = 3n - 8 - 3n$

Simplify:  $4n = -8$

Divide both sides by 4:  $\dfrac{4n}{4} = \dfrac{-8}{4}$

Simplify:  $n = -2$

The solution set is $\{-2\}$. View the **check**.

**You Try It**    **Work through this You Try It problem.**

**Work Exercises 37–45 in this textbook or in the *MyMathLab*®  Study Plan.**

Some equations have **algebraic expressions** on both sides of the equality symbol. We again use the **properties of equality** to move all variables to one side and constants to the other. The only difference here is that we need to apply the **addition property of equality** twice: once for the **constant** and once for the **variable terms**.

### Example 9  Solving Linear Equations with Variables on Both Sides

Use the properties of equality to solve each equation.

a. $7y + 4 = 2y - 6$         b. $2x - 14.5 = 0.5x + 50$

**Solutions**

a. Begin with the original equation:      $7y + 4 = 2y - 6$

Subtract 4 from both sides:    $7y + 4 - 4 = 2y - 6 - 4$  ←  Addition property of equality

Simplify:      $7y = 2y - 10$

Subtract $2y$ from both sides:    $7y - 2y = 2y - 10 - 2y$  ←  Addition property of equality

Simplify:      $5y = -10$

Divide both sides by 5:      $\dfrac{5y}{5} = \dfrac{-10}{5}$  ←  Multiplication property of equality

Simplify:      $y = -2$

The solution set is $\{-2\}$. Check this **solution** in the original equation.

*My video summary*      b. Try solving this equation on your own. Be sure to pay attention to decimal places when combining **coefficients**. View the **answer**, or watch this **video** for the complete solutions to both parts.

**You Try It**    Work through this **You Try It** problem.

Work Exercises 46–52 in this textbook or in the *MyMathLab* Study Plan.

# 2.1 Exercises

In Exercises 1–6, determine if each is a linear equation in one variable. If not, state why.

**You Try It**

**1.** $3x - 2(x + 1)$          **2.** $3x = 2(x + 1)$          **3.** $3x^{-1} = 2(x + 1)$

**4.** $3x^2 = 2(x^2 + 1)$        **5.** $3x = 2(y + 1)$         **6.** $1.3x - 2 + 5x = x - 8.6$

In Exercises 7–12, determine if the given value is a solution to the equation.

**You Try It**

**7.** $2x - 5 = 13 - 4x;\ x = 3$        **8.** $5 - 2(y - 1) = 3y + (y - 5);\ y = 5$

**9.** $z^2 + 2z + 1 = (-1 - z)^2;\ z = -4$        **10.** $\dfrac{1}{2}w - \dfrac{1}{3} = \dfrac{1}{6}w;\ w = 6$

**11.** $\dfrac{2}{3}x - \dfrac{3}{4}(x + 1) = \dfrac{1}{6}x - \dfrac{21}{16};\ x = \dfrac{9}{4}$        **12.** $4.5m - 5 = 3.1(m + 2.6);\ m = 2.5$

In Exercises 13–25, solve each linear equation using the addition property of equality.

**You Try It**

**13.** $a - 3 = 2$          **14.** $c + 4 = 7$          **15.** $y - 2 = -5$

**You Try It**

**16.** $z + \dfrac{2}{3} = \dfrac{5}{4}$         **17.** $10x = 9x + 7$         **18.** $4.8 + 1.5x = 2.5x$

**19.** $7x + \dfrac{3}{4} = 6x$         **20.** $6 - 5x = -4x$         **21.** $4x + 2 = 3x + 5$

**22.** $4 + 3c = 4c - 3$    **23.** $-1 - 6m = 4 - 7m$    **24.** $3z + 2.7 = 8.1 + 2z$

**You Try It**
**25.** $\dfrac{1}{8} - \dfrac{5}{6}w = \dfrac{1}{6}w + \dfrac{9}{8}$

In Exercises 26–36, solve each linear equation using the multiplication property of equality.

**26.** $5x = 10$    **27.** $\dfrac{x}{3} = 9$    **28.** $21 = -3x$

**You Try It**

**29.** $\dfrac{m}{-2} = -16$    **30.** $\dfrac{3}{4}x = 27$    **31.** $-\dfrac{1}{4}z = -3.5$

**32.** $3.2y = 2.4$    **33.** $\dfrac{2}{5}x = \dfrac{1}{3}$    **34.** $\dfrac{3}{2}z = -\dfrac{9}{20}$

**You Try It**

**35.** $17 = 5w$    **36.** $\dfrac{n}{-2} = -\dfrac{3}{8}$

In Exercises 37–52, solve using both the addition and multiplication properties of equality.

**37.** $5x + 2 = 27$    **38.** $2 + 4x = x$    **39.** $7 - 3x = -8$

**You Try It**
**40.** $-12 = 3y + 10$    **41.** $5z = 16 - 3z$    **42.** $-4n + 7 = -6n$

**43.** $3.7x - 4.1 = 3.2x$    **44.** $\dfrac{2}{3}w - 1 = 5$    **45.** $\dfrac{2}{5}y + \dfrac{1}{2} = \dfrac{5}{2}$

**46.** $2x + 1 = 5x - 8$    **47.** $8 + b = 2 - 2b$    **48.** $5 - 4n = 6n + 1$

**You Try It**
**49.** $2z + 6 = 5z + 6$    **50.** $3x - 2.9 = 5.1 - 5x$    **51.** $\dfrac{2}{3}y + 11 = -\dfrac{1}{2}y + 8$

**52.** $0.1x - 14.5 = 0.04x + 10.7$

# 2.2  Solving Linear Equations in One Variable

## THINGS TO KNOW

Before working through this section, be sure you are familiar with the following concepts:

| | | VIDEO | ANIMATION | INTERACTIVE |
|---|---|---|---|---|

**1.** Use the Order of Operations to Evaluate Numeric Expressions (Section 1.4, **Objective 2**)

**You Try It**

**2.** Evaluate Algebraic Expressions (Section 1.5, **Objective 1**)

**You Try It**

**3.** Simplify Algebraic Expressions (Section 1.6, **Objective 2**)

**You Try It**

**4.** Determine If a Given Value Is a Solution to an Equation (Section 2.1, **Objective 2**)

**You Try It**

**5.** Solve Linear Equations Using Both Properties of Equality (Section 2.1, **Objective 5**)

**You Try It**

## OBJECTIVES

**1** Solve Linear Equations Containing Non-Simplified Expressions

**2** Solve Linear Equations Containing Fractions

**3** Solve Linear Equations Containing Decimals; Apply a General Strategy

**4** Identify Contradictions and Identities

**5** Use Linear Equations to Solve Application Problems

---

**OBJECTIVE 1**  **SOLVE LINEAR EQUATIONS CONTAINING NON-SIMPLIFIED EXPRESSIONS**

When **solving** a **linear equation** with **non-simplified expressions** on one or both sides, we simplify each side first before using the **properties of equality**.

### Example 1  Solving Linear Equations Containing Non-Simplified Expressions

Solve: $4x + 7 - 2x = 5 - 3x - 3$

**Solution**  Begin with the original equation:  $4x + 7 - 2x = 5 - 3x - 3$

Combine like terms:  $2x + 7 = 2 - 3x$

Add $3x$ to both sides:  $2x + 7 + 3x = 2 - 3x + 3x$  ← Addition property of equality

Simplify:  $5x + 7 = 2$

Subtract 7 from both sides:  $5x + 7 - 7 = 2 - 7$  ← Addition property of equality

Simplify:  $5x = -5$

Divide both sides by 5:  $\dfrac{5x}{5} = \dfrac{-5}{5}$  ← Multiplication property of equality

Simplify:  $x = -1$

The **solution set** is $\{-1\}$. View the **check**.

**You Try It**  Work through this You Try It problem.

**Work Exercises 1–5 in this textbook or in the MyMathLab® Study Plan.**

If an equation contains **grouping symbols**, we remove the grouping symbols by applying the **distributive property**, **simplify** both sides, and then use the **properties of equality** to solve the equation.

### Example 2  Solving Linear Equations Containing Non-Simplified Expressions and Grouping Symbols

 Solve: $7 - 2(4z - 3) = 3z + 1$

**Solution**

Begin with the original equation:  $7 - 2(4z - 3) = 3z + 1$

Use the **distributive property**:  $7 - 8z + 6 = 3z + 1$  ← $-2(4z - 3) = -2 \cdot 4z - 2 \cdot (-3)$

Combine like terms:  $13 - 8z = 3z + 1$

**2.2**  Solving Linear Equations in One Variable  **2-11**

Both sides of the equation are now **simplified expressions**. Finish solving the equation on your own using the **properties of equality**. Check your **answer**, or watch this **video** for the complete solution.

**You Try It**    Work through this You Try It problem.

**Work Exercises 6–10 in this textbook or in the MyMathLab® Study Plan.**

### Example 3  Solving Equations Containing Non-Simplified Expressions

  Solve: $2(3x - 1) - 5x = 3 - (3x + 1)$

**Solution**  Begin with the original equation:  $2(3x - 1) - 5x = 3 - (3x + 1)$

Use the **distributive property**:  $6x - 2 - 5x = 3 - 3x - 1$

The **grouping symbols** have been removed from both sides of the equation. Simplify each side by **combining like terms**, then finish solving by using the properties of equality. Check your **answer**, or watch this **video** for a fully worked solution.

**You Try It**    Work through this You Try It problem.

**Work Exercises 11–18 in this textbook or in the MyMathLab® Study Plan.**

### OBJECTIVE 2   SOLVE LINEAR EQUATIONS CONTAINING FRACTIONS

When an equation contains **fractions**, we can make the calculations more manageable if we remove the fractions before **combining like terms** or applying the **properties of equality**. To do this, we multiply both sides of the equation by an appropriate **common multiple** of all the **denominators**, usually the **least common denominator (LCD)** of all the fractions.

### Example 4  Solving Equations Containing Fractions

  Solve: $y + \dfrac{2}{3} = \dfrac{1}{5}$

**Solution**  The equation contains fractions with the denominators 3 and 5, so the LCD is 15. We can **clear the fractions** by multiplying both sides of the equation by 15.

Begin with the original equation:  $y + \dfrac{2}{3} = \dfrac{1}{5}$

Multiply both sides by 15:  $15\left(y + \dfrac{2}{3}\right) = 15\left(\dfrac{1}{5}\right)$

Use the **distributive property**:  $15(y) + 15\left(\dfrac{2}{3}\right) = 15\left(\dfrac{1}{5}\right)$

Clear the fractions:  $15y + 10 = 3$

Subtract 10 from both sides:  $15y + 10 - 10 = 3 - 10$

Simplify:  $15y = -7$

Divide both sides by 15:  $\dfrac{15y}{15} = \dfrac{-7}{15}$

Simplify:  $y = -\dfrac{7}{15}$

The **solution set** is $\left\{-\dfrac{7}{15}\right\}$. This is the same result that we obtained in Example 3 of Section 2.1 where we solved the same equation without clearing fractions first.

### Example 5  Solving Equations Containing Fractions

Solve: $\dfrac{w+3}{2} - 4 = w + \dfrac{1}{3}$

**Solution** The denominators in this equation are 2 and 3, so the **LCD** is 6. We can **clear the fractions** by multiplying both sides of the equation by 6.

Begin with the original equation: $\dfrac{w+3}{2} - 4 = w + \dfrac{1}{3}$

Multiply both sides by 6: $6\left(\dfrac{w+3}{2} - 4\right) = 6\left(w + \dfrac{1}{3}\right)$

Use the **distributive property**: $6\left(\dfrac{w+3}{2}\right) - 6(4) = 6(w) + 6\left(\dfrac{1}{3}\right)$

Clear the fractions: $3(w+3) - 24 = 6w + 2 \ \leftarrow$

$$6\left(\dfrac{w+3}{2}\right) = \overset{3}{6}\left(\dfrac{w+3}{\underset{1}{2}}\right)$$
$$= 3(w+3)$$

Use the distributive property: $3w + 9 - 24 = 6w + 2 \ \leftarrow \ 3(w) + 3(3) = 3w + 9$

Simplify: $3w - 15 = 6w + 2$

 *My video summary*  We now have **simplified expressions** on each side of the equation. Finish solving the equation on your own. View the **answer**, or watch this **video** for a complete solution.

**You Try It**  Work through this You Try It problem.

**Work Exercises 19–26 in this textbook or in the *MyMathLab* Study Plan.**

**Note:** When solving an equation that contains fractions, it is usually helpful to clear the fractions first. However, clearing fractions first is not always necessary, as shown in **this popup**.

### Example 6  Solving Linear Equations Containing Fractions

 *My video summary*  Solve: $\dfrac{5x}{2} - \dfrac{7}{8} = \dfrac{3}{4}x - \dfrac{11}{8}$

**Solution** This equation contains fractions with the denominators 2, 4, and 8, so the **LCD** is 8. Begin by multiplying both sides of the equation by 8 to **clear the fractions**. Try solving this equation on your own. Check your **answer** or watch this **video** for a fully worked solution.

**You Try It**  Work through this You Try It problem.

**Work Exercises 27–32 in this textbook or in the *MyMathLab* Study Plan.**

The advantage of clearing fractions is that we can work with integers which are easier to use when **performing operations**. However, equations containing fractions can still be solved without clearing the fractions. View this **popup** to see Example 6 solved without clearing fractions.

OBJECTIVE 3   SOLVE LINEAR EQUATIONS CONTAINING DECIMALS; APPLY A GENERAL STRATEGY

When an equation contains **decimals**, it is often helpful to clear them as we do with fractions. To **clear decimals**, multiply both sides of the equation by an appropriate **power of** 10. Determine which power of 10 to use by looking at the **constants** in the equation and choosing the constant with the greatest number of decimal places. We count those decimal places and then raise 10 to that power. Multiplying both sides of the equation by that power of 10 will usually clear all the decimals.

### Example 7  Solving Linear Equations Containing Decimals

Solve: $1.4x - 3.8 = 6$

**Solution** The **coefficients** 1.4 and $-3.8$ have one decimal place, so the greatest number of decimal places is one. To clear the decimals, we multiply both sides of the equation by $10^1 = 10$.

$$\underbrace{1.4}\, x - \underbrace{3.8} = 6$$

One decimal place · One decimal place

| | |
|---|---|
| Begin with the original equation: | $1.4x - 3.8 = 6$ |
| Multiply both sides by 10: | $10(1.4x - 3.8) = 10(6)$ |
| Apply the **distributive property**: | $14x - 38 = 60$ |
| Add 38 to both sides: | $14x - 38 + 38 = 60 + 38$ |
| Simplify: | $14x = 98$ |
| Divide both sides by 14: | $\dfrac{14x}{14} = \dfrac{98}{14}$ |
| Simplify: | $x = 7$ |

The **solution set** is $\{7\}$. View the **check**.

### Example 8  Solving Linear Equations Containing Decimals

 *My video summary*

Solve: $0.1x + 0.03(7 - x) = 0.05(7)$

**Solution** The **coefficient** 0.1 has one decimal place. The **factors** 0.03 and 0.05 have two decimal places, so the greatest number of decimal places is two.

$$\underbrace{0.1}\, x + \underbrace{0.03}(7 - x) = \underbrace{0.05}(7)$$

One decimal place · Two decimal places · Two decimal places

To **clear the decimals**, we multiply both sides of the equation by $10^2 = 100$.

| | |
|---|---|
| Begin with the original equation: | $0.1x + 0.03(7 - x) = 0.05(7)$ |
| Multiply both sides by 100: | $100[0.1x + 0.03(7 - x)] = 100[0.05(7)]$ |
| Distribute: | $100[0.1x] + 100[0.03(7 - x)] = 100[0.05(7)]$ |
| Multiply: | $10x + 3(7 - x) = 5(7)$ |

 Notice that we only multiplied 100 by the first factor in each term.

Finish solving the equation, then check your **answer**. Watch this **video** for a complete solution.

**You Try It**    **Work through this You Try It problem.**

**Work Exercises 33–40 in this textbook or in the *MyMathLab*® Study Plan.**

Our work so far leads us to the following general strategy for solving linear equations in one variable. If a particular step does not apply to a given equation, then skip it. Remember that clearing fractions or decimals is optional but usually makes equations easier to solve.

---

**A General Strategy for Solving Linear Equations in One Variable**

**Step 1.** Clear all **fractions** by multiplying both sides of the equation by the **LCD. Clear all decimals** by multiplying both sides of the equation by the appropriate **power of** 10.

**Step 2.** Remove grouping symbols using the **distributive property**.

**Step 3.** Simplify each side of the equation by **combining like terms**.

**Step 4.** Use the **addition property of equality** to collect all **variable terms** on one side of the equation and all **constant terms** on the other side.

**Step 5.** Use the **multiplication property of equality** to **isolate the variable**.

**Step 6.** Check that the result **satisfies** the original equation.

---

## OBJECTIVE 4    IDENTIFY CONTRADICTIONS AND IDENTITIES

So far, we have solved equations with exactly one **solution**. However, not every **linear equation in one variable** has a single solution. There are two other cases—no solution and a solution set of all real numbers.

Consider the equation $x = x + 1$. No matter what value is **substituted** for $x$, the resulting value on the right side will always be one greater than the value on the left side. Therefore, the equation can never be true. Trying to solve the equation, we get

$$x = x + 1$$
$$x - x = x + 1 - x$$
$$0 = 1 \quad \longleftarrow \boxed{\text{Variables are gone and statement is false}}$$

No variable terms remain, and a false statement results. We call such an equation a **contradiction**, and it has *no solution*. Its **solution set** is the empty or **null set**, denoted by { } or $\varnothing$, respectively.

Now consider the equation $(x + 3) + (x - 8) = 2x - 5$. The expression on the left side of the equation **simplifies** to the expression on the right side. No matter what value we **substitute** for $x$, the resulting values on both the left and right sides will always be the same. Therefore, the equation is always true regardless of the value for the variable. Trying to solve this equation, we get

$$(x + 3) + (x - 8) = 2x - 5$$
$$x + 3 + x - 8 = 2x - 5$$
$$2x - 5 = 2x - 5$$
$$2x - 5 - 2x = 2x - 5 - 2x$$
$$-5 = -5 \quad \longleftarrow \boxed{\text{Variables are gone and statement is true}}$$

No **variable terms** remain, and a true statement results. We call such an equation an **identity**, and its **solution set** is the set of all real numbers, denoted by $\mathbb{R}$ or $\{x \mid x \text{ is a real number}\}$.

Example 9 shows how to identify **contradictions** and **identities** when solving equations.

### Example 9  Identifying Contradictions and Identities

Determine if the equation is a **contradiction** or an **identity**. State the solution set.

**a.**  $3x + 2(x - 4) = 5x + 7$          **b.** $3(x - 4) = x + 2(x - 6)$

### Solutions

**a.**  Begin with the original equation:  $3x + 2(x - 4) = 5x + 7$

　　　　Use the **distributive property**:  $3x + 2x - 8 = 5x + 7$

　　　　　　　　**Combine like terms**:  $5x - 8 = 5x + 7$

　　　Subtract $5x$ from both sides:  $5x - 8 - 5x = 5x + 7 - 5x$

　　　　　　　　　　**Simplify**:  $-8 = 7$  False

All the **variable terms** drop out, leaving a false statement. Therefore, the equation is a contradiction and has no solution. Its solution set is { } or $\varnothing$.

 *My video summary*      **b.**  Try to complete this problem on your own to verify that it is an identity. View this **popup** for the steps, or watch this **video** for the complete solution.

**You Try It**    Work through this You Try It problem.

**Work Exercises 41–46 in this textbook or in the *MyMathLab*® Study Plan.**

### OBJECTIVE 5   USE LINEAR EQUATIONS TO SOLVE APPLICATION PROBLEMS

We now look at how we can use linear equations to solve application problems. At this point you may wish to review the **Strategy for Solving Linear Equations in One Variable**.

### Example 10  Body Surface Area of Infants and Children

The body surface area and weight of well proportioned infants and children are related by the equation

$$30S = W + 4$$

where $S$ = the body surface area in square meters and $W$ = weight in kilograms. Find the body surface area of a well-proportioned child that weighs 18 kg. *(Source: The Internet Journal of Anesthesiology, Vol. 2, No. 2)*

**Solution**  We are given the child's weight, so we begin by **substituting** 18 for W.

$$30S = (18) + 4$$

**Simplifying** on the right-hand side gives

$$30S = 22.$$

To solve for $S$, we apply the **multiplication property of equality** and divide both sides of the equation by 30.

$$\frac{30S}{30} = \frac{22}{30}$$

$$S = \frac{22}{30} = \frac{11}{15} \approx 0.733$$

The child's body surface area is approximately 0.733 square meters.

### Example 11 Red Meat vs Poultry

 *My video summary*    In the U.S., the average pounds of red meat eaten, $M$, is related to the average pounds of poultry eaten, $P$, by the equation

$$100M = 14{,}000 - 42P.$$

Determine the average amount of poultry eaten if the average amount of red meat eaten is 100.1 pounds. *(Source: U.S. Department of Agriculture)*

**Solution** Substitute 100.1 for $M$ in the equation, then solve the equation for $P$. Check your answer, or watch this video for the complete solution.

**You Try It**    Work through this You Try It problem.

**Work Exercises 47–49 in this textbook or in the *MyMathLab*® Study Plan.**

## 2.2 Exercises

In Exercises 1–46, solve the linear equations.

**You Try It**

**1.** $2x + 3 - 5x = 7 + x$

**2.** $3 + 5x - 2 = 2x + 3 - x$

**3.** $3y + 7 + 2y - 3 = 4y + 1$

**4.** $y + 5 + 2y + 4 = 3 + 3y + 2 - 4y$

**You Try It**

**5.** $3z - 2 - 6z + 1 = 4 - 3z - 7 + z$

**6.** $2(x + 1) + 3x = x - 8$

**7.** $3x + 7(1 - x) = 15$

**8.** $-2(x - 1) + 5x = -2 + x + 7$

**9.** $2x + 6 = 3(x + 2)$

**10.** $2x - 6 + 3x = 2x + 5(x - 2)$

**You Try It**

**11.** $7 + 2(3 - 4z) = 7z + 2(3z - 4)$

**12.** $2(3x - 5) + 1 = (2x - 1) - 3(x - 2)$

**13.** $(4x - 1) + 3(2x + 8) = 5x - 2$

**14.** $(x + 3) - 2(2 - 3x) = 4(x + 4) - 2$

**15.** $3(2x - 1) - 6(x + 1) = 2x + 3$

**16.** $4(5 - 2x) + 1 = 2(x + 3) - 5x$

**17.** $2[3(x + 2) - 2] = x - 2$

**18.** $2x - (x + 1) = -3 + 5(x - 6)$

**You Try It**

**19.** $2y + \dfrac{1}{5} = \dfrac{4}{3}$

**20.** $\dfrac{y}{3} + \dfrac{1}{2} = 3y$

**21.** $\dfrac{1}{3}(y + 1) = \dfrac{2}{3}y$

**22.** $3 - \dfrac{2z}{5} = 5 - \dfrac{2z}{3}$

**23.** $2x + \dfrac{1}{2} = \dfrac{x}{3} + 3$

**24.** $\dfrac{x + 1}{2} + 3 = \dfrac{x + 4}{5}$

**You Try It**

**25.** $2w + \dfrac{7}{3} - \dfrac{w + 2}{3} - 2$

**26.** $2 + \dfrac{3y}{5} - \dfrac{2}{5} + 3y$

**27.** $\dfrac{a}{2} + \dfrac{2}{3} - \dfrac{3a}{4} + \dfrac{4}{5}$

**28.** $\dfrac{2x}{3} + \dfrac{1}{3} + 2 = \dfrac{3x}{4} - \dfrac{3}{2}$

**29.** $\dfrac{2x + 2}{5} + 1 = \dfrac{5x + 1}{3} - 4$

**30.** $\dfrac{5x - 2}{3} + \dfrac{x + 1}{5} = 7$

**31.** $\left(\dfrac{x + 1}{2} + 1\right) + 1 = \dfrac{3x + 3}{4}$

**32.** $3\left[2 - \dfrac{z + 2}{5}\right] + 1 = 5\left(\dfrac{z - 1}{3} + 1\right)$

**33.** $1.5x + 2.4 = 3.2x - 1$

**34.** $2.5x + 1.2 = 6.8$

**You Try It** **35.** $2.57y + 3.21 = 1.5y$

**36.** $5.2z - 7.85 = 9.726$

**37.** $2.26 - 1.12x = 5x - 3.554$

**38.** $2.3(5 + 2x) - 2.5 = 9$

**39.** $0.24x - 0.66 = 1.35x - 0.65x - 2.806$

**40.** $1.5x - 3.5 + 2x = 3.2 + 1.9x - 11.7$

**41.** $3(x + 2) - 2x + 4 = x + 5$

**42.** $5(x + 2) - 2x = 3(x + 1) + 7$

**You Try It** **43.** $-3(x + 1) + x = 2(x + 2) - 7 - 4x$

**44.** $(2x + 5) + (x + 1) = 3(x - 1) + 7$

**45.** $2(x + 3) - 2 = 5(x + 1) - 3x$

**46.** $2\left(\dfrac{x + 1}{3} - 1\right) - x = -\dfrac{x + 1}{3}$

**You Try It** **47. Sliders** The cost, $C$ (in dollars), and the number of mini-burgers (or *sliders*), $N$, are related by the following equation:

$$40C = 27N + 2.$$

Find the number of sliders that can be purchased for $14.90.

**48. Chirp Rate** Temperature, $T$ (in °F) and the *chirp rate of crickets*, $C$, (in chirps per minute) have an actual mathematical relationship as shown in the following equation:

$$T = 50 + \dfrac{C - 40}{4}$$

While sitting outside last night, I heard crickets chirping. If the temperature was about 52 degrees, how many chirps did I hear per minute?

**49. Child Height** The following equation shows the approximate linear relationship between the age (in years), $a$, of a child and the child's height, $h$ (in inches):

$$h = 2.5433a + 28.465$$

If Payton, is 11.5 years old, approximately how tall might she be?
(*Source*: *The Merck Manual of Diagnosis and Therapy*, 15th ed)

# 2.3 Introduction to Problem Solving

## THINGS TO KNOW

Before working through this section, be sure you are familiar with the following concepts:

|  | VIDEO | ANIMATION | INTERACTIVE |

**You Try It** 1. Write Word Statements as Algebraic Expressions (Section 1.6, **Objective 3**)

**You Try It** 2. Solve Linear Equations Using Both Properties of Equality (Section 2.1, **Objective 5**)

**You Try It** 3. Solve Linear Equations Containing Non-Simplified Expressions (Section 2.2, **Objective 1**)

**You Try It** 4. Solve Linear Equations Containing Decimals; Apply a General Strategy (Section 2.2, **Objective 3**)

## OBJECTIVES

**1** Translate Sentences into Equations

**2** Use the Problem-Solving Strategy to Solve Direct Translation Problems

**3** Solve Problems Involving Related Quantities

**4** Solve Problems Involving Consecutive Integers

**5** Solve Problems Involving Value

---

**OBJECTIVE 1**  TRANSLATE SENTENCES INTO EQUATIONS

In **Section 1.6**, we translated *word phrases* into **algebraic expressions**. We can also translate *sentences* into **equations**. Like the **key words** that translate into **arithmetic operations**, there are key words that translate into an equal sign ($=$). See Table 1.

| Key Words That Translate to an Equal Sign | | | |
|---|---|---|---|
| is | was | will be | gives |
| yields | results in | equals | is equal to |
| is equivalent to | is the same as | | |

**Table 1**

Consider the following sentence:

The **product** of 5 and a number is 45.

The key word "product" indicates multiplication, and the key word "is" indicates an equal sign. Letting $x$ represent the unknown number, we can translate:

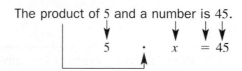

The equation is $5x = 45$.

## Example 1  Translating Sentences into Equations

Translate each sentence into an **equation**. Use $x$ to represent each unknown number.

**a.** Fifty-two less than a number results in $-21$.

**b.** Three-fourths of a number, increased by 8, gives the number.

**c.** The **difference** of 15 and a number is the same as the **sum** of the number and 1.

**d.** If the sum of a number and 4 is multiplied by 2, the result will be 2 less than the **product** of 4 and the number.

## Solutions

**a.** The phrase "less than" indicates subtraction. "Fifty-two less than a number" translates into the **algebraic expression** $x - 52$. "Results in" indicates an equal sign.

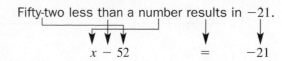

The equation is $x - 52 = -21$.

**b.** "Increased by" indicates addition. Three-fourths of $x$ and 8 are the **terms** being added, which means $\frac{3}{4}x + 8$. The word "gives" means an equal sign, and "the number" means $x$.

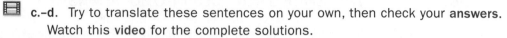

Three-fourths of a number, increased by 8, gives the number.

$$\frac{3}{4} \quad \cdot \quad x \quad + \quad 8 \quad = \quad x$$

The equation is $\frac{3}{4}x + 8 = x$.

 *My video summary*

 **c.–d.** Try to translate these sentences on your own, then check your **answers**. Watch this **video** for the complete solutions.

**You Try It** **Work through this You Try It problem.**

**Work Exercises 1–8 in this textbook or in the *MyMathLab*® Study Plan.**

OBJECTIVE 2 **USE THE PROBLEM-SOLVING STRATEGY TO SOLVE DIRECT TRANSLATION PROBLEMS**

A **mathematical model** uses the language of mathematics to describe a problem. Typically the model is an equation that describes a relationship within an application.

As you proceed through this course, you will learn how to model applied situations by using equations. Let's explore the following problem-solving strategy when translating, modeling, and solving applied problems involving linear equations.

---

**Problem-Solving Strategy for Applications of Linear Equations**

**Step 1.** **Define the Problem.** Read the problem carefully, or multiple times if necessary. Identify what you are trying to find and determine what information is available to help you find it.

**Step 2.** **Assign Variables.** Choose a variable to assign to an unknown quantity in the problem. For example, use $p$ for price. If other unknown quantities exist, express them in terms of the selected variable.

**Step 3.** **Translate into an Equation.** Use the relationships among the known and unknown quantities to form an equation.

**Step 4.** **Solve the Equation.** Determine the value of the variable and use the result to find any other unknown quantities in the problem.

**Step 5.** **Check the Reasonableness of Your Answer.** Check to see if your answer makes sense within the context of the problem. If not, check your work for errors and try again.

**Step 6.** **Answer the Question.** Write a clear statement that answers the question(s) posed.

---

We can use this **problem-solving strategy** to solve problems involving **direct translation**.

**Example 2  Solving Direct-Translation Problems**

Five times a number, increased by 17, is the same as 11 subtracted from the number. Find the number.

**Solution**  Follow the problem-solving strategy.

**Step 1.**  We must find "the number."

**Step 2.**  Let $x$ be the unknown number.

**Step 3.**  "Five times a number, increased by 17" translates to $5x + 17$. The word phrase "is the same as" translates to an equal sign ($=$). The phrase "11 subtracted from the number" translates to $x - 11$.

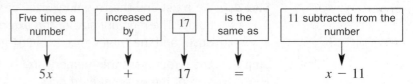

| Five times a number | increased by | 17 | is the same as | 11 subtracted from the number |
|---|---|---|---|---|
| $5x$ | $+$ | $17$ | $=$ | $x - 11$ |

The equation is $5x + 17 = x - 11$.

**Step 4.**

| | |
|---|---|
| Write the equation: | $5x + 17 = x - 11$ |
| Subtract $x$ from both sides: | $5x - x + 17 = x - x - 11$ |
| Simplify: | $4x + 17 = -11$ |
| Subtract 17 from both sides: | $4x + 17 - 17 = -11 - 17$ |
| Simplify: | $4x = -28$ |
| Divide both sides by 4: | $\dfrac{4x}{4} = \dfrac{-28}{4}$ |
| Simplify: | $x = -7$ |

**Step 5.**  Check the number $-7$ in the original sentence to see if it is a reasonable solution.

"5 times $-7$, increased by 17" translates to $5(-7) + 17$.

Simplify: $5(-7) + 17 = -35 + 17 = -18$.

"11 subtracted from $-7$" translates to $-7 - 11$.

Simplify: $-7 - 11 = -18$.

Both phrases simplify to the same result, $-18$, so $-7$ checks.

**Step 6.**  The number is $-7$.

**You Try It**  Work through this You Try It problem.

Work Exercises 9–12 in this textbook or in the *MyMathLab*®  Study Plan.

**Example 3  Solving Direct-Translation Problems**

*My video summary*  Four times the difference of twice a number and 5 results in the number increased by 50. Find the number.

**Solution**  Using the problem-solving strategy, try to solve this problem on your own. View the answer, or watch this video for a detailed solution.

**You Try It**  Work through this You Try It problem.

Work Exercises 13–16 in this textbook or in the *MyMathLab*®  Study Plan.

OBJECTIVE 3   SOLVE PROBLEMS INVOLVING RELATED QUANTITIES

For some problems, we need to find two or more quantities that are related in some way.

### Example 4  Storage Capacity

The storage capacity of Deon's external hard drive is 32 times that of his jump drive, a small portable memory device. Together, his two devices have 264 gigabytes of memory. What is the memory size of each device?

**Solution**  Follow the **problem-solving strategy**.

**Step 1.**  We must find the "memory size" or number of gigabytes that each device holds. The memory size of the external hard drive is 32 times that of the jump drive, and the total capacity is 264 gigabytes.

**Step 2.**  Let $g$ represent the memory size of the jump drive. Then $32g$ is the memory size of the external hard drive.

**Step 3.**  The total size is 264 gigabytes. We add the sizes of the jump drive and external hard drive to get 264.

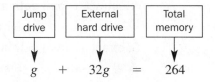

$$g \quad + \quad 32g \quad = \quad 264$$

**Step 4.**    Write the equation:  $g + 32g = 264$

Combine like terms:  $33g = 264$

Divide both sides by 33:  $\dfrac{33g}{33} = \dfrac{264}{33}$

Simplify:  $g = 8$

If $x = 8$, then $32x = 32(8) = 256$.

**Step 5.**  The jump drive holds 8 gigabytes of memory, and the external hard drive holds 256 gigabytes. Because 256 is 32 times 8 and the sum of 8 and 256 is 264, these results make sense.

**Step 6.**  Deon's jump drive has 8 gigabytes of memory, and his external hard drive has 256 gigabytes of memory.

**You Try It**   **Work through this You Try It problem.**

**Work Exercises 17–20 in this textbook or in the *MyMathLab*®  Study Plan.**

### Example 5  Movie Running Times

*My video summary*    Disney's *Toy Story* is 11 minutes shorter than its sequel *Toy Story 2*. *Toy Story 3* is 17 minutes longer than *Toy Story 2*. If the total running time for the three movies is 282 minutes, find the running time of each movie. (*Source*: Disney)

**Solution**

**Step 1.**  We need to find the running time of each movie. The relationship between the quantities is that *Toy Story* is 11 minutes shorter than *Toy Story 2*, and *Toy Story 3* is 17 minutes longer than *Toy Story 2*. Also, the total running time of the three movies is 282 minutes.

**Step 2.** Let $t$ represent the running time of *Toy Story 2*. Then $t - 11$ is the running time for *Toy Story*, and $t + 17$ is the running time for *Toy Story 3*.

**Step 3.** The sum of the three individual running times equals the total running time of 282 minutes.

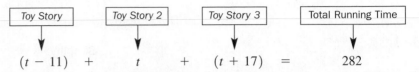

$$(t - 11) \quad + \quad t \quad + \quad (t + 17) \quad = \quad 282$$

Complete the remaining steps of the **problem-solving strategy** to finish this problem on your own. View the **answer**, or watch this video for a complete solution.

**You Try It**  Work through this You Try It problem.

Work Exercises **21** and **22** in this textbook or in the **MyMathLab**® Study Plan.

OBJECTIVE 4  SOLVE PROBLEMS INVOLVING CONSECUTIVE INTEGERS

**Integers** that appear next to each other in an ordered list of all integers are called **consecutive integers**. For example, 7, 8, and 9 are three consecutive integers. Similarly, 10, 12, and 14 are **consecutive even integers**, and 9, 11, and 13 are **consecutive odd integers**.

The **difference** of two consecutive integers is 1 (for example, $9 - 8 = 1$), and the difference of two consecutive even integers or two consecutive odd integers is 2 (for example, $10 - 8 = 2$ and $9 - 7 = 2$). Using these facts, we can establish the general relationships shown in Table 2.

| | Example | General Relationship |
|---|---|---|
| Consecutive integers | 7, 8, 9 | $x, x + 1, x + 2$, where $x$ is an integer |
| Consecutive even integers | 10, 12, 14 | $x, x + 2, x + 4$, where $x$ is an even integer |
| Consecutive odd integers | 9, 11, 13 | $x, x + 2, x + 4$, where $x$ is an odd integer |

**Table 2**

 A common error is to use $x$, $x + 1$, and $x + 3$ to represent consecutive odd integers (because 1 and 3 are odd). This is wrong! Remember consecutive odd integers are 2 units apart.

**Example 6  Solving a Consecutive Integer Problem**

The sum of two **consecutive integers** is 79. Find the two integers.

**Solution**

**Step 1.** We are looking for two consecutive integers with a sum of 79.

**Step 2.** Let $x$ be the first integer. Then $x + 1$ is the next consecutive integer.

**Step 3.** The sum is 79, so we add the two integers to equal 79.

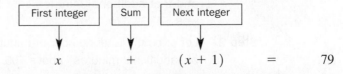

$$x \quad + \quad (x + 1) \quad = \quad 79$$

**Step 4.**    Write the equation:    $x + (x + 1) = 79$

Combine like terms:    $2x + 1 = 79$

Subtract 1 from both sides:    $2x + 1 - 1 = 79 - 1$

Simplify:    $2x = 78$

Divide both sides by 2:    $\dfrac{2x}{2} = \dfrac{78}{2}$

Simplify:    $x = 39$

If $x = 39$, then $x + 1 = 39 + 1 = 40$.

**Step 5.**    Because 39 and 40 are consecutive integers and their sum is 79, the answers make sense.

**Step 6.**    The two consecutive integers are 39 and 40.

**You Try It**    Work through this You Try It problem.

**Work Exercise 23 and 24 in this textbook or in the *MyMathLab*® Study Plan.**

### Example 7  Solving a Consecutive Integer Problem

 *My video summary*    Three **consecutive even integers** add to 432. Find the three integers.

**Solution**  Try to solve this problem on your own. View the **answer**, or watch this **video** for a detailed solution.

**You Try It**    Work through this You Try It problem.

**Work Exercises 25–28 in this textbook or in the *MyMathLab*® Study Plan.**

OBJECTIVE 5    SOLVE PROBLEMS INVOLVING VALUE

Many applications involve finding the cost or value of a set of items. For example, suppose you purchase 3 pastries that cost 69¢ each. Your total cost for the three pastries will be ($0.69)(3) = $2.07. The following equation helps us to solve this problem:

$$\text{Total value} = (\text{Value per item})(\text{Number of items})$$

The "value" in the above equation can represent cost, profit, revenue, earnings, and so forth.

### Example 8  Cell Phone Plan

 *My video summary*    Ethan's cell phone plan costs $34.99 per month for the first 700 minutes, plus $0.35 for each additional minute. If Ethan's bill is $57.39, how many minutes did he use?

**Solution**

**Step 1.**    We need to find the number of minutes Ethan used on his cell phone. We know that the first 700 minutes are included in the monthly cost of $34.99 and that additional minutes cost $0.35 each. We also know the total cost was $57.39.

**Step 2.**    Let $m$ be the total number of minutes Ethan used. Then $m - 700$ is the number of "additional minutes" above 700.

**Step 3.** The cost of the first 700 minutes is $34.99. The cost of the additional minutes is $0.35(m - 700)$. Adding these two amounts results in the overall bill amount of $57.39. So, we have the equation:

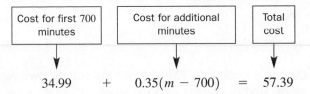

$$34.99 \quad + \quad 0.35(m - 700) \quad = \quad 57.39$$

Finish solving the equation to find the number of minutes Ethan used. Compare your **answer**, or watch this **video** for a fully worked solution.

**You Try It**   Work through this You Try It problem.

**Work Exercises 29–32 in this textbook or in the *MyMathLab*® Study Plan.**

# 2.3 Exercises

In Exercises 1–8, translate each sentence into an equation. Use $x$ to represent each unknown number. Do not solve.

**1.** Thirteen more than a number is 38.

**2.** The sum of twice a number and 16 is equal to three times the number.

**3.** Subtracting 5 from the quotient of a number and 4 gives 31.

**You Try It**

**4.** The product of 7 and the sum of a number and 4 results in the difference of 4 and the number.

**5.** The sum of a number and 2.7, divided by 6, yields the quotient of the number and 9.

**6.** If −7 is decreased by a number, the result will be three-fourths of the number.

**7.** The ratio of a number and 2 is equivalent to the difference of the number and 19.

**8.** Three times the sum of a number and 8, decreased by 10, is the same as five times the difference of the number and 1, increased by 14.

In Exercises 9–16, solve each problem involving direct translation.

**9.** The quotient of a number and 3, decreased by 1, is equal to 37. Find the number.

**10.** Nineteen less than triple a number is −53. Find the number.

**You Try It**

**11.** The sum of a number and 11 is the same as the difference of 17 and the number. Find the number.

**12.** The product 8 and a number, increased by 84, is equivalent to the product of 12 and the sum of the number increased by 4. Find the number.

**13.** If −11 is added to a number, the sum will be 5 times the number. Find the number.

**14.** Five-eighths more than three-fourths of a number results in two times the number.

**You Try It**

**15.** Eight times the difference of twice a number and 9 yields the number increased by 93. Find the number.

**16.** The sum of a number and 80, divided by 6, results in one-half of the number. Find the number.

In Exercises 17–22, solve each problem involving related quantities.

**You Try It**

**17. Cutting a Board**   A 72-inch board is cut into two pieces so that one piece is four times as long as the other. Find the length of each piece.

**18. Inheritance**   An inheritance of $35,000 will be divided between sisters Hannah and Taylor. If Hannah is to receive $3000 less than Taylor, how much does each sister receive?

**19. Skyscrapers**   The Burj Khalifa (in Dubai, United Arab Emirates) is 320 meters taller than Taipei 101 (in Taipei, Taiwan). If the height of the Burj Khalifa is subtracted from twice the height of Taipei 101, the result is 188 meters. How tall is each skyscraper? (*Source*: Infoplease.com)

**20. Movie Classics**   The running time for *Gone with the Wind* is 36 minutes more than twice the running time for *The Wizard of Oz*. If the running time for *Gone with the Wind* is subtracted from triple the running time for *The Wizard of Oz*, the result is 65 minutes. Find the running times for each movie. (*Source*: The Internet Movie Database, IMDb.com)

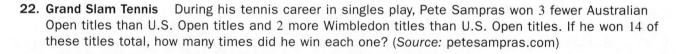
**You Try It**

**21. Most Valuable Player**   During his career, Michael Jordan was named the NBA's MVP one time less than Kareem Abdul-Jabbar. Bill Russell was MVP two more than half the times of Abdul-Jabbar. Together, the three men were named MVP 16 times. How many times were each of the men named MVP? (*Source*: NBA.com)

**22. Grand Slam Tennis**   During his tennis career in singles play, Pete Sampras won 3 fewer Australian Open titles than U.S. Open titles and 2 more Wimbledon titles than U.S. Open titles. If he won 14 of these titles total, how many times did he win each one? (*Source*: petesampras.com)

In Exercises 23–28, solve each problem involving consecutive integers.

**You Try It**

**23.** The sum of two consecutive integers is 691. Find the integers.

**24.** The sum of three consecutive integers is 258. Find the integers.

**25.** The sum of two consecutive odd integers is 116. Find the integers.

**You Try It**

**26.** The sum of three consecutive even integers is 762. Find the integers.

**27. An Open Book**   The page numbers showing on an open book are consecutive integers with a sum of 317. Find the two page numbers.

**28. Office Numbers**   The numbers on the doors of four adjacent offices are consecutive odd integers. If their sum is 880, find the four office numbers.

In Exercises 29–32, solve each problem involving value.

**29. Selling Aluminum**   Doug collects and recycles aluminum cans by selling them to a local salvage yard. If he earns $136.30 by selling aluminum at a price of $0.58 per pound, how many pounds of aluminum did Doug sell?

**You Try It**

**30. Car Rental**   A car rental cost $80 per day, for the first 250 miles, plus $0.20 for each additional mile. If Sonya's rental cost was $116 for a one-day rental, how many miles did she drive?

31. **Working Two Jobs**    John works two jobs. As a security guard he earns $9.50 per hour. As a landscaper he earns $13.00 per hour. One week John worked a total of 40 hours and earned $432.50. How many hours did he work at each job?

32. **Concession Stand**    A basketball concession stand sells sodas for $1.50 each and bags of popcorn for $0.75 each. One night, 36 more sodas were sold than bags of popcorn. If the concession stand made $153 that night, how many sodas and how many bags of popcorn were sold?

# 2.4  Formulas

## THINGS TO KNOW

Before working through this section, be sure you are familiar with the following concepts:

VIDEO          ANIMATION          INTERACTIVE

**You Try It**

1. Evaluate Algebraic Expressions (Section 1.5, **Objective 1**)

**You Try It**

2. Solve Linear Equations Using the Both Properties of Equality (Section 2.1, **Objective 5**)

**You Try It**

3. Solve Linear Equations Containing Non-Simplified Expressions (Section 2.2, **Objective 1**)

**You Try It**

4. Solve Linear Equations Containing Fractions (Section 2.2, **Objective 2**)

**You Try It**

5. Use Linear Equations to Solve Application Problems (Section 2.2, **Objective 5**)

## OBJECTIVES

1  Evaluate a Formula

2  Find the Value of a Non-isolated Variable in a Formula

3  Solve a Formula for a Given Variable

4  Use Geometric Formulas to Solve Applications

OBJECTIVE 1    EVALUATE A FORMULA

> **Definition   Formula**
>
> A **formula** is an **equation** that describes the relationship between two or more **variables**.

Typically formulas apply to physical or financial situations that relate quantities such as length, area, volume, time, speed, money, interest rates, and so on. For example, $d = rt$ is the **distance formula**. It relates three variables: distance $d$, rate (or speed) $r$, and time $t$. If we know values for $r$ and $t$, we can **evaluate the formula** to find the value of $d$.

## Example 1  Evaluating the Distance Formula

A car travels at an average speed (rate) of 55 miles per hour for 4 hours. How far does the car travel?

**Solution** Using the distance formula, we have $r = \dfrac{55 \text{ mi}}{\text{h}}$ and $t = 4 \text{ h}$ and need to find $d$.

Write the distance formula:  $d = rt$

Substitute $r = \dfrac{55 \text{ mi}}{\text{h}}$ and $t = 4\text{h}$:  $d = \dfrac{55 \text{ mi}}{\text{h}} \cdot 4\text{h}$

Simplify:  $d = \dfrac{55 \text{ mi}}{\cancel{\text{h}}} \cdot \dfrac{4 \cancel{\text{h}}}{1} = 220 \text{ mi}$

The car traveled 220 miles.

 When evaluating formulas, the units must be consistent. For example, when evaluating the distance formula, if the unit for rate is meters per second, then the unit for time must be seconds, and the resulting unit for distance will be meters.

**You Try It**   Work through this You Try It problem.

**Work Exercises 1 and 2 in this textbook or in the _MyMathLab_® Study Plan.**

The **Devine Formula** is a popular **formula** in the healthcare profession that uses a person's height to compute his or her ideal body weight. The formula is $w = 110 + 5.06(h - 60)$ for men and $w = 100.1 + 5.06(h - 60)$ for women, where $w$ is ideal body weight, in pounds, and $h$ is height, in inches.

**Note:** This formula is only used for people over 60 inches, or 5 feet, tall.

## Example 2  Computing Ideal Body Weight

 _My video summary_  Evaluate the Devine Formula to find the ideal body weight of each person described.

a.  A man 72 inches tall

b.  A woman 66 inches tall

### Solutions

a.  Write the Devine Formula (for men):  $w = 110 + 5.06(h - 60)$

Substitute $h = 72$:  $w = 110 + 5.06(72 - 60)$

Subtract within parenthases:  $w = 110 + 5.06(12)$

Multiply:  $w = 110 + 60.72$

Add:  $w = 170.72$

The ideal body weight of a man 72 inches tall is 170.72 pounds.

b.  Try to evaluate the Devine Formula (for women) on your own, then check your **answer**. Watch this **video** for complete solutions to both parts.

**You Try It**   Work through this You Try It problem.

**Work Exercises 3–8 in this textbook or in the _MyMathLab_® Study Plan.**

Many **formulas** come from geometry. To review the formulas for **area** and **perimeter**, launch this **popup box**. To review formulas for volume, launch this **popup box**.

### Example 3  Evaluating Geometry Formulas

a. The top of a stainless steel sink is shaped like a square with each side measuring $15\frac{3}{4}$ inches long. How many inches of aluminum molding will be required to surround the outside of the sink? Hint: Use the **formula** for the **perimeter of a square**.

$15\frac{3}{4}$ inches

$15\frac{3}{4}$ inches

b. A yield sign has the shape of a triangle with a base of 3 feet and a height of 2.6 feet. Find the area of the sign. Hint: Use the formula for the **area of a triangle**.

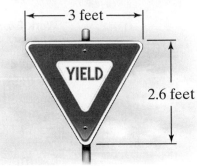

3 feet

2.6 feet

### Solutions

a. We must find the **perimeter** of the sink. The **formula** for the perimeter of a square is $P = 4s$. We are given $s = 15\frac{3}{4}$ inches.

Write the formula:    $P = 4s$

Substitute $s = 15\frac{3}{4}$ in:    $A = 4\left(15\frac{3}{4}\text{ in}\right)$

Simplify:    $A = 4\left(\frac{63}{4}\text{ in}\right) = 63\text{ in}$

63 inches of aluminum molding are needed.

 *My video summary*        b. Try working this problem on your own by using the formula $A = \frac{1}{2}bh$. View the answer, or watch this video for a detailed solution.

**You Try It**    Work through this You Try It problem.

Work Exercises 9–12 in this textbook or in the **MyMathLab**® Study Plan.

OBJECTIVE 2    FIND THE VALUE OF A NON-ISOLATED VARIABLE IN A FORMULA

Most **formulas** are solved for one of their variables. This means that one variable is **isolated** on one side of the equation. For example, the formula $d = rt$ is solved for $d$. We have **evaluated formulas** to find values of isolated variables. Now let's find values of non-isolated variables.

### Example 4  Finding the Value of a Non-isolated Variable

The **perimeter** of a rectangle is given by the formula $P = 2l + 2w$. If $P = 84$ cm and $l = 26$ cm, find $w$.

**Solution**  We want to find the width $w$ of a rectangle with perimeter $P = 84$ and length $l = 26$ cm.

Write the original formula:    $P = 2l + 2w$

Substitute $P = 84$ cm and $l = 26$ cm:    $84\text{ cm} = 2(26\text{ cm}) + 2w$

**2.4**  Formulas    2-29

| | |
|---|---|
| Simplify: | $84\text{ cm} = 52\text{ cm} + 2w$ |
| Subtract 52 cm from both sides: | $84\text{ cm} - 52\text{ cm} = 52\text{ cm} + 2w - 52\text{ cm}$ |
| Simplify: | $32\text{ cm} = 2w$ |
| Divide both sides by 2: | $\dfrac{32\text{ cm}}{2} = \dfrac{2w}{2}$ |
| Simplify: | $16\text{ cm} = w$ |

The width of the rectangle is 16 cm.

**You Try It**    Work through this You Try It problem.

**Work Exercises 13–16 in this textbook or in the *MyMathLab*® Study Plan.**

A common banking formula, $I = Prt$, is used for computing **simple interest**. In this formula, $I$ is the *simple interest* earned on an investment or paid for a loan; $P$ is the *principal*, or the amount that is invested or borrowed; $r$ is the *interest rate* in decimal form; and $t$ is the *time* that the money is invested or borrowed.

### Example 5  Using the Simple Interest Formula

 *My video summary*

Paige has invested $15,000 in a **certificate of deposit** (CD) that pays 4% simple interest annually. If she earns $750 in interest when the CD matures, how long has Paige invested the money?

**Solution**  We know $P = \$15{,}000$, $r = 4\% = 0.04$, and $I = \$750$. To find $t$, we substitute these values into the formula $I = Prt$ and solve for $t$. Try to finish this problem on your own. View the **answer**, or watch this **video** for the complete solution.

**You Try It**    Work through this You Try It problem.

**Work Exercise 17 in this textbook or in the *MyMathLab*® Study Plan.**

A **formula** that relates the two most common measures of temperature is $F = \dfrac{9}{5}C + 32$, where $F$ represents degrees Fahrenheit and $C$ represents degrees Celsius.

### Example 6  Finding Equivalent Temperatures

 *My video summary*

During the month of February, the average high temperature in Montreal, QC is $-4.6°C$ while the average high temperature in Phoenix, AZ is $70.7°F$. (*Source:* World Weather Information Service)

a.  What is the equivalent Fahrenheit temperature in Montreal?

b.  What is the equivalent Celsius temperature in Phoenix?

**Solutions**  Try working these problems on your own, then check your **answers**. Watch this **video** for complete solutions.

**You Try It**    Work through this You Try It problem.

**Work Exercises 18–22 in this textbook or in the *MyMathLab*® Study Plan.**

## OBJECTIVE 3   SOLVE A FORMULA FOR A GIVEN VARIABLE

Sometimes it is helpful to **solve a formula for a given variable** in terms of the other **variables**. This means that we write the **formula** with the desired variable **isolated** on one side of the equal sign, resulting in all other variables and **constants** on the other side of the equal sign. For example, the distance formula $d = rt$ can be solved for $t$:

$$\text{Write the distance formula:} \quad d = rt$$

$$\text{Divide both sides by } r: \quad \frac{d}{r} = \frac{rt}{r}$$

$$\text{Simplify:} \quad \frac{d}{r} = t$$

To solve a formula for a given variable, we follow the **general strategy for solving linear equations** from Section 2.2.

 When we solve a formula for a given variable, the appearance of the formula changes, but the relationships among the variables do not.

### Example 7   Solving a Formula for a Given Variable

Solve each formula for the given variable.

**a.** Selling price: $S = C + M$ for $M$

**b.** Area of a triangle: $A = \dfrac{1}{2}bh$ for $b$

**c.** **Perimeter** of a rectangle: $P = 2l + 2w$ for $l$

### Solutions

**a.**
$$\text{Write the original formula:} \quad S = C + M$$
$$\text{Subtract } C \text{ from both sides:} \quad S - C = C + M - C$$
$$\text{Simplify:} \quad S - C = M$$

The formula $S - C = M$, or $M = S - C$, is solved for $M$.

**b.**
$$\text{Write the original formula:} \quad A = \frac{1}{2}bh$$

$$\text{Multiply both sides by 2 to clear the fraction:} \quad 2(A) = 2\left(\frac{1}{2}bh\right)$$

$$\text{Simplify:} \quad 2A = bh$$

$$\text{Divide both sides by } h: \quad \frac{2A}{h} = \frac{bh}{h}$$

$$\text{Simplify:} \quad \frac{2A}{h} = b$$

The formula $b = \dfrac{2A}{h}$ is solved for the variable $b$.

  **c.** Try to solve this formula for $l$ on your own. View the **answer**, or watch this **video** to see a fully worked solution.

 For a formula to be solved for a given variable, the given variable must be isolated on one side of the equation and must be the only variable of its type in the equation.

**You Try It**   **Work through this You Try It problem.**

**Work exercises 23–30 in this textbook or in the *MyMathLab*® Study Plan.**

## OBJECTIVE 4   USE GEOMETRIC FORMULAS TO SOLVE APPLICATIONS

We have seen that **formulas** can be very useful when solving applications. Some applications require work beyond a single formula. Let's look at some applied problems that involve geometric formulas.

### Example 8   Filling a Swimming Pool

An above-ground pool is shaped like a **circular cylinder** with a **diameter** of 28 ft and a **depth** of 4.5 ft. If $1 \text{ ft}^3 \approx 7.5$ gal, how many gallons of water will the pool hold? Use $\pi \approx 3.14$ and round to the nearest thousand gallons.

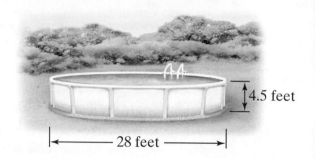

4.5 feet

28 feet

**Solution** We must find the **volume** of the pool in cubic feet and convert to gallons. The formula for the volume of a cylinder is $V = \pi r^2 h$. The diameter is 28 ft, so the **radius** is half of the diameter, or 14 ft.

Write the volume formula:   $V = \pi r^2 h$

Substitute $r = 14$ ft, $h = 4.5$ ft, and $\pi \approx 3.14$:   $V \approx (3.14)(14 \text{ ft})^2(4.5 \text{ ft})$

Evaluate the exponent:   $V \approx (3.14)(196 \text{ ft}^2)(4.5 \text{ ft})$

Multiply:   $V \approx 2769.48 \text{ ft}^3$

To find the number of gallons, we multiply the volume by $\dfrac{7.5 \text{ gal}}{1 \text{ ft}^3}$:

$$2769.48 \text{ ft}^3 \cdot \frac{7.5 \text{ gal}}{1 \text{ ft}^3} \approx 20{,}771.1 \text{ gal}$$

Rounding to the nearest thousand gallons, the pool will hold approximately 21,000 gallons.

**You Try It**   Work through this You Try It problem.

Work Exercises 31 and 32 in this textbook or in the **MyMathLab** Study Plan.

### Example 9   Finding the Cost of a New Floor

*My video summary*   Terrence wants to have a new floor installed in his living room, which measures 20 ft by 15 ft. Extending out 3 ft from one wall is a fireplace in the shape of a **trapezoid** with base lengths of 4 ft and 8 ft.

**a.** Find the **area** that needs flooring.

**b.** If the flooring costs $5.29 per square foot, how much will Terrence pay for the new floor? (Assume there is no wasted flooring.)

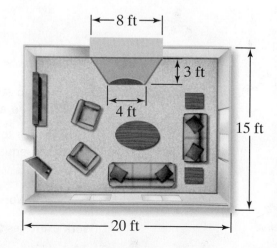

8 ft, 3 ft, 4 ft, 15 ft, 20 ft

## Solutions

a. The **area** that needs new flooring is the area of the room (rectangle) minus the area of the fireplace (trapezoid). The area formula for a rectangle is $A = lw$, and the area formula for a trapezoid is $A = \frac{1}{2}h(a + b)$, so

$$\text{Area needing flooring} = \underbrace{lw}_{\text{Area of Rectangle}} - \underbrace{\frac{1}{2}h(a + b)}_{\text{Area of Trapezoid}}.$$

Substitute $l = 20$ ft, $w = 15$ ft, $h = 3$ ft, $a = 4$ ft, and $b = 8$ ft to finish part a, then complete all of part b on your own. View the **answers**, or watch this **video** for detailed solutions.

**You Try It**   Work through this You Try It problem.

**Work Exercises 33 and 34 in this textbook or in the *MyMathLab*®  Study Plan.**

### Example 10  Constructing a Roundabout Intersection

*My video summary*   The roundabout intersection with the top view shown in the figure will be constructed using concrete pavement 9 inches thick. How many cubic yards of concrete will be needed for the roundabout? Use $\pi \approx 3.14$.

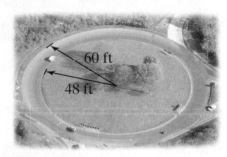

**Solution**  First, we must find the volume of concrete needed, in cubic yards. This problem involves finding the volume of two **cylinders**: an outer cylinder with a **radius** of 60 feet and an inner cylinder with a radius of 48 feet. Both cylinders have heights of 9 inches. The volume of concrete needed is the difference in the volumes of these two cylinders:

Volume of concrete = Volume of outer cylinder − Volume of inner cylinder

Try to finish solving this problem on your own. Remember to use consistent units when computing the volumes. (Hint: 1 yd = 3 ft = 36 in.) View the **answer**, or watch this **video** for a detailed solution.

**You Try It**   Work through this You Try It problem.

**Work Exercises 35–36 in this textbook or in the *MyMathLab*®  Study Plan.**

# 2.4  Exercises

In Exercises 1 and 2, use the distance formula $d = rt$.

**You Try It**

**1.** An airplane travels for 2.5 hours at an average rate of 140 miles per hour. How far does the plane travel?

**2.** If a cheetah runs at a rate of 16 meters per second for 7 seconds, how far will it travel?

In Exercises 3 and 4, use the appropriate **Devine Formula** to compute the ideal body weight of each person described.

**You Try It**

**3.** A man 70 inches tall

**4.** A woman 62 inches tall

In Exercises 5–8, use the formula provided to answer each question.

**5. Good Tip** $T = 0.2C$ is a formula for computing a 20% tip $T$ in a restaurant for a meal that costs $C$ dollars. Find the tip for a meal costing $75.

**6. Retail Price** The formula $R = W + pW$ gives the retail price $R$ of an item with a wholesale price $W$ that is marked up $p$ percent (in decimal form). Determine the retail price of a pair of jeans with a wholesale price of $35.40 that is marked up 40%.

**7. High Blood Pressure** The formula $p = 1.068a - 18.39$ can be used to find the percent $p$ of men at age $a$, in years, with high blood pressure. What percent of 35 year-old men have high blood pressure? Remember to convert from decimals to percent for your answer. Round to the nearest tenth of a percent. (*Source:* Based on data from the British Heart Foundation)

**8. Horse Power** The formula $H = \dfrac{VIE}{746}$ gives the horsepower $H$ produced by an electrical motor with voltage $V$ in volts, current $I$ in amps, and percent efficiency $E$ in decimal form. What is the horsepower of a 240-volt motor pulling 40 amps and having 86% efficiency? Remember to convert percents to decimals in your calculations. Round $H$ to the nearest tenth.

In Exercises 9–12, evaluate the appropriate geometry formula to answer each question.

**You Try It**

**9. Road Work Ahead** The roadwork ahead sign shown has the shape of a square (rotated to look like a diamond). A construction worker wants to outline the edge of the sign with reflective tape. What length of reflective tape will be needed?

**10. NBA Basketball Court** The dimensions of an official NBA basketball court are shown in the figure. Find both the perimeter and area of the court. (*Source:* NBA.com)

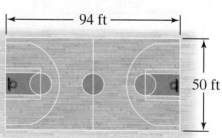

**11. Table Top** The top of a multi-purpose table is the shape of the trapezoid shown. Find both the perimeter and area of the table top.

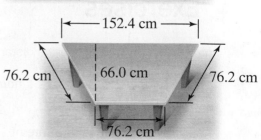

**12. Volume of a Printer**   The laser printer shown is approximately a rectangular solid. What is the volume of the printer?

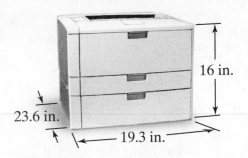

16 in.

23.6 in.

19.3 in.

In Exercises 13–16, find the value of the unknown variable.

**You Try It**

**13. Profit**   Find $R$ if $P = R - C$, $P - \$2150$, and $C = \$1525$.

**14. Area of a Rectangle**   Find the value of $w$ if $A = lw$, $A = 48$ m$^2$, and $l = 4$ m.

**15. Retail Price**   Find the value of $W$ if $R = W + pW$, $R = \$31.85$, and $p = 30\% = 0.3$.

**16. Area of a Trapezoid**   Find the value of $a$ if $A = \dfrac{1}{2}h(a + b)$, $A = 210$ mm$^2$, $h = 14$ mm, and $b = 18$ mm.

In Exercises 17–22, use the appropriate formula to answer each question.

**17. Simple Interest Principal**   How much principal should be invested in a savings account paying 6% simple interest if you want to earn \$450 in interest in 2 years.

**You Try It**

**18. Record Heat**   The warmest temperature officially recorded on Earth is $136°F$, which occurred on September 13, 1922 in Al 'Aziziyah, Libya. (*Source:* World Meteorological Organization) What is the equivalent temperature in Celsius? Round to the nearest tenth.

**19. Time for a Trip**   How long will it take Cory to drive 210 miles if he averages 60 miles per hour on the trip?

**20. NBA Ball Size**   The circumference of an official NBA basketball is 29.5 inches. Find the diameter of the ball. Use $\pi \approx 3.14$. Round to the nearest tenth. (*Source:* NBA.com)

**21. Sail Height**   A triangular sail for a boat has an area of $120$ ft$^2$. If the base of the sail is 16 ft, find the height.

?

16 ft

**22. Popcorn Box**   The popcorn box shown has a volume of $123$ in$^3$. What is the height of the box?

?

6 in.

2.5 in.

**2.4** Formulas   **2-35**

In Exercises 23–30, solve the formula for the given variable.

You Try It

**23.** $C = 2\pi r$ for $r$

**24.** $P = a + b + c$ for $a$

**25.** $Ax + By = C$ for $y$

**26.** $A = P + Prt$ for $t$

**27.** $E = I(r + R)$ for $R$

**28.** $A = \dfrac{1}{2}h(a + b)$ for $a$

**29.** $F = \dfrac{9}{5}C + 32$ for $C$

**30.** $E = \dfrac{I - P}{I}$ for $P$

In Exercises 31–36, solve each application. When necessary, use $\pi \approx 3.14$.

You Try It

**31. Rain Barrel** The rain barrel shown is a right circular cylinder. If $1\text{ ft}^3 \approx 7.5$ gallons, how many gallons of rainwater can the barrel hold? Round to the nearest tenth.

←1.75 ft →

3 ft

**32. Martini Glass** The martini glass shown has the shape of cone (on a stem). If $1\text{ cm}^3 = 1$ mL, how many milliliters of liquid can the glass hold? Round to the nearest tenth.

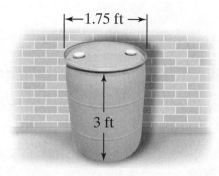

←9.6 cm→

6.4 cm

You Try It

**33. Carpeting a Bedroom** Lauren's bedroom measures 12 feet by 15 feet, with a closet that measures 6 feet by 3 feet. If carpet costs $4.75 per square foot, how much will it cost to carpet Lauren's room and closet?

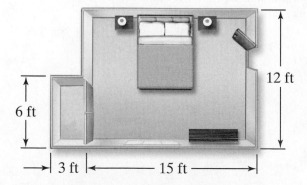

6 ft

12 ft

→ 3 ft ←──── 15 ft ────→

**34. Sodding a Lawn**   If sod costs $0.45 per square foot, how much will it cost to sod the lawn shown in the figure? Hint: Sod is not needed where the house and driveway are located.

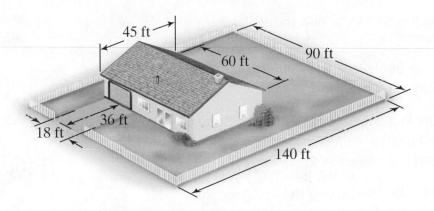

45 ft

60 ft

90 ft

36 ft

18 ft

140 ft

**You Try It**

**35. Building a Sidewalk**   A concrete sidewalk 5 feet wide and 6 inches thick will be built around the square playground shown. How many cubic yards of concrete will be needed for the sidewalk?

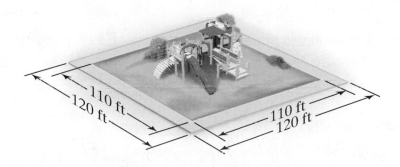

110 ft
120 ft

110 ft
120 ft

**36. Resurfacing a Driveway**   A semicircular driveway 10 feet wide will be resurfaced with asphalt. See the figure. If the 2-inch thick asphalt costs $1.95 per square foot, find the cost of the driveway.

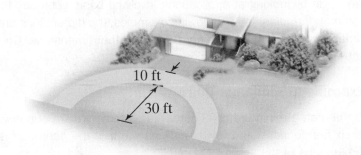

10 ft

30 ft

# 2.5 Geometry and Uniform Motion Problem Solving

## THINGS TO KNOW

Before working through this section, be sure you are familiar
with the following concepts:

| | | | VIDEO | ANIMATION | INTERACTIVE |
|---|---|---|---|---|---|

**You Try It**    **1.** Translate Sentences into Equations
(Section 2.3, **Objective 1**)    🎞️

**You Try It**    **2.** Use the Problem-Solving Strategy to Solve
Direct Translation Problems
(Section 2.3, **Objective 2**)    🎞️

**You Try It**    **3.** Solve Problems Involving Related Quantities
(Section 2.3, **Objective 3**)    🎞️

**You Try It**    **4.** Evaluate a Formula
(Section 2.4, **Objective 1**)    🎞️

**You Try It**    **5.** Solve a Formula for a Given Variable
(Section 2.4, **Objective 3**)    🎞️

**You Try It**    **6.** Use Geometric Formulas to Solve Applications
(Section 2.4, **Objective 4**)    🎞️

## OBJECTIVES

**1** Solve Problems Involving Geometry Formulas

**2** Solve Problems Involving Angles

**3** Solve Problems Involving Uniform Motion

---

OBJECTIVE 1    SOLVE PROBLEMS INVOLVING GEOMETRY FORMULAS

We begin by looking at applications involving basic geometry formulas. **Review** some of the
common formulas. In the following examples, the dimensions are given in terms of a common
variable, which may or may not be one of the dimensions. Once a value for the variable is
known, the individual dimensions can be found.

### Example 1   Miniature Golf

A green on a miniature golf course has a rectangular boundary. The length of the boundary
is six feet longer than twelve times its width. If the perimeter is 103 feet, what are the
dimensions of the green?

**Solution** We use the **problem-solving
steps.**

**Step 1.**    We must find both the length
and the width of the miniature golf
course green. The perimeter is
103 feet, and the length is 6 feet
longer than 12 times the width.

We will use the formula for the **perimeter of a rectangle** to help us solve this problem.

**Step 2.** If we let $w$ represent the width of the green, then the length is $12w + 6$.

**Step 3.** Use the formula for the perimeter of a rectangle.

$$2l + 2w = P$$

Substituting 103 for $P$ (the perimeter) and $12w + 6$ for $l$ we have

$$2(12w + 6) + 2w = 103$$

**Step 4.** Solve the equation for $w$ and then use the result to find the length.

| | |
|---:|:---|
| Begin with the original equation: | $2(12w + 6) + 2w = 103$ |
| **Distribute:** | $24w + 12 + 2w = 103$ |
| Combine like terms: | $26w + 12 = 103$ |
| Subtract 12 from both sides: | $26w + 12 - 12 = 103 - 12$ |
| Simplify: | $26w = 91$ |
| Divide both sides by 26: | $\dfrac{26w}{26} = \dfrac{91}{26}$ |
| Simplify: | $w = \dfrac{7}{2}$ or 3.5 |

**Step 5.** The width of the green is 3.5 feet. Substituting 3.5 for $w$, we get $l = 12(3.5) + 6 = 48$ feet. This gives a perimeter of $P = 2(48) + 2(3.5) = 103$ feet, as given in the original problem.

**Step 6.** The width of the green is 3.5 feet, and the length is 48 feet.

## Example 2 Canoe Paddle

*My video summary*

The blade of a canoe paddle is in the shape of an **isosceles triangle** so that two sides have the same length. The two common sides are each 4 inches longer than twice the length of the third side. If the **perimeter** is 48 inches, find the lengths of the sides of the blade.

## Solution

**Step 1.** We need to find the lengths of the sides of a canoe paddle blade. We know that the blade is an isosceles triangle, the common side is 4 inches longer than twice the length of the third side, and the perimeter of the blade is 48 inches. We will use the perimeter formula for a triangle to solve this problem.

**Step 2.** If we let $a$ represent the length of the third side, then the length of each common side is $2a + 4$.

**Step 3.** Start with the formula for the perimeter of a triangle.

$$a + b + c = P$$

Substituting 48 for $P$ (the perimeter) and $2a + 4$ for $b$ and $c$ (the two common sides), we have the equation:

$$a + (2a + 4) + (2a + 4) = 48$$

**Step 4.** Finish solving the equation to find the length of the third side, then use that result to find the length of the common side. When finished, check your **answer**, or watch this **video** for a fully worked solution.

### Example 3  Fresno Favorite

 *My video summary*

The Triangle Drive In has been a local favorite for hamburgers in Fresno, CA since 1963. It is located on a triangular-shaped lot. One side of the lot is 4 meters longer than the shortest side, and the third side is 32 meters less than twice the length of the shortest side. If the perimeter of the lot is 180 meters, find the length of each side.

**Solution** Follow the **problem-solving steps** to solve this problem on your own. View the **answer**, or watch this **video** for a fully worked solution.

**You Try It**   Work through this You Try It problem.

Work Exercises 1–5 in this textbook or in the *MyMathLab*® Study Plan.

### Example 4  Bathtub Surround

 *My video summary*

A bathtub is surrounded on three sides by a vinyl wall enclosure. The height of the enclosure is 20 inches less than three times the width, and the length is 5 inches less than twice the width. If the sum of the length, width, and height is 185 inches, what is the **volume** of the enclosed space?

### Solution

**Step 1.**   We need to find the volume of the enclosed space. We know that the height is 20 inches less than three times the width, the length is 5 inches less than twice the width, and the sum of the length, width, and height is 185 inches. We will need to use the **formula** for the volume of a **rectangular solid** to compute the volume.

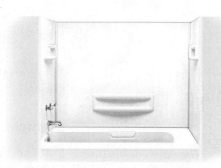

**Step 2.**   If we let $w$ represent the width of the enclosure, then the height is $h = 3w - 20$ and the length is $l = 2w - 5$.

**Step 3.**   Start with the **sum** of the three dimensions:

$$l + w + h = 185 \quad \leftarrow \text{Given sum}$$

Substituting $3w - 20$ for $h$ and $2w - 5$ for $l$, we have the equation:

$$(2w - 5) + w + (3w - 20) = 185$$

**Step 4.**   Finish solving the equation to find the width of the enclosure. Use the width to find the height and length and then the volume of the enclosed space. When finished, view the **answer**, or watch this **video** for a fully worked solution.

**You Try It**   Work through this You Try It problem.

Work Exercises 6 and 7 in this textbook or in the *MyMathLab*® Study Plan.

OBJECTIVE 2   SOLVE PROBLEMS INVOLVING ANGLES

Geometry problems sometimes involve *complementary* and *supplementary* angles. The **measures** of two **complementary angles** add to 90°. The measures of two **supplementary angles** add to 180°. Thus, if an angle measures $x$ degrees, its **complement** is given by $90 - x$ and its **supplement** is given by $180 - x$. Complementary angles form **right angles**, while supplementary angles form a straight line.

### Example 5  Solve Problems Involving Complementary Angles

Find the measure of each **complementary angle** in the
following figure.

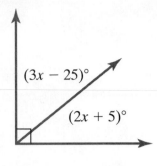

**Solution**  We use the **problem-solving steps.**

**Step 1.**  We must find the measure of each angle. We know
that one angle measure is $3x - 25$ degrees, the
other is $2x + 5$ degrees, and the sum of the
measures of complementary angles is $90°$.

**Step 2.**  The variable $x$ is the unknown quantity. Each angle
measure is given in terms of $x$.

**Step 3.**  Because the sum of the measures of complementary angles is $90°$, we write

$$\overset{\text{1st angle}}{\overbrace{(3x - 25)}} + \overset{\text{2nd angle}}{\overbrace{(2x + 5)}} = 90.$$

**Step 4.**  Solve the equation for $x$ and then use the result to find the **measure** of each
angle.

| | |
|---:|:---|
| Begin with the original equation: | $(3x - 25) + (2x + 5) = 90$ |
| Remove grouping symbols: | $3x - 25 + 2x + 5 = 90$ |
| Combine like terms: | $5x - 20 = 90$ |
| Add 20 to both sides: | $5x - 20 + 20 = 90 + 20$ |
| Simplify: | $5x = 110$ |
| Divide both sides by 5: | $\dfrac{5x}{5} = \dfrac{110}{5}$ |
| Simplify. | $x = 22$ |

**Step 5.**  Substituting 22 for $x$, we get $3(22) - 25 = 41°$ for the measure of one angle
and $2(22) + 5 = 49°$ for the measure of the second. The sum of the two angle
measures is $90°$, as required. ($41° + 49° = 90°$)

**Step 6.**  The angle measures are $41°$ and $49°$.

**You Try It**  Work through this You Try It problem.

**Work Exercises 8 and 9 in this textbook or in the *MyMathLab*®  Study Plan.**

### Example 6  Solve Problems Involving Supplementary Angles

*My video summary*  Find the measure of each **supplementary angle** in the following figure.

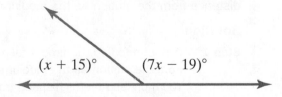

**Solution**

**Step 1.**  We want to find the measure of each angle. We know that one angle measure is
$x + 15$ degrees, the other is $7x - 19$ degrees, and the sum of the measures of
supplementary angles is $180°$.

**Step 2.**  In the diagram, the variable $x$ is the unknown quantity. Each angle measure is
given in terms of $x$.

**Step 3.** Because the sum of the measures of supplementary angles is 180°, we can write

$$\overbrace{(x + 15)}^{\text{1st angle}} + \overbrace{(7x - 19)}^{\text{2nd angle}} = 180.$$

**Step 4.** Finish solving the equation to find the angle measures. When finished, view the **answer**, or watch this **video** for a fully worked solution.

**You Try It** **Work through this You Try It problem.**

**Work Exercises 10 and 11 in this textbook or in the *MyMathLab*® Study Plan.**

A well-known property of triangles is that the **sum** of the **measures** of the three **angles** equals 180°. We can use this fact as a **formula** to solve problems involving triangles.

### Example 7  Triangle Park

 Triangle Park in Lexington, KY, has a roughly triangular shape such that the smallest angle measures 10 degrees less than the middle-sized angle. The largest angle measures 30 degrees less than twice the middle-sized angle. Find the measures of all three angles.

**Solution** Work through the **problem-solving steps** to solve this problem on your own. View the **answer**, or watch this **video** for a fully worked solution.

**You Try It** **Work through this You Try It problem.**

**Work Exercises 12–16 in this textbook or in the *MyMathLab*® Study Plan.**

OBJECTIVE 3   SOLVE PROBLEMS INVOLVING UNIFORM MOTION

Some problems involve two **rates**. This means that there will also be two distances and two times. These are called **uniform motion** problems. In these types of problems, the **equations** formed will typically set two quantities equal to each other (e.g. distance traveled) or add two quantities together (e.g. time of travel).

### Example 8  High Speed Trains

In January 2010, the U.S. government announced plans for the development of high-speed rail projects. A medium-fast passenger train leaves a station traveling 100 mph. Two hours later, a high-speed passenger train leaves the same station traveling 180 mph on a different track. How long will it take the high-speed train to be the same distance from the station as the medium-fast passenger train?

### Solution

**Step 1.** We need to find the time it takes the high-speed train to be the same distance from the station as the medium-fast train. The medium-fast train travels at 100 mph, and the high-speed train travels at 180 mph. The medium-fast train travels 2 hours more than the high-speed train. We also know that the distance traveled by each train will be the same.

**Step 2.** If we let $t$ = the travel time for the high-speed train, then the travel time for the medium-fast train will be $t + 2$ (two hours more).

**Step 3.** Because the distance traveled is the same, we can write

$$\text{distance}_{\text{med-fast}} = \text{distance}_{\text{high-speed}}$$

Using the **distance formula**, $d = rt$, we write

$$(\text{rate})(\text{time})_{\text{med-fast}} = (\text{rate})(\text{time})_{\text{high-speed}}$$

**Substituting** the given rates and our **expressions** for the times from Step 2, we get

$$(100)(t + 2) = (180)(t)$$

Sometimes it is helpful to first summarize the given information in a table. View a **table** for this example.

**Step 4.** Solve the equation for $t$, the travel time for the high-speed train.

| | |
|---:|:---|
| Begin with the original equation: | $(100)(t + 2) = (180)(t)$ |
| Distribute: | $100t + 200 = 180t$ |
| Subtract $100t$ from both sides: | $100t + 200 - 100t = 180t - 100t$ |
| Simplify: | $200 = 80t$ |
| Divide both sides by 80: | $\dfrac{200}{80} = \dfrac{80t}{80}$ |
| Simplify: | $2.5 = t$ |

**Step 5.** Substituting 2.5 for $t$, we get $(2.5) + 2 = 4.5$ hours for the travel time of the medium-fast train.

$$\left.\begin{array}{l} 100(4.5) = 450 \text{ miles} \\ 180(2.5) = 450 \text{ miles} \end{array}\right\} \text{Same distance}$$

The distance traveled by each train is the same, as described in the original problem.

**Step 6.** The high-speed train will be the same distance away from the station after 2.5 hours.

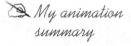 *My animation summary*

 Watch this **animation** for an example where we add distances.

## Example 9  Travel Mileage

 *My video summary*

▤ Brennan provides in-home healthcare in a rural county and gets reimbursed for mileage. On one particular day he spent 4 hours driving to visit patients. His average speed is 50 mph on the highway but then slows to 30 mph when driving through towns. If he traveled five times as far on the highway as through towns, how far did he travel that day?

### Solution

**Step 1.** We must find the total distance Brennan traveled on the day in question. We know that the total time is four hours, his rate on the highway is 50 mph, and his rate through towns is 30 mph. We also know that the distance he travels on the highway is five times the distance traveled through towns.

**Step 2.** If we let $d$ = distance traveled in towns, then the distance traveled on the highway is $5d$ (five times the distance in towns).

**Step 3.** The total travel time is the sum of the time spent driving on the highway and the time spent driving through towns.

$$\text{time}_{\text{highway}} + \text{time}_{\text{town}} = \text{time}_{\text{total}}$$

Solving the **distance formula** for $t$ gives us $t = \dfrac{d}{r}$, or time $= \dfrac{\text{distance}}{\text{rate}}$. So we rewrite our equation as

$$\left(\frac{\text{distance}}{\text{rate}}\right)_{\text{highway}} + \left(\frac{\text{distance}}{\text{rate}}\right)_{\text{town}} = \text{time}_{\text{total}}$$

Substituting the given rates, total time, and our expressions for the distances, we get

$$\frac{5d}{50} + \frac{d}{30} = 4.$$

View a **table summary**.

**Step 4.** Solve the equation for $d$ (the distance traveled in towns) on your own and then answer the question. Remember that his total distance is the sum of his distance on the highway and his distance in towns. When finished, view the **answer**, or watch this **video** for a fully worked solution.

**You Try It**   Work through this You Try It problem.

**Work Exercises 17–20 in this textbook or in the *MyMathLab*® Study Plan.**

When working with the distance formula, the units for rate and time must be consistent. For example, if the unit for rate is miles per hour, then the unit for time must be hours.

# 2.5 Exercises

**You Try It**

**1. Banquet Table**   The length of a banquet table is 2 feet longer than 5 times its width. If the perimeter is 52 feet, what are the dimensions of the table?

**2. Solar Panel**   A rectangular solar panel has a length that is 12 inches shorter than 3 times its width. If the perimeter of the panel is 136 inches, what are the dimensions of the panel?

**3. Lacrosse Field**   The length of a lacrosse field is 10 yards less than twice its width, and the perimeter is 340 yards. The defensive area of the field is $\frac{7}{22}$ of the total field area. Find the defensive area of the lacrosse field.

**4. Patio Garden**   A man has a rectangular garden. One length of the garden lies along a patio wall. However, the rest of the garden is enclosed by 36 feet of fencing. If the length of the garden is twice its width, what is the area of the garden?

**5. Triangular Deck**   A triangular-shaped deck has one side that is 5 feet longer than the shortest side and a third side that is 5 feet shorter than twice the length of the shorter side. If the perimeter of the deck is 60 feet, what are the lengths of the three sides?

**You Try It**

**6. Brick Size**   The width of a brick is half the length, which is 1 inch less than four times the height. If the sum of the three dimensions is 14.25 inches, find the volume of the brick.

**7. Mailing a Package**   The sum of the length, width, and height of a rectangular package is 60 inches. If the length is 2 inches longer than three times the height and the width is 2 inches shorter than twice the height, what is the volume of the package?

**You Try It**

**8. Complementary Angles**   Find the measures of the angles in the following diagram.

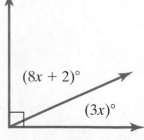

9. **Complementary Angles**   Two complementary angles are $(2x - 2)°$ and $(6x + 4)°$. Find their measures.

10. **Supplementary Angles**   Find the measures of the angles in the following diagram.

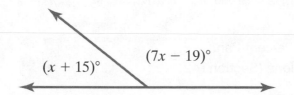

$(x + 15)°$

$(7x - 19)°$

11. **Supplementary Angles**   Angle $A$ has the property that five times its complement is equal to twice its supplement. Find the measure of angle $A$.

12. **Triangle Angles**   The largest angle of a triangle is 5 times the size of the smallest, and the middle angle is 3 times the size of the smallest. What are the three angle measures?

13. **Isosceles Triangle**   An isosceles triangle contains two equal angles. Each of these angles is five degrees larger than twice the smallest angle. What are the measures of the three angles?

14. **Triangular Sunshade**   The smallest angle of a triangular sunshade measures 10 degrees less than the middle-sized angle. The largest angle measures 10 degrees more than twice the middle-sized angle. What are the angle measures?

15. **Support Cable**   A platform on the back of an RV is anchored by two support cables. Each cable is attached to the front of the platform and the back wall of the RV. See the diagram below. If angle $B$ measures 30° less than twice the measure of angle $A$, what is the measure of angle $B$ (the angle between the cable and the wall)?

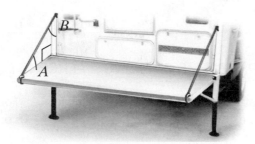

16. **Triangular Flag**   The three angles of a triangular flag are $x$, twice the complement of $x$, and one-fourth the supplement of $x$. Find the angle measures.

17. **Jogging Time**   David leaves his house, jogging down the street at 6 mph. Once David has gone $\frac{1}{4}$ mile, Jacob comes out and follows him at a rate of 6.5 mph. How long will it take Jacob to catch David?

18. **Hiking Speed**   Landon can climb a certain hill at a rate that is 2.5 mph slower than his rate coming down the hill. If it takes him 2 hours to climb the hill and 45 minutes to come down the hill, what is his rate coming down?

19. **Road Trip**   Perry goes for a 3-hour trip through towns and on country roads. If he averages 55 mph on country roads and 35 mph through towns, and if he travels four times as far on country roads as he does through towns, what is the total length of his trip?

20. **Biathlon**   Francesca plans to compete in a biathlon that involves running and cycling. During one training session, she covered a total distance of 70 miles in three hours. If she ran at a rate of 6 mph and cycled at a rate of 30 mph, how long did she spend cycling?

# 2.6 Percent and Mixture Problem Solving

## THINGS TO KNOW

Before working through this section, be sure you are familiar with the following concepts:

VIDEO ANIMATION INTERACTIVE

**You Try It**
1. Convert Between Percents and Decimals or Fractions (Section R.2, **Objective 4**)

**You Try It**
2. Solve Linear Equations Containing Decimals; Apply a General Strategy (Section 2.2, **Objective 3**)

**You Try It**
3. Use the Problem-Solving Strategy to Solve Direct Translation Problems (Section 2.3, **Objective 2**)

**You Try It**
4. Solve Problems Involving Related Quantities (Section 2.3, **Objective 3**)

## OBJECTIVES

**1** Solve Problems by Using a Percent Equation

**2** Solve Percent Problems Involving Discounts, Markups, and Sales Tax

**3** Solve Percent of Change Problems

**4** Solve Mixture Problems

........................................................................................

### OBJECTIVE 1 SOLVE PROBLEMS BY USING A PERCENT EQUATION

Recall that the word "of" indicates multiplication. For example, the sentence "thirty percent of sixty is eighteen" translates into $30\% \cdot 60 = 18$ or $0.3 \cdot 60 = 18$.

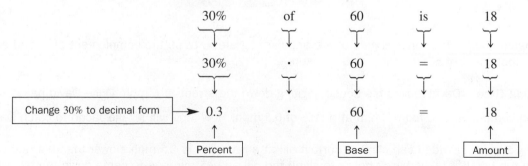

In the sentence, 30% is the *percent*, 60 is the *base*, and 18 is the *amount*. Using these labels, we can write the **general equation for percents**.

---
**General Equation for Percents**

$$\text{Percent} \cdot \text{Base} = \text{Amount}$$

---

When we translate **percent** problems into **equations**, the word *percent* or the % symbol identifies the percent. The **base** represents one whole and typically follows the word *of*. The amount is the part compared to the whole and is normally isolated from the percent and the base on one side of the word *is* (or a different **key word meaning equals**).

 When using the **general equation for percents**, the percent must be **written as a decimal** or **fraction** in order to solve the problem. Typically, we will use decimal form.

## Example 1 Using the Percent Equation

 *My video summary*    Use equations to solve each percent problem.

**a.** 32 is 40% of what number?

**b.** 145% of 78 is what number?

**c.** 8.2 is what percent of 12.5?

**Solutions** Using the general equation for percents, we translate each sentence into an equation and solve.

**a.** Translate into an equation: 
$$\underset{\downarrow}{32} \quad \underset{\downarrow}{\text{is}} \quad \underset{\downarrow}{40\%} \quad \underset{\downarrow}{\text{of}} \quad \underset{\downarrow}{\text{what number}}$$
$$32 = 40\% \cdot x$$

Convert 40% to a decimal:    $32 = 0.4 \cdot x$

Divide both sides by 0.4:    $\dfrac{32}{0.4} = \dfrac{0.4 \cdot x}{0.4}$

Simplify:    $80 = x$

So, 32 is 40% of 80.

**b.** Translate into an equation: 
$$\underset{\downarrow}{145\%} \quad \underset{\downarrow}{\text{of}} \quad \underset{\downarrow}{78} \quad \underset{\downarrow}{\text{is}} \quad \underset{\downarrow}{\text{what number}}$$
$$145\% \cdot 78 = x$$

Try to solve this equation on your own to finish the problem. View the **answer**, or watch this **video** for fully-worked solutions to all three parts.

**c.** Try solving this problem your own, then check your **answer**. Watch this **video** for fully worked solutions to all three parts.

**You Try It**    **Work through this You Try It problem.**

**Work Exercises 1–6 in this textbook or in the *MyMathLab*®  Study Plan.**

## Example 2 Bleach Concentration

6% of a 128 fluid-ounce bottle of bleach is sodium hypochlorite. How many fluid ounces of sodium hypochlorite are in the bottle?

**Solution** We follow the **problem-solving steps**.

**Step 1.**    We need to find the *amount* of sodium hypochlorite in the bottle of bleach. We know the *percent*, 6%, and the *base*, 128 fluid ounces.

**Step 2.**    Let $x$ be the number of fluid ounces of sodium hypochlorite.

**Step 3.**    Translate into an equation:
$$\underset{\downarrow}{6\%} \quad \underset{\downarrow}{\text{of}} \quad \underset{\downarrow}{128 \text{ fl oz}} \quad \underset{\downarrow}{\text{is}} \quad \underset{\downarrow}{\text{sodium hypochlorite}}$$
$$\underset{\uparrow}{6\%} \quad \cdot \quad \underset{\uparrow}{128} \quad = \quad \underset{\downarrow}{x}$$

| Percent | Base | Amount |
|---------|------|--------|

**Step 4.** Change 6% to decimal form: $0.06 \cdot 128 = x$

Multiply: $7.68 = x$

**Step 5.** To check, we divide the amount, 7.68 by the base, 128, to see if we get the correct percent, 6%: $7.68 \div 128 = 0.06 = 6\%$. The result checks.

**Step 6.** The bottle contains 7.68 fluid ounces of sodium hypochlorite.

**You Try It**   Work through this You Try It problem.

Work Exercises 7–10 in this textbook or in the *MyMathLab*®  Study Plan.

OBJECTIVE 2   SOLVE PERCENT PROBLEMS INVOLVING DISCOUNTS, MARKUPS, AND SALES TAX

During a sale, a store may take a certain percent off the price of merchandise. We can calculate the **amount** of this **discount** by using a **percent equation**. We then *subtract* the discount from the original price to find the new price.

---

**Computing Discounts**

Discount = Percent · Original price

New price = Original price − Discount

---

### Example 3   Going Out of Business Sale

A furniture store is going out of business and cuts all prices by 55%. What is the sale price of a sofa with an original price of $1199?

**Solution**  Compute the discount and subtract it from the original price.

Discount = Percent · Original price = $55\% \cdot \$1199 = 0.55 \cdot \$1199 = \$659.45$

Sale price = Original price − Discount = $\$1199 - \$659.45 = \$539.55$

The sale price of the sofa is $539.55.

**You Try It**   Work through this You Try It problem.

Work Exercises 11–13 in this textbook or in the *MyMathLab*®  Study Plan.

**Markups** work like discounts, except that they are *added* to the original price.

---

**Computing Markups**

Markup = Percent · Original price

New price = Original price + Markup

---

### Example 4   College Book Store

  A college book store sells all textbooks at a 30% **markup** over its cost. If the price marked on a biology textbook is $124.28, what was the cost of the book to the store? Round to the nearest cent.

**Solution**  Follow the **problem-solving strategy**.

**Step 1.**   We need to find the original price of the book to the bookstore. We know the markup is 30% and the selling price is $124.28.

**Step 2.** Let $x$ be the original price of the book to the bookstore.

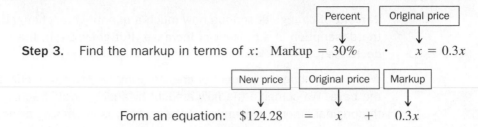

**Step 3.** Find the markup in terms of $x$: Markup = 30% · $x = 0.3x$

Form an equation: $124.28 = x + 0.3x$

Try to finish this problem by completing the **problem-solving steps**. View the **answer**, or watch this **video** for a complete solution.

**You Try It** **Work through this You Try It problem.**

**Work Exercises 14 and 15 in this textbook or in the *MyMathLab*® Study Plan.**

**Sales tax** is handled like markups. Multiply the tax rate (a percent) by the purchase price to determine the sales tax amount. Then add the sales tax to the purchase price to get the overall price.

---

**Computing Sales Tax**

$$\text{Sales tax} = \text{Tax rate} \cdot \text{Purchase price}$$
$$\text{Overall price} = \text{Purchase price} + \text{Sales tax}$$

---

## Example 5  Buying Jeans

Charlotte bought a pair of jeans priced at $51.99. When sales tax was added, she paid an overall price of $55.37. What was the tax rate? Round to the nearest tenth of a percent.

**Solution**  Follow the **problem-solving steps**.

**Step 1.** We need to find the tax rate, a percent. We know that the purchase price is $51.99 and the overall (total) price is $55.37.

**Step 2.** Let $x$ be the tax rate.

**Step 3.** Find the sales tax in terms of $x$: Sales tax = $x$ · $51.99 = 51.99x$

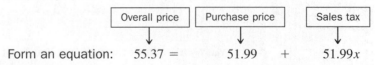

Form an equation: $55.37 = 51.99 + 51.99x$

**Step 4.** Subtract 51.99 from both sides: $3.38 = 51.99x$

Divide both sides by 51.99: $0.065 \approx x$

Change to a percent: $6.5\% \approx x$

**Step 5.** View the **check**.

**Step 6.** The tax rate is $6.5\%$.

**You Try It** **Work through this You Try It problem.**

**Work Exercises 16–18 in this textbook or in the *MyMathLab*® Study Plan.**

## OBJECTIVE 3    SOLVE PERCENT OF CHANGE PROBLEMS

**Percent of change** describes how much a quantity has changed. If the quantity goes up, the description is a **percent of increase**. If it goes down, the description is a **percent of decrease**.

When computing the **amount** of change from the **percent** of change, the original amount is the **base**. We compute the new amount by adding or subtracting the amount of change to the original amount, depending if the quantity is increasing or decreasing.

**Percent of Increase**

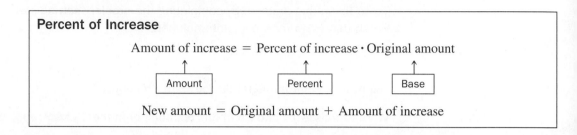

New amount = Original amount + Amount of increase

### Example 6  Enrollment Increase

Last year, 16,528 students attended City Community College. This year, enrollment increased by 3.2%. How many students attend City Community College this year? Round to the nearest whole student.

**Solution**  Follow the **problem-solving steps.**

**Step 1.**  We must find the current enrollment of City Community College. Last year's enrollment was 16,528, and the percent of increase is 3.2%.

**Step 2.**  Let $x$ be the **amount of increase** in enrollment. The current enrollment is 16,528 + $x$.

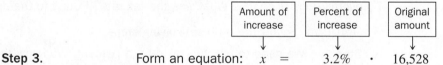

**Step 3.**              Form an equation:    $x$  =        3.2%    ·    16,528

**Step 4.**    Change 3.2% to decimal form:   $x$  =   0.032 · 16,528

Multiply:   $x$  ≈   529

So, 16,528 + $x$ = 16,528 + 529 = 17,057 students.

**Step 5.**    View the check.

**Step 6.**    This year's enrollment at City Community College is 17,057 students.

**You Try It**    Work through this You Try It problem.

**Work Exercises 19–21 in this textbook or in the *MyMathLab*®** Study Plan.

**Percent of Increase**

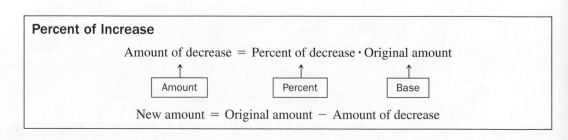

New amount = Original amount − Amount of decrease

### Example 7 General Motors Reorganization

 *My video summary*

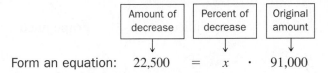 Prior to reorganization in 2010, General Motors (GM) had 91,000 U.S. employees. After the reorganization, GM had 68,500 U.S. employees. By what percent did the number of U.S. employees decrease? (*Source:* General Motors)

#### Solution

**Step 1.** We must find the **percent of decrease**. The original number of U.S. employees was 91,000 (before reorganization), and the new number was 68,500 (afterwards).

**Step 2.** Let $x$ be the percent of decrease in the number of U.S. employees.

**Step 3.** The **amount of decrease** is $91,000 - 68,500 = 22,500$ employees.

| Amount of decrease | Percent of decrease | Original amount |
|:---:|:---:|:---:|
| ↓ | ↓ | ↓ |

Form an equation:   $22,500 \quad = \quad x \quad \cdot \quad 91,000$

Try to finish this problem by completing the **problem-solving steps**. View the **answer**, or watch this **video** for a complete solution.

**You Try It**   **Work through this You Try It problem.**

**Work Exercises 22–24 in this textbook or in the *MyMathLab*® Study Plan.**

 When calculating the **percent of change**, be sure to use the original amount as the **base**, not the new amount.

OBJECTIVE 4   SOLVE MIXTURE PROBLEMS

Suppose that a 50-pound bag of lawn fertilizer has a 12% nitrogen **concentration**. This means that out of all the components in the fertilizer, 12% is pure nitrogen. The 50-pound bag contains a total of $0.12(50) = 6$ pounds of pure nitrogen.

We can find the amount of a particular component in a **mixture** by multiplying its concentration (in decimal form) by the amount of mixture. The *amount of component* is the **amount** in the **general equation for percents**, the *concentration* is the **percent**, and the *amount of mixture* is the **base**.

---

**Mixture Problem Equation**

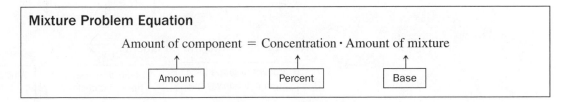

Amount of component = Concentration · Amount of mixture

| Amount | Percent | Base |
|:---:|:---:|:---:|

---

We can also write this in the following equivalent form:

$$\text{Concentration} = \frac{\text{Amount of component}}{\text{Amount of mixture}}$$

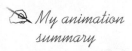 *My animation summary*

With mixture problems, remember that the total *amount* of a component does not change when two or more substances are mixed together, but the *concentration* of that component might change. Watch this **animation** for an illustration.

### Example 8 Organic Juice Mix

*My video summary*

An organic cranberry-grape juice is 40% grape juice, while an organic fruit cocktail juice is 10% grape juice. If 8 ounces of the cranberry-grape juice are mixed with 22 ounces of the fruit cocktail juice, what is the mixed juice's **concentration** of grape juice?

**Solution** Follow the **problem-solving steps.**

**Step 1.** 8 ounces of cranberry-grape juice (40% grape) are mixed with 22 ounces of fruit cocktail juice (10% grape) to form $8 + 22 = 30$ ounces of a third juice. We must find the concentration of grape juice in the mix. See Figure 2.

| **Cranberry-Grape** | | **Fruit Cocktail** | | **Mixture** |
|:---:|:---:|:---:|:---:|:---:|
| 40% | | 10% | | $x\%$ |
| grape juice | | grape juice | | grape juice |

**Figure 2**   8 oz   +   22 oz   =   30 oz

**Step 2.** Let $x$ be the concentration (percent) of grape juice in the mix.

**Step 3.** The **amount** of grape juice in the cranberry-grape juice is

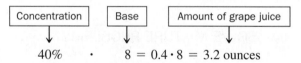

| Concentration | Base | Amount of grape juice |
|:---:|:---:|:---:|
| 40% $\cdot$ | 8 | $= 0.4 \cdot 8 = 3.2$ ounces |

The amount of grape juice in the fruit cocktail juice is

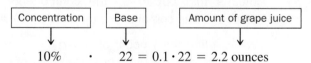

| Concentration | Base | Amount of grape juice |
|:---:|:---:|:---:|
| 10% $\cdot$ | 22 | $= 0.1 \cdot 22 = 2.2$ ounces |

The amount of grape juice in the mixture is

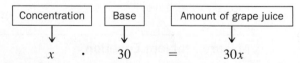

| Concentration | Base | Amount of grape juice |
|:---:|:---:|:---:|
| $x$ $\cdot$ | 30 | $= 30x$ |

Now we can write the equation:

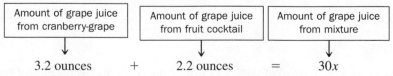

| Amount of grape juice from cranberry-grape | | Amount of grape juice from fruit cocktail | | Amount of grape juice from mixture |
|:---:|:---:|:---:|:---:|:---:|
| 3.2 ounces | + | 2.2 ounces | = | $30x$ |

Try to finish this problem on your own. View the **answer,** or watch this **video** to see a detailed solution.

**You Try It**   **Work through this You Try It problem.**

**Work Exercises 25 and 26 in this textbook or in the *MyMathLab*®️ Study Plan.**

### Example 9  Alcohol Concentration

 *My video summary*       How many milliliters of a 25% alcohol solution must be mixed with 10 mL of a 60% alcohol solution to result in a mixture that is 30% alcohol?

### Solution

**Step 1.**  We mix the two solutions together to result in a third solution. We know that an unknown amount of a 25% alcohol solution will be mixed with 10 mL of a 60% alcohol solution to result in a 30% alcohol solution.

**Step 2.**  Let $x$ be the unknown amount of the 25% alcohol solution (in mL). When we mix this amount with 10 mL of a 60% alcohol solution, the resulting amount of a 30% alcohol solution is $x + 10$ (in mL). See Figure 3.

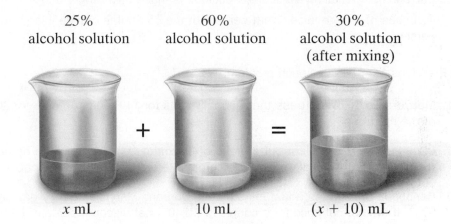

25% alcohol solution        60% alcohol solution        30% alcohol solution (after mixing)

**Figure 3**        $x$ mL        10 mL        $(x + 10)$ mL

**Step 3.**

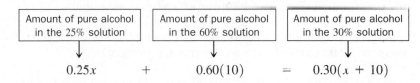

| Amount of pure alcohol in the 25% solution | | Amount of pure alcohol in the 60% solution | | Amount of pure alcohol in the 30% solution |
|---|---|---|---|---|
| $0.25x$ | $+$ | $0.60(10)$ | $=$ | $0.30(x + 10)$ |

Try to finish this problem on your own. View the **answer**, or watch this **video** to see a detailed solution.

**You Try It**    Work through this You Try It problem.

Work Exercises 27–30 in this textbook or in the *MyMathLab*® Study Plan.

 The **concentration** of a mixture must always be between the concentrations of the two mixed solutions. Do you see **why**?

## 2.6  Exercises

In Exercises 1–6, use an equation to solve each percent problem.

**You Try It**

**1.** 21 is 75% of what number?          **2.** What number is 38% of 155?

**3.** 52.7 is what percent of 62?          **4.** What number is 214% of 86.5?

**5.** 2.5% of what number is 91?          **6.** What percent of 55 is 88?

In Exercises 7–10, use a percent equation to solve each percent problem.

**You Try It**

**7. Sculpture**   A 240-pound sculpture is made by using a cast that is 90% copper and 10% tin. How many pounds of copper are in the sculpture?

**8. Ethanol Production**   Twenty-six percent of all U.S. grain crops grown in 2008 were used to produce ethanol, which amounts to 104 million tons of grain. (*Source:* Earth Policy Institute) How much grain was grown in the U.S. in 2008?

**9. Mobile Devices**   In March 2010, a total of 234 million people age 13 or older in the U.S. used mobile devices. Of these, 51.2 million people used devices manufactured by Motorola. (*Source:* comScore.com) What percent of the population used Motorola devices? Round to the nearest tenth of a percent.

**10. Bartending**   A Harvey Wallbanger cocktail is made by mixing 1 ounce of vodka, $\frac{1}{2}$ ounce of Galliano, and 4 ounces of orange juice. What percent of the cocktail is orange juice? Round to the nearest tenth of a percent.

In Exercises 11–18, solve each problem.

**You Try It**

**11. Mattress Sale**   A mattress that normally sells for $499.90 is on sale for 30% off. What is the sale price of the mattress?

**12. Clearance Sale**   To clear out a brand of discontinued power tools, a home supply store marks down the tools by 25%. If the sale price for a cordless power drill is $82.35, what was the original price?

**13. Caribbean Cruise**   As a last minute deal, Don and Mary booked a 7-day Caribbean cruise for a total of $737. If the normal price for a couple is $1340, what discount percent did Don and Mary receive?

**You Try It**

**14. Seafood Restaurant**   An oceanside restaurant prices its fresh fish using a 90% markup over its cost. If it sells fresh red snapper for $19.95, what was the restaurant's cost for this fish?

**15. Fundraiser**   An athletic booster club purchased bottles of soda for $0.80 each and sold them during games for $2.00 each. What was the markup percent?

**16. Buying a Computer**   The purchase price of a laptop computer is $529.50. If the sales tax rate is 7.5%, find the overall price of the laptop. Round to the nearest cent.

**You Try It**

**17. Buying Jewelry**   Trevor bought a diamond pendant with a purchase price of $324. When sales tax was added, he paid an overall price of $349.92. What was the tax rate?

**18. Buying a Game System**   Destiny bought a game system for an overall price of $239.03. If the tax rate is 6%, what was the purchase price of the game system?

In Exercises 19–24, solve each percent of change problem.

**You Try It**

**19. Tuition Increase**   A state university's board of directors approved a 7.5% tuition increase for next year. If the current annual tuition is $9485, what will the tuition be next year? Round to the nearest whole dollar.

**20. Population Increase**   According to the U.S. Census Bureau, the population of the Palm Coast metro area in Florida increased by 84% to 92,000 people between 2000 and 2010. What was the population of the Palm Coast metro area in 2000?

**21. Life Expectancy**   The life expectancy for a newborn child in the U.S. increased from 73.7 years in 1980 to 78.3 years in 2010. What is the percent of increase for this time period? Round to the nearest tenth of a percent. (*Source:* U.S. Census Bureau)

**22. Daycare Enrollment**   Following an investigation into finance fraud, a daycare center's enrollment decreased by 30% to 56 students. What was the daycare center's enrollment before the investigation?

**23. GM Plant Closings**   Before the 2010 reorganization, GM had 47 U.S. plants. After the reorganization, GM had 34 U.S. plants. By what percent did the number of U.S. plants decrease? Round to the nearest tenth of a percent. (*Source: General Motors*)

**24. The Biggest Loser**   During Season 8 of NBC's reality show *The Biggest Loser*, the winner Danny Cahill lost 55.6% of his body weight. If he weighed 430 pounds at the start of the show, what was his weight at the end of the show?

In Exercises 25–30, solve each mixture problem.

**25. Mixing Alcohol**   Suppose 8 pints of a 12% alcohol solution is mixed with 2 pints of a 60% alcohol solution. What is the concentration of alcohol in the new 10-pint mixture?

**26. Mixing Fertilizer**   The granular fertilizer 12-12-12 is composed of 12% nitrogen, 12% phosphate, and 12% potassium. Similarly, the fertilizer 16-20-0 is composed of 16% nitrogen, 20% phosphate, and 0% potassium. If a gardener mixes a 10 pound bag of 12-12-12 with a 15 pound bag of 16-20-0, what are the concentrations of nitrogen, phosphate, and potassium in the mixture? Express the answers as percents.

**27. Mixing Cranberry Juice**   How much of a 75% cranberry juice drink must be mixed with 3 liters of a 20% cranberry juice drink to result in a mixture that is 50% cranberry juice?

**28. Salad Dressing**   A cook needs to mix a regular ranch dressing containing 46% fat with a light ranch dressing containing 26% fat to result in 5 cups of a dressing containing 32% fat. How many cups of each kind of dressing should be mixed?

**29. Diluting Acid**   How many cups of pure water should be mixed with 1.5 cups of a 30% acid solution to dilute it into a 2% acid solution?

**30. Fuel Mixture**   A two-cycle engine mechanic has 2 gallons of a fuel mixture that is 96% gasoline and 4% oil. How many gallons of gasoline should be added to form a fuel mixture that is 99% gasoline and 1% oil?

# 2.7  Linear Inequalities in One Variable

## THINGS TO KNOW

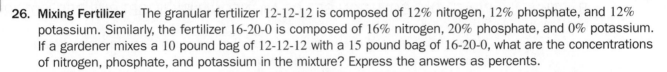

Before working through this section, be sure you are familiar with the following concepts:

|  |  | VIDEO | ANIMATION | INTERACTIVE |
|---|---|---|---|---|
| **1.** | Use the Order of Operations to Evaluate Numeric Expressions (Section 1.4, **Objective 2**) | ▣ |  | ▣ |
| **2.** | Evaluate Algebraic Expressions (Section 1.5, **Objective 1**) | ▣ |  |  |
| **3.** | Simplify Algebraic Expressions (Section 1.6, **Objective 2**) |  |  | ▣ |
| **4.** | Solve Linear Equations Containing Non-simplified Expressions (Section 2.2, **Objective 1**) | ▣ |  |  |
| **5.** | Use the Problem-Solving Strategy to Solve Direct Translation Problems (Section 2.3, **Objective 2**) | ▣ |  |  |
| **6.** | Solve Problems Using Geometry Formulas (Section 2.5, **Objective 1**) | ▣ |  |  |

## OBJECTIVES

**1** Write the Solution Set of an Inequality in Set-Builder Notation

**2** Graph the Solution Set of an Inequality on a Number Line

**3** Use Interval Notation to Express the Solution Set of an Inequality

**4** Solve Linear Inequalities in One Variable

**5** Solve Three-Part Inequalities

**6** Use Linear Inequalities to Solve Application Problems

---

OBJECTIVE 1    WRITE THE SOLUTION SET OF AN INEQUALITY IN SET-BUILDER NOTATION

*My interactive video summary*

An equal sign is used in **equations** to show that two quantities are equal. We use **inequality symbols** in **inequalities** to show that two quantities are unequal.

Watch this **interactive video** to practice distinguishing between equations and inequalities.

While most equations have a *finite* number of solutions, most **inequalities** have an *infinite* number of solutions. For example, the inequality $x > -2$ has infinitely many solutions because there are infinitely many values of $x$ that are greater than $-2$. View this **popup** for an explanation of the difference between "finite" and "infinite."

The collection of all solutions to an inequality forms the **solution set** of the inequality. When a solution set contains an infinite number of values, we cannot list each solution, so we use **set-builder notation** to express the solution set. The set of all numbers greater than $-2$ is written in set-builder notation as follows.

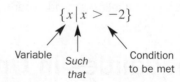

$$\{x \mid x > -2\}$$

Variable    *Such that*    Condition to be met

The variable appears to the left of the vertical bar. The vertical bar means "such that," and the expression to the right of the vertical bar is the condition that must be met for a number to be part of the solution set.

### Example 1   Writing the Solution Set of an Inequality in Set-Builder Notation

Write the **solution set** of each inequality in **set-builder notation**.

**a.** $x < 6$        **b.** $y \geq -3$        **c.** $2 < m \leq 9$

### Solutions

**a.** To be part of the solution set, the value for $x$ must be less than 6. We write the solution set as $\{x \mid x < 6\}$ and read this as "the set of all values for $x$ such that $x$ is less than 6."

**b.** We write the solution set as $\{y \mid y \geq -3\}$ and read this as "the set of all values for $y$ such that $y$ is greater than or equal to $-3$."

**c.** To be part of the solution set, the value for $m$ must be greater than 2 but less than or equal to 9. We write the solution set as $\{m \mid 2 < m \leq 9\}$ and read this as "the set of all values for $m$ such that 2 is less than $m$ and $m$ is less than or equal to 9."

**You Try It**   Work through this **You Try It** problem.

**Work Exercises 1–4 in this textbook or in the *MyMathLab*®** Study Plan.

OBJECTIVE 2   GRAPH THE SOLUTION SET OF AN INEQUALITY ON A NUMBER LINE

To graph the **solution set** on a number line, we use an open circle (O) to indicate that a value *is not* included in the solution set and a closed circle (●) to indicate that a value *is* included in the solution set. For example, the graph of $\{x|x > -2\}$ is shown below.

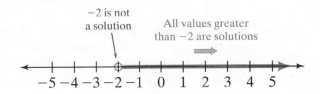

### Example 2  Graphing the Solution Set of an Inequality on a Number Line

Graph each solution set on a number line.

a. $\{x|x \leq 0\}$    b. $\{x|-2 \leq x < 4\}$    c. $\{x|x > -1\}$

d. $\{x|3 < x < 7\}$    e. $\{x|-1 \leq x \leq 5\}$    f. $\{x|x$ is any real number$\}$

### Solutions

a. We translate this solution set as "the set of all values for $x$ such that $x$ is less than or equal to 0." Because the inequality is **non-strict**, we place a closed circle at 0 to show that 0 is a solution. Then we shade the number line to the left to show that all values less than 0 are also solutions.

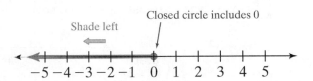

b. The solution set reads as "the set of all values for $x$ such that $-2$ is less than or equal to $x$ and $x$ is less than 4." The inequality on the left is non-strict, so we place a closed circle at $-2$ to show that $-2$ is a solution. The inequality on the right is **strict**, so we place an open circle at 4 to show that 4 is not a solution and then shade the number line between the two circles to indicate that all values between $-2$ and 4 are also solutions.

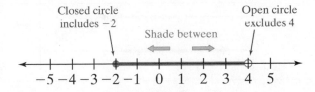

c.–f. View this **popup** for the remaining solution sets and their graphs.

**You Try It**   Work through this You Try It problem.

Work Exercises 5–9 in this textbook or in the *MyMathLab* Study Plan.

OBJECTIVE 3   USE INTERVAL NOTATION TO EXPRESS THE SOLUTION SET
OF AN INEQUALITY

We can also use **interval notation** to express the **solution set** of an inequality. First, each solution set has a **lower bound** and an **upper bound**, separated by a comma, which make up the **endpoints** of the interval. An interval has the form:

lower bound, upper bound

Then we indicate if the endpoints are included in the interval. A parenthesis— '(' for the lower bound or ')' for the upper bound — shows that the endpoint is not included in the solution set. This is like using an open circle when graphing on a number line. A square bracket— '[' for the lower bound or ']' for the upper bound — shows that the endpoint is included in the solution set. This is like using a closed circle when graphing on a number line.

In **Example 2b**, the solution set $\{x|-2 \leq x < 4\}$ is written as $[-2, 4)$ in interval notation. We use a square bracket on $-2$ because it *is* included in the solution set, and we use a parenthesis on 4 because it *is not* included in the solution set.

 A parenthesis is always used for $-\infty$ and $\infty$ because these are infinity symbols, not numbers. They indicate that the interval is **unbounded** in a particular direction.

 Watch this **interactive video** to help you determine the lower and upper bounds of an interval or solution set.

Table 1 summarizes the three ways of expressing intervals used in this text: **number line graph**, **interval notation**, and **set-builder notation**. Typically we use interval notation and graphs when expressing **solution sets** for inequalities.

| Graph | Interval Notation | Set-Builder Notation |
|---|---|---|
| ○——○ <br> a    b | $(a, b)$ | $\{x|a < x < b\}$ |
| ●——● <br> a    b | $[a, b,]$ | $\{x|a \leq x \leq b\}$ |
| ○——● <br> a    b <br> ●——○ <br> a    b | $(a, b]$ <br><br> $[a, b)$ | $\{x|a < x \leq b\}$ <br><br> $\{x|a \leq x < b\}$ |
| ○—————→ <br> a <br> ←—————○ <br>         b | $(a, \infty)$ <br><br> $(-\infty, b)$ | $\{x|x > a\}$ <br><br> $\{x|x < b\}$ |
| ●—————→ <br> a <br> ←—————● <br>         b | $[a, \infty)$ <br><br> $(-\infty, b]$ | $\{x|x \geq a\}$ <br><br> $\{x|x \leq b\}$ |

**Table 1**

_(eText Screens 2.7-1–2.7-29)_

### Example 3 Using Interval Notation to Express the Solution Set of an Inequality

Write each **solution set** using **interval notation**.

a. $\{x\,|\,x \geq -2\}$    b. $\{x\,|\,0 < x \leq 6\}$    c. $x$ is less than 4

d. $x$ is between $-1$ and 5, **inclusive**

e. $\{x\,|\,x \text{ is any real number}\}$    f. $\{x\,|\,8 > x \geq -3\}$

### Solutions

a.  This solution set has a **lower bound** of $-2$ and no **upper bound**. The interval notation is $[-2, \infty)$. We use a square bracket on $-2$ to show that $-2$ is included in the solution set. We use a parenthesis on $\infty$ to show that there is no upper bound.

b.  There is a lower bound of 0 and an upper bound of 6. The interval notation is $(0, 6]$. We use a parenthesis on 0, the lower bound, to show that 0 is not included in the solution set, and a square bracket on 6, the upper bound, to show that 6 is included in the solution set.

_My video summary_  **c.–f.** Try working these parts on your own. View the **answers**, or watch this **video** for complete solutions.

**You Try It**   Work through this You Try It problem.

**Work Exercises 10–14 in this textbook or in the _MyMathLab_® Study Plan.**

CAUTION  When writing a solution set in **set-builder notation** or **interval notation**, the values should increase from left to right.

OBJECTIVE 4   SOLVE LINEAR INEQUALITIES IN ONE VARIABLE

The inequality $3x + 2 \leq -6$ is a **linear inequality in one variable**.

---

**Definition**

A **linear inequality in one variable** is an inequality that can be written in the form $ax + b < c$, where $a, b,$ and $c$ are real numbers and $a \neq 0$.

**Note:** The inequality symbol $<$ can be replaced with $>, \leq, \geq,$ or $\neq$.

---

Why do we require $a \neq 0$ in our definition? View this **popup** for an explanation.

Solving linear inequalities is similar to solving linear equations. Our goal is to **isolate** the variable on either side of the inequality symbol, then form the **solution set** based on the resulting inequality.

Similar to solving equations, we have an **addition property of inequality**.

---

**Addition Property of Inequality**

Let $a, b,$ and $c$ be real numbers.

$$\text{If } a < b, \text{ then } a + c < b + c \text{ and } a - c < b - c.$$

Adding or subtracting the same quantity from both sides of an inequality results in an **equivalent inequality**.

**Note:** The inequality symbol $<$ can be replaced with $>, \leq, \geq,$ or $\neq$.

---

## Example 4 Solving a Linear Inequality in One Variable Using the Addition Property of Inequality

Solve each inequality using the **addition property of inequality**. Write the **solution set** in **interval notation** and graph it on a number line.

**a.** $x + 5 > 4$ **b.** $y - 3 \leq 1$

## Solutions

**a.** Our approach is similar to the one used when solving linear equations. First, we **isolate** the variable on one side of the inequality symbol.

Begin with the original inequality: $x + 5 > 4$

Subtract 5 from both sides: $x + 5 - 5 > 4 - 5$

Simplify: $x > -1$

The graph of the solution set appears on the following number line:

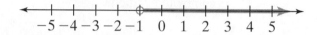

In interval notation, the solution set is $(-1, \infty)$.

**Check** With an infinite number of values in the solution set, we cannot check every solution. So, we pick one **test value** from the solution set as a "check." Let's choose $x = 0$ as our test value and **substitute** it for $x$ in the original inequality to see if a true statement results.

Begin with the original inequality: $x + 5 > 4$

Substitute 0 for $x$: $0 + 5 \overset{?}{>} 4$

Simplify: $5 > 4$ True

Because the resulting statement, $5 > 4$, is true, $x = 0$ is a solution to the inequality.

 Checking one solution, or even several solutions, does not guarantee that the solution set of the inequality is correct. It only guarantees that the tested values are solutions. However, it is still good practice to check one or two test values.

 **b.** Apply the **addition property of inequality**. View the **answer**, or watch this **video** for a complete solution.

**You Try It** Work through this You Try It problem.

Work Exercises 15–18 in this textbook or in the *MyMathLab*® Study Plan.

**Equations** and **inequalities** have similar addition properties. However, there is an important difference between the two when we multiply or divide. Consider the following examples.

True statement: $5 > 3$ True statement: $5 > 3$

Multiply both sides by 4: $4(5) > 4(3)$ Multiply both sides by $-4$: $-4(5) > -4(3)$

Simplify: $20 > 12$ True Simplify: $-20 > -12$ False

When we multiply (or divide) both sides of a true inequality statement by a positive number, the resulting inequality is also a true statement. However, when we multiply (or divide) both sides of the same true inequality statement by a **negative** number, the resulting inequality is a false statement.

In order to make the second inequality true, we must switch the direction of the inequality. View this **illustration**.

$$-20 > -12 \quad \text{False} \qquad\qquad -20 < -12 \quad \text{True}$$

Switch the inequality directions

This leads us to the **multiplication property of inequality**.

---

**Multiplication Property of Inequality**

Let $a$, $b$, and $c$ be real numbers.

$$\text{If } a < b \text{ and } c > 0, \text{ then } ac < bc \text{ and } \frac{a}{c} < \frac{b}{c}.$$

Multiplying or dividing both sides of an inequality by a positive number $c$ results in an equivalent inequality.

$$\text{If } a < b \text{ and } c < 0, \text{ then } ac > bc \text{ and } \frac{a}{c} > \frac{b}{c}.$$

Multiplying or dividing both sides of an inequality by a negative number $c$, and switching the direction of the inequality, results in an equivalent inequality.

**Note:** The inequality symbol $<$ can be replaced with $<$, $\leq$, $\geq$, or $\neq$.

---

## Example 5  Solving a Linear Inequality in One Variable Using the Multiplication Property of Inequality

Solve each inequality. Graph the solution set on a number line and write the solution set in **interval notation**.

**a.** $-5x < 15$ **b.** $\dfrac{m}{3} \geq 2$

### Solutions

**a.** Apply the **multiplication property of inequality** to isolate the variable.

Begin with the original inequality: $\qquad -5x < 15$

Divide both sides by $-5$: $\qquad \dfrac{-5x}{-5} > \dfrac{15}{-5}$

Reverse the direction
of the inequality

Simplify: $\qquad x > -3$

The graph of the solution set appears on the following number line:

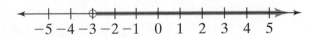

In interval notation, the solution set is $(-3, \infty)$.

 *My video summary*  **b.** Apply the **multiplication property of inequality**. View the **answer**, or watch this video for a complete solution.

**You Try It**  Work through this **You Try It** problem.

**Work Exercises 19–22 in this textbook or in the *MyMathLab*®  Study Plan.**

Our approach to solving **linear inequalities in one variable** is similar to our approach to **solving linear equations** in one variable. Let's apply the following guidelines.

---

**Guidelines for Solving Linear Inequalities in One Variable**

**Step 1.** Clear all fractions from the inequality by multiplying both sides by the **LCD**. (If LCD $< 0$, reverse the direction of the inequality) Clear all decimals by multiplying both sides by the appropriate power of 10.

**Step 2.** Remove grouping symbols using the **distributive property**.

**Step 3.** Simplify each side of the inequality by **combining like terms**.

**Step 4.** Use the **addition property of inequality** to move all variable terms to one side of the inequality and all constant terms to the other side.

**Step 5.** Use the **multiplication property of inequality** to isolate the variable.

**Step 6.** Write the solution set in **interval notation** (or **set-builder notation**) and graph it on a number line.

---

View this summary of the **inequality properties**.

## Example 6  Solving a Linear Inequality in One Variable

Solve each inequality. Write the **solution set** in **interval notation** and graph it on a number line.

**a.**  $-3x + 2 \leq 8$          **b.**  $6x - 3 > 4x + 9$

### Solutions

**a.** Follow the **guidelines** for solving linear inequalities in one variable.

$$\begin{aligned}
\text{Begin with the original inequality:} \quad & -3x + 2 \leq 8 \\
\text{Subtract 2 from both sides:} \quad & -3x + 2 - 2 \leq 8 - 2 \\
\text{Simplify:} \quad & -3x \leq 6 \\
\text{Divide both sides by } -3: \quad & \frac{-3x}{-3} \geq \frac{6}{-3} \quad \leftarrow \text{Remember to switch the direction} \\
\text{Simplify:} \quad & x \geq -2
\end{aligned}$$

The graph of the solution set appears on the following number line:

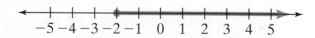

In interval notation, the solution set is $[-2, \infty)$.

 *My video summary*  **b.** Try solving this inequality on your own. View the **answer**, or watch this **video** for a complete solution.

**You Try It**    Work through this **You Try It** problem.

**Work Exercises 23–26** in this textbook or in the *MyMathLab* Study Plan.

### Example 7 Solving a Linear Inequality in One Variable

Solve the inequality $5x - 4 < 2(x + 3) + x$. Write the solution set in **interval notation**, and graph it on a number line.

**Solution** Follow the **guidelines** for solving linear inequalities in one variable.

| | |
|---|---|
| Begin with the original inequality: | $5x - 4 < 2(x + 3) + x$ |
| Distribute: | $5x - 4 < 2x + 6 + x$ |
| Combine like terms: | $5x - 4 < 3x + 6$ |
| Subtract $3x$ from both sides: | $5x - 4 - 3x < 3x + 6 - 3x$ |
| Simplify: | $2x - 4 < 6$ |
| Add 4 to both sides: | $2x - 4 + 4 < 6 + 4$ |
| Simplify: | $2x < 10$ |
| Divide both sides by 2: | $\dfrac{2x}{2} < \dfrac{10}{2}$ |
| Simplify: | $x < 5$ |

The graph of the solution set:

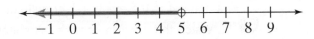

In interval notation, the solution set is $(-\infty, 5)$.

### Example 8 Solving a Linear Inequality in One Variable

 Solve the inequality $4 + 2(3 - x) > 3(2x + 7) + 5$. Write the solution set in **set-builder notation** and graph it on a number line.

**Solution** Try to solve this inequality on your own. View the **answer**, or watch this **video** to see the complete solution.

**You Try It** Work through this **You Try It** problem.

Work Exercises 27–30 in this textbook or in the *MyMathLab*® Study Plan.

### Example 9 Solving a Linear Inequality Containing Fractions

Solve the inequality $\dfrac{n}{3} - 4 > -\dfrac{n}{6} + 1$ and write the solution set in **interval notation**.

**Solution** View the **answer**, or watch this **video** to see the detailed solution.

**You Try It** Work through this **You Try It** problem.

Work Exercises 31–36 in this textbook or in the *MyMathLab*® Study Plan.

When solving inequalities involving fractions, some students like to clear fractions right away. This is helpful, but not required. The same solution set will result whether or not fractions are cleared immediately.

As with equations, a linear inequality can have no solution or all real numbers as its solution. Remember that a **contradiction** has no solution, and an **identity** has the set of all real numbers as its solution.

### Example 10 Solving Special Cases of Linear Inequalities

Solve the following inequalities. Write each solution set in **interval notation**.

**a.** $10 - 2(x + 1) > -5x + 3(x + 8)$    **b.** $2(5 - x) - 2 < 3(x + 3) - 5x$

**Solutions**

a. We use the **properties of inequalities** to isolate the variable on one side of the inequality.

$$\begin{aligned}
\text{Begin with the original ineqality:} &\quad 10 - 2(x + 1) > -5x + 3(x + 8)\\
\text{Use the distributive property:} &\quad 10 - 2x - 2 > -5x + 3x + 24\\
\text{Simplify:} &\quad -2x + 8 > -2x + 24\\
\text{Add } 2x \text{ to both sides:} &\quad -2x + 8 + 2x > -2x + 24 + 2x\\
\text{Simplify:} &\quad 8 > 24 \quad \text{False}
\end{aligned}$$

Because the final statement is a contradiction, the inequality has no solution. The solution set is the empty { } or null set, $\varnothing$.

 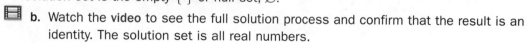 **b.** Watch the **video** to see the full solution process and confirm that the result is an identity. The solution set is all real numbers.

**You Try It**   **Work through this You Try It problem.**

**Work Exercises 35 and 36 in this textbook or in the *MyMathLab*® Study Plan.**

### OBJECTIVE 5   SOLVE THREE-PART INEQUALITIES

Consider the following application: A Bing internet search in June 2010 showed the range in price of iPad apps from $0.99 to $14.99. We can use the **three-part inequality** $0.99 \le P \le 14.99$ to show this relationship. Let $P$ be the price of an iPad app.

To solve a three-part inequality, we use the **properties of inequalities** to isolate the variable between the two inequality symbols. Remember that *what we do to one part, we must do to all three parts* in order to write an equivalent inequality.

### Example 11 Solving a Three-Part Inequality

Solve the inequality $4 \le 3x - 2 < 7$. Write this solution set in **interval notation** and then graph it on a number line.

**Solution**   Begin with the original inequality: $\quad 4 \le 3x - 2 < 7$

Add 2 to all three parts: $\quad 4 + 2 \le 3x - 2 + 2 < 7 + 2$

Simplify: $\quad 6 \le 3x < 9$

Divide all parts by 3: $\quad \dfrac{6}{3} \le \dfrac{3x}{3} < \dfrac{9}{3}$

Simplify: $\quad 2 \le x < 3$ ← Variable isolated between inequality symbols

The graph of this solution set is shown below:

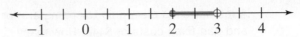

The solution set in interval notation is $[2, 3)$.

 When solving **three-part inequalities**, remember that what you do to one part of the inequality must be done to **all three parts** in order to obtain an equivalent inequality.

**You Try It**   Work through this You Try It problem.

Work Exercises 37–40 in this textbook or in the *MyMathLab*® Study Plan.

### Example 12  Solving a Three-Part Inequality

   Solve the inequality $-1 < \dfrac{2x + 1}{3} < 1$ and write its solution set in **interval notation**.

**Solution**   Begin with the original inequality:   $-1 < \dfrac{2x + 1}{3} < 1$

Multiply all parts by 3 to clear the fraction:   $3(-1) < 3 \cdot \dfrac{2x + 1}{3} < 3(1)$

Simplify:   $-3 < 2x + 1 < 3$

Finish solving the inequality on your own. View the **answer**, or watch this **video** for the complete solution.

**You Try It**   Work through this You Try It problem.

Work Exercises 41–44 in this textbook or in the *MyMathLab*® Study Plan.

OBJECTIVE 6   USE LINEAR INEQUALITIES TO SOLVE APPLICATION PROBLEMS

At this point, you may wish to review the **strategy** for solving application problems with linear equations. We present a variation of this strategy here using **mathematical models** for linear inequalities.

---

**Strategy for Solving Application Problems Involving Linear Inequalities**

Step 1.   **Define the Problem.** Read the problem carefully, or multiple times if necessary. Identify what you are trying to find and determine what information is available to help you find it.

Step 2.   **Assign Variables.** Choose a variable to assign to an unknown quantity in the problem. For example, use $p$ for price. If other unknown quantities exist, express them in terms of the selected variable.

Step 3.   **Translate into an Inequality.** Use the relationships among the known and unknown quantities to form an **inequality**.

Step 4.   **Solve the Inequality.** Determine the **solution set** of the inequality.

Step 5.   **Check the Reasonableness of Your Answer.** Check to see if your results make sense within the context of the problem. If not, check your work for errors and try again.

Step 6.   **Answer the Question.** Write a clear statement that answers the question(s) posed.

---

## Example 13  Making a Profit

A retailer sells electronic toy hamsters for $10. He pays $2 for each hamster wholesale and has fixed costs of $440. How many hamsters must he sell in order to make a **profit**? Solve the inequality $R > C$ with $R$ as his **revenue** and $C$ as his **cost**.

### Solution

**Step 1.**  We want to find the number of electronic hamsters that must be sold in order to make a profit. The retail price is $10 each, and the wholesale cost to the retailer is $2 each. The fixed costs are $440.

**Step 2.**  Let the variable $x$ represent the number of hamsters that are sold. The retailer's revenue, $R$, is $10x$ because the number sold times the price equals the revenue. The retailer's cost, $C$, is the sum of the fixed costs and the wholesale costs per toy. Therefore, $C$ is $2x + 440$.

**Step 3.**  We can write the following inequality:

$$\overbrace{10x}^{\text{Revenue, } R} > \overbrace{2x + 440}^{\text{Cost, } C}$$

**Step 4.**  Begin with the original inequality: $\qquad\qquad 10x > 2x + 440$

Subtract $2x$ from both sides: $\quad 10x - 2x > 2x + 440 - 2x$

Simplify: $\qquad\qquad\qquad\qquad\qquad 8x > 440$

Divide both sides by 8: $\qquad\qquad\qquad \dfrac{8x}{8} > \dfrac{440}{8}$

Simplify: $\qquad\qquad\qquad\qquad\qquad x > 55$

**Step 5.**  Because $55(10) = 550$ and $2(55) + 440 = 550$, the retailer's cost and revenue are equal if 55 hamsters are sold. To make a profit, the retailer needs to sell more than 55 electronic hamsters, so this answer is reasonable.

**Step 6.**  The retailer needs to sell more than 55 (at least 56) electronic hamsters to make a profit.

**You Try It**  Work through this You Try It problem.

Work Exercises 45 and 46 in this textbook or in the **_MyMathLab_**® Study Plan.

## Example 14  Incredible Birthday Party

Lesley is planning a birthday party for her son at Incredible Pizza. The party will cost $43 plus $15 for each guest. If she does not want to spend more than $300, what is the largest number of guests that can attend the party?

### Solution

**Step 1.**  We want to find the number of guests that can attend the birthday party. We know that the cost is $43 plus $15 per guest and Lesley does not want to spend more than $300.

**Step 2.**  Let the variable $x$ represent the number of guests. The cost of the party is $43 + 15x$ because there is a fixed cost of $43 and a per guest cost of $15.

**Step 3.**  Because Lesley has a maximum amount she wants to spend, the inequality will involve a "less than or equal to" symbol.

$$\overbrace{43 + 15x}^{\text{Party cost}} \le \overbrace{300}^{\text{Maximum amount}}$$

Watch this **video** to see the rest of the solution completed in detail. There can be at most 17 guests at the party.

 Because it is not feasible for a fractional guest to attend, we must round down the number of guests.

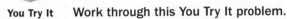

**You Try It**  Work through this You Try It problem.

Work Exercises 47 and 48 in this textbook or in the *MyMathLab*® Study Plan.

# 2.7 Exercises

In Exercises 1–4, write the solution set in set-builder notation.

**1.** $x < -3$

**2.** $-1 < y < 5$

**You Try It**
**3.** $z$ is any real non-negative number

**4.** $x$ is smaller than 8 and greater than 5

In Exercises 5–9, graph the solution set on a number line.

**5.** $\{x|x > 0\}$

**6.** $\{x|-3 \le x < 2\}$

**You Try It**
**7.** $\{x|x \le 5\}$

**8.** $\{x|2 \le x \le 5\}$

**9.** $\{x|-3 < x < 1\}$

In Exercises 10–14, write the solution set in interval notation.

**10.** $\{x|x \ge -4\}$

**11.** $\{y|-5 < y < -2\}$

**12.** $\{y|0 < y\}$

**You Try It**
**13.** $t$ is larger than 3 but not larger than 10

**14.** $w$ is nonnegative

In Exercises 15–30, solve the inequality. Graph the solution set and write it in interval notation.

**You Try It**
**15.** $x + 3 < 5$

**16.** $x - 1 \ge 4$

**17.** $x + 3 \le 2x + 1$

**18.** $3x + 7 > 2x - 5$

**19.** $5x > 10$

**20.** $\dfrac{m}{-4} \ge 1$

**You Try It**
**21.** $\dfrac{y}{2} \le -3$

**22.** $-6x < 24$

**23.** $2x + 1 > 5x + 7$

**24.** $15 - x < 3x - 1$

**You Try It**
**25.** $3x - 2 < 8 + x$

**26.** $5x - 2 < 3x + 8$

**27.** $4x - 5 > 3(x + 2) + 1$

**28.** $2x - 1 \ge 5(x + 1) + 3$

**29.** $5 + (x - 3) > 3 - (x - 5)$

**30.** $2 + 3(4 + x) < 2(7x + 3) - 3$

**You Try It**

In Exercises 31–36, solve the inequality. Write the solution set in interval notation.

**You Try It**

**31.** $\dfrac{m}{5} - 2 + m > -\dfrac{m}{2} + \dfrac{7}{5}$

**32.** $\dfrac{3}{5}x + 1 - 2x \le \dfrac{x}{2} - \dfrac{9}{10}$

**33.** $2x + \dfrac{1}{2} > \dfrac{2}{3}x - 1$

**34.** $2.5x - 4 \ge 3.1 + 2.25x$

**You Try It**

**35.** $5x - (2x - 5) \ge 3x + 7$

**36.** $3(x - 1) + 2x < 2 + 5(x + 1)$

In Exercises 37–44, solve the inequality. Graph the solution set and write it in interval notation.

**37.** $-2 < 2x + 1 < 7$

**38.** $-4 < 1 - 3x < -1$

**You Try It**

**39.** $1 < 5x + 6 \le 16$

**40.** $6.4 \le 3.5x - 2 < 14.1$

**41.** $1 \le \dfrac{2x + 1}{5} < 3$

**42.** $-1 \le \dfrac{3 - 3x}{-3} \le 3$

**You Try It**

**43.** $-4 \le \dfrac{3(x - 1) + 4}{2} < 2$

**44.** $2 < \dfrac{8 - 2x}{5} < 4$

In Exercises 45–48, solve each application problem.

**45. Flea Market**   Nancy sells handcrafted bracelets at a flea market for $7. If her monthly fixed costs are $675 and each bracelet costs her $2.75 to make, how many bracelets must she sell in a month to make a profit?

**46. Car Rental**   A rental car company offers a rental plan of $75 per day, plus $0.12 per mile. An alternate rental plan is for $60 per day, plus $0.20 per mile. For what mileage is the first plan cheaper than the second?

**47. Wedding Reception**   Beverly is planning her wedding reception at the Missouri Botanical Gardens. The rental cost is $2000 plus $76 per person for catering. How many guests can she invite if her reception budget is $10,000? (*Source:* www.mobot.org)

**48. Wireless Plan**   Suppose Verizon Wireless offers a Nationwide 450 Talk & Test monthly plan that includes 450 anytime minutes and unlimited nights and weekends for $60. Each additional anytime minute (or fraction of a minute) costs the user $0.45. If Tori subscribes to this plan, how many anytime minutes can she use each month while keeping her total cost to no more than $100 (before taxes)? (*Source:* www.verizonwireless.com)

CHAPTER THREE

# Graphs of Linear Equations and Inequalities in Two Variables

## 3.1 The Rectangular Coordinate System

### THINGS TO KNOW

Before working through this section, be sure you are familiar with the following concepts:

VIDEO      ANIMATION      INTERACTIVE

**You Try It**

1. Plot Real Numbers on a Number Line
   (Section 1.1, **Objective 2**)

### OBJECTIVES

1 Read Line Graphs

2 Identify Points in the Rectangular Coordinate System

3 Plot Ordered Pairs in the Rectangular Coordinate System

4 Create Scatter Plots

OBJECTIVE 1    READ LINE GRAPHS

Graphs are often used to show relationships between two **variables**. Figure 1 shows the selling price of MasterCard, Inc. stock at the start of each month from May 2006 (when it began trading publicly) until June 2010. The graph shows us how the value of the stock has changed over time.

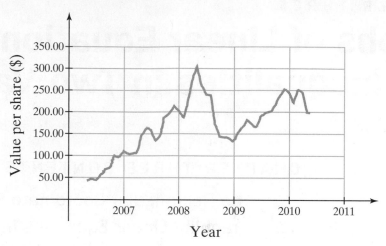

**Figure 1** MasterCard, Inc. stock value
*Source:* TD Ameritrade

Figure 1 is an example of a *line graph*. **Line graphs** consist of a series of points (•) that are connected by **line segments**.

## Example 1  Finding Temperatures on a Line Graph

My video summary  The following **line graph** shows the average daily temperature in St. Louis, MO, for each month.

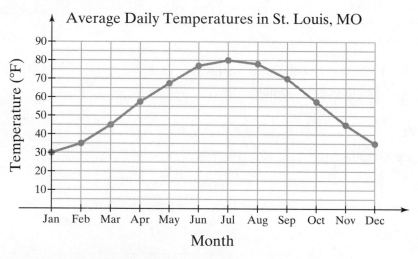

*Source:* National Climatic Data Center

**a.** What is the average daily temperature in February?

**b.** What is the average daily temperature in November?

**c.** In what month is the average daily temperature 70°F?

**d.** Which month has the highest average daily temperature? What is the average daily temperature for that month?

**e.** In what months are the average daily temperatures above 65°F?

## Solutions

a. The **red** point on the graph below corresponds to February. Note that the point for February is about halfway between the tick marks for 30°F and 40°F, which means that the average daily temperature for February is about 35°F.

b. The **green** point on the graph below corresponds to November. Because that point lies about halfway between the tick marks for 40°F and 50°F, the average daily temperature is about 45°F.

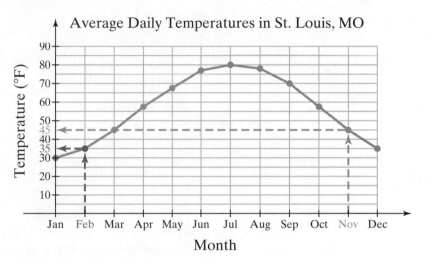

*Source:* National Climatic Data Center

**c.–e.** Try answering these questions on your own. View the **answers**, or watch this **video** for detailed solutions to all five parts.

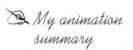

**You Try It**    **Work through this You Try It problem.**

**Work Exercises 1 and 2 in this textbook or in the *MyMathLab*® Study Plan.**

**OBJECTIVE 2**    IDENTIFY POINTS IN THE RECTANGULAR COORDINATE SYSTEM

*My animation summary*

Line graphs like **Figure 1** use points in two dimensions to show how two **variables** are related. We now extend this method of identifying points by defining the *rectangular coordinate system*. Watch this **animation** for an overview.

The **rectangular coordinate system**, also known as the **Cartesian coordinate system** in honor of its inventor **René Descartes**, consists of two **perpendicular** real **number lines** called the **coordinate axes**. The horizontal axis is called the ***x*-axis**, and the vertical axis is called the ***y*-axis**. The two axes intersect at a point called the **origin**.

The **plane** represented by this system is called the **coordinate plane**, also known as the **Cartesian plane** or ***xy*-plane**. The axes divide the plane into four regions called **quadrants**. The upper-right region is *Quadrant I*, the upper-left region is *Quadrant II*, the lower-left region is *Quadrant III*, and the lower-right region is *Quadrant IV*. See Figure 2.

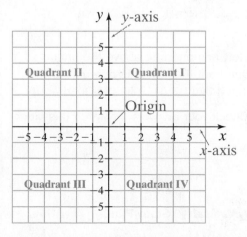

**Figure 2**

The rectangular coordinate system

Each position, or **point**, on the **coordinate plane** can be identified using an **ordered pair** of numbers in the form $(x, y)$. For example, the point shown in Figure 3 is identified by the ordered pair $(3, 2)$. The first number 3 is called the **x-coordinate** or **abscissa** and indicates the point is located 3 units to the right of the origin. The second number 2 is called the **y-coordinate** or **ordinate** and indicates the point is located 2 units above the **origin**.

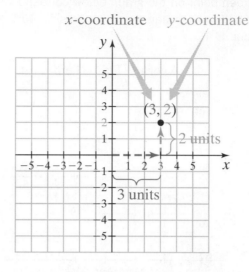

**Figure 3**
The ordered pair $(3, 2)$

In Figure 3, both the $x$-coordinate and $y$-coordinate are positive $(x > 0, y > 0)$ because the point is located in Quadrant I. Launch this **popup box** to determine the signs for $x$- and $y$-coordinates of points that lie in the other three **quadrants** or on the **axes**.

## Example 2  Identifying Points

*My video summary*   Use an **ordered pair** to identify each point on the **coordinate plane** shown. State the **quadrant** or axis where each point lies.

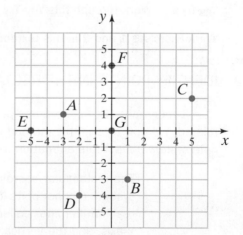

**Solution** Point $A$ is positioned three units to the left and one unit above the **origin**, so it is identified by the ordered pair $(-3, 1)$. It lies in Quadrant II. Point $B$ is one unit to the right and three units below the origin, so it is identified by $(1, -3)$. It lies in Quadrant IV.

Try to finish identifying these points on your own. Check your **answers**, or watch this **video** for a detailed solution.

**You Try It**   Work through this You Try It problem.

Work Exercises 3 and 4 in this textbook or in the *MyMathLab*® Study Plan.

The order of the pair of **coordinates** in an **ordered pair** is just as important as the coordinates themselves. For example, compare the ordered pairs and the positions of points $A$ and $B$ in **Example 2**. Both ordered pairs include the numbers 1 and $-3$, but they are in reverse order: $A(-3, 1)$ versus $B(1, -3)$. Point $A$ lies in Quadrant II, whereas point $B$ lies in Quadrant IV.

## OBJECTIVE 3  PLOT ORDERED PAIRS IN THE RECTANGULAR COORDINATE SYSTEM

We **plot**, or **graph**, an **ordered pair** by placing a point (•) at its location on the **coordinate plane**.

### Example 3  Plotting Ordered Pairs

*My video summary*

Plot each ordered pair on the coordinate plane. State the **quadrant** or **axis** where each point lies.

**a.** $(2, 4)$      **b.** $(4, -5)$      **c.** $(0, -2)$

**d.** $(-3, -4)$      **e.** $\left(-\dfrac{7}{2}, \dfrac{5}{2}\right)$      **f.** $(1.5, 0)$

**Solutions**

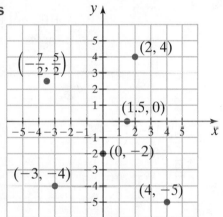

Point $(2, 4)$ lies in Quadrant I.

Point $(4, -5)$ lies in Quadrant IV.

Point $(0, -2)$ lies on the $y$-axis.

Point $(-3, -4)$ lies in Quadrant III.

Point $\left(-\dfrac{7}{2}, \dfrac{5}{2}\right)$ lies in Quadrant II.

Point $(1.5, 0)$ lies on the $x$-axis.

Watch this **video** for a detailed solution.

**You Try It**  Work through this You Try It problem.

Work Exercises 5–10 in this textbook or in the *MyMathLab*® Study Plan.

## OBJECTIVE 4  CREATE SCATTER PLOTS

**Ordered pairs** can be used to study relationships between variables. For example, a doctor studying height and weight can use ordered pairs of the form (*height, weight*) to record data for each patient. Data represented as ordered pairs are called **paired data**. A **scatter plot** is created by graphing paired data as points on the **coordinate plane**. Scatter plots can reveal patterns in the paired data.

### Example 4  Ethanol Industry

The table to the right shows the number of U.S. ethanol plants operating in the month of January for the years 2000–2010. List ordered pairs in the form (*year*, *number of plants*). Create a scatter plot of the paired data. Do the paired data show a trend? If so, what is the trend?

| Year | Number of Plants |
|------|------------------|
| 2000 | 54 |
| 2001 | 56 |
| 2002 | 61 |
| 2003 | 68 |
| 2004 | 72 |
| 2005 | 81 |
| 2006 | 95 |
| 2007 | 110 |
| 2008 | 139 |
| 2009 | 170 |
| 2010 | 189 |

*Source:* Renewable Fuels Association

**Solution**  The ordered pairs are $(2000, 54)$, $(2001, 56)$, $(2002, 61)$, $(2003, 68)$, $(2004, 72)$, $(2005, 81)$, $(2006, 95)$, $(2007, 110)$, $(2008, 139)$, $(2009, 170)$, and $(2010, 189)$.

Because the paired data consist of all positive numbers, we need only **Quadrant I** of the **coordinate plane**. We mark the given years along the *x*-axis. For the *y*-axis, we let the grid marks represent multiples of 10 ethanol plants. (**Note:** We labeled only the multiples of 20 to avoid clutter.) Plotting the ordered pairs gives us the graph shown.

The scatter plot shows that the number of U.S. ethanol plants has grown steadily from 2000 to 2010, with faster growth after 2005.

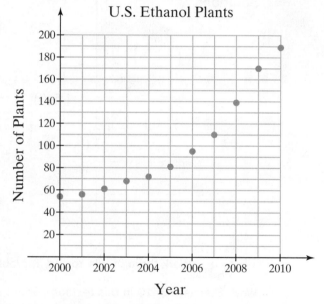

**You Try It**  Work through this You Try It problem.

Work Exercises 11–14 in this textbook or in the *MyMathLab*®  Study Plan.

**Scatter plots** and **line graphs** typically involve only Quadrant I because real data primarily involve non-negative numbers.

## 3.1 Exercises

In Exercises 1 and 2, use the line graph provided to answer the questions.

**You Try It**    **1. Twisters**  The following line graph shows the average number of tornadoes in Oklahoma each month.

   **a.** To the nearest whole number, what is the average number of tornadoes in March?

   **b.** What month has an average of about 8 tornadoes?

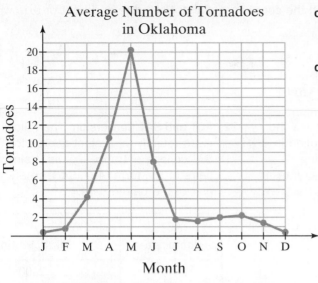

Average Number of Tornadoes in Oklahoma

*Source:* National Weather Service, NOAA

c. Which month has the highest average number of tornadoes? To the nearest whole number, what is the average number of tornadoes this month?

d. In what months are the average numbers of tornadoes below 1?

**2. Florida** The following line graph shows the population of Florida since 1900.

Population of Florida

*Source:* U.S. Census Bureau

a. To the nearest million, what was the population of Florida in 1960?

b. To the nearest million, what was the population in 2000?

c. In what year was the population about 13 million people?

d. Estimate the increase in population from 1900 to 2010.

In Exercises 3 and 4, use an ordered pair to identify each point on the given coordinate plane. State the quadrant or axis where each point lies.

**You Try It**

**3.**

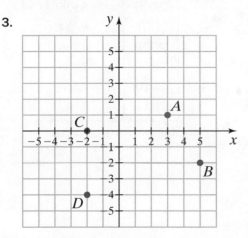

**4.**

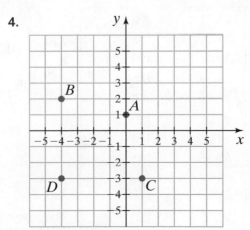

**3.1** The Rectangular Coordinate System   3-7

In Exercises 5–10, plot each ordered pair on the coordinate plane. State the quadrant or axis where each point lies.

**You Try It**

**5.** $(-4, 1)$

**6.** $(0, -3)$

**7.** $(-6, -2)$

**8.** $(2, 5)$

**9.** $(3.5, 0)$

**10.** $\left(\dfrac{1}{2}, -\dfrac{5}{2}\right)$

**11. Price of Gasoline** The following table gives the average price per gallon of regular gasoline in the U.S. for the years 2000–2009. Using ordered pairs of the form (*year, price*), create a scatter plot of the paired data.

| Year | Price per gallon (dollars) |
|------|----------------------------|
| 2000 | 1.48 |
| 2001 | 1.42 |
| 2002 | 1.35 |
| 2003 | 1.56 |
| 2004 | 1.85 |
| 2005 | 2.27 |
| 2006 | 2.57 |
| 2007 | 2.80 |
| 2008 | 3.25 |
| 2009 | 2.35 |

*Source:* Energy Information Administration

**12. ACT and SAT Conversion** The following table gives roughly equivalent scores for the ACT and SAT college entrance exams. Using ordered pairs of the form (*ACT, SAT*), create a scatter plot of the paired data.

| ACT Composite | SAT Critical Reading, Math, Writing |
|---------------|--------------------------------------|
| 36 | 2400 |
| 33 | 2190 |
| 30 | 1980 |
| 27 | 1820 |
| 24 | 1650 |
| 21 | 1500 |
| 18 | 1290 |
| 15 | 1060 |
| 12 | 780 |

*Source:* InLikeMe.com

**You Try It**

**13. Maximum Heart Rate** *Maximum heart rate* is the highest number of times a person's heart can beat in one minute of exercise. The following table gives the average maximum heart rate for ages 20–70. Using ordered pairs of the form (*age, maximum heart rate*), create a scatter plot of the paired data. Do the paired data show a trend? If so, what is the trend?

| Age | Average Maximum Heart Rate (beats per minute) |
|-----|-----------------------------------------------|
| 20 | 200 |
| 25 | 195 |
| 30 | 190 |
| 35 | 185 |
| 40 | 180 |
| 45 | 175 |
| 50 | 170 |
| 55 | 165 |
| 60 | 160 |
| 65 | 155 |
| 70 | 150 |

*Source:* American Heart Association

**14. College Tuition** The following table shows annual in-state tuition and fees for the Connecticut State University system for the years 2005–2011. Using ordered pairs of the form (*year, tuition & fees*), create a scatter plot of the paired data. Do the paired data show a trend? If so, what is the trend?

| Year | Tuition & Fees (dollars) |
|------|--------------------------|
| 2005 | 5,611 |
| 2006 | 5,936 |
| 2007 | 6,284 |
| 2008 | 6,736 |
| 2009 | 7,180 |
| 2010 | 7,566 |
| 2011 | 8,043 |

*Source:* State of Connecticut Department of Higher Education

# 3.2 Graphing Linear Equations in Two Variables

## THINGS TO KNOW

Before working through this section, be sure you
are familiar with the following concepts:

VIDEO     ANIMATION     INTERACTIVE

 **You Try It**
**1.** Determine If a Given Value Is a
Solution to an Equation (Section 2.1,
**Objective 2**)

 **You Try It**
**2.** Solve Linear Equations Using Both
Properties of Equality (Section 2.1,
**Objective 5**)

 **You Try It**
**3.** Plot Ordered Pairs in the Rectangular
Coordinate System (Section 3.1, **Objective 3**)

## OBJECTIVES

**1** Determine If an Ordered Pair Is a Solution to an Equation

**2** Determine the Unknown Coordinate of an Ordered Pair Solution

**3** Graph Linear Equations by Plotting Points

**4** Find $x$- and $y$-Intercepts

**5** Graph Linear Equations Using Intercepts

**6** Use Linear Equations to Model Data

**7** Graph Horizontal and Vertical Lines

---

OBJECTIVE 1    DETERMINE IF AN ORDERED PAIR IS A SOLUTION TO AN EQUATION

In **Section 2.1** we learned that a **solution** to an **equation in one variable** is a value that, when substituted for the **variable**, makes the equation true. In this section, we expand on this idea and consider equations that contain two variables. For example, we may wish to find solutions to the equation $x + y = 4$.

Solutions to equations in two variables require *two* values—one value for each variable. For the equation $x + y = 4$, $x = 1$ and $y = 3$ form a solution because a true statement results if we **substitute** the values for the variables.

$$\text{Begin with the original equation:} \quad x + y = 4$$
$$\text{Substitute 1 for } x \text{ and 3 for } y: \quad (1) + (3) \stackrel{?}{=} 4$$
$$\text{Simplify:} \quad 4 = 4 \quad \text{True}$$

We can write this solution as the **ordered pair** $(1, 3)$, where the first number is the $x$-value and the second number is the $y$-value.

> **Solution to an Equation in Two Variables**
>
> A **solution to an equation in two variables** is an ordered pair of values that, when substituted for the variables, makes the equation true.

### Example 1 Determining If an Ordered Pair Is a Solution to an Equation

Determine if each **ordered pair** is a **solution** to the equation $x + 2y = 8$.

**a.** $(-2, 5)$       **b.** $(2, 6)$       **c.** $\left(-11, \dfrac{3}{2}\right)$       **d.** $(0, 4)$

### Solutions

**a.** Substitute $-2$ for $x$ and $5$ for $y$. Work through the simplification to determine if the ordered pair is a solution.

$$
\begin{aligned}
\text{Begin with the original equation:} \qquad & x + 2y = 8 \\
\text{Substitute } -2 \text{ for } x \text{ and } 5 \text{ for } y: \qquad & (-2) + 2(5) \overset{?}{=} 8 \\
\text{Simplify:} \qquad & -2 + 10 \overset{?}{=} 8 \\
& 8 = 8 \quad \text{True}
\end{aligned}
$$

The final statement is true, so $(-2, 5)$ is a solution to the equation.

**b.**
$$
\begin{aligned}
\text{Begin with the original equation:} \qquad & x + 2y = 8 \\
\text{Substitute } 2 \text{ for } x \text{ and } 6 \text{ for } y: \qquad & (2) + 2(6) \overset{?}{=} 8 \\
\text{Simplify:} \qquad & 2 + 12 \overset{?}{=} 8 \\
& 14 = 8 \quad \text{False}
\end{aligned}
$$

The final statement is not true, so $(2, 6)$ is not a solution to the equation.

*My video summary*  **c–d.** Substitute the values for the variables and work through the simplification to determine if each ordered pair is a solution. View the **answers**, or watch this **video** to see the details.

**You Try It**    Work through this **You Try It** problem.

Work Exercises 1–3 in this textbook or in the *MyMathLab*® Study Plan.

OBJECTIVE 2    DETERMINE THE UNKNOWN COORDINATE OF AN ORDERED PAIR SOLUTION

In Example 1, we determined if a given ordered pair was a **solution** to an equation in two variables. But how do we find such ordered pair solutions if we know only one **coordinate**? Substituting the given coordinate for the corresponding variable will result in an equation in one variable that we can solve as we did in **Section 2.1** or **Section 2.2**. The resulting values for both variables together form an **ordered pair solution** to the equation.

### Example 2 Determining Unknown Coordinates

*My interactive video summary* 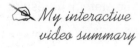 Find the unknown coordinate so that each ordered pair **satisfies** $2x - 3y = 15$.

**a.** $(6, ?)$       **b.** $(?, 7)$       **c.** $\left(-\dfrac{5}{2}, ?\right)$

### Solutions

**a.** To find the $y$-coordinate when the $x$-coordinate is $6$, we substitute $6$ for $x$ and solve for $y$.

$$
\begin{aligned}
\text{Begin with the original equation:} \qquad & 2x - 3y = 15 \\
\text{Substitute } 6 \text{ for } x: \qquad & 2(6) - 3y = 15 \\
\text{Simplify:} \qquad & 12 - 3y = 15 \; \leftarrow \text{Linear equation in one variable} \\
\text{Subtract } 12 \text{ from both sides:} \qquad & -3y = 3 \\
\text{Divide both sides by } -3: \qquad & y = -1 \; \leftarrow \text{Missing coordinate}
\end{aligned}
$$

We combine the two coordinates to form the solution. The ordered pair $(6, -1)$ is a solution to the equation $2x - 3y = 15$.

b. Begin with the original equation:      $2x - 3y = 15$

Substitute 7 for $y$:      $2x - 3(7) = 15$

Simplify:      $2x - 21 = 15$      ← Linear equation in one variable

Finish solving for $x$ to find the $x$-coordinate when the $y$-coordinate is 7. Once completed, view the **answer** or watch this **interactive video** to check your work.

c. Substitute the $x$-coordinate into the equation and solve for $y$ to determine the $y$-coordinate. Once finished, view the **answer** or watch this **interactive video** to check your work for all parts.

**You Try It**    **Work through this You Try It problem.**

**Work Exercises 4-6 in this textbook or in the *MyMathLab*® Study Plan.**

OBJECTIVE 3    GRAPH LINEAR EQUATIONS BY PLOTTING POINTS

From **Example 1**, we see that an equation in two variables can have more than one solution. The ordered pairs $(-2, 5)$ and $(0, 4)$ are two of the solutions to the equation $x + 2y = 8$. Building on **Example 2**, we can find additional solutions to this equation by selecting different values for $x$ and determining the corresponding values for $y$. Or, we can select values for $y$ and determine the corresponding values for $x$.

| $x$ | $y$ |
|-----|-----|
| 2 | ? |
| ? | 2 |
| 8 | ? |
| ? | -1 |

Substitute given values →

$(2) + 2y = 8$
$x + 2(2) = 8$
$(8) + 2y = 8$
$x + 2(-1) = 8$

Solve for the remaining variable →

| $x$ | $y$ |
|-----|-----|
| 2 | 3 |
| 4 | 2 |
| 8 | 0 |
| 10 | -1 |

Solutions →

$(2, 3)$
$(4, 2)$
$(8, 0)$
$(10, -1)$

The **graph of an equation in two variables** includes all points whose **coordinates** are solutions to the equation. So a graph is a visual display of the solution set for an equation. To make such a graph, we can **plot** several points that **satisfy** the equation. Then we connect the points with a line or smooth curve.

In Figure 4a, we plot each of the **ordered pair solutions** we have found for $x + 2y = 8$.

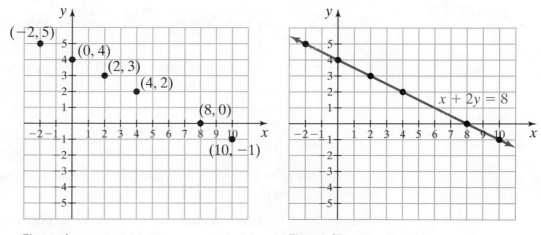

**Figure 4a**                    **Figure 4b**

There is a pattern to the plotted points. They all appear to lie on the same line. Connecting the points with a line as in Figure 4b completes the **graph** of the equation $x + 2y = 8$. For every point on this line, the coordinates of the point give an ordered pair solution to the equation.

The equations from **Example 1**, $x + 2y = 8$, and from **Example 2**, $2x - 3y = 15$, are both examples of **linear equations in two variables**. The graphs of these equations are always straight lines.

---

**Definition    Linear Equation in Two Variables (Standard Form)**

A **linear equation in two variables** is an equation that can be written in the standard form $Ax + By = C$, where $A, B,$ and $C$ are real numbers, and $A$ and $B$ are not both equal to 0.

---

A line can be completely determined by two points. So, to graph a linear equation in two variables, we find two **ordered pair solutions**, plot the corresponding points, and connect the points with a line. In practice, we will generally find a third solution to serve as a **check**. If the corresponding third point also lies on the **graph**, we can feel comfortable with our solution.

### Example 3  Graphing Linear Equations by Plotting Points

Graph $3x - y = 2$ by plotting points.

**Solution**  First, we determine some **ordered pair solutions** to the equation. Select a value for one of the variables and then solve for the remaining variable.

Let $x = 1$:

| | |
|---|---|
| Original equation: | $3x - y = 2$ |
| Substitute 1 for $x$: | $3(1) - y = 2$ |
| Simplify: | $3 - y = 2$ |
| Subtract 3 from both sides: | $-y = -1$ |
| Divide both sides by $-1$: | $y = 1$ |

Let $y = -2$:

| | |
|---|---|
| Original equation: | $3x - y = 2$ |
| Substitute $-2$ for $y$: | $3x - (-2) = 2$ |
| Simplify: | $3x + 2 = 2$ |
| Subtract 2 from both sides: | $3x = 0$ |
| Divide both sides by 3: | $x = 0$ |

When $x = 1$, we get $y = 1$ and when $y = -2$ we get $x = 0$. So, the ordered pairs $(1, 1)$ and $(0, -2)$ are solutions to the equation and are points on the graph.

The two points are enough to sketch the graph, but we want to find a third point as a **check**.

Let $x = -1$:

| | |
|---|---|
| Original equation: | $3x - y = 2$ |
| Substitute $-1$ for $x$: | $3(-1) - y = 2$ |
| Simplify: | $-3 - y = 2$ |
| Add 3 to both sides: | $-y = 5$ |
| Divide both sides by $-1$: | $y = -5$ |

The ordered pairs $(-1, -5)$, $(1, 1)$, and $(0, -2)$ are all solutions to the equation $3x - y = 2$ and points on the graph for this equation.

We plot the three points in Figure 5a. Connecting the points with a line, the complete graph of $3x - y = 2$ is given in Figure 5b.

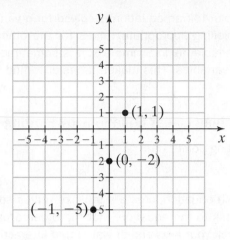

Figure 5a

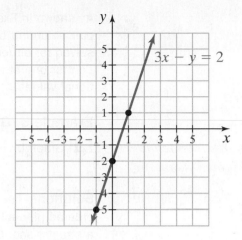

Figure 5b

**You Try It**    Work through this You Try It problem.

**Work Exercises 7 and 8 in this textbook or in the *MyMathLab*® Study Plan.**

### Example 4   Graphing Linear Equations by Plotting Points

Graph $y = -\dfrac{1}{3}x + 2$ by plotting points.

**Solution** The equation is solved for $y$. We can create a table of values by selecting appropriate values for $x$ and determining the corresponding values for $y$. Because the **coefficient** on $x$ has a **denominator** of 3, we will select three distinct values for $x$ that are **multiples** of 3. This will allow us to **clear the fraction** in the equation and avoid fractions as coordinates in our solutions.

Using $x = -3$, $x = 0$, and $x = 3$, the resulting **ordered pair solutions** are $(-3, 3), (0, 2)$, and $(3, 1)$, respectively. View the **work**.

The points are plotted in **Figure 6a** and the complete graph is shown in **Figure 6b**.

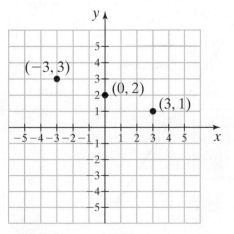

Figure 6a

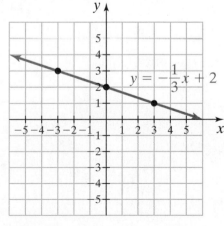

Figure 6b

**You Try It**    Work through this You Try It problem.

**Work Exercises 9 and 10 in this textbook or in the *MyMathLab*® Study Plan.**

As shown in Example 4, if an equation is **solved for a variable**, we can easily find ordered pair solutions by picking appropriate values for the other variable. If an equation is not solved for either variable, as in **Example 3**, it is sometimes helpful to first solve the equation for one of the variables. This makes it easier to find **integer** solutions and avoid fractional values.

### Example 5 Graphing Linear Equations by Plotting Points

 Graph by plotting points.

*My interactive video summary*

a. $y = 2x$          b. $3x + 2y = 5$

**Solutions** For each equation, create a **table of values** so you have at least three distinct **ordered pair solutions** to the equation. Plot the ordered pairs and connect the points with a straight line. Check your **answers**, or watch this **interactive video** for complete solutions.

**You Try It** Work through this You Try It problem.

Work Exercises 11–16 in this textbook or in the *MyMathLab*® Study Plan.

OBJECTIVE 4    FIND $x$- AND $y$-INTERCEPTS

Look at **Figure 4b**. Notice that the graph of $x + 2y = 8$ crosses the $y$-axis at the point $(0, 4)$ and also crosses the $x$-axis at the point $(8, 0)$. The points where a graph crosses or touches the **axes** are called its **intercepts**. The graph of $x + 2y = 8$ has two intercepts: $(0, 4)$ and $(8, 0)$. If a graph never crosses an axis, then it will have no intercepts.

A **$y$-intercept** is the **$y$-coordinate** of a point where a graph crosses or touches the **$y$-axis**. Its corresponding ordered pair would be $(0, y)$. An **$x$-intercept** is the **$x$-coordinate** of a point where a graph crosses or touches the **$x$-axis**. Its corresponding ordered pair would be $(x, 0)$. So for the graph of $x + 2y = 8$, the $y$-intercept is 4 and the $x$-intercept is 8.

Figure 7 illustrates examples of $x$- and $y$-intercepts.

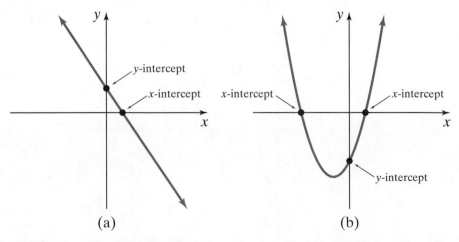

(a)          (b)

**Figure 7** $x$- and $y$-Intercepts

**Note:** When the type of intercept is not specified, we must list the intercept as an ordered pair. If the type of intercept is specified, then we can list only the coordinate of the intercept.

## Example 6  Finding *x*- and *y*-Intercepts

Find the **intercepts** of the graph shown in Figure 8. What are the *x*-**intercepts**? What are the *y*-**intercepts**?

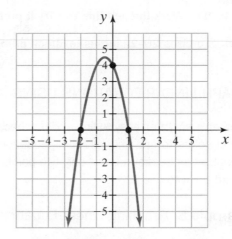

**Figure 8**

**Solution** The graph crosses the *y*-axis at $(0, 4)$ and crosses the *x*-axis at $(-2, 0)$ and $(1, 0)$. The *x*-intercepts are $-2$ and $1$. The *y*-intercept is $4$. The intercepts are $(0, 4)$, $(-2, 0)$, and $(1, 0)$.

**You Try It**  **Work through this You Try It problem.**

**Work Exercises 17–19 in this textbook or in the *MyMathLab*®  Study Plan.**

To find *x*-**intercepts** of the graph of an equation, we let $y = 0$ and solve for $x$ because every point on the *x*-axis has a *y*-coordinate of $0$. To find *y*-**intercepts**, we let $x = 0$ and solve for $y$ because every point on the *y*-axis has an *x*-coordinate of $0$.

---

**Finding *x*- and *y*-Intercepts of a Graph Given an Equation**

- To find an *x*-intercept, let $y = 0$ and solve for $x$.
- To find a *y*-intercept, let $x = 0$ and solve for $y$.

---

## Example 7  Finding *x*- and *y*-Intercepts

Find the *x*- and *y*-intercepts for the graph of each equation.

**a.**  $2x + y = 4$          **b.**  $4x = 3y + 8$

### Solutions

**a.** To find the *x*-intercept, we let $y = 0$ and solve for $x$. To find the *y*-intercept, we let $x = 0$ and solve for $y$.

*x-intercept:*

$$\text{Let } y = 0: \quad 2x + (0) = 4$$
$$\text{Simplify:} \qquad 2x = 4$$
$$\text{Divide both sides by 2:} \qquad x = 2$$

The *x*-intercept is 2.

*y-intercept:*

$$\text{Let } x = 0: \quad 2(0) + y = 4$$
$$\text{Simplify:} \qquad y = 4$$

The *y*-intercept is 4.

*My video summary*

▦ **b.** Try finding the intercepts for this equation on your own. Check your **answers**, or watch this **video** for the complete solution.

**You Try It**  Work through this You Try It problem.

Work Exercises 20–22 in this textbook or in the **MyMathLab**® Study Plan.

OBJECTIVE 5   GRAPH LINEAR EQUATIONS USING INTERCEPTS

**Intercepts** are often easy to find and plot because one of the **coordinates** is 0. This makes them useful points to find when graphing equations. Because the graph of a **linear equation in two variables** can be drawn using only two points, we can use the $x$- and $y$-intercepts together with a third point to check.

### Example 8  Graphing Linear Equations Using Intercepts

Graph $3x - 2y = 6$ using intercepts.

**Solution**  To find the $x$-intercept, we let $y = 0$ and solve for $x$. To find the $y$-intercept, we let $x = 0$ and solve for $y$. As a check, we randomly let $x = 4$ and find the corresponding $y$-value.

|  |  |  |
|---|---|---|
| *x-intercept:* | Let $y = 0$: | $3x - 2(0) = 6$ |
|  | Simplify: | $3x = 6$ |
|  | Divide both sides by 3: | $x = 2$ |

The $x$-intercept is 2, so the corresponding point is $(2, 0)$.

|  |  |  |
|---|---|---|
| *y-intercept:* | Let $x = 0$: | $3(0) - 2y = 6$ |
|  | Simplify: | $-2y = 6$ |
|  | Divide both sides by $-2$: | $y = -3$ |

The **y-intercept** is $-3$, so the corresponding point is $(0, -3)$.

| *Check point:* | Let $x = 4$: | $3(4) - 2y = 6$ |
|---|---|---|
|  | Simplify: | $12 - 2y = 6$ |
|  | Subtract 12 from both sides: | $-2y = -6$ |
|  | Divide both sides by $-2$: | $y = 3$ |

The corresponding check point is $(4, 3)$.

We plot the three points and connect the points with a straight line. The resulting graph is shown in Figure 9.

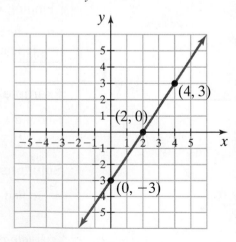

Figure 9

**You Try It**  Work through this You Try It problem.

Work Exercises 23–30 in this textbook or in the **MyMathLab**® Study Plan.

The graph of a **linear equation** always has at least one intercept, but it may have two. When the graph has only one **intercept**, we need to find at least one more point in order to graph the equation.

### Example 9 Graphing Linear Equations Using Intercepts

*My video summary*  Graph $2x = 5y$ using intercepts.

**Solution** Find the *x*-intercept by letting $y = 0$ and solving for $x$. Find the *y*-intercept by letting $x = 0$ and solving for $y$.

| | |
|---|---|
| Let $y = 0$: $\quad 2x = 5(0)$ | Let $x = 0$: $\quad 2(0) = 5y$ |
| Simplify: $\quad 2x = 0$ | Simplify: $\quad 0 = 5y$ |
| Divide by 2: $\quad x = 0$ | Divide by 5: $\quad 0 = y$ |
| The *x*-intercept is 0. | The *y*-intercept is 0. |

The graph has only one intercept: $(0, 0)$. Select two more values for $x$ and find the corresponding *y*-values to plot two additional points on the graph. Use the points to graph the equation. Check your **answer**, or watch this **video** for a detailed solution.

**You Try It**   Work through this You Try It problem.

**Work Exercises 31 and 32 in this textbook or in the *MyMathLab*® Study Plan.**

Compare the equations in **Examples 8** and **9**. How can we tell when the graph of a linear equation in two variables will go through the origin? **Find out.**

OBJECTIVE 6   USE LINEAR EQUATIONS TO MODEL DATA

Linear equations can be used to **model** real-world applications. In these situations, the **coordinates** of the points on the graph have meaning in the context of the situation.

### Example 10 Disappearing Drive-Ins

*My video summary*  The number of U.S. drive-in theaters can be modeled by the linear equation $y = -7.5x + 435$, where $x$ is the number of years after 2000. (*Source:* United Drive-In Theater Owners Association, 2009)

**a.** Sketch the graph of the equation for the year 2000 and beyond.

**b.** Find the missing coordinate for the ordered pair solution $(?, 390)$.

**c.** Interpret the point from part (b).

**d.** Find and interpret the *y*-intercept.

**e.** What does the *x*-intercept represent in this problem?

**Solutions** Try to work this problem on your own using earlier examples to guide you. View the **answers**, or watch this **video** for the complete solution.

**You Try It**   Work through this You Try It problem.

**Work Exercises 37 and 38 in this textbook or in the *MyMathLab*® Study Plan.**

## OBJECTIVE 7   GRAPH HORIZONTAL AND VERTICAL LINES

The equations $x = a$ and $y = b$ are considered **linear equations in two variables** because they can be written in standard form. The equation $x = a$ can be written as $x + 0y = a$ and its graph is a **vertical line**. The equation $y = b$ can be written as $0x + y = b$ and its graph is a **horizontal line**.

### Example 11  Graphing Horizontal Lines

Graph $y = 4$.

**Solution** The equation $y = 4$ can be written as $0x + y = 4$ to show that it is a linear equation in two variables. For any value **substituted** for $x$ in the equation, $y$ will always equal 4. Therefore, every point on the graph of the equation will have a $y$-coordinate of 4. If we choose $x$-values of $-2, 0$, and 3, the corresponding ordered pair solutions are $(-2, 4)$, $(0, 4)$, and $(3, 4)$. The resulting graph, shown in Figure 10, is a horizontal line with $y$-**intercept** 4 and no $x$-**intercept** because the graph does not cross the $x$-axis.

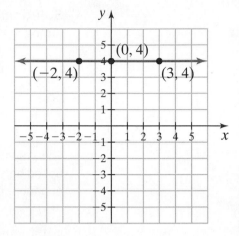

**Figure 10**

### Example 12  Graphing Vertical Lines

*My video summary*   Graph $x = -3$.

**Solution** The equation $x = -3$ can be written as $x + 0y = -3$ to show that it is a **linear equation in two variables**. For any value **substituted** for $y$ in the equation, $x$ will always equal $-3$. Therefore, every point on the graph of the equation will have an $x$-**coordinate** of $-3$. Choose three values for $y$ and find the corresponding points. Plot the points and connect them with a line. View the **answer**, or watch this **video** to see a detailed solution.

**You Try It**   Work through this You Try It problem.

Work Exercises 33–36 in this textbook or in the *MyMathLab*® Study Plan.

## 3.2  Exercises

In Exercises 1–3, determine if the given ordered pairs are solutions to the equations.

**You Try It**

**1.** $4x + 3y = 17$

a. $(3, 0)$

b. $(2, 3)$

c. $\left(-\dfrac{1}{4}, 6\right)$

**2.** $y = \dfrac{3}{4}x - 2$

a. $(-8, -8)$

b. $(4, -1)$

c. $\left(2, -\dfrac{1}{2}\right)$

**3.** $2x = 3y + 2$

a. $\left(\dfrac{1}{2}, 1\right)$

b. $(-2, -2)$

c. $\left(-\dfrac{2}{3}, 0\right)$

In Exercises 4–6, find the unknown coordinate so that each ordered pair satisfies the given equation.

**You Try It**

**4.** $5x + 4y = 10$

   **a.** $(2, ?)$

   **b.** $\left(?, -\dfrac{5}{4}\right)$

**5.** $-3x = 4y + 3$

   **a.** $\left(?, -\dfrac{3}{4}\right)$

   **b.** $(-2, ?)$

**6.** $y = -\dfrac{1}{3}x + 4$

   **a.** $(6, ?)$

   **b.** $(?, -5)$

In Exercises 7–16, graph each equation by plotting points.

**You Try It**

**7.** $2x + y = 6$

**9.** $y = \dfrac{3}{4}x - 4$

**8.** $4x - y = 2$

**10.** $y = -\dfrac{2}{3}x + 3$

**You Try It**

**11.** $y = 2x$

**13.** $3x - 2y = 5$

**15.** $1.2x - 0.5y = 2.4$

**12.** $y = -\dfrac{3}{2}x$

**14.** $2x + 5y = 3$

**16.** $1.6x + 1.2y = 2.4$

In Exercises 17–19, find the *x*- and *y*-intercepts of each graph.

**You Try It**

**17.**

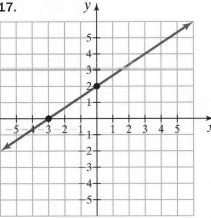

**18.**

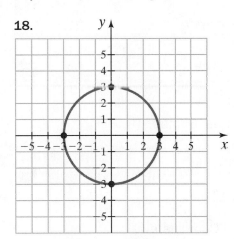

**19.**

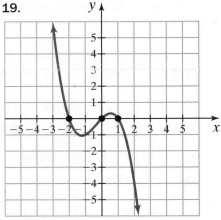

In Exercises 20–22, find the *x*- and *y*-intercepts of the graph of the given equation.

**20.** $3x + y = -2$

**21.** $y = 4x - 2$

**22.** $y = 0.4x + 1.4$

**You Try It**

In Exercises 23–32, graph the linear equation using intercepts.

**23.** $4x + y = 8$

**You Try It**

**26.** $y = x - 3$

**29.** $x + 3y = \dfrac{3}{4}$

**24.** $x - 2y = -6$

**27.** $y = -x - \dfrac{5}{2}$

**30.** $y = -\dfrac{3}{5}x - 2$

**25.** $-9x + 3y = 9$

**28.** $y = 3.5x + 1.4$

**31.** $y = -2x$

**32.** $3y = 4x$

**You Try It**

In Exercises 33–36, graph the given line.

**33.** $y = 3$

**34.** $x = -2$

**35.** $x - 2.5 = 0$

**36.** $y + 2 = 0$

**You Try It**

**37. Car Loan** Natalie took out a loan for a 2011 Ford Fusion with 0% financing and a monthly payment of $335. The balance remaining on the loan, $y$, is given by the linear equation $y = -335x + 24{,}120$, where $x$ is the number of months that Natalie has made payments.

**You Try It**

  a. Sketch the graph of the equation.

  b. Find the missing coordinate for the ordered pair solution $(25, ?)$.

  c. Interpret the point from part (b).

  d. Find and interpret the $y$-intercept.

  e. What does the $x$-intercept represent in this problem?

**38. Video Stores** The number of Blockbuster video stores in the U.S. can be approximated by the linear equation $y = -318x + 4935$, where $x$ is the number of years after 2005. (*Source:* www.blockbuster.com)

  a. Sketch the graph of the equation.

  b. Find the missing coordinate for the ordered pair solution $(8, ?)$.

  c. Interpret the point from part (b).

  d. Find and interpret the $y$-intercept.

  e. What does the $x$-intercept represent in this problem?

# 3.3 Slope

## THINGS TO KNOW

Before working through this section, be sure you are familiar with the following concepts:

|  | VIDEO | ANIMATION | INTERACTIVE |

**You Try It** 1. Write Fractions in Simplest Form (Section R.1, **Objective 2**)

**You Try It** 2. Find the Opposite of a Real Number (Section 1.1, **Objective 3**)

**You Try It** 3. Evaluate a Formula (Section 2.4, **Objective 1**)

**You Try It** 4. Determine the Unknown Coordinate of an Ordered Pair Solution (Section 3.2, **Objective 2**)

**You Try It** 5. Graph Linear Equations by Plotting Points (Section 3.2, **Objective 3**)

**You Try It** 6. Graph Horizontal and Vertical Lines (Section 3.2, **Objective 7**)

## OBJECTIVES

**1** Find the Slope of a Line Given Two Points

**2** Find the Slopes of Horizontal and Vertical Lines

**3** Graph a Line Using the Slope and a Point

**4** Find and Use the Slopes of Parallel and Perpendicular Lines

**5** Use Slope in Applications

## OBJECTIVE 1   FIND THE SLOPE OF A LINE GIVEN TWO POINTS

In **Section 3.2**, we graphed **linear equations in two variables** by **plotting points**. Although the graph of every linear equation is a straight line, there can be many differences between the graphs. A key feature of a line is its **slant** or **steepness**. For example, looking from left to right at the graphs in Figure 11, the lines in (a) and (b) slant upward (or rise), whereas the lines in (c) and (d) slant downward (or fall). Also, the line in (a) has a steeper upward slant than the line in (b), and the line in (c) has a steeper downward slant than the line in (d).

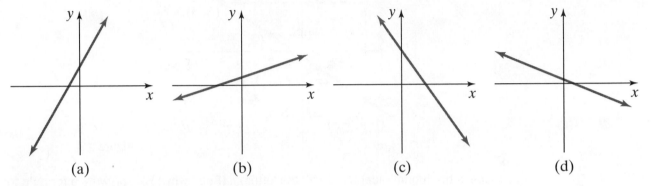

(a)     (b)     (c)     (d)

**Figure 11**

We measure the slant or steepness of a line using *slope*.

---

**Definition   Slope**

The **slope** of a line is the **ratio** of the vertical change in $y$, or **rise**, to the horizontal change in $x$, or **run**.

$$\text{slope} = \frac{\text{vertical change}}{\text{horizontal change}} = \frac{\text{change in } y}{\text{change in } x} = \frac{\text{rise}}{\text{run}}$$

---

The line graphed in Figure 12 goes through the points $(2, 1)$ and $(5, 3)$. The **rise** of the line is the **difference** between the two $y$-coordinates: $3 - 1 = 2$, and the **run** of the line is the difference between the corresponding two $x$-coordinates: $5 - 2 = 3$. The **slope** of the line is the ratio of rise to run:

$$\text{slope} = \frac{\text{rise}}{\text{run}} = \frac{2}{3}.$$

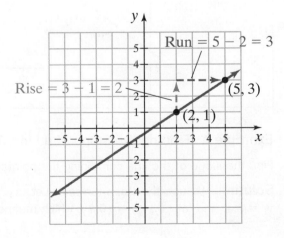

**Figure 12**

A slope of $\frac{2}{3}$ means that for every horizontal change of 3 units, there is a corresponding vertical change of 2 units. Because the slope is the same everywhere on a line, it does not matter which two points on the line are used to find its slope, as shown in Figure 13.

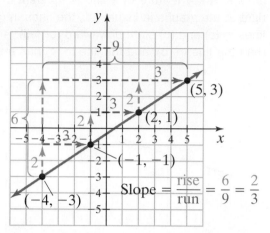

**Figure 13**

Slope is traditionally identified with the letter $m$. (Read **why**.) We can write a **formula** for finding slope using **subscript notation** to identify the two general points on the line, $(x_1, y_1)$ and $(x_2, y_2)$. We read $x_1$ as "$x$ sub one" and $y_2$ as "$y$ sub two."

**Slope Formula**

Given two points, $(x_1, y_1)$ and $(x_2, y_2)$, on the graph of a line, the **slope $m$** of the line containing the two points is given by the formula

$$m = \frac{\text{change in } y}{\text{change in } x} = \frac{\text{rise}}{\text{run}} = \frac{y_2 - y_1}{x_2 - x_1},$$

where $x_1 \neq x_2$.

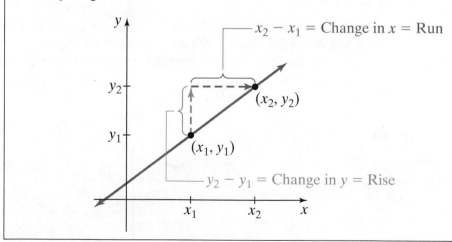

## Example 1 Finding the Slope of a Line Given Two Points

Find the **slope** of the line containing the points $(-2, -4)$ and $(0, 1)$.

**Solution** We use the **slope formula**. Let $(x_1, y_1) = (-2, -4)$, which means $x_1 = -2$ and $y_1 = -4$, and let $(x_2, y_2) = (0, 1)$, which means $x_2 = 0$ and $y_2 = 1$.

Write the slope formula:    $m = \dfrac{y_2 - y_1}{x_2 - x_1}$

Substitute $x_1 = -2$, $y_1 = -4$, $x_2 = 0$, and $y_2 = 1$:    $= \dfrac{1 - (-4)}{0 - (-2)}$

Simplify:    $= \dfrac{1 + 4}{0 + 2}$

$= \dfrac{5}{2}$

The slope of the line is $\dfrac{5}{2}$. This means that for every horizontal change, or **run**, of 2 units, there is a corresponding vertical change, or **rise**, of 5 units. The slope of this line is illustrated in Figure 14.

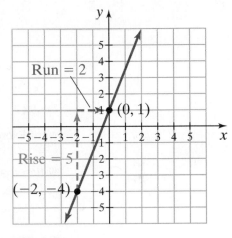

**Figure 14**

**You Try It**    **Work through this You Try It problem.**

**Work Exercises 1 and 2 in this textbook or in the *MyMathLab*® Study Plan.**

 It does not matter which point is called $(x_1, y_1)$ and which point is called $(x_2, y_2)$. Just be consistent. Whichever $y$-coordinate goes first in the **numerator** of the **slope formula**, the corresponding $x$-coordinate must go first in the **denominator**. This **alternate solution** shows that letting $(x_1, y_1) = (0, 1)$ and $(x_2, y_2) = (-2, -4)$ in **Example 1** gives the same result $m = \dfrac{5}{2}$.

## Example 2  Finding the Slope of a Line Given Two Points

Find the **slope** of the line containing the points $(-2, 4)$ and $(1, -3)$.

**Solution**  We use the slope formula. Let $(x_1, y_1) = (-2, 4)$ and $(x_2, y_2) = (1, -3)$.

Write the slope formula:    $m = \dfrac{y_2 - y_1}{x_2 - x_1}$

Substitute $x_1 = -2$, $y_1 = 4$, $x_2 = 1$, and $y_2 = -3$:    $= \dfrac{-3 - 4}{1 - (-2)}$

Simplify:    $= \dfrac{-3 + (-4)}{1 + 2}$

$= \dfrac{-7}{3} = -\dfrac{7}{3}$

The slope of the line is $-\dfrac{7}{3}$. For every horizontal change of 3 units, there is a corresponding vertical change of $-7$ units. This slope is illustrated in Figure 15.

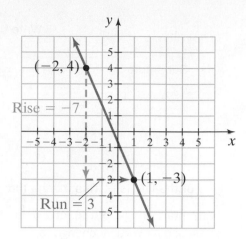

Figure 15

You Try It    Work through this You Try It problem.

Work Exercises 3 and 4 in this textbook or in the *MyMathLab*® Study Plan.

Notice that the line in **Figure 14** slants upward, or rises, from left to right and has a positive **slope**, whereas the line in **Figure 15** slants downward, or falls, from left to right and has a negative slope.

---

**Positive versus Negative Slope**

A line that slants upward, or rises, from left to right has a positive slope.
A line that slants downward, or falls, from left to right has a negative slope.

---

### Example 3  Finding the Slope of a Line Given Two Points

*My video summary*    Find the slope of the line containing the given points. Simplify if possible.

**a.** $(-6, -1)$ and $(4, 5)$          **b.** $(1, 5)$ and $(3, -1)$

**Solutions**  Use the **slope** formula.

**a.**  Let $(x_1, y_1) = (-6, -1)$ and $(x_2, y_2) = (4, 5)$.

$$\text{Write the slope formula:} \quad m = \frac{y_2 - y_1}{x_2 - x_1}$$

$$\text{Substitute } x_1 = -6, y_1 = -1, x_2 = 4, \text{ and } y_2 = 5: \quad = \frac{5 - (-1)}{4 - (-6)}$$

$$\text{Simplify:} \quad = \frac{5 + 1}{4 + 6} = \frac{6}{10} = \frac{3}{5}$$

See this **visual representation** of this slope.

**b.**  Try to find this slope on your own, then check your **answer**.
    Watch this **video** for detailed solutions to both parts of this example.

You Try It    Work through this You Try It problem.

Work Exercises 5–8 in this textbook or in the *MyMathLab*® Study Plan.

OBJECTIVE 2    FIND THE SLOPES OF HORIZONTAL AND VERTICAL LINES

Let's look at the **slopes** of lines that do not slant upward or downward from left to right.

### Example 4   Finding the Slope of a Line Given Two Points

*My video summary*

Find the slope of the line containing the given points. Simplify if possible.

**a.** $(-3, 2)$ and $(1, 2)$          **b.** $(4, 2)$ and $(4, -5)$

**Solutions** Use the **slope formula.**

**a.** Let $(x_1, y_1) = (-3, 2)$ and $(x_2, y_2) = (1, 2)$.

Write the slope formula: $\quad m = \dfrac{y_2 - y_1}{x_2 - x_1}$

Substitute $x_1 = -3, y_1 = 2, x_2 = 1$, and $y_2 = 2$: $\quad = \dfrac{2 - 2}{1 - (-3)}$

Simplify: $\quad = \dfrac{0}{4}$

$\quad = 0$

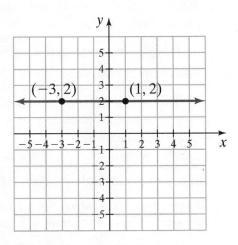

**Figure 16** Slope 0

The slope of the line is 0. Figure 16 shows that the line through these points is horizontal. It does not slant upward or downward.

**b.** Let $(x_1, y_1) = (4, 2)$ and $(x_2, y_2) = (4, -5)$.

Write the slope formula: $\quad m = \dfrac{y_2 - y_1}{x_2 - x_1}$

Substitute $x_1 = 4, y_1 = 2, x_2 = 4$, and $y_2 = -5$: $\quad = \dfrac{-5 - 2}{4 - 4}$

Simplify: $\quad = \dfrac{-7}{0}$

Division by 0 is undefined, so this line has an *undefined slope*. Figure 17 shows that the graph of the line through these points is vertical. It does not slant upward nor downward.

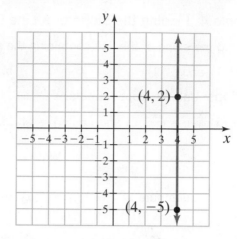

**Figure 17**
Undefined slope

Watch this **video** for detailed solutions to both parts of this example.

**You Try It**   **Work through this You Try It problem.**

**Work Exercises 9 and 10 in this textbook or in the *MyMathLab*® Study Plan.**

**Horizontal lines** have no vertical change. This makes the **numerator** in the slope formula 0, which makes the slope of any horizontal line 0. Similarly, **vertical lines** have no horizontal change. This makes the **denominator** of the slope formula 0, which makes the slope of any vertical line undefined.

---

**Slopes of Horizontal and Vertical Lines**

All horizontal lines (which have equations of the form $y = b$) have **slope 0**.
All vertical lines (which have equations of the form $x = a$) have **undefined slope**.

---

Avoid using the term "no slope." Technically, "no slope" means undefined slope (vertical line), but it can easily be confused with zero slope (horizontal line). Therefore, it is better to clearly state "zero slope" or "undefined slope" and avoid "no slope."

Figure 18 summarizes the relationship between the slope and the graph of a linear equation.

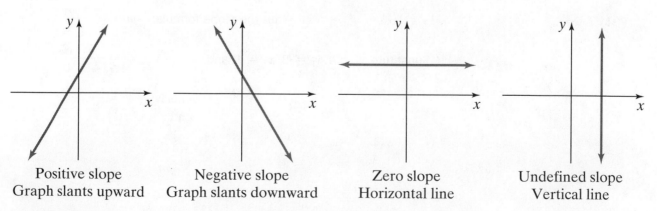

Positive slope
Graph slants upward

Negative slope
Graph slants downward

Zero slope
Horizontal line

Undefined slope
Vertical line

**Figure 18** Types of lines

OBJECTIVE 3    GRAPH A LINE USING THE SLOPE AND A POINT

In **Section 3.2**, we found that only two points are needed to sketch the graph of a line. Given one point on a line and the **slope** of the line, we can also sketch the graph. We **plot** the given point and then use the **rise** and the **run** to find a second point.

### Example 5  Graphing a Line Using the Slope and a Point

Graph the line that has slope $m = \dfrac{3}{2}$ and passes through the point $(1, -2)$.

**Solution**  Note that the slope is positive, so the line will slant upward from left to right. The slope is $m = \dfrac{3}{2}$, so rise = 3 and run = 2. Plot the given point $(1, -2)$. From this point, move up (rise) 3 units and move right (run) 2 units. This brings us to a second point on the line, $(3, 1)$. Draw the line between $(1, -2)$ and $(3, 1)$, as shown in Figure 19.

**Note:** If desired, additional points on the line can be plotted using the slope. Launch this **popup box** to see.

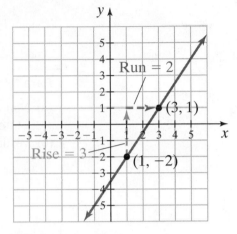

**Figure 19**

**You Try It**    Work through this You Try It problem.

Work Exercises 11 and 12 in this textbook or in the *MyMathLab*® Study Plan.

### Example 6  Graphing a Line Using the Slope and a Point

*My video summary*    Graph the line that has slope $m = -3$ and passes through the point $(2, -1)$.

**Solution**  The slope is negative, so the line will slant downward from left to right. The slope is $m = -3 = \dfrac{-3}{1} = \dfrac{3}{-1}$. We can use either rise = $-3$ and run = 1, or rise = 3 and run = $-1$. Try to graph this line on your own. View the **answer**, or watch this **video** for a detailed solution.

**You Try It**    Work through this You Try It problem.

Work Exercises 13–16 in this textbook or in the *MyMathLab*® Study Plan.

OBJECTIVE 4    FIND AND USE THE SLOPES OF PARALLEL AND PERPENDICULAR LINES

**Parallel lines** are two lines in the **coordinate plane** that never touch (have no points in common). The distance between two parallel lines remains the same across the plane. In Figure 20, lines $l_1$ and $l_2$ are parallel. Notice that both lines have the same **slope** $m_1 = m_2 = \dfrac{2}{3}$. If the slope of either line changed, then the two lines would no longer be parallel.

Notice also that $l_1$ and $l_2$ have different $y$-intercepts. If two lines have equal slopes and different $y$-intercepts, then the lines are parallel.

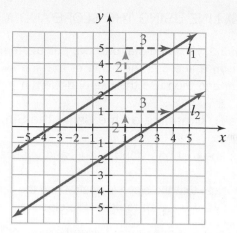

**Figure 20** Parallel lines

---

**Parallel Lines**

**(a)** Non-vertical lines are parallel **if and only if** they have equal slopes and different $y$-intercepts.

**(b)** Vertical lines are parallel if they have different $x$-intercepts.

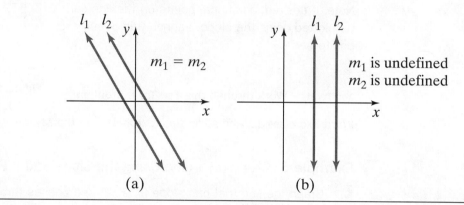

(a)       (b)

---

Two lines that **intersect** at a 90° (or right) angle are called **perpendicular lines**. In Figure 21, lines $l_1$ and $l_2$ are perpendicular. Notice that $l_1$ has **slope** $m_1 = 2$, whereas $l_2$ has slope $m_2 = -\dfrac{1}{2}$. If the slope of either line were changed, then the two lines would no longer be perpendicular. Notice also that the **product** of the slopes is $2\left(-\dfrac{1}{2}\right) = -1$. This is true for the slopes of perpendicular lines. We say that the slopes of perpendicular lines are **opposite reciprocals**, which means that the slopes have opposite signs and their **absolute values** are **reciprocals**.

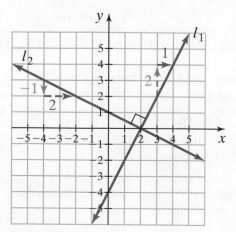

**Figure 21** Perpendicular lines

**Perpendicular Lines**

(a) Two non-vertical lines are perpendicular if and only if the product of their slopes is $-1$. Their slopes are opposite reciprocals.

(b) A vertical line and a horizontal line are perpendicular to each other.

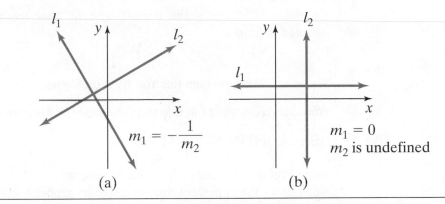

$$m_1 = -\frac{1}{m_2}$$

(a)

$$m_1 = 0$$
$m_2$ is undefined

(b)

## Example 7  Finding Slopes of Parallel and Perpendicular Lines

Line $l_1$ has a slope of $m_1 = -\dfrac{4}{5}$.

a. If line $l_2$ is **parallel** to line $l_1$, what is its slope?

b. If line $l_3$ is **perpendicular** to line $l_1$, what is its slope?

### Solutions

a. Parallel lines have equal slopes, so the slope of line $l_2$ is equal to the slope of line $l_1$.

   Line $l_2$ has a slope of $m_2 = -\dfrac{4}{5}$.

b. Perpendicular lines have slopes that are **opposite reciprocals**, so the slope of line $l_3$ is the opposite reciprocal of $m_1 = -\dfrac{4}{5}$. Changing the sign and finding the reciprocal,

   we get $\dfrac{5}{4}$. Note that $\left(-\dfrac{4}{5}\right)\left(\dfrac{5}{4}\right) = -1$. Line $l_3$ has a slope of $m_3 = \dfrac{5}{4}$.

**You Try It**    Work through this **You Try It** problem.

Work Exercises 17–20 in this textbook or in the ***MyMathLab***® Study Plan.

## Example 8  Graphing Parallel and Perpendicular Lines

*My video summary*     Figure 22 shows the graph of a line $l_1$.

a. Graph a line $l_2$ that is **parallel** to $l_1$ and passes through the point $(3, -2)$.

b. Graph a line $l_3$ that is **perpendicular** to $l_1$ and passes through the point $(3, -2)$.

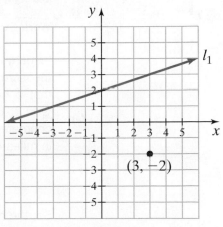

**Figure 22**

**Solutions** From looking at the graph, we can see that the slope of $l_1$ is $m_1 = \frac{1}{3}$. Because $l_2$ is parallel to $l_1$, its slope will be the same, $m_2 = \frac{1}{3}$. Because $l_3$ is perpendicular to $l_1$, its slope will be $m_3 = -3$, the **opposite reciprocal** of $m_1 = \frac{1}{3}$. Use these slopes with the point $(3,-2)$ to graph lines $l_2$ and $l_3$. View the **answers**, or watch this **video** for detailed solutions.

**You Try It**    Work through this **You Try It** problem.

**Work Exercises 21–24 in this textbook or in the *MyMathLab*® Study Plan.**

OBJECTIVE 5    USE SLOPE IN APPLICATIONS

**Slope** has a wide range of real-world applications. For example, in road construction, the slope of a road in percent form is called its **grade**. A road with a 6% grade means that the slope is $\frac{6}{100}$. For every 100 feet of horizontal change, the road rises 6 feet vertically. The slope of a roof is called its **pitch**. The pitch of the following roof is $\frac{4}{9}$.

$$\frac{6}{100} = 6\% \text{ grade}$$

100 ft    6 ft    9 ft    4 ft

**Example 9  Finding the Grade of a Wheelchair Ramp**

A standard wheelchair ramp should rise no more than 1 foot vertically for every 12 feet horizontally. Find the grade of this ramp. Round to the nearest tenth of a percent. (*Source:* Americans with Disabilities Act Accessibility Guidelines (ADAAG))

**Solution** The grade of the ramp is its slope expressed as a percent.

$$\text{grade} = m = \frac{\text{rise}}{\text{run}} = \frac{1}{12} \approx 0.083 \text{ or } 8.3\%$$

The grade of a wheelchair ramp should be about 8.3%.

**You Try It**    Work through this **You Try It** problem.

**Work Exercises 25 and 26 in this textbook or in the *MyMathLab*® Study Plan.**

In many applications, the **slope** of a line is called the **average rate of change** because we can measure the change in one variable with respect to another variable. For example, suppose that a car's gas tank is full with $x_1 = 16$ gallons of gasoline when the odometer reads $y_1 = 35,652$ miles. Later, when the tank is one-fourth full with $x_2 = 4$ gallons of gasoline, the odometer reads $y_2 = 35,976$ miles. We can use slope to measure the average rate of change in miles driven with respect to gallons of gasoline used:

$$m = \frac{y_2 - y_1}{x_2 - x_1} = \frac{\text{change in miles}}{\text{change in gasoline}} = \frac{35,976 - 35,652}{16 - 4} = \frac{324}{12} = 27$$

So, the car traveled 27 miles per 1 gallon of gasoline used, on average.

### Example 10  Using Slope as the Average Rate of Change

 *My video summary*

 The average tuition and fees for U.S. public two-year colleges were $2130 in 1999. The average tuition and fees were $2540 in 2009. Find and interpret the slope of the line connecting the points $(1999, 2130)$ and $(2009, 2540)$. (*Source*: College Board, *Trends in College Pricing 2009*)

**Solution**  Let $(x_1, y_1) = (1999, 2130)$ and $(x_2, y_2) = (2009, 2540)$. Use the **slope formula** to find $m = 41$. Between 1999 and 2009, the average tuition and fees for two-year colleges increased by $41 per year. Watch this **video** for a detailed solution.

**You Try It**  Work through this You Try It problem.

Work Exercises 27–30 in this textbook or in the **MyMathLab** Study Plan.

## 3.3 Exercises

In Exercises 1–10, find the slope of the line containing the given points. Simplify if possible.

**You Try It**

**1.** $(-1, 4)$ and $(3, 7)$

**2.** $(1, -2)$ and $(3, 2)$

**You Try It**

**3.** $(-3, 5)$ and $(2, -1)$

**4.** $(-1, 7)$ and $(5, 3)$

**5.** $(5, 0)$ and $(-1, -3)$

**6.** $(2, 13)$ and $(8, 5)$

**You Try It**

**7.**

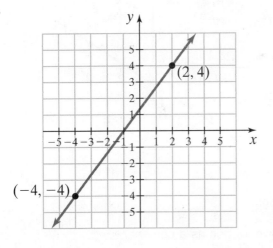

**8.**

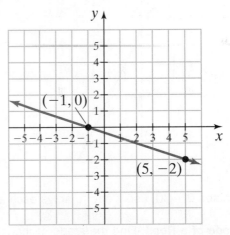

**You Try It**

**9.** $(4, -2)$ and $(-1, -2)$

**10.** $(5, -3)$ and $(5, 4)$

In Exercises 11–16, graph the line given the slope and a point on the line.

**You Try It**

**11.** $m = \dfrac{1}{3}; (-2, -2)$

**12.** $m = 2; (0, 4)$

**13.** $m = -\dfrac{5}{2}; (3, 1)$

**You Try It**

**14.** $m = -\dfrac{3}{4}; (0, 0)$

**15.** $m = 0; (4, -3)$

**16.** Undefined slope; $(-2, 5)$

In Exercises 17–20, for the given slope of line $l_1$, **(a)** find the slope of a line $l_2$ parallel to $l_1$, and **(b)** find the slope of a line $l_3$ perpendicular to $l_1$.

**You Try It**

**17.** $m = \dfrac{2}{3}$        **18.** $m = -\dfrac{7}{4}$        **19.** $m = 6$        **20.** $m = 0$

In Exercises 21 and 22, graph a line $l_2$ that is parallel to the given line $l_1$ and passes through the given point.

**You Try It**

**21.**

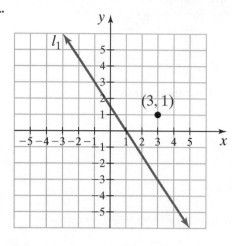

**22.**

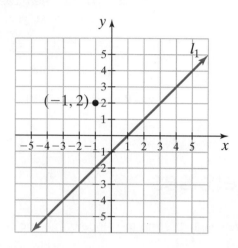

In Exercises 23 and 24, graph a line $l_3$ that is perpendicular to the given line $l_1$ and passes through the given point.

**23.**

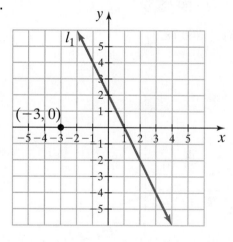

**24.**

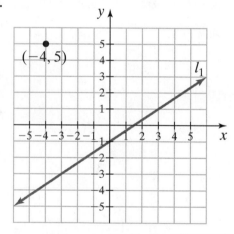

In Exercises 25–30, use slope to solve each application.

**You Try It**

**25. Grade of a Road** Find the grade of the road shown, written as a percent.

4 m

32 meters

26. **Pitch of a Roof** Find the pitch of the roof shown, written as a fraction in simplest form.

**You Try It**

27. **Tuition and Fees** The average tuition and fees for U.S. public four-year colleges and universities were $3930 in 1994. The average tuition and fees were $7020 in 2009. Find and interpret the slope of the line connecting the points $(1994, 3930)$ and $(2009, 7020)$. (*Source:* College Board, *Trends in College Pricing 2009*)

28. **Car Value** In 2005, Julie purchased a new SUV for $32,000. In 2010, she sold the SUV for $18,000. Find and interpret the slope of the line connecting the points $(2005, 32000)$ and $(2010, 18000)$.

29. **Autism Prevalence** In 2000, there were 93,650 cases of autism in the U.S. among people ages 3 to 22. In 2008, there were 337,795 cases of autism among this same U.S. age group. Find and interpret the slope of the line connecting the points $(2000, 93650)$ and $(2008, 337795)$. (*Source:* ThoughtfulHouse.org)

30. **Video MP4 Players** An 8-gigabyte video MP4 player costs $79.99 in a store. A 16-gigabyte video MP4 player of the same brand cost $99.99 in the same store. Find and interpret the slope of the line connecting the points $(8, 79.99)$ and $(16, 99.99)$.

# 3.4 **Equations of Lines**

## THINGS TO KNOW

Before working through this section, be sure you are familiar with the following concepts:

| | | VIDEO | ANIMATION | INTERACTIVE |

**You Try It**

1. Solve Linear Equations Using Both Properties of Equality (Section 2.1, **Objective 5**)

**You Try It**

2. Solve a Formula for a Given Variable (Section 2.4, **Objective 3**)

**You Try It**

3. Find the Slope of a Line Given Two Points (Section 3.3, **Objective 1**)

**You Try It**

4. Graph a Line Using the Slope and a Point (Section 3.3, **Objective 3**)

## OBJECTIVES

1 Determine the Slope and $y$-Intercept from a Linear Equation

2 Use the Slope-Intercept Form to Graph a Linear Equation

3 Write the Equation of a Line Given Its Slope and $y$-Intercept

4 Write the Equation of a Line Given Its Slope and a Point on the Line

5 Write the Equation of a Line Given Two Points

**6** Determine the Relationship Between Two Lines

**7** Write the Equation of a Line Parallel or Perpendicular to a Given Line

**8** Use Linear Equations to Solve Applications

OBJECTIVE 1   DETERMINE THE SLOPE AND $y$-INTERCEPT FROM A LINEAR EQUATION

In **Section 3.2**, we graphed the **linear equation** $x + 2y = 8$ by **plotting points** as shown in Figure 23a. In Section 3.3, we learned how to use the graph to determine the **slope** of the line $m = \dfrac{\text{rise}}{\text{run}} = -\dfrac{1}{2}$, as shown in Figure 23b. Notice that the graph crosses the $y$-axis at the point $(0, 4)$, so the $y$-**intercept** is 4.

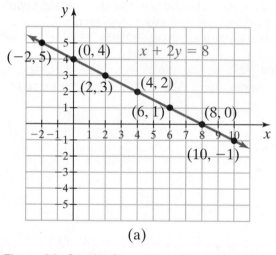
(a)

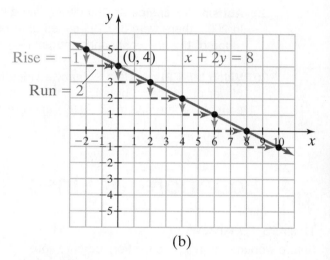
(b)

**Figure 23** Graph of $x + 2y = 8$

We can determine the **slope** and $y$-**intercept** directly from the equation by first solving the equation for $y$.

| | |
|---|---|
| Original equation: | $x + 2y = 8$ |
| Subtract $x$ from both sides: | $2y = -x + 8$ |
| Divide both sides by 2: | $\dfrac{2y}{2} = \dfrac{-x}{2} + \dfrac{8}{2}$ |
| Simplify: | $y = -\dfrac{1}{2}x + 4$ |

Notice that, in this form, the slope, $m = -\dfrac{1}{2}$, is the **coefficient** of $x$ and the $y$-intercept, $b = 4$, is the **constant term**. This result is true in general.

> **Slope-Intercept Form**
>
> A **linear equation** in two **variables** of the form
>
> $$y = mx + b$$
>
> is written in **slope-intercept form**, where $m$ is the slope of the line and $b$ is the $y$-intercept.

Be careful not to assume that the coefficient of $x$ is always the slope. This is true when the equation is in slope-intercept form but is not true in general.

### Example 1 Determining the Slope and $y$-Intercept from a Linear Equation

Find the **slope** and **$y$-intercept** of the given line.

**a.** $2x + y = 3$ **b.** $4x - 3y = 6$

### Solutions

**a.** Put the equation in slope-intercept form by solving for $y$.

Begin with the original equation: $2x + y = 3$

Subtract $2x$ from both sides: $y = -2x + 3 \leftarrow y = mx + b$ form

Comparing this result to the slope-intercept form, we see that the coefficient of $x$ is $-2$ and the constant term is 3. Therefore, the slope is $m = -2$, and the $y$-intercept is $b = 3$.

**b.** Begin with the original equation: $4x - 3y = 6$

Subtract $4x$ from both sides: $-3y = -4x + 6$

Divide both sides by $-3$: $\dfrac{-3y}{-3} = \dfrac{-4x}{-3} + \dfrac{6}{-3}$

Simplify: $y = \dfrac{4}{3}x - 2 \leftarrow y = mx + b$ form

Now determine the slope and $y$-intercept on your own, then view the **answers**.

**You Try It** Work through this You Try It problem.

Work Exercises 1–4 in this textbook or in the *MyMathLab*® Study Plan.

It is possible that either $m$ or $b$ (or both) are equal to 0, as shown in Example 2.

### Example 2 Determining the Slope and $y$-Intercept from a Linear Equation

Find the **slope** and **$y$-intercept** of the given line.

*My video summary* 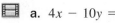 **a.** $4x - 10y = 0$ **b.** $y = 4$

**Solutions** Write each equation in the form $y = mx + b$ to determine the slope and $y$-intercept, then view the **answers**. For a detailed solution, watch this **video**.

**You Try It** Work through this You Try It problem.

Work Exercises 5–8 in this textbook or in the *MyMathLab*® Study Plan.

Not every linear equation in two variables can be written in **slope-intercept form**. Think about what type of line cannot have its equation written in this form. View this **example** of such a line.

OBJECTIVE 2 USE THE SLOPE-INTERCEPT FORM TO GRAPH A LINEAR EQUATION

*My animation summary* 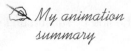 Recall that the $y$-intercept, $b$, corresponds to the point $(0, b)$ on the graph. Given the equation of a nonvertical line, we can use the **slope-intercept form** to determine the slope of the line and the point on the line corresponding to the y-intercept. Watch this **animation**, which illustrates the concept, and then work through Example 3.

### Example 3 Graphing a Linear Equation Using Slope-Intercept Form

Graph the equation $y = \frac{3}{5}x - 2$ using the slope and y-intercept.

**Solution** The equation is written in slope-intercept form. The slope is $m = \frac{3}{5}$, and the y-intercept is $b = -2$. We plot the point $(0, -2)$ and use the slope to obtain a second point on the graph. Then we can complete the graph by connecting the points with a straight line.

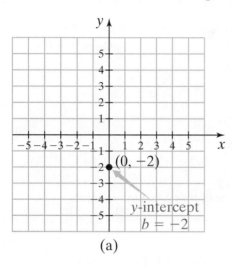

(a)

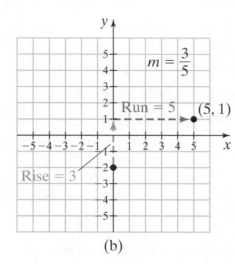

(b)

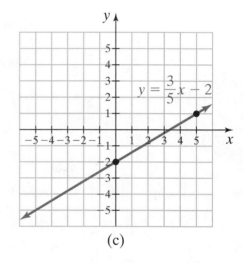

(c)

**You Try It** Work through this You Try It problem.

Work Exercises 9–11 in this textbook or in the **MyMathLab**® Study Plan.

### Example 4 Graphing a Linear Equation Using Slope-Intercept Form

✎ *My video summary* ▦ Graph the equation $2x + 3y = 9$ using the **slope** and **y-intercept**.

**Solution** We start by writing the equation in **slope-intercept form**, $y = -\frac{2}{3}x + 3$.

The slope is $m = -\frac{2}{3}$ and the y-intercept is $b = 3$, so the point $(0, 3)$ is on the graph of the equation. Use this point and the slope to sketch the graph of the equation. Compare your graph to the **answer**, or watch this **video** for a detailed solution.

**You Try It** Work through this You Try It problem.

Work Exercises 12–14 in this textbook or in the **MyMathLab**® Study Plan.

OBJECTIVE 3    WRITE THE EQUATION OF A LINE GIVEN ITS SLOPE AND y-INTERCEPT

Given the **slope** and **y-intercept** of a line, we can write the equation of the line by **substituting** the corresponding values into the **slope-intercept form**.

**Example 5** **Writing an Equation of a Line Given Its Slope and *y*-Intercept**

Write an equation of the line with the given slope and *y*-intercept.

**a.** slope $-4$; *y*-intercept 3

**b.** slope $\dfrac{2}{5}$; *y*-intercept $-7$

### Solutions

**a.** We have $m = -4$ and $b = 3$. Substitute these values into the slope-intercept form.

$$y = mx + b \qquad y = -4x + 3$$

The equation of the line is $y = -4x + 3$.

**b.** We have $m = \dfrac{2}{5}$ and $b = -7$. Substitute these values into the slope-intercept form.

$$y = mx + b \qquad y = \dfrac{2}{5}x + (-7)$$

The equation of the line is $y = \dfrac{2}{5}x - 7$.

**You Try It**    **Work through this You Try It problem.**

**Work Exercises 15–20 in this textbook or in the *MyMathLab*® Study Plan.**

OBJECTIVE 4    **WRITE THE EQUATION OF A LINE GIVEN ITS SLOPE AND A POINT ON THE LINE**

Given the **slope** of a nonvertical line and a point on the line (other than the *y*-**intercept**), we can still use the **slope-intercept form** to write the equation of the line. The given point $(x, y)$ is an **ordered pair solution** to the equation, so must satisfy the equation. We use these **coordinates** and the given slope to solve for the *y*-intercept, *b*.

**Example 6** **Writing the Equation of a Line Given Its Slope and a Point on the Line**

Write the equation of the line that has slope 2 and passes through the point $(4,7)$.

**Solution** We are given $m = 2$ and $(x, y) = (4, 7)$. Substitute these values into the slope-intercept form and solve for *b*.

$$\begin{aligned}
\text{Slope-intercept form:} \quad & y = mx + b \\
\text{Substitute 2 for } m, 4 \text{ for } x, \text{ and } 7 \text{ for } y: \quad & 7 = (2)(4) + b \\
\text{Simplify:} \quad & 7 = 8 + b \\
\text{Subtract 8 from both sides:} \quad & -1 = b
\end{aligned}$$

Now use the slope and *y*-intercept to write the equation $y = 2x - 1$.

We can use an alternate form to write the equation of a line given the **slope** and a point on the line. This form is called the **point-slope form**.

> **Point-Slope Form**
>
> Given the slope $m$ of a line and a point $(x_1, y_1)$ on the line, the **point-slope form** of the equation of the line is given by
>
> $$y - y_1 = m(x - x_1).$$

### Example 7 Writing the Equation of a Line Given Its Slope and a Point on the Line

 Use the point-slope form to determine the equation of the line that has slope $-\dfrac{3}{4}$ and passes through the point $(2, -5)$. Write the equation in **slope-intercept form**.

*My interactive video summary*

**Solution** We are given $m = -\dfrac{3}{4}$ and $(x_1, y_1) = (2, -5)$. **Substitute** these values into the **point-slope form.**

$$\text{Point-slope form:} \qquad y - y_1 = m(x - x_1)$$

$$\text{Substitute } -\frac{3}{4} \text{ for } m,\ 2 \text{ for } x_1, \text{ and } -5 \text{ for } y_1: \quad y - (-5) = \left(-\frac{3}{4}\right)(x - (2))$$

$$\text{Simplify:} \qquad y + 5 = -\frac{3}{4}(x - 2)$$

Finish solving the equation for $y$ and write it in **slope-intercept form**. Check your **answer**, or watch this **interactive video** to see a detailed solution using either point-slope form or slope-intercept form.

**You Try It**    Work through this You Try It problem.

Work Exercises 21–26 in this textbook or in the **MyMathLab**® Study Plan.

Whether we use the slope-intercept form or the point-slope form, the work is basically the same to determine the equation of the line. To see **Example 6** worked out using the **point-slope form**, view this popup.

### Example 8 Writing the Equation of a Line Given Its Slope and a Point on the Line

Write the equation of a line that passes through the point $(-3, 2)$ and has the given **slope**.

**a.** $m = 0$                  **b.** undefined slope

### Solutions

**a.** **Horizontal lines** have a slope of $0$, and the equation of a horizontal line is given by $y = b$ where $b$ is the $y$-**intercept**. Recall that every point on the graph of a horizontal line has the same $y$-**coordinate**, $b$. The line passes through the point $(-3, 2)$, which has a $y$-coordinate of $2$. Therefore, the equation of the line passing through $(-3, 2)$ with slope $0$ is $y = 2$.

**b.** Similarly, **vertical lines** have undefined slope, and the equation of a vertical line is given by $x = a$ where $a$ is the $x$-**intercept**. Recall that every point on the graph of a vertical line has the same $x$-**coordinate**, $a$. The line passes through the point $(-3, 2)$, which has an $x$-coordinate of $-3$. Therefore, the equation of the line passing through $(-3, 2)$ with undefined slope is $x = -3$.

**You Try It**   **Work through this You Try It problem.**

**Work Exercises 27 and 28 in this textbook or in the *MyMathLab*® Study Plan.**

OBJECTIVE 5    WRITE THE EQUATION OF A LINE GIVEN TWO POINTS

Recall that two points determine a line. Given two points on a line, we can find the equation of the line by first finding the slope of the line and then using the slope and either given point to determine the equation.

**Example 9  Writing the Equation of a Line Given Two Points**

 Write the equation of the line passing through the points $(-4, 1)$ and $(2, 4)$. Write your answer in **slope-intercept form**.

**Solution** We are given two points, but we don't know the slope. We can compute the slope by using the **slope formula**:

$$m = \frac{4 - 1}{2 - (-4)} = \frac{4 - 1}{2 + 4} = \frac{3}{6} = \frac{1}{2}.$$

Now that we know the slope, we can use it together with *either* of the two given points to write the equation of the line. Remember that you can use either the slope-intercept form or the **point-slope form** to determine the equation. View the **answer** or watch this **video** to see a detailed solution.

**You Try It**   **Work through this You Try It problem.**

**Work Exercises 29–32 in this textbook or in the *MyMathLab*® Study Plan.**

We summarize the slope formula and the different forms for equations of lines in Table 1.

| | |
|---|---|
| $m = \dfrac{y_2 - y_1}{x_2 - x_1}$ | **Slope** <br> Average rate of change |
| $y - y_1 = m(x - x_1)$ | **Point-Slope Form** <br> Slope is $m$, and $(x_1, y_1)$ is a point on the line. |
| $y = mx + b$ | **Slope-Intercept Form** <br> Slope is $m$, and $y$-intercept is $b$. |
| $Ax + By = C$ | **Standard Form** <br> $A$, $B$, and $C$ are real numbers, with <br> $A$ and $B$ not both zero. |
| $y = b$ | **Horizontal Line** <br> Slope is zero, and $y$-intercept is $b$. |
| $x = a$ | **Vertical Line** <br> Slope is undefined, and $x$-intercept is $a$. |

**Table 1**  Slope and equations of lines

## OBJECTIVE 6   DETERMINE THE RELATIONSHIP BETWEEN TWO LINES

**Slopes** and **y-intercepts** can be used to determine the relationship between two lines. This will be helpful when we solve systems of linear equations in Chapter 4.

Recall that **parallel lines** have the same slope but different y-intercepts, while perpendicular lines have **opposite-reciprocal** slopes (product of the slopes is $-1$). **Coinciding lines** have the same slope *and* the same y-intercept. Coinciding lines appear to be the same line because their graphs lie on top of each other. Two lines with different slopes will intersect. Two intersecting lines that are not perpendicular are called **only intersecting** lines.

In Table 2 we summarize how the slope and y-intercept allow us to determine the relationship between two lines.

|  | Parallel Lines | Coinciding Lines | Intersecting and Perpendicular Lines | Only Intersecting Lines (not Perpendicular) |
|---|---|---|---|---|
| **Slopes are** | Same | Same | opposite reciprocals $\left(m_1 \cdot m_2 = -1 \text{ or } m_1 = -\dfrac{1}{m_2}\right)$ | Different (but $m_1 \cdot m_2 \neq -1$) |
| **y-Intercepts are** | Different | Same | Same or Different | Same or Different |

**Table 2**   Relationship between two lines

### Example 10  Determining the Relationship Between Two Lines

*My interactive video summary*

For each pair of lines, determine if the lines are **parallel**, **perpendicular**, **coinciding**, or **only intersecting**.

**a.** $3y = -2x + 7$
$3x - 2y = 8$

**b.** $y = -3x + 1$
$6x + 2y = 2$

**c.** $4x - 5y = 15$
$y = \dfrac{4}{5}x + 1$

**d.** $3x - 4y = 2$
$x + 2y = -12$

### Solutions

**a.** Begin by writing each equation in **slope-intercept** form.

Original equation:   $3y = -2x + 7$

Divide both sides by 3:   $y = -\dfrac{2}{3}x + \dfrac{7}{3}$   $\leftarrow y = mx + b$

Original equation:   $3x - 2y = 8$

Subtract $3x$ from both sides:   $-2y = -3x + 8$

Divide both sides by $-2$:   $y = \dfrac{3}{2}x - 4$   $\leftarrow y = mx + b$

In slope-intercept form, we see that the two lines have **opposite-reciprocal** slopes, $\left(-\dfrac{2}{3}\right)\left(\dfrac{3}{2}\right) = -1$, so the lines are perpendicular.

**b–d.** View the **answers** or watch this **interactive video** for a more detailed solution to all parts.

**You Try It**   Work through this You Try It problem.

Work Exercises 33–38 in this textbook or in the *MyMathLab*® Study Plan.

OBJECTIVE 7    WRITE THE EQUATION OF A LINE PARALLEL OR PERPENDICULAR TO A GIVEN LINE

When writing the equation of a **parallel line** or **perpendicular line**, we must remember that parallel lines have the same slope and perpendicular lines have **opposite-reciprocal** slopes.

### Example 11   Writing the Equation of a Perpendicular Line or a Parallel Line

Write the equation of the line that passes through the point $(6, -5)$ and is

**a.** perpendicular to $6x - 2y = -1$.      **b.** parallel to $y = -2x + 4$.

### Solutions

**a.** We are given a point on the graph, but we do not know the slope. However, we know that the line is perpendicular to $6x - 2y = -1$, which gives us information about the slope. Perpendicular lines have opposite-reciprocal slopes, so we first need to determine the slope of the given line.

$$\text{Original equation:} \quad 6x - 2y = -1$$

$$\text{Subtract } 6x \text{ from each side:} \quad -2y = -6x - 1$$

$$\text{Divide both sides by } -2\text{:} \quad y = 3x + \frac{1}{2} \quad \leftarrow \frac{-6}{-2} = 3; \frac{-1}{-2} = \frac{1}{2}$$

The slope of the given line is 3, or $\frac{3}{1}$. The opposite reciprocal is $-\frac{1}{3}$, so the slope of our line is $-\frac{1}{3}$. Now that we know the slope and a point on the graph, we can use the **slope-intercept form** or **point-slope form** to determine the equation.

$$\text{Slope-intercept form:} \quad y = mx + b$$

$$\text{Substitute } -\frac{1}{3} \text{ for } m, 6 \text{ for } x, -5 \text{ for } y\text{:} \quad -5 = \left(-\frac{1}{3}\right)(6) + b$$

$$\text{Simplify:} \quad -5 = -2 + b$$

$$\text{Add 2 to both sides:} \quad -3 = b$$

The equation of the line is $y = -\frac{1}{3}x - 3$.

(View the **alternate steps** using point-slope form.)

 *My video summary*      **b.** The equation is already written in slope-intercept form. Determine the slope of the given line. Then use this slope and the given point to determine the equation of the parallel line. See the answer in this **popup**, or see the complete solution in this **video**.

**You Try It**    **Work through this You Try It problem.**

**Work Exercises 39–42 in this textbook or in the *MyMathLab*®️ Study Plan.**

Table 3 summarizes how to write the equation of a line from given information.

| If you are given ... | Then do this ... |
|---|---|
| **Slope** and *y*-intercept | Plug values for *m* and *b* directly into the **slope-intercept form** to write the equation. |
| Slope and a point | Use the slope-intercept form to find *b* and then write the equation in slope-intercept form. |
| Two points | Find the slope. Then use this slope and either point in the slope-intercept form to find *b*. |
| Point and slope = 0 (horizontal line) | Use the *y*-coordinate of the point to write *y = b*. |
| Point and undefined slope (vertical line) | Use the *x*-coordinate of the point to write *x = a*. |
| Point and **parallel** line in equation form | Find the slope of the given line. Use this slope and the point in the slope-intercept form to find *b*. |
| Point and **perpendicular** line in equation form | Find the slope of the given line. Find its **opposite reciprocal** to find the slope of the perpendicular line. Using this as the slope and the given point, find *b* by using the slope-intercept form. |

**Table 3** Writing equations of lines from given information

 Table 3 emphasizes the use of the slope-intercept form because it is generally more useful (view this **popup** for an explanation). Remember that given the slope and a point, we could also use the **point-slope form** if desired.

OBJECTIVE 8    USE LINEAR EQUATIONS TO SOLVE APPLICATIONS

 *My video summary*   Many real-life applications can be **modeled** using **linear equations**. Work through Example 12, or watch the **video** solution.

### Example 12  Football Attendance

If attendance at professional football games is 17 million in a given year, then the corresponding attendance at college football games is 31 million. Increasing attendance at professional football games to 25 million increases attendance at college football games to 55 million. (*Source: Statistical Abstract, 2010*)

a. Assume that the relationship between professional football attendance (in millions) and college football attendance (in millions) is linear. Find the equation of the line that describes this relationship. Write your answer in **slope-intercept form**.

b. Use your equation from part (a) to estimate the attendance at college football games if the attendance at professional football games is 21 million.

### Solutions

a. We start by writing two **ordered pairs** of the form $(x, y) = $ (**pro atd, college atd**), where each coordinate represents millions of people. From the problem statement, the first ordered pair is $(17, 31)$ and the second is $(25, 55)$.

Next we use the two ordered pairs to determine the slope.

$$m = \frac{55 - 31}{25 - 17} = \frac{24}{8} = 3$$

We then use the slope-intercept form to determine the $y$-intercept, $b$. Either of the two ordered pairs can be used. We will use $(17, 31)$.

$$\text{Slope-intercept form:} \quad y = mx + b$$
$$\text{Substitute 3 for } m, \text{ 17 for } x, \text{ and 31 for } y: \quad 31 = (3)(17) + b$$
$$\text{Simplify:} \quad 31 = 51 + b$$
$$\text{Subtract 51 from both sides:} \quad -20 = b$$

Using the slope $m = 3$ and $y$-intercept $b = -20$, the equation is $y = 3x - 20$.

**b.** To estimate the attendance at college football games if the attendance at professional games is 21 million, we substitute 21 for $x$ and solve for $y$.

$$\text{Slope-intercept form:} \quad y = 3x - 20$$
$$\text{Substitute 21 for } x: \quad y = 3(21) - 20$$
$$\text{Simplify:} \quad y = 63 - 20$$
$$y = 43$$

The estimated attendance at college football games is 43 million.

**You Try It**    Work through this You Try It problem.

Work Exercises 43–45 in this textbook or in the *MyMathLab*® Study Plan.

# 3.4  Exercises

In Exercises 1–8, determine the slope and $y$-intercept for each equation.

**You Try It**

**1.** $3x + y = 9$        **2.** $-4x + y = 7$        **3.** $6x + 5y = -4$        **4.** $2x - 5y = 3$

**5.** $5x + 4y = 0$        **6.** $3x - 8y = 0$        **7.** $y = 10$        **8.** $x = -2$

**You Try It**  In Exercises 9–14, graph the equation using the slope and $y$-intercept.

**9.** $y = 3x + 2$        **10.** $y = \dfrac{4}{3}x - 1$        **11.** $y = -\dfrac{1}{2}x$

**You Try It**

**12.** $2x + 3y = 4$        **13.** $5x - 3y = -12$        **14.** $1.2x + 1.5y = 4.5$

**You Try It**  In Exercises 15–20, write the equation of the line with the given slope and $y$-intercept.

**15.** $m = 4, b = 8$        **16.** $m = -2, b = -7.2$        **17.** slope $= \dfrac{1}{3}$, $y$-intercept $= 4$

**You Try It**

**18.** slope $= \dfrac{3}{4}$, $y$-intercept $= \dfrac{1}{2}$        **19.** slope $= 0$, $y$-intercept $= \dfrac{5}{4}$        **20.** slope $= -3.5$, $y$-intercept $= 0$

In Exercises 21–26, write the equation of the line that has the given slope and passes through the given point. Write the equation in slope-intercept form.

**21.** $m = -3; (2, 5)$        **22.** $m = \dfrac{3}{5}; (10, -3)$        **23.** slope $= -\dfrac{2}{3}; (-3, 11)$

**You Try It**

**24.** slope $= 3.5; (-5, 2.4)$

**25.** slope $= \dfrac{2}{3}; (6, 4)$

**26.** $m = -\dfrac{1}{4}; \left(\dfrac{8}{3}, -\dfrac{5}{2}\right)$

**27.** $m = 0; (4, 1)$

**28.** undefined slope; $(-5, 9)$

You Try It

In Exercises 29–32, write the equation of the line passing through the given points. Write your answer in slope-intercept form.

**29.** Passing through $(2, 6)$ and $(4, 9)$

**30.** Passing through $(-1, 2)$ and $(2, -7)$

You Try It

**31.** Passing through $(0, 0)$ and $\left(-3, \dfrac{2}{3}\right)$

**32.** Passing through $(3, 5)$ and $(-6, 5)$

In Exercises 33–38, determine if the two lines are parallel, perpendicular, coinciding, or only intersecting.

**33.** $y = -2x + 9$
$\quad\; 8x + 4y = 1$

**34.** $5x - 3y = 4$
$\quad\; 3x - 5y = -7$

**35.** $-8x + 2y = 13$
$\quad\;\; x + 4y = -3$

You Try It

**36.** $3x - 4y = 28$

$\quad\; y = \dfrac{3}{4}x - 7$

**37.** $y = -6$
$\quad -2y = -6$

**38.** $x = 4$
$\quad y = 4$

**39.** Write the equation of the line perpendicular to $y = \dfrac{1}{6}x + 4$ that passes through the point $(-3, 7)$.

You Try It

**40.** Write the equation of the line perpendicular to $7x - 5y = -2$ that passes through the point $(-7, 4)$.

**41.** Write the equation of the line parallel to $y = \dfrac{1}{6}x + 4$ that passes through the point $(-8, 1)$.

**42.** Write the equation of the line parallel to $7x - 5y = -2$ that passes through the point $(10, -2)$.

**43. Mortgage Rates** If the average interest rate for a 15-year fixed rate mortgage is 5.00%, then the corresponding 30-year fixed rate is 5.85%. If the 15-year rate decreases to 4.00%, then the 30-year rate decreases to 4.55%. (*Source: mortgagenewsdaily.com*).

You Try It

    **a.** Assume that the relationship between the 15-year fixed rate (%) and the 30-year fixed rate (%) is linear. Find the equation of the line that describes this relationship. Write your answer in slope-intercept form.

    **b.** Use your equation from part (a) to estimate the 30-year fixed rate if the 15-year fixed rate is 4.8%.

**44. Hard Drive Size** Jasmine is building a new computer. She notices that for $55, she can get 250 GB of hard drive space, but for $75, she can get 500 GB of space.

    **a.** Assume that the relationship between price ($) and hard drive space (GB) is linear. Find the equation of the line that describes this relationship. Write your answer in slope-intercept form.

    **b.** Use your equation from part (a) to estimate the hard drive space that Jasmine can get for $85.

**45. e-Reader Sales** An e-reader manufacturer expects to sell 50,000 units each year if the selling price is $250. If the selling price is reduced to $180, then the manufacturer expects to sell 140,000 units each year.

    **a.** Assume that the relationship between selling price ($) and units sold (1000s) is linear. Find the equation of the line that describes this relationship. Write your answer in slope-intercept form.

# 3.5 Linear Inequalities in Two Variables

## THINGS TO KNOW

Before working through this section, be sure you are familiar with the following concepts:

|  | | VIDEO | ANIMATION | INTERACTIVE |

**You Try It**  1. Graph the Solution Set of an Inequality on a Number Line (Section 2.7, **Objective 2**)

**You Try It**  2. Solve Linear Inequalities in One Variable (Section 2.7, **Objective 4**)   🎞

**You Try It**  3. Use Linear Inequalities to Solve Application Problems (Section 2.7, **Objective 6**)   🎞

**You Try It**  4. Identify Points in the Rectangular Coordinate System (Section 3.1, **Objective 2**)   🎞   ↗

**You Try It**  5. Graph Horizontal and Vertical Lines (Section 3.2, **Objective 7**)   🎞

**You Try It**  6. Use the Slope-Intercept Form to Graph a Linear Equation (Section 3.4, **Objective 2**)   🎞   ↗

**You Try It**  7. Use Linear Equations to Solve Applications (Section 3.4, **Objective 8**)   🎞

## OBJECTIVES

**1** Determine If an Ordered Pair Is a Solution to a Linear Inequality in Two Variables

**2** Graph a Linear Inequality in Two Variables

**3** Solve Applications Involving Linear Inequalities in Two Variables

---

OBJECTIVE 1   DETERMINE IF AN ORDERED PAIR IS A SOLUTION TO A LINEAR INEQUALITY IN TWO VARIABLES

In **Section 2.7**, we solved **linear inequalities in one variable**. The **solution set** for such an inequality is the set of all values that make the inequality true. Typically, we graph the solution set of a linear inequality in one variable on a **number line**.

In this section, we solve **linear inequalities in two variables**. A linear inequality in two variables looks like a **linear equation in two variables** except that an **inequality symbol** replaces the equal sign.

> **Definition   Linear Inequality in Two Variables**
>
> A **linear inequality in two variables** is an inequality that can be written in the form $Ax + By < C$, where $A$, $B$, and $C$ are real numbers, and $A$ and $B$ are not both equal to zero.
> **Note:** The inequality symbol "$<$" can be replaced with $>$, $\leq$, or $\geq$.

An **ordered pair** is a **solution to a linear inequality in two variables** if, when substituted for the variables, it makes the inequality true.

### Example 1  Determining If an Ordered Pair Is a Solution to a Linear Inequality in Two Variables

 *My video summary*  ▤  Determine if the given **ordered pair** is a **solution to the inequality** $2x - 3y < 6$.

a. $(-1, -2)$          b. $(4, -1)$          c. $(6, 2)$

**Solutions**  We substitute the $x$- and $y$-coordinates for the **variables** and simplify. If the resulting statement is true, then the ordered pair is a solution to the inequality.

a.  Begin with the original inequality:  $\qquad 2x - 3y < 6$

Substitute $-1$ for $x$ and $-2$ for $y$:  $2(-1) - 3(-2) \overset{?}{<} 6$

Simplify:  $\qquad\qquad -2 + 6 \overset{?}{<} 6$

$\qquad\qquad\qquad\qquad\qquad 4 < 6 \quad$ True

The final statement is true, so $(-1, -2)$ is a solution to the inequality.

**b–c.**  Try to determine if these ordered pairs are solutions on your own. View the **answers**, or watch this **video** for detailed solutions to all three parts.

▲

**You Try It**    Work through this You Try It problem.

**Work Exercises 1–4 in this textbook or in the *MyMathLab*®  Study Plan.**

### OBJECTIVE 2  GRAPH A LINEAR INEQUALITY IN TWO VARIABLES

Let's look for all solutions to a **linear inequality in two variables**. First, we focus on the **linear equation** related to the inequality. To find solutions to $x + y > 2$ or $x + y < 2$, we would look at the equation $x + y = 2$. For an **ordered pair** to **satisfy** this equation, the sum of its **coordinates** must be 2. So, $(-2, 4), (1, 1), (2, 0)$, and $(4, -2)$ are all examples of solutions to the equation. (View the **details**.) Plotting and connecting these points gives the line shown in Figure 24.

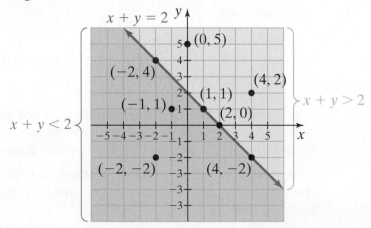

**Figure 24**

Now let's look to see which **ordered pairs** satisfy the inequalities $x + y > 2$ or $x + y < 2$. Notice that the line in **Figure 24** divides the **coordinate plane** into two **half-planes**, an *upper half-plane* shaded blue and a *lower half-plane* shaded pink.

Choose any point that lies in the upper half-plane, such as $(4, 2)$. The sum of its **coordinates** is $4 + 2 = 6$, which is larger than 2. Choose any other point in this region, such as $(0, 5)$. The sum of its coordinates is $0 + 5 = 5$, which is also larger than 2.

If we choose any point in the upper half-plane, the sum of its coordinates will be larger than 2. This means that any ordered pair from the upper half-plane is a **solution to the inequality** $x + y > 2$. The area shaded blue in **Figure 24** represents the set of all **ordered pair solutions** to the inequality $x + y > 2$.

Repeat this process for the lower half-plane to see that the area shaded pink in **Figure 24** represents the set of all ordered pair solutions to the inequality $x + y < 2$. (Launch this **popup box** to see the process.)

The linear equation $x + y = 2$ acts as a **boundary line** that separates the solutions of the two linear inequalities $x + y > 2$ and $x + y < 2$.

Based on this information, we can define **steps for graphing linear inequalities in two variables**.

---

**Steps for Graphing Linear Inequalities in Two Variables**

**Step 1.** Find the **boundary line** for the inequality by replacing the **inequality symbol** with an equal sign and graphing the resulting equation. If the inequality is **strict**, graph the boundary using a dashed line. If the inequality is **non-strict**, graph the boundary using a solid line.

**Step 2.** Choose a **test point** that does not belong to the boundary line and determine if it is a solution to the inequality.

**Step 3.** If the test point is a solution to the inequality, then shade the **half-plane** that contains the test point. If the test point is not a solution to the inequality, then shade the half-plane that does not contain the test point. The shaded area represents the set of all **ordered pair solutions** to the inequality.

---

**Note:** When graphing a **linear inequality in two variables** on the **coordinate plane**, using a dashed line is similar to using an open circle when graphing a **linear inequality in one variable** on a **number line**. Using a solid line is similar to using a solid circle. Do you see **why?**

### Example 2  Graphing a Linear Inequality in Two Variables

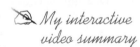 

 Graph each inequality.

**a.** $3x - 4y \leq 8$    **b.** $y > 3x$    **c.** $y < -2$

**Solutions**

**a.** We follow the **three-step process**.

**Step 1.** The **boundary line** is $3x - 4y = 8$. We use a solid line as the graph because the inequality is **non-strict**. See Figure 25(a).

**Step 2.** We choose the **test point** $(0, 0)$ and check to see if it **satisfies** the inequality.

Begin with the original inequality:    $3x - 4y \leq 8$

Substitute $x = 0$ and $y = 0$:    $3(0) - 4(0) \overset{?}{\leq} 8$

Simplify:    $0 \leq 8$  True

The point $(0, 0)$ is a solution to the inequality.

**Step 3.** Because the test point is a solution to the inequality, we shade the **half-plane** that contains $(0, 0)$. See Figure 25(b).
The shaded region, including the boundary line, represents all ordered pair solutions to the inequality $3x - 4y \leq 8$.

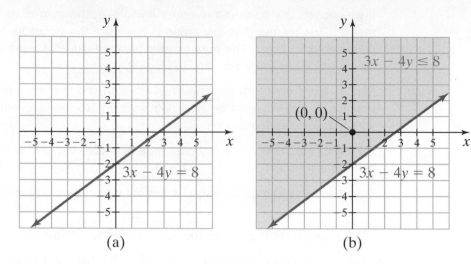

(a)                              (b)                        **Figure 25**

**b.** For $y > 3x$, the **boundary line** is $y = 3x$. Because the inequality is **strict**, we graph the boundary line using a dashed line. Choose a **test point** and complete the graph. See the **answer**, or watch this **interactive video** for a detailed solution.

**c.** Try to graph this inequality $y < -2$ on your own. Note that the boundary line is the **horizontal line** $y = -2$. See the **answer**, or watch this **interactive video** for a detailed solution.

 A test point cannot belong to the boundary line. Do you see **why**?

**You Try It**    **Work through this You Try It problem.**

**Work Exercises 5–12 in this textbook or in the *MyMathLab*® Study Plan.**

OBJECTIVE 3    SOLVE APPLICATIONS INVOLVING LINEAR INEQUALITIES IN TWO VARIABLES

Sometimes real-life applications can be **modeled by linear inequalities in two variables**.

### Example 3  A Piggy Bank

*My video summary*    A piggy bank contains only nickels and dimes with a total value of less than $9. Let $n$ = the number of nickels and $d$ = the number of dimes.

**a.** Write an inequality describing the possible numbers of coins in the bank.

**b.** Graph the inequality. Because $n$ and $d$ must be **whole numbers**, restrict the graph to **Quadrant I**.

**c.** Could the piggy bank contain 90 nickels and 60 dimes?

### Solutions

**a.** The value of $n$ nickels, in cents, is $5n$, and the value of $d$ dimes is $10d$, so the total value of the coins in the bank is $5n + 10d$. The total value is less than $9, or 900¢, so the inequality is

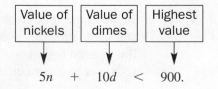

$$5n \ +\ 10d \ < \ 900.$$

**b.** Our two variables are $n$ and $d$, so label the horizontal axis $n$ and the vertical axis $d$. Follow the **steps for graphing linear inequalities in two variables** to obtain the graph in **Figure 26**. Watch this **video** for a detailed solution.

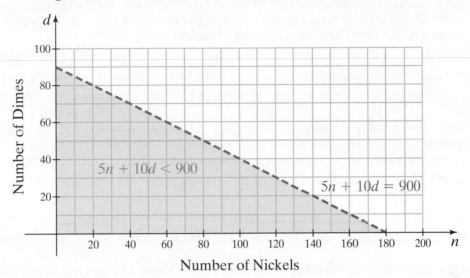

Figure 26

The number of nickels and dimes must be **whole numbers**, so only those points in the shaded region that have whole number coordinates are acceptable solutions to this application.

**c.** Substitute $(n, d) = (90, 60)$ into the inequality to see if a true statement results.

$$\text{Write the inequality:} \qquad 5n + 10d < 900$$
$$\text{Substitute } n = 90 \text{ and } d = 60: \quad 5(90) + 10(60) \overset{?}{<} 900$$
$$\text{Simplify:} \qquad\qquad 1050 < 900 \quad \text{False}$$

The piggy bank cannot contain 90 nickels and 60 dimes.

**You Try It**  Work through this You Try It problem.

Work Exercises 13–16 in this textbook or in the *MyMathLab*® Study Plan.

# 3.5 Exercises

In Exercises 1–4, determine if each ordered pair is a solution to the given inequality.

**You Try It**

**1.** $2x - y > 5$
   **a.** $(6, 8)$
   **b.** $(3, -4)$
   **c.** $(3, 4)$

**2.** $4x + 3y \le 24$
   **a.** $(8, -2)$
   **b.** $(-5, 10)$
   **c.** $(3, 4)$

**3.** $y < \dfrac{4}{3}x - 2$
   **a.** $(-6, -5)$
   **b.** $(3, 0)$
   **c.** $\left(\dfrac{3}{2}, \dfrac{1}{2}\right)$

**4.** $3x - 5y \ge 45$
   **a.** $(7, -5)$
   **b.** $(-2, -10)$
   **c.** $(7.5, -4.5)$

In Exercises 5–12, graph each inequality.

**You Try It**

**5.** $3x + y \le 6$

**6.** $5x - 3y > 15$

**7.** $-x + y \ge 0$

**8.** $x + 2y < 0$

**9.** $y < -\dfrac{3}{2}x + 4$

**10.** $y \ge -2x$

**11.** $y > -1$

**12.** $x \le 3$

In Exercises 13–16, solve each application by using an inequality.

**You Try It**

13. **Growing Crops** In Illinois the annual cost to grow corn is about $450 per acre, and the annual cost to grow soybeans is about $300 per acre (not counting the cost of land). Suppose a small farmer has at most $90,000 to spend on growing corn and soybeans this year. Let $c$ = acres of corn and $s$ = acres of soybeans. (*Source: Farmdoc*, University of Illinois)

  a. Write an inequality that describes the possible acres of corn and soybeans this farmer can afford to grow.

  b. Graph the inequality. Because $c$ and $s$ must be non-negative, restrict the graph to Quadrant I.

  c. Can this farmer afford to grow 125 acres of corn and 125 acres of soybeans?

14. **Transporting Crops** An agriculture company needs to ship its crops. Corn weighs about 50 pounds per bushel, while soybeans weigh about 60 pounds per bushel. A river barge can safely carry up to 3,000,000 pounds of cargo. Let $c$ = bushels of corn and $s$ = bushels of soybeans. (*Source:* powderandbulk.com)

  a. Write an inequality that describes the bushels of corn and soybeans that can be shipped together on this barge.

  b. Graph the inequality. Because $c$ and $s$ must be non-negative, restrict the graph to Quadrant I.

  c. Can the company ship 25,000 bushels of corn and 25,000 bushels of soybeans on this barge?

15. **Coin Collection** A coin collection consists of only dimes and quarters with a value of more than $12. Let $d$ = the number of dimes and $q$ = the number of quarters.

  a. Write an inequality that describes the possible numbers of dimes and quarters in the collection.

  b. Graph the inequality. Because $d$ and $q$ must be whole numbers, restrict the graph to Quadrant I.

  c. Could this collection consist of 35 dimes and 45 quarters?

16. **Car Rental** While on a business trip, Bruce pays $50 per day to rent a car, plus $0.20 per mile driven. Bruce's company will reimburse him at most $200 for the entire rental. Let $d$ = number of days and $m$ = number of miles.

  a. Write an inequality that describes Bruce's car rental limitations.

  b. Graph the inequality. Because $d$ and $m$ must be non-negative, restrict the graph to Quadrant I.

  c. If Bruce keeps the car for 3 days and drives 400 miles, will he be within his budget?

CHAPTER FOUR

# Systems of Linear Equations and Inequalities

# 4.1 Solving Systems of Linear Equations by Graphing

## THINGS TO KNOW

Before working through this section, be sure you are familiar with the following concepts:

VIDEO　　ANIMATION　　INTERACTIVE

**You Try It**  1. Determine If an Ordered Pair Is a Solution to an Equation (Section 3.2, **Objective 1**)

**You Try It**  2. Graph a Line Using the Slope and a Point (Section 3.3, **Objective 3**)

**You Try It**  3. Use the Slope-Intercept Form to Graph a Linear Equation (Section 3.4, **Objective 2**)

**You Try It**  4. Determine the Relationship between Two Lines (Section 3.4, **Objective 6**)

## OBJECTIVES

1 Determine If an Ordered Pair Is a Solution to a System of Linear Equations in Two Variables
2 Determine the Number of Solutions to a System Without Graphing
3 Solve Systems of Linear Equations by Graphing

OBJECTIVE 1    DETERMINE IF AN ORDERED PAIR IS A SOLUTION TO A SYSTEM
OF LINEAR EQUATIONS IN TWO VARIABLES

Figure 1 shows that the approximate revenue for digital music is projected to increase steadily through 2014, while revenue for physical music (CDs and vinyl) is projected to decrease. The **x-coordinate** of the point where the two lines **intersect** gives the year when revenues are expected to be equal for both music types, and the **y-coordinate** of the intersection point gives the expected revenue when the values are equal for both music types.

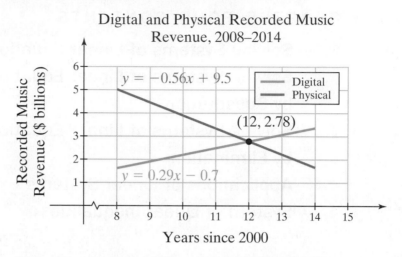

Digital and Physical Recorded Music
Revenue, 2008–2014

*Source:* Forrester Research Internet Music Forecast, 12/09

**Figure 1**

Figure 1 illustrates the graph of a **system of linear equations in two variables**

---

**Definition    System of Linear Equations in Two Variables**

**A system of linear equations in two variables** is a collection of two or more **linear equations in two variables** considered together.

---

**Note:** Although a system of linear equations in two variables can contain more than two equations, we will only consider cases involving two equations in this text.

The system graphed in Figure 1 is given by the equations $\begin{cases} y = 0.29x - 0.7 \\ y = -0.56x + 9.5 \end{cases}$.

The following examples also represent systems of linear equations in two variables.

$$\begin{cases} 5x - y = -9 \\ 3x + 4y = 15 \end{cases} \qquad \begin{cases} \dfrac{1}{3}a - \dfrac{2}{5}b = 6 \\ -\dfrac{1}{6}a + \dfrac{4}{3}b = \dfrac{3}{7} \end{cases} \qquad \begin{cases} 10p + 3q = 28 \\ q = -5 \end{cases}$$

Within a given system of linear equations in two variables, the same two **variables** will be used. For example, the variables $x$ and $y$ are used in both equations of the first system, while the variables $a$ and $b$ are used in the second system.

From **Section 3.2** we know that an **ordered pair** is a **solution** to an equation in two variables if a true statement results when **substituting** the values for the variables. Each ordered pair solution is a point on the graph of the equation. The **intersection point** in **Figure 1**, $(12, 2.78)$, lies on the graph of both equations, so it is a solution to both equations. Substituting 12 for $x$ and 2.78 for $y$ in either equation results in a true statement (view the **checks**). The ordered pair $(12, 2.78)$ is an example of a **solution to a system of linear equations**.

---

> **Definition    Solution to a System of Linear Equations in Two Variables**
>
> A **solution to a system of linear equations** in two variables is an ordered pair that, when substituted for the variables, makes all equations in the system true.

 To determine if an ordered pair is a solution to a system, we check to see if the ordered pair makes *both* equations true when substituted for the variables. It is not enough to check only one equation in the system.

### Example 1  Determining If an Ordered Pair Is a Solution to a System of Linear Equations in Two Variables

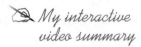 *My interactive video summary*

 Determine if each ordered pair is a solution to the following system:

$$\begin{cases} 2x + 3y = 12 \\ x + 2y = 7 \end{cases}$$

**a.** $(-3, 6)$          **b.** $(3, 2)$

**Solutions**  Check each ordered pair to see if it makes both equations true. If both equations are true, then the ordered pair is a **solution** to the system. If either equation is false, then the ordered pair is not a solution to the system.

**a.**          Begin with the          First Equation          Second Equation
                original equations:          $2x + 3y = 12$          $x + 2y = 7$

Substitute $-3$ for $x$ and
6 for $y$ in each equation:  $2(-3) + 3(6) \overset{?}{=} 12$     $(-3) + 2(6) \overset{?}{=} 7$

Finish simplifying both equations to see if $(-3, 6)$ is a solution to the system  View the **answer**, or watch this **interactive video** for the full solution.

**b.** Substitute 3 for $x$ and 2 for $y$ in each equation. Simplify to see if the resulting equations are true. View the **answer**, or watch this **interactive video** for the full solution.

**You Try It**    Work through this You Try It problem.

**Work Exercises 1–5 in this textbook or in the *MyMathLab*® Study Plan.**

---

OBJECTIVE 2    DETERMINE THE NUMBER OF SOLUTIONS TO A SYSTEM WITHOUT GRAPHING

Solutions to **systems of linear equations** are the **intersection points** of the graphs of the equations. When two linear equations are graphed, one of three possible outcomes will occur.

1. The two lines intersect at one point. See Figure 2(a). The system has one solution.
2. The two lines are **parallel** and do not intersect at all. See Figure 2(b). The system has no solution.
3. The two lines **coincide** and have an **infinite** number of intersection points. See Figure 2(c). The system has an infinite number of solutions.

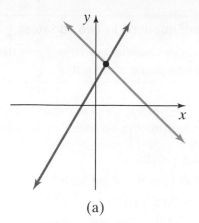

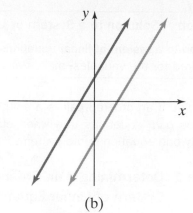

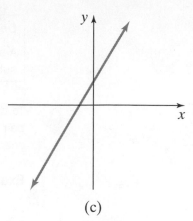

|(a)|(b)|(c)|

**Intersecting Lines**
One solution
Two lines are different, having one common point.

**Parallel Lines**
No solution
Two lines are different, having no common points.

**Coinciding Lines**
Infinitely many solutions
Two lines are the same, having infinitely many common points.

**Figure 2**

A system with at least one solution is **consistent**. A system without a solution is **inconsistent**. When the equations in a system are coinciding lines, the system is **dependent**. When the equations in a system are different lines, the system is **independent**.

In **Section 3.4** we saw how to use the **slope** and **y-intercept** to determine the relationship between two lines. We can use this same information to find the number of **solutions** to a system of linear equations without actually graphing the system.

| Intersecting lines (different slopes) | → | One solution |
|---|---|---|
| Parallel lines (same slope, different y-intercepts) | → | No solution |
| Coinciding lines (same slope, same y-intercept) | → | Infinite number of solutions |

### Example 2  Determining the Number of Solutions to a System Without Graphing

Determine the number of solutions to each system without graphing.

a. $\begin{cases} y = 3x - 4 \\ 6x + 3y = 8 \end{cases}$
b. $\begin{cases} 2x - 4y = \dfrac{8}{3} \\ 3x - 6y = 4 \end{cases}$
c. $\begin{cases} 5x - 2y = 3 \\ -\dfrac{5}{2}x + y = 7 \end{cases}$

**Solutions**

a. Write each equation in **slope-intercept form** so that the slope is easy to find.

$$\begin{cases} y = 3x - 4 \\ 6x + 3y = 8 \end{cases}$$

First equation:  $y = 3x - 4$  ← Slope-intercept form

Second Equation:  $6x + 3y = 8$

Subtract $6x$ from both sides:  $3y = -6x + 8$

Divide both sides by 3:  $y = \dfrac{-6x + 8}{3} \leftarrow \dfrac{-6x + 8}{3} = \dfrac{-6x}{3} + \dfrac{8}{3}$

Simplify:  $y = -2x + \dfrac{8}{3} \leftarrow$ Slope-intercept form

The two **slopes**, 3 and $-2$, are different. Therefore, the lines will **intersect**, and the system will have one solution.

**b.** The equations in slope-intercept form are both $y = \dfrac{1}{2}x - \dfrac{2}{3}$ (view the **details**). This means that the graphs of the equations are coinciding lines and there are an infinite number of solutions. All **ordered pairs** that satisfy the equation $y = \dfrac{1}{2}x \quad \dfrac{2}{3}$ are solutions to the system.

*✎ My video summary*   **c.** Watch this **video** to confirm that the system will have no solution because the two lines are **parallel**.

**You Try It**   Work through this **You Try It** problem.

Work Exercises 6–11 in this textbook or in the *MyMathLab*® Study Plan.

OBJECTIVE 3   SOLVE SYSTEMS OF LINEAR EQUATIONS BY GRAPHING

In this chapter, we look at three methods for solving **systems of linear equations** in two variables: *graphing, substitution,* and *elimination*.

If the system has *no solution* or an *infinite number of solutions*, then there is no need to graph the system since we already know the solution set. We will look at these special cases more in the next two sections. For now, we will focus on **consistent systems** with one solution (intersecting lines).

To solve systems of linear equations by graphing, we use a three-step process.

---

**Solving Systems of Linear Equations in Two Variables by Graphing**

**Step 1.**   Graph the two equations on the same set of **axes**.

**Step 2.**   If the lines intersect, find the **coordinates** of the **intersection point**. The ordered pair is the **solution to the system**.

**Step 3.**   Check the ordered-pair solution in <u>both</u> of the original equations.

**Note:**   If the lines are **parallel**, then the system has no solution. If the lines **coincide**, then the system has infinitely many solutions.

---

It may not be possible to identify the exact intersection point by graphing, so it is essential to check the ordered-pair solution in both equations.

**Example 3  Solving Systems of Linear Equations by Graphing**

Solve the following system by graphing:

$$\begin{cases} y = 2x + 1 \\ y = -x + 4 \end{cases}$$

**Solution** Note that the **slopes**, 2 and $-1$, are different, so there will be one solution to the system. We follow the **three-step process.**

**Step 1.** Graph each line.

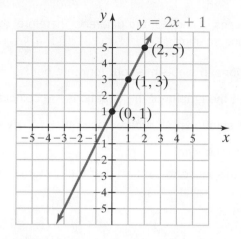

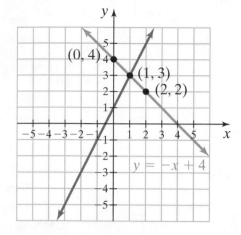

**Step 2.** Find the **coordinates** of the intersection point.

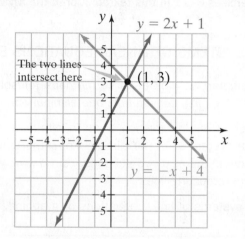

**Figure 3**

Graph of system $\begin{cases} y = 2x + 1 \\ y = -x + 4 \end{cases}$

The **intersection point** is $(1, 3)$.

**Step 3.** Check $(1, 3)$ in both equations to see if it is the **solution** to the system.

|  | First Equation | Second Equation |
|---|---|---|
| Original equation: | $y = 2x + 1$ | $y = -x + 4$ |
| Substitute 1 for $x$ and 3 for $y$: | $(3) \overset{?}{=} 2(1) + 1$ | $(3) \overset{?}{=} -(1) + 4$ |
| Simplify: | $3 = 3$ True | $3 = 3$ True |

The **ordered pair** checks in both equations, so $(1, 3)$ is the solution to the system.

### Example 4  Solving Systems of Linear Equations by Graphing

*My video summary*  Solve the following system by graphing:

$$\begin{cases} 3x + y = -2 \\ x + y = 2 \end{cases}$$

**Solution** We follow the **three-step process.**

**Step 1.** Graph each line, as shown in Figure 4.

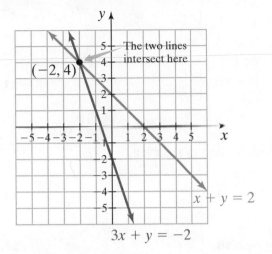

**Figure 4**

Graph of system $\begin{cases} 3x + y = -2 \\ x + y = 2 \end{cases}$

**Step 2.** The **intersection point** is $(-2, 4)$.

**Step 3.** Check $(-2, 4)$ in both equations to see if it is the **solution** to the system.

Watch this **video** to see the fully worked solution.

**You Try It**   Work through this You Try It problem.

Work Exercises **12–21** in this textbook or in the *MyMathLab*® Study Plan.

# 4.1 **Exercises**

In Exercises 1–5, determine if each ordered pair is a solution to the given system.

**You Try It**

**1.** $\begin{cases} 3x + 2y = 1 \\ 5x + 3y = 1 \end{cases}$    **a.** $(-1, 2)$    **b.** $(5, -7)$

**2.** $\begin{cases} 6x - y = 1 \\ -x + 4y = 19 \end{cases}$    **a.** $(1, 5)$    **b.** $(3, 17)$

**3.** $\begin{cases} 0.7x - 0.2y = 0.8 \\ 0.4x + 0.8y = 3.2 \end{cases}$    **a.** $(6, 14)$    **b.** $(2, 3)$

**4.** $\begin{cases} x + 2y = 6.3 \\ 5x - 7y = -4 \end{cases}$    **a.** $\left(\dfrac{2}{5}, -\dfrac{1}{3}\right)$    **b.** $(1.3, 2.5)$

**5.** $\begin{cases} \dfrac{2}{3}x - y = -2 \\ \dfrac{5}{4}x + \dfrac{1}{2}y = \dfrac{23}{4} \end{cases}$    **a.** $(3, 4)$    **b.** $(2, -3)$

In Exercises 6–11, determine the number of solutions to the given system without graphing.

**You Try It**

**6.** $\begin{cases} y = 7x + 2 \\ y = 2x + 7 \end{cases}$

**7.** $\begin{cases} y = 3x + 7 \\ 6x - 2y = 9 \end{cases}$

**8.** $\begin{cases} 2x - 5y = 15 \\ y = \dfrac{2}{5}x - 3 \end{cases}$

**9.** $\begin{cases} 5x - 8y = 4 \\ 2y - 1.25x = 1 \end{cases}$

**10.** $\begin{cases} 2x + 7y = 1 \\ 7x + 2y = 1 \end{cases}$

**11.** $\begin{cases} y = \dfrac{3}{4}x - 1 \\ x = \dfrac{4}{3}y + \dfrac{4}{3} \end{cases}$

In Exercises 12–21, solve each system by graphing.

**You Try It**

**12.** $\begin{cases} y = 3x - 1 \\ y = 2x + 1 \end{cases}$

**13.** $\begin{cases} 4x + 2y = -8 \\ -7x + y = 5 \end{cases}$

**14.** $\begin{cases} 3x - 4y = 0 \\ 9x - 4y = 24 \end{cases}$

**15.** $\begin{cases} 2x - 6y = -6 \\ 2x + 3y = 12 \end{cases}$

**16.** $\begin{cases} 2x - 3y = -9 \\ x + 0.5y = -2.5 \end{cases}$

**17.** $\begin{cases} x + 0.5y = 6 \\ 1.8x + 1.5y = 12 \end{cases}$

**18.** $\begin{cases} 9x - 2y - 8 = 0 \\ y = 2x + 1 \end{cases}$

**19.** $\begin{cases} 4x - \dfrac{5}{3}y = -15 \\ -5y = x + 20 \end{cases}$

**20.** $\begin{cases} 7x + \dfrac{7}{2}y = 28 \\ \dfrac{3}{2}x + \dfrac{5}{4}y = 5 \end{cases}$

**21.** $\begin{cases} 3x - \dfrac{7}{3}y = \dfrac{14}{3} \\ 6x - \dfrac{7}{2}y = \dfrac{35}{2} \end{cases}$

# 4.2 Solving Systems of Linear Equations by Substitution

## THINGS TO KNOW

Before working through this section, be sure you are familiar with the following concepts:

| | VIDEO | ANIMATION | INTERACTIVE |
|---|---|---|---|

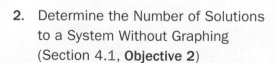
**You Try It**

**1.** Determine If an Ordered Pair Is a Solution to a System of Linear Equations in Two Variables (Section 4.1, **Objective 1**)     

**You Try It**

**2.** Determine the Number of Solutions to a System Without Graphing (Section 4.1, **Objective 2**)     

**You Try It**

**3.** Solve Systems of Linear Equations by Graphing (Section 4.1, **Objective 3**)

## OBJECTIVES

**1** Solve Systems of Linear Equations by Substitution

**2** Solve Special Systems by Substitution

---

OBJECTIVE 1    SOLVE SYSTEMS OF LINEAR EQUATIONS BY SUBSTITUTION

In this section, we present the first of two algebraic methods for solving systems of linear equations in two variables. Consider the following system:

$$\begin{cases} 2x + 3y = 14 \\ y = x - 2 \end{cases}$$

Recall that when solving a system, we are looking for the **intersection point(s)** of the two lines in the system. At these points, both equations have the same $x$-values and the same $y$-values. If the second equation says that $y$ equals $x - 2$, this means that $y$ must equal $x - 2$ in the first equation as well.

$$2x + 3y = 14 \qquad\qquad\qquad y = x - 2$$

$y$ must equal $x - 2$ in both equations

Replacing $y$ in the first equation by $x - 2$ gives us an equation we can solve for $x$.

| | |
|---|---|
| First equation: | $2x + 3y = 14$ |
| Replace $y$ with $x - 2$: | $2x + 3(x - 2) = 14$ |
| Distribute: | $2x + 3x - 6 = 14$ |
| Simplify: | $5x - 6 = 14$ |
| Add 6 to both sides: | $5x = 20$ |
| Divide both sides by 5: | $x = 4$ |

This value for $x$ can then be used to determine the corresponding value for $y$. We can choose either original equation to find $y$. Here we substitute 4 for $x$ in the second equation.

| | |
|---|---|
| Second equation: | $y = x - 2$ |
| Replace $x$ with 4: | $y = (4) - 2$ |
| Simplify: | $y = 2$ |

The **solution** to this system is the ordered pair $(4, 2)$.

The **substitution method** involves solving one of the equations for one **variable**, substituting the resulting expression into the other equation, and then solving for the remaining variable. The substitution method can be summarized in four steps.

---

**Solving Systems of Linear Equations in Two Variables by Substitution**

**Step 1.**  Choose an equation and solve for one variable in terms of the other variable.

**Step 2.**  Substitute the expression from Step 1 into the other equation.

**Step 3.**  Solve the equation in one variable from Step 2.

**Step 4.**  Substitute the solution from Step 3 into one of the original equations to find the value of the other variable.

---

It is always a good idea to check the ordered-pair solution by **substituting** the $x$- and $y$-values into *both* original equations to see if both equations result in true statements.

## Example 1 Solving Systems of Linear Equations by Substitution

Use the **substitution method** to solve the following system:

$$\begin{cases} 4x + 2y = 10 \\ y = 3x - 10 \end{cases}$$

### Solution

**Step 1.** The second equation, $y = 3x - 10$, is already solved for $y$.

**Step 2.** Substitute $3x - 10$ for $y$ in the first equation.

First equation: $\qquad\qquad 4x + 2y = 10$

Substitute $3x - 10$ for $y$: $\quad 4x + 2(\overbrace{3x - 10}^{y}) = 10$

**Step 3.** Solve for $x$.

$$
\begin{aligned}
\text{Equation from Step 2:} \quad & 4x + 2(3x - 10) = 10 \\
\text{Distribute:} \quad & 4x + 6x - 20 = 10 \\
\text{Simplify:} \quad & 10x - 20 = 10 \\
\text{Add 20 to both sides:} \quad & 10x = 30 \\
\text{Divide both sides by 10:} \quad & x = \frac{30}{10} = 3
\end{aligned}
$$

**Step 4.** Find $y$ by substituting $x = 3$ into one of the original equations.

$$
\begin{aligned}
\text{Original second equation:} \quad & y = 3x - 10 \\
\text{Substitute 3 for } x: \quad & y = 3(3) - 10 \\
\text{Multiply:} \quad & y = 9 - 10 \\
\text{Simplify:} \quad & y = -1
\end{aligned}
$$

The solution to this system is the **ordered pair** $(3, -1)$. Check this solution in both original equations. View this **popup** to see the check.

Figure 5 shows the graph of the system in Example 1. We see that the solution $(3, -1)$ is the **intersection point** of the two lines in the system.

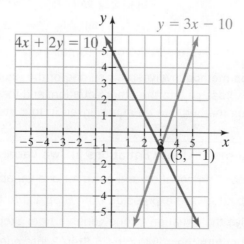

**Figure 5**

Graph of system $\begin{cases} 4x + 2y = 10 \\ y = 3x - 10 \end{cases}$

**You Try It**   Work through this **You Try It** problem.

**Work Exercises 1–8** in this textbook or in the **MyMathLab**® Study Plan.

### Example 2 Solving Systems of Linear Equations by Substitution

*My video summary* Use the **substitution method** to solve the following system:

$$\begin{cases} 3x + y = -6 \\ 2x - \dfrac{1}{3}y = 2 \end{cases}$$

**Solution**

**Step 1.** Choose an equation and solve for one of the variables. If possible, choose a variable that has a **coefficient** of 1. In this example, it is easiest to solve the first equation for $y$.

Begin with the first equation:   $3x + y = -6$

Subtract $3x$ from both sides:   $y = -3x - 6$

**Step 2.** Substitute $-3x - 6$ for $y$ in the second equation.

Second equation:   $2x - \dfrac{1}{3}y = 2$

Substitute $-3x - 6$ for $y$:   $2x - \dfrac{1}{3}\overbrace{(-3x - 6)}^{y} = 2$

**Step 3.** Solve for $x$.

Equation from Step 2:   $2x - \dfrac{1}{3}\overbrace{(-3x - 6)}^{y} = 2$

Distribute:   $2x + x + 2 = 2 \leftarrow -\frac{1}{3}(-3x) - \frac{1}{3}(-6) = x + 2$

Simplify:   $3x + 2 = 2$

Subtract 2 from both sides:   $3x = 0$

Divide both sides by 3:   $x = \dfrac{0}{3} = 0$

Use this value for $x$ to find the value for $y$ and complete the solution to the system. View the **answer**, or watch this **video** for a complete solution.

### Example 3 Solving Systems of Linear Equations by Substitution

*My video summary* Solve the following system:

$$\begin{cases} 4x + 3y = 7 \\ x + 9y = -1 \end{cases}$$

**Solution** Try solving this system on your own. View the **answer**, or watch this **video** for the fully worked solution.

**You Try It**   Work through this **You Try It** problem.

Work Exercises 9–16 in this textbook or in the ***MyMathLab*®** Study Plan.

## Example 4  Solving Systems of Linear Equations by Substitution

*My video summary*  Use the **substitution** method to solve the following system:

$$\begin{cases} 6x - 3y = -33 \\ 2x + 4y = 4 \end{cases}$$

**Solution** In this system, we do not see any variables with a **coefficient** of 1 or −1. We can select either equation and solve for either variable. Notice that if we divide both sides of the first equation by 3, we get

$$\frac{6x}{3} - \frac{3y}{3} = \frac{-33}{3} \rightarrow 2x - y = -11.$$

And if we divide both sides of the second equation by 2, we get

$$\frac{2x}{2} + \frac{4y}{2} = \frac{4}{2} \rightarrow x + 2y = 2.$$

Try solving this system on your own. View the **answer**, or watch this **interactive video** for a fully worked solution.

**You Try It**  Work through this You Try It problem.

Work Exercises 17–20 in this textbook or in the *MyMathLab*® Study Plan.

OBJECTIVE 2  SOLVE SPECIAL SYSTEMS BY SUBSTITUTION

Recall that a **system of linear equations in two variables** will always result in one of three possible situations. We see these situations graphically in Figure 6.

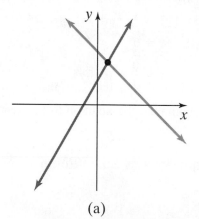

(a)

**Intersecting Lines**
One solution
Consistent, Independent
Two lines are different, having one common point. The lines have different slopes.

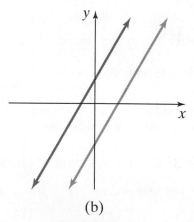

(b)

**Parallel Lines**
No solution
Inconsistent, Independent
Two lines are different, having no common points. The lines have the same slope but different *y*-intercepts.

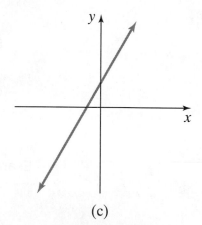

(c)

**Coinciding Lines**
Infinitely many solutions
Consistent, Dependent
Two lines are the same, having infinitely many common points. The lines have the same slope and same *y*-intercept.

**Figure 6**

When solving a system algebraically, we recognize these three situations as follows:

1. **A value for both variables exists.** This is the "one solution" found for **intersecting lines**. The system is **independent** and **consistent**.

2. **A contradiction occurs.** This is the "no solution" found for **parallel lines**. The system is independent and **inconsistent**.

3. **An identity results.** This is the "infinite solutions" found for **coinciding lines**. The system is **dependent** and consistent.

So far we have focused on solving systems with one solution. We now consider systems with no solution or an infinite number of solutions.

### Example 5  Solving a Dependent System by Substitution

Use the **substitution method** to solve the following system:

$$\begin{cases} 2x + 10y = 8 \\ x + 5y = 4 \end{cases}$$

### Solution

**Step 1.** Solve the second equation for $x$.

$$\text{Begin with the second equation:} \quad x + 5y = 4$$
$$\text{Subtract } 5y \text{ from both sides:} \quad x = -5y + 4$$

**Step 2.** Substitute $-5y + 4$ for $x$ in the first equation.

$$\text{First equation:} \quad 2x + 10y = 8$$
$$\text{Substitute } -5y + 4 \text{ for } x: \quad 2(-5y + 4) + 10y - 8$$

**Step 3.** Solve for $y$.

$$\text{Distribute:} \quad -10y + 8 + 10y = 8$$
$$\text{Combine like terms:} \quad 8 = 8 \quad \text{True}$$

Our equation in Step 3 simplifies to $8 = 8$, which is an **identity**, so the equation is true for any value of $y$. For each $y$, there is a corresponding value for $x$ such that $x = -5y + 4$. So, the system has an infinite number of solutions and is a **dependent system**. Any ordered pair that is a solution to one of the equations is a solution to both equations.

**Figure 6** shows that the graphs of the two equations in the system are **coinciding lines**. The **solution to the system** is the set of all ordered pairs that lie on the graph of either equation.

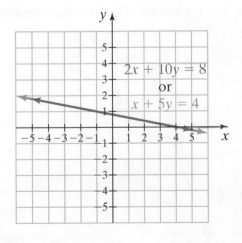

**Figure 7**

Graph of system $\begin{cases} 2x + 10y = 8 \\ x + 5y = 4 \end{cases}$

Because there are an **infinite** number of solutions to **dependent systems**, it is not possible to write all the solutions. Instead, we can use **set-builder notation**, including one of the equations from the system (or any **equivalent equation**). The solution set for the system in **Example 5** can be written as $\{(x, y) | x + 5y = 4\}$. We read this as "the set of all ordered pairs $(x, y)$, such that $x + 5y = 4$." We can also write the solution set equivalently as $\{(x, y) | 2x + 10y = 8\}$ by using the first equation, or as $\left\{ (x, y) \middle| y = -\dfrac{1}{5}x + \dfrac{4}{5} \right\}$ by writing either equation in **slope-intercept form**. For convenience, we will agree to use slope-intercept form when writing the equation of the common line in a dependent system.

Although dependent systems have an infinite number of ordered-pair solutions, not all ordered pairs are solutions. The solution set consists only of those ordered pairs whose points lie on the coinciding lines.

## Example 6 Solving an Inconsistent System by Substitution

Use the **substitution method** to solve the following system:

$$\begin{cases} 3x - y = -1 \\ -12x + 4y = 8 \end{cases}$$

### Solution

**Step 1.** Solve the first equation for $y$.

$$\begin{aligned} \text{First equation:} \quad & 3x - y = -1 \\ \text{Subtract } 3x \text{ from both sides:} \quad & -y = -3x - 1 \\ \text{Multiply both sides by } -1: \quad & y = 3x + 1 \end{aligned}$$

**Step 2.** Substitute $3x + 1$ for $y$ in the second equation.

$$\begin{aligned} \text{Second equation:} \quad & -12x + 4y = 8 \\ \text{Substitute } 3x + 1 \text{ for } y: \quad & -12x + 4(3x + 1) = 8 \end{aligned}$$

**Step 3.** Solve for $x$.

$$\begin{aligned} \text{Distribute:} \quad & -12x + 12x + 4 = 8 \\ \text{Combine like terms:} \quad & 4 = 8 \quad \text{False} \end{aligned}$$

The equation in Step 3 simplifies to $4 = 8$, which is a **contradiction**. This means that the equation is never true. Therefore, the system has no solution and is **inconsistent**. The solution set is $\varnothing$ or $\{\ \}$.

Figure 8 shows that the graphs of the two equations in the system are **parallel lines**.

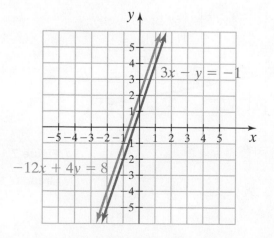

**Figure 8**

Graph of system $\begin{cases} 3x - y = -1 \\ -12x + 4y = 8 \end{cases}$

**You Try It**    Work through this You Try It problem.

**Work Exercises 21–24 in this textbook or in the MyMathLab® Study Plan.**

**Example 7  Solving Special Systems by Substitution**

 Use the **substitution method** to solve the following systems:

*My interactive video summary*

a. $\begin{cases} \dfrac{1}{4}x + y = 5 \\ x + 4y = 8 \end{cases}$

b. $\begin{cases} -2.4x + 1.5y = -3 \\ 0.8x - 0.5y = 1 \end{cases}$

**Solutions** Try solving these systems on your own. View the **answers**, or watch this interactive video for fully worked solutions.

**You Try It**    Work through this You Try It problem.

**Work Exercises 25–28 in this textbook or in the MyMathLab® Study Plan.**

# 4.2  Exercises

In Exercises 1–20, solve each system by substitution.

**You Try It**
1. $\begin{cases} y = 3x - 2 \\ 3x + 5y = 44 \end{cases}$

2. $\begin{cases} 2x - 3y = 17 \\ x = 2y + 10 \end{cases}$

3. $\begin{cases} x = 4y - 3 \\ 5x - 3y = 2 \end{cases}$

4. $\begin{cases} 5x - 2y = -1 \\ y = 3x \end{cases}$

5. $\begin{cases} 6x + y = 9 \\ y = 3 - 3x \end{cases}$

6. $\begin{cases} y = 2x - 5 \\ y = 5x - 2 \end{cases}$

7. $\begin{cases} \dfrac{2}{3}x + \dfrac{5}{2}y = -3 \\ x = \dfrac{2}{3}y + \dfrac{13}{3} \end{cases}$

8. $\begin{cases} y = \dfrac{3}{7}x + \dfrac{1}{7} \\ \dfrac{1}{2}x + \dfrac{3}{8}y = \dfrac{11}{8} \end{cases}$

**You Try It**

9. $\begin{cases} 2x + y = 7 \\ 4x - 7y = 5 \end{cases}$

10. $\begin{cases} x + y = 5 \\ 3x - y = -13 \end{cases}$

11. $\begin{cases} x - 4y = 6 \\ 7x - 5y = -4 \end{cases}$

12. $\begin{cases} 5x - 3y = -18 \\ -2x + y = 7 \end{cases}$

13. $\begin{cases} 5x - 2y = 0 \\ 3x + y = 0 \end{cases}$

14. $\begin{cases} x - 2y = 7 \\ 3x + 2y = -3 \end{cases}$

15. $\begin{cases} 2x + \dfrac{3}{2}y = 1 \\ x + 3y = 7 \end{cases}$

16. $\begin{cases} 1.5x + y = 1.1 \\ 2.4x - 3y = 8.4 \end{cases}$

**You Try It**

17. $\begin{cases} 2x + 4y = 6 \\ 5x + 2y = -1 \end{cases}$

18. $\begin{cases} 3x + 6y = \dfrac{19}{2} \\ \dfrac{8}{5}x + 3y = 5 \end{cases}$

19. $\begin{cases} 1.7x + 3.2y = 0.22 \\ 2.1x - 2.1y = 6.93 \end{cases}$

20. $\begin{cases} \dfrac{1}{2}x + \dfrac{5}{3}y = \dfrac{2}{3} \\ 3x - 2y = \dfrac{8}{5} \end{cases}$

In Exercises 21–28, solve each special system by substitution. If the system is dependent, write the solution using set-builder notation with the common line expressed in slope-intercept form.

**You Try It**

**21.** $\begin{cases} 6x + 2y = 18 \\ y = 8 - 3x \end{cases}$

**22.** $\begin{cases} 2x + 6y = 18 \\ x + 3y = 9 \end{cases}$

**23.** $\begin{cases} 5x - 15y = -10 \\ -x + 3y = 2 \end{cases}$

**24.** $\begin{cases} -2x + y = 11 \\ 6x - 3y = 5 \end{cases}$

**You Try It**

**25.** $\begin{cases} x - 3y = -1 \\ -2x + 6y = 2 \end{cases}$

**26.** $\begin{cases} y = 4x + 3 \\ y = 4x - 3 \end{cases}$

**27.** $\begin{cases} 3.4x - 5.1y = -1.7 \\ -2.4x + 3.6y = 1.2 \end{cases}$

**28.** $\begin{cases} \dfrac{2}{3}x + \dfrac{7}{3}y = \dfrac{13}{3} \\ \dfrac{4}{5}x + \dfrac{14}{5}y = \dfrac{18}{5} \end{cases}$

# 4.3 Solving Systems of Linear Equations by Elimination

## THINGS TO KNOW

Before working through this section, be sure you are familiar with the following concepts:

| | VIDEO | ANIMATION | INTERACTIVE |
|---|---|---|---|

**You Try It**

**1.** Determine If an Ordered Pair Is a Solution to a System of Linear Equations in Two Variables (Section 4.1, **Objective 1**)

**You Try It**

**2.** Determine the Number of Solutions to a System Without Graphing (Section 4.1, **Objective 2**)

**You Try It**

**3.** Solve Systems of Linear Equations by Graphing (Section 4.1, **Objective 3**)

**You Try It**

**4.** Solve Systems of Linear Equations by Substitution (Section 4.2, **Objective 1**)

**You Try It**

**5.** Solve Special Systems by Substitution (Section 4.2, **Objective 2**)

## OBJECTIVES

**1** Solve Systems of Linear Equations by Elimination

**2** Solve Special Systems by Elimination

OBJECTIVE 1     SOLVE SYSTEMS OF LINEAR EQUATIONS BY ELIMINATION

The **elimination method** for solving a **system of linear equations** in two variables involves adding the two equations together in a way that will *eliminate* one of the **variables**. This algebraic method is based on the following logic.

> **Logic for the Elimination Method**
>
> If $A = B$ and $C = D$, then $A + C = B + D$.

This means that if two true equations are added, then the result will be a third true equation. Because equations are added, the elimination method is also known as the **addition method**.

### Example 1  Solving Systems of Linear Equations by Elimination

Solve the following system.

$$\begin{cases} x + y = 8 \\ x - y = -2 \end{cases}$$

**Solution**  Notice that the first equation includes "$+y$" while the second equation includes "$-y$." Let's see what happens when we add these two equations together.

$$\begin{array}{r} x + y = 8 \\ x - y = -2 \\ \hline 2x + 0 = 6 \end{array}$$

The $y$-variable is eliminated. By solving the resulting equation for $x$, we can find the $x$-coordinate of the **solution to the system**.

Rewrite the equation: $\quad 2x + 0 = 6$

Simplify: $\qquad\qquad 2x = 6$

Divide both sides by 2: $\qquad x = 3$

The value of $x$ is 3. We can now substitute $x = 3$ into one of the original equations to find the value of $y$.

Begin with the first original equation: $\qquad x + y = 8$

Substitute 3 for $x$: $\qquad 3 + y = 8$

Subtract 3 from both sides: $\quad 3 + y - 3 = 8 - 3$

Simplify: $\qquad\qquad\qquad y = 5$

The solution is $(3, 5)$. Now we can check $(3, 5)$ in both of the original equations.

|  | First Equation | Second Equation |
|---|---|---|
| Begin with each original equation: | $x + y = 8$ | $x - y = -2$ |
| Substitute $x = 3$ and $y = 5$: | $3 + 5 \stackrel{?}{=} 8$ | $3 - 5 \stackrel{?}{=} -2$ |
| Simplify: | $8 = 8$ True | $-2 = -2$ True |

The ordered pair checks, so $(3, 5)$ is the solution to the system.

Figure 9 shows the graph of the system. We see that the solution $(3, 5)$ is the **intersection point** of the two lines in the system.

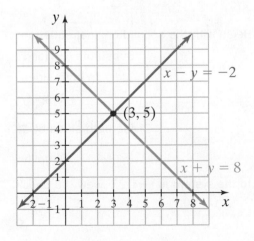

**Figure 9**

Graph of system $\begin{cases} x + y = 8 \\ x - y = -2 \end{cases}$

**You Try It    Work through this You Try It problem.**

**Work Exercises 1–6 in this textbook or in the *MyMathLab*® Study Plan.**

In **Example 1**, the $y$-variable was eliminated simply by adding the two equations together. Unfortunately, a variable will not always be eliminated so easily. Consider the following system:

$$\begin{cases} x - y = -4 \\ x + 2y = 5 \end{cases}$$

Adding these two equations as a first step does not eliminate a variable:

$$\begin{array}{r} x - \phantom{2}y = -4 \\ x + 2y = 5 \\ \hline 2x + \phantom{2}y = 1 \end{array}$$

In cases like this, we must write **equivalent equations** in the system before adding so that a variable will be eliminated.

To eliminate a variable, the **coefficients** of the variable in the two equations must be **opposites**. We can make this happen by multiplying one or both of the equations by an appropriate nonzero **constant**.

### Example 2  Solving Systems of Linear Equations by Elimination

Solve the following system.

$$\begin{cases} x - y = -4 \\ x + 2y = 5 \end{cases}$$

**Solution**  Multiplying both sides of the first equation by 2 will make the coefficient of its $y$-variable $-2$, which is the **opposite** of the coefficient of the $y$-variable in the second equation. Add the result to the second equation in order to eliminate the $y$-variable.

$$x - \phantom{2}y = -4 \xrightarrow{\text{Multiply by 2}} 2(x - y) = 2(-4) \xrightarrow{\text{Simplify}} 2x - 2y = -8$$

$$x + 2y = 5 \xrightarrow{\text{No change}} x + 2y = 5 \xrightarrow{\text{No change}} \underline{x + 2y = 5}$$

$$\text{Add the new equations:} \quad 3x \phantom{+2y} = -3$$

$$\text{Solve for } x \text{ by dividing both sides by 3:} \quad x = -1$$

The value of $x$ is $-1$. Substitute $x = -1$ into one of the original equations to find the value of $y$.

$$\begin{aligned} \text{Begin with the second original equation:} & \quad x + 2y = 5 \\ \text{Substitute } -1 \text{ for } x: & \quad -1 + 2y = 5 \\ \text{Add 1 to both sides:} & \quad -1 + 2y + 1 = 5 + 1 \\ \text{Simplify:} & \quad 2y = 6 \\ \text{Divide both sides by 2:} & \quad \frac{2y}{2} = \frac{6}{2} \\ \text{Simplify:} & \quad y = 3 \end{aligned}$$

The **solution to the system** is $(-1, 3)$. View the **check**.

**You Try It    Work through this You Try It problem.**

Work Exercises 7–9 in this textbook or in the **MyMathLab** Study Plan.

The elimination method can be summarized in five steps.

---

**Solving Systems of Linear Equations in Two Variables by Elimination**

**Step 1.** Choose a variable to eliminate.

**Step 2.** If necessary, multiply one or both equations by an appropriate nonzero **constant** so that the sum of the **coefficients** of one of the variables is zero.

**Step 3.** Add the two equations from Step 2 together to obtain an equation in one variable.

**Step 4.** Solve the equation in one variable from Step 3.

**Step 5.** Substitute the value found in Step 4 into one of the original equations to solve for the other variable.

---

CAUTION It is good practice to check the solution. Do this by substituting the ordered pair into the original equation that was not used in Step 5.

### Example 3  Solving Systems of Linear Equations by Elimination

*My video summary*  Use the **elimination method** to solve the following system.

$$\begin{cases} x - 3y = -9 \\ 5x + 4y = -7 \end{cases}$$

**Solution**

**Step 1.** We choose to eliminate the $x$-variable because the **coefficients** of the $x$-variables will be **opposites** if we multiply the first equation by $-5$.

**Step 2.** Multiply both sides of the first equation by $-5$.

$$x - 3y = -9 \xrightarrow{\text{Multiply by } -5} -5(x - 3y) = -5(-9) \xrightarrow{\text{Simplify}} -5x + 15y = 45$$

$$5x + 4y = -7 \xrightarrow{\text{No change}} 5x + 4y = -7 \xrightarrow{\text{No change}} 5x + 4y = -7$$

**Step 3.** Add the two new equations together.

$$\begin{array}{r} -5x + 15y = 45 \\ 5x + 4y = -7 \\ \hline 19y = 38 \end{array}$$

**Step 4.** Solve the resulting equation for $y$.

Rewrite the resulting equation from Step 3:   $19y = 38$

Divide both sides by 19:     $y = 2$

**Step 5.** Find $x$ by substituting $y = 2$ into one of the original equations.

Begin with the original first equation:     $x - 3y = -9$

Substitute 2 for $y$:     $x - 3(2) = -9$

Simplify:     $x - 6 = -9$

Add 6 to both sides:   $x - 6 + 6 = -9 + 6$

Simplify:     $x = -3$

The **solution to the system** is $(-3, 2)$

Since we used the original first equation in Step 5, we use the original second equation to check the final answer. View the **check**, or watch this **video** for a fully worked solution.

**You Try It**   **Work through this You Try It problem.**

**Work Exercises 10–12 in this textbook or in the *MyMathLab*® Study Plan.**

In **Example 3**, we had to multiply one equation by a nonzero constant to eliminate a variable. To solve some systems, we must multiply both equations by nonzero constants. Consider the next example.

### Example 4  Solving Systems of Linear Equations by Elimination

*My video summary*   Use the **elimination method** to solve the following system.

$$\begin{cases} -3x + 4y = 7 \\ 5x + 6y = 1 \end{cases}$$

### Solution

**Step 1.** For this problem, we choose to eliminate the $x$-variable. The **coefficients** of the $x$-variables will be **opposites** if we multiply the first equation by 5 and the second equation by 3.

**Step 2.** Multiply both sides of the first equation by 5, and multiply both sides of the second equation by 3.

$$-3x + 4y = 7 \xrightarrow{\text{Multiply by 5}} 5(-3x + 4y) = 5(7) \xrightarrow{\text{Simplify}} -15x + 20y = 35$$

$$5x + 6y = 1 \xrightarrow{\text{Multiply by 3}} 3(5x + 6y) = 3(1) \xrightarrow{\text{Simplify}} 15x + 18y = 3$$

**Step 3.** Add the two new equations together.

$$\begin{array}{r} -15x + 20y = 35 \\ 15x + 18y = 3 \\ \hline 38y = 38 \end{array}$$

**Step 4.** Solve the resulting equation for $y$.

Rewrite the resulting equation from Step 3:   $38y = 38$

Divide both sides by 38:   $y = 1$

**Step 5.** Try to finish solving this problem on your own. See the **answer**, or watch this **video** for the complete solution.

### Example 5  Solving Systems of Linear Equations by Elimination

*My video summary*   Use the **elimination method** to solve the following system.

$$\begin{cases} 5x - 6y = 20 \\ 4x + 9y = 16 \end{cases}$$

**Solution**  Try solving this system on your own. View the **answer**, or watch this **video** for a fully worked solution.

**You Try It**   **Work through this You Try It problem.**

**Work Exercises 13–18 in this textbook or in the *MyMathLab*® Study Plan.**

If the equations in a system include fractions, then it is usually helpful to **clear the fractions** before proceeding. It may also be helpful to **clear any decimals** from a system.

### Example 6 Solving Systems of Linear Equations Involving Fractions

*My video summary*  Use the **elimination method** to solve the following system.

$$\begin{cases} x - \dfrac{3}{5}y = \dfrac{4}{5} \\ \dfrac{1}{2}x + 3y = -\dfrac{9}{5} \end{cases}$$

**Solution** We begin by clearing the fractions from the equations. Multiply both sides of the first equation by the **LCD** 5, and multiply both sides of the second equation by the LCD 10.

$$x - \frac{3}{5}y = \frac{4}{5} \xrightarrow{\text{Multiply by 5}} 5\left(x - \frac{3}{5}y\right) = 5\left(\frac{4}{5}\right) \xrightarrow{\text{Simplify}} 5x - 3y = 4$$

$$\frac{1}{2}x + 3y = -\frac{9}{5} \xrightarrow{\text{Multiply by 10}} 10\left(\frac{1}{2}x + 3y\right) = 10\left(-\frac{9}{5}\right) \xrightarrow{\text{Simplify}} 5x + 30y = -18$$

Try to finish this problem on your own by solving the resulting system.

$$\begin{cases} 5x - 3y = 4 \\ 5x + 30y = -18 \end{cases}$$

See the **answer**, or watch this **video** for a fully worked solution.

**You Try It** **Work through this You Try It problem.**

**Work Exercises 19–22 in this textbook or in the *MyMathLab*® Study Plan.**

---

OBJECTIVE 2   SOLVE SPECIAL SYSTEMS BY ELIMINATION

Now we look at **inconsistent** and **dependent systems**. Like the **substitution method**, when solving by **elimination**, an inconsistent system will lead to a **contradiction** and a dependent system will lead to an **identity**.

### Example 7 Solving Inconsistent and Dependent Systems by Elimination

Use the elimination method to solve each system.

a. $\begin{cases} 3x + y = 6 \\ 6x + 2y = 4 \end{cases}$    b. $\begin{cases} 2x - 8y = 6 \\ 3x - 12y = 9 \end{cases}$

**Solution**

a. **Step 1.** For this problem, we choose to eliminate the $x$-variable. The **coefficients** of the $x$-variables will be **opposites** if we multiply the first equation by $-2$.

   **Step 2.** Multiply both sides of the first equation by $-2$.

$$3x + y = 6 \xrightarrow{\text{Multiply by } -2} -2(3x + y) = -2(6) \xrightarrow{\text{Simplify}} -6x - 2y = -12$$

$$6x + 2y = 4 \xrightarrow{\text{No change}} 6x + 2y = 4 \xrightarrow{\text{No change}} 6x + 2y = 4$$

**Step 3.** Add the two new equations together.

$$-6x - 2y = -12$$
$$6x + 2y = 4$$
$$\overline{\phantom{6x + 2y}}$$
$$0 = -8 \quad \text{False}$$

Both variables are eliminated, and we are left with the **contradiction** $0 = -8$. The system has no solution and is **inconsistent**. The solution set is $\varnothing$ or $\{\}$. Figure 10 shows that the graphs of the two linear equations in this system are **parallel**.

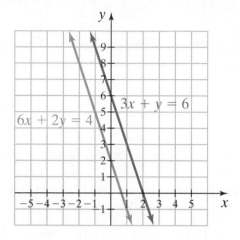

**Figure 10**

Graph of system $\begin{cases} 3x + y = 6 \\ 6x + 2y = 4 \end{cases}$

*My video summary*  **b.** Try to solve this system on your own. View the **answer**, or watch this **video** for a complete solution.

**You Try It**  Work through this You Try It problem.

**Work Exercises 23–26 in this textbook or in the *MyMathLab*® Study Plan.**

# 4.3 Exercises

In Exercises 1–22, solve each system by elimination.

**You Try It**

**1.** $\begin{cases} x + y = 8 \\ x - y = 4 \end{cases}$

**2.** $\begin{cases} x + 4y = 1 \\ -x + 2y = 5 \end{cases}$

**3.** $\begin{cases} x - 5y = -3 \\ x + 5y = 17 \end{cases}$

**4.** $\begin{cases} -4x + 3y = -4 \\ 4x + 5y = -28 \end{cases}$

**5.** $\begin{cases} -3x + 4y = -15 \\ 2x - 4y = 10 \end{cases}$

**6.** $\begin{cases} 10x - 9y = 11 \\ 2x + 9y = -5 \end{cases}$

**You Try It**

**7.** $\begin{cases} x + 3y = 4 \\ 2x - y = -6 \end{cases}$

**8.** $\begin{cases} 2x - 5y = 13 \\ 6x - 5y = 29 \end{cases}$

**9.** $\begin{cases} 8x - 3y = -17 \\ 2x + y = 1 \end{cases}$

**You Try It 10.** $\begin{cases} 4x - y = 10 \\ x - 3y = -25 \end{cases}$

**11.** $\begin{cases} 10x - 3y = 45 \\ 2x - 5y = -13 \end{cases}$

**12.** $\begin{cases} 7x - 2y = -22 \\ 5x + 6y = 14 \end{cases}$

**13.** $\begin{cases} 2x + 5y = 3 \\ 3x + 2y = -12 \end{cases}$

**14.** $\begin{cases} 3x + 5y = -20 \\ -5x + 4y = -16 \end{cases}$

**15.** $\begin{cases} 2x + 3y = 13 \\ 5x + 2y = -6 \end{cases}$

**You Try It 16.** $\begin{cases} 4x - 5y = 51 \\ 3x + 7y = -37 \end{cases}$

**17.** $\begin{cases} 9x - 2y = 0 \\ 5x + 3y = 0 \end{cases}$

**18.** $\begin{cases} 4x + 9y = 1 \\ 6x + 6y = -1 \end{cases}$

**19.** $\begin{cases} \dfrac{1}{4}x - y = \dfrac{9}{8} \\ x + \dfrac{3}{5}y = -\dfrac{1}{10} \end{cases}$ **You Try It**

**20.** $\begin{cases} \dfrac{1}{3}x + \dfrac{1}{2}y = -\dfrac{1}{3} \\ \dfrac{1}{2}x + \dfrac{1}{3}y = -\dfrac{4}{3} \end{cases}$

**21.** $\begin{cases} 0.3x + 0.2y = 1 \\ 0.1x - 0.3y = 1.8 \end{cases}$

**22.** $\begin{cases} 0.5x - 0.4y = 1.5 \\ 0.1x - 0.2y = 0.3 \end{cases}$

In Exercises 23–26, solve each special system by elimination. If the system is dependent, write the solution using set-builder notation with the common line expressed in slope-intercept form.

 **You Try It**

**23.** $\begin{cases} 2x - 6y = 12 \\ -3x + 9y = -18 \end{cases}$

**24.** $\begin{cases} -x - 5y = 7 \\ 8x + 40y = -25 \end{cases}$

**25.** $\begin{cases} 6x - 10y = 40 \\ -9x + 15y = 45 \end{cases}$

**26.** $\begin{cases} 6x - 12y = 9 \\ -4x + 8y = -6 \end{cases}$

# 4.4 Applications of Linear Systems

## THINGS TO KNOW

Before working through this section, be sure you are familiar with the following concepts:

|  |  | VIDEO | ANIMATION | INTERACTIVE |
|---|---|---|---|---|

 **You Try It**
**1.** Solve Problems Involving Related Quantities (Section 2.3, **Objective 3**)

 **You Try It**
**2.** Solve Problems Using Geometry Formulas (Section 2.5, **Objective 1**)

 **You Try It**
**3.** Solve Problems Involving Angles (Section 2.5, **Objective 2**)

 **You Try It**
**4.** Solve Problems Involving Uniform Motion (Section 2.5, **Objective 3**)

 **You Try It**
**5.** Solve Mixture Problems (Section 2.6, **Objective 4**)

 **You Try It**
**6.** Solve Systems of Linear Equations by Substitution (Section 4.2, **Objective 1**)

 **You Try It**
**7.** Solve Systems of Linear Equation by Elimination (Section 4.3, **Objective 1**)

## OBJECTIVES

**1** Solve Related Quantity Applications Using Systems

**2** Solve Geometry Applications Using Systems

**3** Solve Uniform Motion Applications Using Systems

**4** Solve Mixture Applications Using Systems

## OBJECTIVE 1  SOLVE RELATED QUANTITY APPLICATIONS USING SYSTEMS

Sometimes we can use an **equation in one variable** to solve application problems such as those in **Section 2.3**. However, it is often easier to use two variables and create a **system of linear equations**. The following steps are based on the **problem-solving strategy** for applications of linear equations from Section 2.3.

---

**Problem-Solving Strategy for Applications Using Systems of Linear Equations**

**Step 1.**  **Define the Problem.**  Read the problem carefully; multiple times if necessary. Identify what you need to find and determine what information is available to help you find it.

**Step 2.**  **Assign Variables.**  Choose variables that describe each unknown quantity.

**Step 3.**  **Translate into a System of Equations.**  Use the relationships among the known and unknown quantities to form a system of equations.

**Step 4.**  **Solve the System.**  Use **graphing**, **substitution**, or **elimination** to solve the system.

**Step 5.**  **Check the Reasonableness of Your Answers.**  Check to see if your answers make sense within the context of the problem. If not, check your work for errors and try again.

**Step 6.**  **Answer the Question.**  Write a clear statement that answers the question(s).

---

### Example 1  Storage Capacity

The storage capacity of Deon's external hard drive is 32 times that of his jump drive, a small portable memory device. Together, his two devices have 264 gigabytes of memory. What is the memory size of each device?

**Solution**  Follow the **problem-solving strategy** for applications using systems.

**Step 1.** We must find the "memory size" or number of gigabytes that each device holds. The memory size of the external hard drive is 32 times that of the jump drive, and the total capacity is 264 gigabytes.

**Step 2.** Let $e$ = the memory size of the external hard drive and $j$ = the memory size of the jump drive. Note that we use the descriptive variables $e$ for external and $j$ for jump because they are easier to remember than other letters such as $x$ or $y$.

**Step 3.** Because the memory of the external hard drive is 32 times that of the jump drive, we have the equation $e = 32j$.

Also, the two devices have a total of 264 gigabytes of memory, so we have the equation $e + j = 264$.

The two equations together give the following **system:**

$$\begin{cases} e = 32j \\ e + j = 264 \end{cases}$$

**Step 4.** Because the first equation is solved for $e$, we will solve the system using **substitution**.

| | |
|---|---|
| Rewrite the second equation: | $e + j = 264$ |
| Substitute $32j$ for $e$: | $(32j) + j = 264$ |
| Simplify: | $33j = 264$ |
| Divide both sides by 33: | $j = 8$ |

To find $e$, substitute 8 for $j$ in the first equation.

Rewrite the first equation:  $e = 32j$

Substitute 8 for $j$:  $e = 32(8)$

Simplify:  $e = 256$

**Step 5.** The jump drive holds 8 gigabytes of memory, and the external hard drive holds 256 gigabytes. Because 256 is 32 times 8 and the sum of 8 and 256 is 264, these results make sense.

**Step 6.** Deon's jump drive holds 8 gigabytes of memory, and his external hard drive holds 256 gigabytes of memory.

Did the problem in Example 1 look familiar? Look back to the solution shown in **Example 4 in Section 2.3.** Now compare that solution process with the one just shown in Example 1. Which approach do you prefer? Algebra problems often can be solved in more than one way!

### Example 2  Ages of a Brother and Sister

*My interactive video summary*

 The **sum** of the ages of Ben and his younger sister Annie is 18 years. The **difference** of their ages is 4 years. What is the age of each child?

**Solution**  We follow the **problem-solving strategy** for applications using systems.

**Step 1.** We need to find the ages of Ben and Annie. We know that the sum of their ages is 18 years and the difference of their ages is 4 years.

**Step 2.** Let $B$ = Ben's age and $A$ = Annie's age.

**Step 3.** The sum of their ages is 18 years, so we have the equation $B + A = 18$.

Also, the difference of their ages is 4 years. Because Ben is the older child, we have the equation $B - A = 4$.

The two equations together give the following **system**:

$$\begin{cases} B + A = 18 \\ B - A = 4 \end{cases}$$

**Step 4.** Finish solving this system of equations on your own. View the **answer**, or watch this **interactive video** for a fully worked solution.

**You Try It**    Work through this You Try It problem.

Work Exercises 1–6 in this textbook or in the *MyMathLab*® Study Plan.

---

OBJECTIVE 2    SOLVE GEOMETRY APPLICATIONS USING SYSTEMS

In **Section 2.5,** we solved geometry problems using **formulas**. We can also use **systems of equations** to solve such applications.

### Example 3  A Calculator Display Panel

*My interactive video summary*

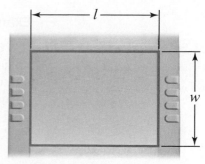 The display panel of a graphing calculator has the shape of a rectangle with a **perimeter** of 264 millimeters. If the length of the display panel is 18 millimeters longer than the width, find its dimensions.

## Solution

**Step 1.** We need to find the length and width of the display panel. We know that the perimeter is 264 mm and the length is 18 mm longer than the width.

**Step 2.** Let $l$ = length and $w$ = width.

**Step 3.** The perimeter of a rectangle is given by the formula $P = 2l + 2w$, so we have the equation $2l + 2w = 264$.

Also, the length is 18 mm longer than the width, so we have the equation $l = w + 18$.

The two equations together give the following system:

$$\begin{cases} 2l + 2w = 264 \\ l = w + 18 \end{cases}$$

**Step 4.** Finish solving this system of equations on your own. See the **answer**, or watch this **interactive video** for a fully worked solution.

**You Try It**    Work through this You Try It problem.

**Work Exercises 7–9 in this textbook or in the *MyMathLab*® Study Plan.**

Geometry problems sometimes involve *complementary* and *supplementary* angles. The measures of two **complementary angles** add to 90°. The measures of two **supplementary angles** add to 180°. Complementary angles form a **right angle**, and supplementary angles form a **straight angle**.

### Example 4  Supplementary Angles

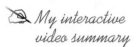

 Find the measures of two **supplementary angles** if the measure of the larger angle is 20 degrees less than three times the measure of the smaller angle.

**Solution** Follow the **problem-solving strategy** for applications using systems and try to solve this problem on your own. View the **answer**, or watch this **interactive video** for a detailed solution.

**You Try It**    Work through this You Try It problem.

**Work Exercises 10–12 in this textbook or in the *MyMathLab*® Study Plan.**

OBJECTIVE 3    SOLVE UNIFORM MOTION APPLICATIONS USING SYSTEMS

In **Section 2.5** we solved **uniform motion** applications involving only one variable. We can use a **system of equations** to solve similar types of problems when there are two variables.

### Example 5  Finding Different Times

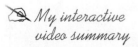

 Shawn is training for the Dirty Duo running-and-bicycling race. During a three-hour training session, his total distance cycling and running was 33 miles. If he cycled at a rate of 18 miles per hour and ran at a rate of 6 miles per hour, how much time did he spend doing each activity?

**Solution** We apply our **problem-solving strategy** for applications using systems.

**Step 1.** We want to find the time spent cycling and the time spent running. We know the total distance is 33 miles and the total time is 3 hours. We also know that Shawn can cycle 18 miles in one hour and run 6 miles in one hour.

**Step 2.** We know the time traveled and the distance traveled but not the individual times. We can define two variables to represent the times.

$$x = \text{time spent cycling}$$
$$y = \text{time spent running}$$

**Step 3.** This is a **uniform motion** problem, so we will need the **distance formula**, $d = r \cdot t$. The distance traveled for each activity is given by the product of the person's speed and time traveled. Because the total distance is 33 miles, we can write the following equation:

$$\underbrace{18x}_{\text{Distance cycling}} + \underbrace{6y}_{\text{Distance running}} = \underbrace{33}_{\text{Total distance}}$$

And, because the total time was three hours, we can write the following second equation:

$$x + y = 3$$

Writing the two equations together gives the system:

$$\begin{cases} 18x + 6y = 33 \\ x + y = 3 \end{cases}$$

**Step 4.** Solve this system by **substitution** or **elimination**. Complete Steps 5 and 6 on your own. View the answer, or watch this interactive video for a detailed solution.

**You Try It** Work through this You Try It problem.

Work Exercises 13–14 in this textbook or in the *MyMathLab*® Study Plan.

Some **uniform motion** problems involve one motion that works with or against another. Walking up the down escalator or paddling upstream are examples of two motions working against each other. Kicking a football with the wind or walking in the direction of motion on a moving sidewalk are examples of two motions working together. For these problems, it is important to remember that when motions work together the **rates** are added, but when they work against each other, the rates are subtracted.

### Example 6 Finding Different Rates

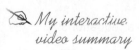
*My interactive video summary*

A jet plane travels 1950 miles in 3.9 hours going with the wind. On the return trip, the plane must fly into the wind and the travel time increases to 5 hours. Find the speed of the jet plane in still air and the speed of the wind. Assume the wind speed is the same for both trips.

**Solution** We apply our **problem-solving strategy** for applications using systems.

**Step 1.** We want to find the speed of the plane in still air and the speed of the wind. We know that the time of travel is 3.9 hours with the wind and 5 hours against (into) the wind. We also know that the plane travels 1950 miles each way.

**Step 2.** We know the time traveled and the distance traveled but not the speeds. We can define two variables to represent the speeds.

$$r = \text{speed of the plane in still air}$$

$$w = \text{wind speed}$$

**Step 3.** This is a **uniform motion** problem, so we will need the **distance formula**, $d = r \cdot t$. The distance traveled by the plane is given by the product of the plane's speed and time traveled. When flying with the wind, the plane's speed is found by adding the speed of the plane in still air to the wind speed. This allows us to write the following equation:

$$\underbrace{1950}_{\substack{\text{Distance} \\ \text{plane} \\ \text{traveled}}} = \underbrace{(r + w)}_{\substack{\text{Plane} \\ \text{speed with} \\ \text{the wind}}} \cdot \underbrace{(3.9)}_{\substack{\text{Time of} \\ \text{travel with} \\ \text{the wind}}}$$

We also know that when flying into (or against) the wind, we find the plane's speed by subtracting the speed of the wind from the speed of the plane in still air. The distance traveled is the same, so we get the following second equation:

$$\underbrace{1950}_{\substack{\text{Distance} \\ \text{plane} \\ \text{traveled on} \\ \text{return trip}}} = \underbrace{(r - w)}_{\substack{\text{Plane} \\ \text{speed} \\ \text{against} \\ \text{the wind}}} \cdot \underbrace{(5)}_{\substack{\text{Time of} \\ \text{travel} \\ \text{against} \\ \text{the wind}}}$$

Writing the two equations together gives the system:

$$\begin{cases} 1950 = (r + w)(3.9) \\ 1950 = (r - w)(5) \end{cases}$$

**Step 4.** Solve the system by **substitution** or **elimination**. Complete Steps 5 and 6 on your own. View the **answer**, or watch this **interactive video** for a detailed solution.

**You Try It**    **Work through this You Try It problem.**

**Work Exercises 15–16 in this textbook or in the *MyMathLab*® Study Plan.**

## OBJECTIVE 4    SOLVE MIXTURE APPLICATIONS USING SYSTEMS

### Example 7  Mailing Packages

*My interactive video summary*

A shipping company delivered 160 packages one day. The cost of regular delivery is $6.50, and the cost for express delivery is $17.50. Total shipping **revenue** for the day was $1513. How many of each kind of delivery were made?

**Solution**  We apply our **problem-solving strategy** for systems of linear equations.

**Step 1.** We want to find the number of regular deliveries and the number of express deliveries. We know that the total number of deliveries was 160 with a total revenue of $1513. We also know that regular deliveries cost $6.50 and overnight deliveries cost $17.50.

**Step 2.** We know the total number of deliveries and total revenue, but not the number of each delivery type. We define two variables to represent the number of each type of delivery.

$$R = \text{number of regular deliveries in one day}$$

$$E = \text{number of express deliveries in one day}$$

**Step 3.** For mixture problems, *totals* are good places to start when trying to write equations in a system. The first sentence tells us about the *total* number of deliveries.

A shipping company delivered 160 packages one day.

The total number of deliveries is the sum of the numbers for each delivery type. Therefore, our first equation is

$$\underbrace{R}_{\text{Regular deliveries}} + \underbrace{E}_{\text{Express deliveries}} = \underbrace{160}_{\text{Total deliveries}}$$

The second and third sentences tell us about the *total* **revenue.**

The cost of regular delivery is $6.50, and the cost for express delivery is $17.50. Total shipping revenue for the day was $1513.

Revenue is found by multiplying the number of packages delivered by the charge per package. The total revenue is the sum of the revenue from each delivery type. This will be our second equation.

$$\underbrace{6.50R}_{\substack{\text{Regular delivery} \\ \text{revenue}}} + \underbrace{17.50E}_{\substack{\text{Express delivery} \\ \text{revenue}}} = \underbrace{1513}_{\text{Total revenue}}$$

Writing the two equations together gives the system:

$$\begin{cases} R + E = 160 \\ 6.50R + 17.50E = 1513 \end{cases}$$

**Step 4.** Solve this system by **substitution** or **elimination**. Complete Steps 5 and 6 on your own. View the **answer**, or watch this **interactive video** for a detailed solution.

**You Try It** **Work through this You Try It problem.**

**Work Exercises 17–19 in this textbook or in the *MyMathLab*®** Study Plan.

## Example 8  Mixing Solutions

*My interactive video summary*

A chemist needs eight liters of a 50% alcohol solution but only has a 30% solution and an 80% solution available. How many liters of each solution should be mixed to form the needed solution?

**Solution** We apply our **problem-solving strategy** for applications using systems.

**Step 1.** We want to find the number of liters of a 30% alcohol solution and the number of liters of an 80% alcohol solution that must be mixed to form 8 liters of a 50% alcohol solution.

**Step 2.** We know the total number of liters needed for the mixture, but not the number of liters for each type of solution. We define two variables to represent the number of liters of each type.

$$x = \text{liters of 30\% alcohol solution}$$
$$y = \text{liters of 80\% alcohol solution}$$

**Step 3.** Since we know the total number of liters, our first equation deals with the number of liters.

$$x + y = 8$$

The second equation deals with the percents, but we will consider the amount of alcohol contained in each solution. The total amount of alcohol in the mixture will equal the sum of the alcohol from each solution. If we multiply each percent by the number of liters of that solution type, we can find the number of liters of alcohol in each solution. In words we have

$$\underbrace{(\text{percent})(\text{liters of }30\%)}_{\text{Alcohol in 30\% solution}} + \underbrace{(\text{percent})(\text{liters of }80\%)}_{\text{Alcohol in 80\% solution}} = \underbrace{(\text{percent})(\text{total liters of }50\%)}_{\text{Alcohol in 50\% mixture}}$$

Using the given percents and the defined variables, we get our second equation:

$$0.3(x) + 0.8(y) = 0.5(8)$$

or

$$0.3x + 0.8y = 4$$

So the two equations in our system are

$$\begin{cases} x + y = 8 \\ 0.3x + 0.8y = 4 \end{cases}.$$

**Step 4.** Solve this system by **substitution** or **elimination**. Complete Steps 5 and 6 on your own. View the **answer**, or watch this **interactive video** for a detailed solution.

**You Try It**    Work through this You Try It problem.

Work Exercises 20–22 in this textbook or in the *MyMathLab*® Study Plan.

### Example 9  Counting Calories

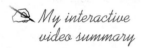 *My interactive video summary*

 Logan and Payton went to Culver's for lunch. Logan ate two Butterburgers with cheese and a small order of fries for a total of 1801 calories. Payton ate one Butterburger with cheese and two small orders of fries for a total of 1313 calories. How many calories are in a Culver's Butterburger with cheese? How many calories are in a small order of fries? (*Source*: Culver's Nutritional Guide, September 2010)

**Solution**  Apply the **problem-solving strategy** for applications using systems and try solving this problem on your own. View the **answer**, or watch this **interactive video** for the complete solution using either **substitution** or **elimination**.

**You Try It**    Work through this You Try It problem.

Work Exercises 23–25 in this textbook or in the *MyMathLab*® Study Plan.

# 4.4  Exercises

In Exercises 1–25, use a system of linear equations to solve each application.

**You Try It**

1. **Two Brothers**   Clayton is 2 years younger than his brother, Josh. If the sum of their ages is 26, how old is each boy?

2. **Number Problem**   The sum of two numbers is 121, while the difference of the two numbers is 15. What are the two numbers?

3. **Weights of Friends** Isaiah weighs 20 pounds more than his friend, Geoff. If the sum of their weights is 340 pounds, how much does each man weigh?

4. **Coin Collection** Susan has a collection of 50 nickels and dimes. If the number of nickels is four times the number of dimes, how many nickels and how many dimes does she have?

5. **Heights of Landmarks** The height of the St. Louis Arch is 19 feet more than twice the height of the Statue of Liberty (from the base of its pedestal). If the difference in their heights is 324.5 feet, find the heights of the two landmarks. (*Sources:* gatewayarch.com and statueofliberty.org)

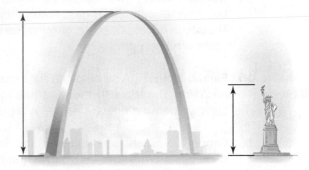

6. **River Lengths** The Missouri River is 200 miles longer than the Mississippi River. If the sum of their lengths is 4880 miles, what is the length of each river? (*Source:* U.S. Geological Survey)

**You Try It**

7. **TV Dimensions** The flat screen of a 3D television is 49 centimeters longer than it is wide. If the perimeter of the screen is 398 centimeters, find its dimensions.

8. **Tennis Court** A doubles tennis court has a perimeter of 228 feet. If six times the length of the court equals thirteen times the width, what are its dimensions?

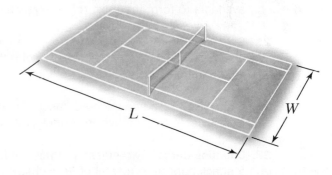

9. **Isosceles Triangle** Recall that an isosceles triangle has two sides with equal lengths. The perimeter of an isosceles triangle is 80 centimeters. If the non-equal side is 10 centimeters shorter than each of the two equal sides, find the lengths of the three sides.

**You Try It**

10. **Supplementary Angles** Two angles are supplementary. The measure of the larger angle is 8 degrees more than three times the measure of the smaller angle. Find the measure of each angle.

11. **Complementary Angles** Two angles are complementary. The measure of the larger angle is 35 degrees less than four times the measure of smaller angle. Find the measure of each angle.

12. **Right Triangle** Recall that a right triangle contains a 90-degree angle. Also recall that the measures of the three angles in any triangle add to 180 degrees. In the figure shown, the measure of angle $y$ is 6 degrees larger than twice the measure of angle $x$. Find the measures of angles $x$ and $y$.

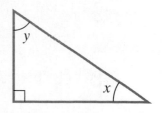

You Try It

**13. Aquathon** Nikola competed in an aquathon (swimming and running) competition. He swam at a rate of 3 km/hr and ran at a rate of 10 km/hr for a total distance traveled of 11.5 km. If he completed the race in 1.5 hours, how long did he take to complete each part of the race?

**14. Traveling Apart** Two friends, Thaddeus and Ian, start at the same location and drive in opposite directions, but leave at different times. When they are 620 miles apart, their combined travel time is 10 hours. If Thaddeus drives at a rate of 60 miles per hour and Ian drives at a rate of 65 miles per hour, how long had each been driving?

You Try It

**15. Puddle Jumper** A small plane flies roundtrip between Wichita, KS, and Columbus, OH. The trip is 784 miles each way. The trip with the wind (the wind going in the same direction) takes 3.5 hours, while the trip against the wind takes 4 hours. What is the speed of the plane in still air? What is the speed of the wind?

**16. Canoe Trip** Nicholas paddles his canoe downstream from the Lodge to Big Bend in 4 hours and then back upstream to the Lodge in 6 hrs. If the distance from the Lodge to Big Bend is 18 miles, find Nicholas' speed in still water and the speed of the current.

You Try It

**17. Soda Mix** Elodie is buying cans of soda for the school picnic. Her choices are Coke at $0.50 per can and Dr. Pepper at $0.55 per can. She wants to provide a mixture of choices but has a limited budget. If she buys 271 cans and spends $141, how many cans of each flavor did she buy?

**18. Pancake Breakfast** A youth club held a pancake breakfast to raise money for a trip. Tickets were $3 for children and $5.50 for adults. If 279 tickets were sold and the group took in $1322, find the number of adults and the number of children that attended the breakfast.

**19. Trail Mix** Chloe purchased some dried mango and dried kiwi to make some trail mix. The dried mango cost $3.86 per pound and the dried kiwi cost $4.60 per pound. If she spent $17.29 for 4 pounds of dried fruit, how many pounds of each fruit did she have?

You Try It

**20. Acid Solution** A chemist needs 3 liters of a 12% acid solution. He has a 10% solution and a 20% solution available to form the mixture. How much of each should be used to form the 12% solution?

**21. Gold Jewelry** A jeweler wants to make gold chains that are 60% gold. She has a supply of 18-carat gold (75% gold) and 12-carat gold (50% gold) that can be melted to form the desired alloy. How much 18-carat gold and how much 12-carat gold should be used to form 350 ounces of the 60% alloy?

**22. Fruit Juice** To be considered "nectar" a drink must contain 40% fruit juice. Find how much pure (100%) fruit juice and how much fruit drink (10% juice) must be mixed to make 9 gallons of "nectar."

You Try It

**23. Counting Carbs** Two glasses of milk and 3 snack bars have a total of 61 carbs, and 1 glass of milk and 5 snack bars have a total of 76 carbs. How many carbs are in one glass of milk? How many carbs are in one snack bar?

**24. Pastry Purchase** Bill and Mary Ann went to the Viola bakery. Bill bought 5 danishes and 7 filled donuts for $13.30. Mary Ann bought 6 of each for $13.08. What is the price of each type of pastry?

**25. Counting Calories** At In-N-Out Burger, 6 cheeseburgers w/onion and 3 orders of french fries contain 4080 calories. 3 cheeseburgers w/onion and 2 orders of french fries contain 2240 calories. How many calories are in a cheeseburger w/onion? How many calories are in one order of french fries? (*Source*: In-N-Out Burger Nutritional Facts, September 2010)

# 4.5 System of Linear Inequalities

## THINGS TO KNOW

Before working through this section, be sure you are familiar with the following concepts:

|  |  | VIDEO | ANIMATION | INTERACTIVE |

**You Try It**

1.  Determine If an Ordered Pair Is a Solution to a Linear Inequality in Two Variables (Section 3.5, **Objective 1**)

**You Try It**

2.  Graph a Linear Inequality in Two Variables (Section 3.5, **Objective 2**)

**You Try It**

3.  Solve Applications Involving Linear Inequalities in Two Variables (Section 3.5, **Objective 3**)

## OBJECTIVES

**1** Determine If an Ordered Pair Is a Solution to a System of Linear Inequalities in Two Variables

**2** Graph Systems of Linear Inequalities

**3** Solve Applications Involving Systems of Linear Inequalities

---

OBJECTIVE 1  DETERMINE IF AN ORDERED PAIR IS A SOLUTION TO A SYSTEM OF LINEAR INEQUALITIES IN TWO VARIABLES

In **Section 3.5**, we learned about **linear inequalities in two variables**. In this section, we explore *systems of linear inequalities in two variables*.

> **Definition**  **System of Linear Inequalities in Two Variables**
>
> A **system of linear inequalities in two variables** is a collection of two or more linear inequalities in two variables considered together.

The following three examples all represent systems of linear inequalities in two variables.

$$\begin{cases} 2x - 3y < 6 \\ 3x + 2y > 12 \end{cases} \qquad \begin{cases} q \geq -\dfrac{5}{2}p + 6 \\ q \leq 3 \end{cases} \qquad \begin{cases} y < -x + 5 \\ x \geq 0 \\ y \geq 0 \end{cases}$$

Notice that a system of linear inequalities looks like a **system of linear equations**, except **inequality symbols** replace the equal signs.

> **Definition**  **Solution to a System of Linear Inequalities in Two Variables**
>
> A **solution to a system of linear inequalities** in two variables is an **ordered pair** that, when substituted for the variables, makes all inequalities in the system true.

### Example 1  Determining If an Ordered Pair Is a Solution to a System of Linear Inequalities in Two Variables

 *My video summary*

Determine if each ordered pair is a **solution** to the following **system of inequalities**.

$$\begin{cases} 2x + y \geq -3 \\ x - 4y \leq 12 \end{cases}$$

**a.** $(4, 2)$      **b.** $(2, -5)$      **c.** $(0, -3)$

**Solutions** We substitute the *x*- and *y*-coordinates for the variables in each inequality and **simplify**. If the resulting statements are both true, then the ordered pair is a solution to the system.

**a.**

|  | First inequality | Second inequality |
|---|---|---|
| Write the original inequalities: | $2x + y \geq -3$ | $x - 4y \leq 12$ |
| Substitute 4 for *x* and 2 for *y*: | $2(4) + 2 \overset{?}{\geq} -3$ | $4 - 4(2) \overset{?}{\leq} 12$ |
| Multiply: | $8 + 2 \overset{?}{\geq} -3$ | $4 - 8 \overset{?}{\leq} 12$ |
| Add or subtract: | $10 \geq -3$  True | $-4 \leq 12$  True |

The ordered pair makes both inequalities true, so $(4, 2)$ is a solution to the system of inequalities.

**b.–c.** Try working through these parts on your own, then check your **answers**. Watch this **video** for detailed solutions to all three parts.

**You Try It**  Work through this You Try It problem.

Work Exercises 1–3 in this textbook or in the **MyMathLab**® Study Plan.

## OBJECTIVE 2  GRAPH SYSTEMS OF LINEAR INEQUALITIES

The **graph of a system of linear inequalities in two variables** is the **intersection** of the graphs of each inequality in the system. This graph represents the set of all **solutions to the system of inequalities**.

 *My animation summary*

To graph a system of linear inequalities, graph each inequality in the system and find the region they all have in common, if any. This common region is called the **solution region**. Watch this **animation** for an overview of graphing systems of linear inequalities, using the

system $\begin{cases} 2x - 3y \leq 9 \\ 2x - y > -1 \end{cases}$.

---

**Steps for Graphing Systems of Linear Inequalities**

**Step 1.** Use the **steps for graphing linear inequalities in two variables** from **Section 3.5** to graph each linear inequality in the system on the same **coordinate plane**.

**Step 2.** Determine the region where the shaded areas overlap, if any. This region represents the set of all solutions to the system of inequalities.

---

 Remember to use a dashed **boundary line** when the inequality is **strict** and a solid boundary line when the inequality is **non-strict**.

**Note:** For the remainder of this section, we will not find or label the exact **intersection point** of boundary lines. If desired, however, these points could be found by using the **substitution** or **elimination methods** for solving **systems of linear equations**.

### Example 2 Graphing Systems of Linear Inequalities

 *My video summary*  Graph the system of linear inequalities from **Example 1**.

$$\begin{cases} 2x + y \geq -3 \\ x - 4y \leq 12 \end{cases}$$

**Solution** Follow the steps for graphing systems of linear inequalities.

**Step 1. Graph $2x + y \geq -3$.** The inequality is **non-strict**, so we use a solid line to graph its **boundary line** $2x + y = -3$. We use the **test point** $(0, 0)$. Do you see **why**? Because $2(0) + 0 \geq -3$ is true, we shade the **half-plane** that contains $(0, 0)$. See **Figure 11(a)**.

**Graph $x - 4y \leq 12$.** The inequality is non-strict, so we use a solid line to graph its boundary line $x - 4y = 12$. The test point $(0, 0)$ satisfies the inequality because $0 - 4(0) \leq 12$ is true, so we shade the half-plane that contains $(0, 0)$. See **Figure 11(b)**.

**Step 2.** The graph of the system of inequalities is the area where the two inequalities overlap. This is the darkest shaded region in **Figure 11(c)**. Any point that falls in this darkest region is a **solution to the system**.

Watch this **video** for a detailed solution.

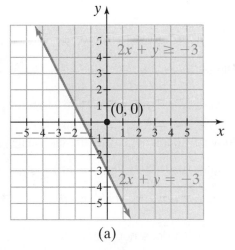

(a)

Graph of the linear inequality
$2x + y \geq -3$

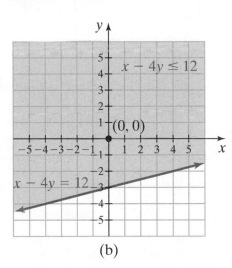

(b)

Graph of the linear inequality
$x - 4y \leq 12$

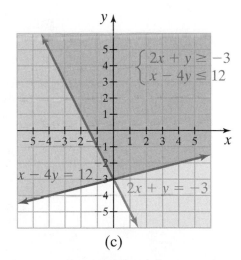

(c)

Graph of the system
$\begin{cases} 2x + y \geq -3 \\ x - 4y \leq 12 \end{cases}$

**Figure 11**

Only the points that fall in the darkest-shaded region of Figure 11(c) are **solutions to the system** in **Example 2**. This includes the points on the adjacent sections of the two solid boundary lines $2x + y = -3$ and $x - 4y = 12$, including their intersection point.

**You Try It** Work through this You Try It problem.

Work Exercises 4–6 in this textbook or in the *MyMathLab*® Study Plan.

## Example 3  Graphing Systems of Linear Inequalities

 *My video summary*     Graph the system of linear inequalities.

$$\begin{cases} x + y < 4 \\ x - 2y < -2 \end{cases}$$

**Solution** Both inequalities in this system are **strict**, so the **boundary lines** $x + y = 4$ and $x - 2y = -2$ are dashed lines. **Graph the inequalities** on the same **coordinate plane** and find the region they have in common. View the **graph**, or watch this **video** for a detailed solution.

The points on the two dashed boundary lines $x + y = 4$ and $x - 2y = -2$, including the intersection point, are not solutions to this system.

**You Try It**   **Work through this You Try It problem.**

**Work Exercises 7–9 in this textbook or in the *MyMathLab*® Study Plan.**

A system of linear inequalities typically has an infinite number of solutions, but a system of inequalities with no solution, called an **inconsistent system of inequalities**, is also possible.

## Example 4  Identifying Inconsistent Systems of Linear Inequalities

*My video summary*    Graph the system of linear inequalities.

$$\begin{cases} y \leq -\dfrac{1}{3}x - 3 \\ y > -\dfrac{1}{3}x + 2 \end{cases}$$

**Solution** Follow the **steps for graphing systems of linear inequalities.**

**Step 1. Graph $y \leq -\dfrac{1}{3}x - 3$.** The inequality is **non-strict**, so we use a solid line to graph

its **boundary line** $y = -\dfrac{1}{3}x - 3$. The **test point** $(0, 0)$ does not **satisfy** the inequality, so we shade the **half-plane** that does not contain $(0, 0)$. See **Figure 12(a)**.

**Graph $y > -\dfrac{1}{3}x + 2$.** The inequality is strict, so we use a dashed line to graph

its boundary line $y = -\dfrac{1}{3}x + 2$. The test point $(0, 0)$ does not satisfy the inequality, so we shade the half-plane that does not contain $(0, 0)$. See **Figure 12(b)**.

**Step 2.** The boundary lines $y = -\dfrac{1}{3}x - 3$ and $y = -\dfrac{1}{3}x + 2$ are **parallel**, and the shaded

regions are on opposite sides of these parallel lines, so there is no shared region. See **Figure 12(c)**. There are no ordered pairs that satisfy both inequalities, so the system has no solution and is **inconsistent**. We can use the null symbol $\varnothing$ or empty set $\{\ \}$ to indicate this.

Watch this **video** for a more detailed solution.

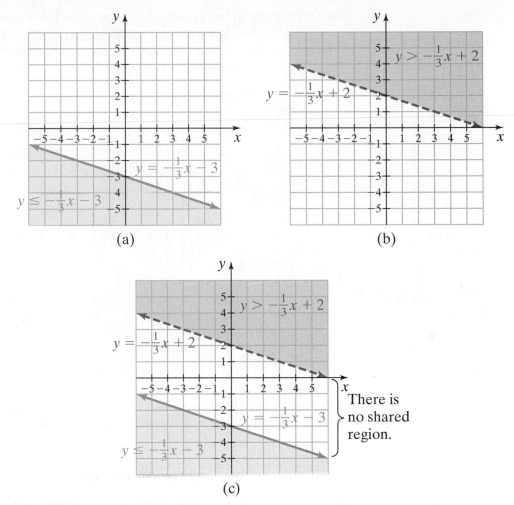

(a)

(b)

(c)

**Figure 12**

**You Try It**   Work through this You Try It problem.

**Work Exercises 10–12** in this textbook or in the **MyMathLab**® **Study Plan.**

Some systems of linear inequalities involve **horizontal** or **vertical** boundary lines.

### Example 5  Graphing Systems of Linear Inequalities

*My video summary*   Graph the system of linear inequalities.

$$\begin{cases} x - 3y > 6 \\ x \geq 1 \end{cases}$$

**Solution**  Follow the steps for graphing systems of linear inequalities.

Step 1.  **Graph** $x - 3y > 6$**.** The inequality is **strict,** so we use a dashed line to graph its
   **boundary line** $x - 3y = 6$**.** The **test point** $(0, 0)$ does not **satisfy** the inequality
   because $0 - 2(0) > 6$ is false, so we shade the **half-plane** that does not contain
   $(0, 0)$. See **Figure 13(a).**

   **Graph** $x \geq 1$**.** The inequality is **non-strict,** so we use a solid line to graph its
   vertical boundary line $x = 1$. The test point $(0, 0)$ does not satisfy the inequality
   because $0 \geq 1$ is false, so we shade the half-plane that does not contains $(0, 0)$.
   See **Figure 13(b).**

**Step 2.** The graph of the system of inequalities is the area where the two inequalities overlap. This is the darkest shaded region in Figure 13(c). Any point that falls in this darkest region, including the adjacent section of the solid boundary line $x = 1$, is a **solution to the system**.

Watch this **video** for a detailed solution.

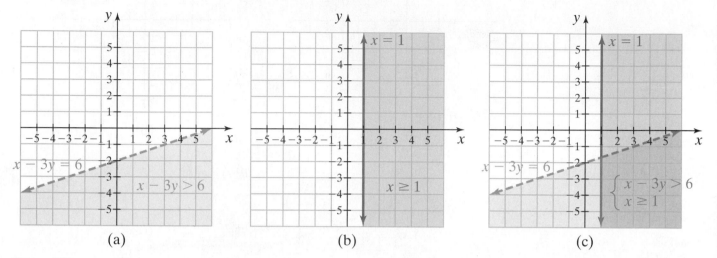

(a)          (b)          (c)

**Figure 13**

The points on the dashed boundary line $x - 3y = 6$, including the intersection point with the solid boundary line $x = 1$, are not solutions to this system.

**You Try It**    **Work through this You Try It problem.**

Work Exercises 13–15 in this textbook or in the **MyMathLab**® Study Plan.

Consider the system $\begin{cases} x \geq 0 \\ y \geq 0 \end{cases}$. What is its **solution region**? See this **explanation**.

Sometimes a **system of linear inequalities** contains more than two inequalities.

### Example 6   Graphing Systems with More Than Two Linear Inequalities

*My video summary*   Graph the system of linear inequalities.

$$\begin{cases} x + y < 4 \\ y > 0 \\ 2x - y > -4 \end{cases}$$

**Solution**   Follow the **steps for graphing systems of linear inequalities**.

**Step 1.** **Graph $x + y < 4$.** The inequality is **strict**, so we use a dashed line to graph its **boundary line** $x + y = 4$. Check a convenient **test point** to confirm that we shade the **half-plane** below this boundary line. View this **graph**.

**Graph $y > 0$.** The inequality is strict, so we use a dashed line to graph its horizontal boundary line $y = 0$. Test a convenient point to confirm that we shade above this boundary line. (Note that the test point in this case cannot be $(0, 0)$.) View this **graph**.

**Graph $2x - y > -4$.** The inequality is strict, so we use a dashed line to graph its boundary line $2x - y = -4$. Test a convenient point to confirm that the half plane below this boundary line is shaded. View this **graph**.

**Step 2.** The graph of the system of inequalities is the area where the three inequalities overlap. This is the shaded region in Figure 14. Any point that falls in this shaded region is a **solution to the system**.

Watch this **video** for a detailed solution.

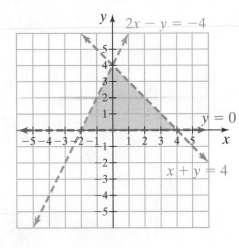

**Figure 14**

$$\text{Graph of } \begin{cases} x + y < 4 \\ y > 0 \\ 2x - y > -4 \end{cases}$$

The points on the dashed boundary lines, including the intersection points, are not solutions to this system.

**You Try It**   **Work through this You Try It problem.**

**Work Exercises 16–18 in this textbook or in the *MyMathLab*®  Study Plan.**

OBJECTIVE 3   SOLVE APPLICATIONS INVOLVING SYSTEMS OF LINEAR INEQUALITIES

Let's look at a real life application that is **modeled** by **system of linear inequalities**.

**Example 7  Planning a Barbeque**

*My video summary*   Savannah is planning a barbeque for her family and friends. She will spend $150 or less to buy hamburger patties that cost $3 per pound and boneless chicken breast that cost $5 per pound. To limit waste, she will purchase at most 40 pounds of meat all together. Also, the amount of hamburger and chicken purchased must be **non-negative**. A system of linear inequalities that models this situation is

$$\begin{cases} 3h + 5c \le 150 \\ h + c \le 40 \\ h \ge 0 \\ c \ge 0 \end{cases}$$

where $h$ = pounds of hamburger patties and $c$ = pounds of chicken breast.

a.  Graph the system of linear inequalities.

b.  Can Savannah purchase 20 pounds of hamburger patties and 15 pounds of chicken breast for the barbeque?

c.  Can Savannah purchase 10 pounds of hamburger patties and 30 pounds of chicken breast for the barbeque?

**Solution**

a. Label the horizontal axis $h$ and the vertical axis $c$. The inequalities $h \geq 0$ and $c \geq 0$ restrict the graph to **Quadrant I.** Follow the **steps for graphing systems of linear inequalities** to result in the graph in Figure 15. Watch this **video** for a detailed solution.

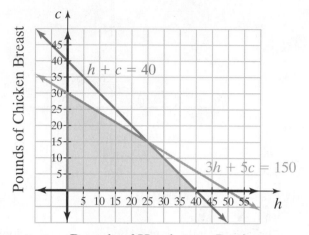

Pounds of Hamburger Patties          **Figure 15**

b. Yes. The ordered pair $(h, c) = (20, 15)$ lies within the shaded region in Figure 15, so Savannah can purchase 20 pounds of hamburger patties and 15 pounds of chicken.

c. No. The ordered pair $(h, c) = (10, 30)$ lies outside of the shaded region in Figure 15, so Savannah cannot purchase 10 pounds of hamburger patties and 30 pounds of chicken.

**You Try It    Work through this You Try It problem.**

Work Exercises 19–22 in this textbook or in the *MyMathLab*® Study Plan.

# 4.5  **Exercises**

In Exercises 1–3, determine if each ordered pair is a solution to the given system of linear inequalities in two variables.

**You Try It**

**1.** $\begin{cases} x + y \leq 10 \\ x - y \geq -2 \end{cases}$

    a. $(1, -5)$
    b. $(4, 6)$
    c. $(-3, 7)$

**2.** $\begin{cases} 2x - 3y > 6 \\ 5x + 2y < -1 \end{cases}$

    a. $(2, -2)$
    b. $(1, -3)$
    c. $(0, -4)$

**3.** $\begin{cases} 2x - 5y \leq 10 \\ x - y > -6 \\ 3x + 4y < 12 \end{cases}$

    a. $(-1, -2)$
    b. $(3, 2)$
    c. $(-5, -4)$

In Exercises 4–18, graph each system of linear inequalities in two variables.

**You Try It**

**4.** $\begin{cases} y > 2x - 1 \\ y \leq x + 2 \end{cases}$

**5.** $\begin{cases} x + y \geq 2 \\ x - y \leq 4 \end{cases}$

**6.** $\begin{cases} 3x - 2y \leq -2 \\ x + 2y > 6 \end{cases}$

**You Try It**

**7.** $\begin{cases} 3x + y < 2 \\ x + 2y < 8 \end{cases}$

**8.** $\begin{cases} y < 2x \\ y > -3x \end{cases}$

**9.** $\begin{cases} 3x - 2y \leq -2 \\ x + 2y \geq 6 \end{cases}$

**You Try It**

**10.** $\begin{cases} y < x - 2 \\ y \geq x + 4 \end{cases}$

**11.** $\begin{cases} 2x + y \leq 4 \\ y \geq -2x - 1 \end{cases}$

**12.** $\begin{cases} 3x - 4y > 12 \\ 6x - 8y < -32 \end{cases}$

**You Try It**

**13.** $\begin{cases} 4x - y > 2 \\ y \geq -1 \end{cases}$

**14.** $\begin{cases} x \geq -1 \\ y \leq 3 \end{cases}$

**15.** $\begin{cases} \dfrac{1}{2}x - \dfrac{1}{3}y \leq 1 \\ x \geq -1 \end{cases}$

**You Try It**

**16.** $\begin{cases} x + y \leq 5 \\ x \geq 0 \\ y \geq 0 \end{cases}$

**17.** $\begin{cases} x + y > -2 \\ x - y < 2 \\ y < 2 \end{cases}$

**18.** $\begin{cases} x + y < 6 \\ 3x - y > -2 \\ x \geq 0 \\ y \geq 0 \end{cases}$

In Exercises 19–22, solve each application by using a system of linear inequalities in two variables.

**You Try It**

**19. Investing** Judy has $50,000 to invest. Her financial advisor recommends that she place at least $30,000 in mutual funds and no more than $20,000 in bonds. A system of linear inequalities that models this situation is

$$\begin{cases} m + b \leq 50{,}000 \\ m \geq 30{,}000 \\ b \leq 20{,}000 \\ b \geq 0 \end{cases}$$

where $m$ = the dollar amount invested in mutual funds and $b$ = the dollar amount invested in bonds.
**a.** Graph the system of linear inequalities.
**b.** Can Judy invest $20,000 in mutual funds and $20,000 in bonds?
**c.** Can Judy invest $38,000 in mutual funds and $12,000 in bonds?

**20. Getting Cash** For an upcoming shopping trip, Elaine will withdraw cash from her checking account which contains $100. She will want the cash in only five-dollar and ten-dollar bills, and she will carry no more than 12 bills all together. A system of linear inequalities that models this situation is

$$\begin{cases} 5f + 10t \leq 100 \\ f + t \leq 12 \\ f \geq 0 \\ t \geq 0 \end{cases}$$

where $f$ = the number $5 bills and $t$ = the number of $10 bills Elaine receives.
**a.** Graph the system of linear inequalities.
**b.** Can Elaine receive 4 five-dollar bills and 8 ten-dollar bills?
**c.** Can Elaine receive 7 five-dollar bills and 3 ten-dollar bills?

**21. Growing Crops** In Illinois, the annual cost to grow corn is about $450 per acre, and the annual cost to grow soybeans is about $300 per acre (not counting the cost of land). A small farmer has at most 150 acres for growing corn and soybeans. His budget for growing these crops is at most $60,000. A system of linear inequalities that models this situation is

$$\begin{cases} c + s \leq 150 \\ 450c + 300s \leq 60{,}000 \\ c \geq 0 \\ s \geq 0 \end{cases}$$

where $c$ = acres of corn and $s$ = acres of soybeans. (*Source: Farmdoc, University of Illinois*)
**a.** Graph the system of linear inequalities.
**b.** Can this farmer grow 100 acres of corn and 50 acres of soybeans?
**c.** Can this farmer grow 120 acres of corn and 30 acres of soybeans?

22. **Dimensions of a Garden**   Austin is planning to build a rectangular garden. Its perimeter will be at most 36 feet, and its length will be no more than twice its width. A system of linear inequalities that models this situation is

$$\begin{cases} 2l + 2w \leq 36 \\ l \leq 2w \\ l \geq w \end{cases}$$

where $l$ = the length and $w$ = the width of the garden.
a. Graph the system of linear inequalities.
b. Can Austin's garden be 15 feet long and 10 feet wide?
c. Can Austin's garden be 10 feet long and 8 feet wide?

CHAPTER FIVE

# Exponents and Polynomials

## CHAPTER FIVE CONTENTS

# 5.1  Exponents

## THINGS TO KNOW

Before working through this section, be sure you are familiar with the following concepts:

VIDEO          ANIMATION          INTERACTIVE

**You Try It**

1.  Evaluate Exponential Expressions
    (Section 1.4, **Objective 1**)

## OBJECTIVES

**1** Simplify Exponential Expressions Using the Product Rule

**2** Simplify Exponential Expressions Using the Quotient Rule

**3** Use the Zero-Power Rule

**4** Use the Power-to-Power Rule

**5** Use the Product-to-Power Rule

**6** Use the Quotient-to-Power Rule

**7** Simplify Exponential Expressions Using a Combination of Rules

OBJECTIVE 1    SIMPLIFY EXPONENTIAL EXPRESSIONS USING THE PRODUCT RULE

An **exponential expression** is a **constant** or **algebraic expression** that is raised to a **power**. The constant or algebraic expression makes up the **base**, and the power is the **exponent** on the base. In **Section 1.4**, we learned that an **exponent** can be used to show repeated multiplication. For example, $3 \cdot 3 \cdot 3 \cdot 3$ can be written as $3^4$.

$$\underbrace{3 \cdot 3 \cdot 3 \cdot 3}_{4 \text{ factors of } 3} = 3^{\underset{\uparrow}{4}} \leftarrow \text{Exponent}$$

Base

The exponent 4 indicates that there are four **factors** of the base 3.

The same is true when the base is a **variable** or an algebraic expression.

$$\underbrace{x \cdot x \cdot x \cdot x \cdot x}_{5 \text{ factors of } x} = x^{5} \leftarrow \text{Exponent} \qquad \underbrace{(2y) \cdot (2y) \cdot (2y)}_{3 \text{ factors of } 2y} = (2y)^{3} \leftarrow \text{Exponent}$$

Base                                     Base

The exponent 5 means there are five factors of the base $x$, and the exponent 3 means there are three factors of the base $2y$.

When an algebraic expression such as $2y$ is the base of an exponential expression, we must use parentheses (or other **grouping symbols**) to show that the entire expression $2y$ is raised to the third power, not just the variable $y$.

$$(2y) \cdot (2y) \cdot (2y) = (2y)^{3} \neq 2y^{3} = 2 \cdot y \cdot y \cdot y$$

If we multiply **exponential expressions** with the same base, such as $x^{3} \cdot x^{4}$, we write

$$x^{3} \cdot x^{4} = \underbrace{(x \cdot x \cdot x)}_{3 \text{ factors of } x} \cdot \underbrace{(x \cdot x \cdot x \cdot x)}_{4 \text{ factors of } x}$$

$$= \underbrace{x \cdot x \cdot x \cdot x \cdot x \cdot x \cdot x}_{7 \text{ factors of } x}$$

$$= x^{7}.$$

Multiplying three **factors** of $x$ by four factors of $x$ means we are really multiplying seven factors of $x$, which is the sum of the two initial exponents 3 and 4:

$$x^{3} \cdot x^{4} = x^{3+4} = x^{7}$$

This result suggests the following rule.

---

**The Product Rule for Exponents**

When multiplying exponential expressions with the same base, add the exponents and keep the common base.

$$a^{m} \cdot a^{n} = a^{m+n}$$

---

 When a factor has no written **exponent**, it is understood to be 1. For example, $x = x^{1}$.

### Example 1  Using the Product Rule for Exponents

 Use the **product rule** to simplify each expression.

**a.** $5^{4} \cdot 5^{6}$      **b.** $x^{5} \cdot x^{7}$      **c.** $y^{3} \cdot y$      **d.** $b^{3} \cdot b^{5} \cdot b^{4}$

**Solutions**

**a.** The two exponential expressions $5^{4}$ and $5^{6}$ have the same base 5, so we add the exponents:

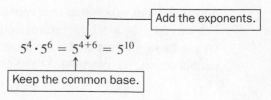

**b.** $x^5 \cdot x^7 = x^{5+7} = x^{12}$

**c.–d.** Try to simplify these expressions on your own. View the **answers**, or watch this **video** for detailed solutions to all four parts of this example.

**You Try It** Work through this You Try It problem.

Work Exercises **1–4** in this textbook or in the *MyMathLab*® Study Plan.

When using the product rule for exponents, do not multiply the bases. Instead, keep the common base and add the exponents. For example,

|  Incorrect  |  Correct  |
|-------------|-----------|
| $2^8 \cdot 2^6 = 4^{14}$ | $2^8 \cdot 2^6 = 2^{14}$ |

When the **exponential expressions** being multiplied involve more than one base, we first use the **commutative** and **associative** properties of multiplication to group **like bases** together and then apply the **product rule**.

### Example 2 Using the Product Rule for Exponents

 Simplify using the product rule.

**a.** $(4x^2)(7x^3)$ **b.** $(m^4n^2)(m^3n^6)$ **c.** $(-3a^5b^3)(-8a^2b)$

**Solutions**

**a.**
| | |
|---|---|
| Begin with the original expression: | $(4x^2)(7x^3)$ |
| Rearrange factors to group like bases: | $= 4 \cdot 7 \cdot x^2 \cdot x^3$ |
| Multiply constants; apply product rule: | $= 28 \cdot x^{2+3}$ |
| Simplify: | $= 28x^5$ |

**b.**
| | |
|---|---|
| Begin with the original expression: | $(m^4n^2)(m^3n^6)$ |
| Rearrange factors to group like bases: | $= m^4 \cdot m^3 \cdot n^2 \cdot n^6$ |
| Apply the product rule: | $= m^{4+3} \cdot n^{2+6}$ |
| Simplify: | $= m^7n^8$ |

**c.** Try to simplify this expression on your own. View the **answer**, or watch this **video** for detailed solutions to all three parts of this example.

**You Try It** Work through this You Try It problem.

Work Exercises **5–8** in this textbook or in the *MyMathLab*® Study Plan.

OBJECTIVE 2 SIMPLIFY EXPONENTIAL EXPRESSIONS USING THE QUOTIENT RULE

When dividing **exponential expressions** with the same nonzero **base**, such as $\dfrac{x^5}{x^2}$ with $x \neq 0$, we can expand each expression and divide out common factors:

$$\text{Expand:} \quad \frac{x^5}{x^2} = \frac{x \cdot x \cdot x \cdot x \cdot x}{x \cdot x} \quad \begin{array}{l} \leftarrow \text{5 factors of } x \\ \leftarrow \text{2 factors of } x \end{array}$$

$$\text{Divide out common factor:} \quad = \frac{\cancel{x} \cdot \cancel{x} \cdot x \cdot x \cdot x}{\cancel{x} \cdot \cancel{x}}$$

$$\text{Simplify:} \quad = x \cdot x \cdot x$$

$$= x^3$$

*My video summary*

Notice that both factors of $x$ from the **denominator** divide out, leaving three remaining factors of $x$ in the **numerator**. This result is the same if we subtract the initial exponents 5 and 2:

$$\frac{x^5}{x^2} = x^{5-2} = x^3$$

As long as the base is not 0, this result is true in general.

---

**The Quotient Rule for Exponents**

When dividing **exponential expressions** with the same non-zero **base**, subtract the **denominator** exponent from the **numerator** exponent and keep the common base.

$$\frac{a^m}{a^n} = a^{m-n} \quad (a \neq 0)$$

---

### Example 3  Using the Quotient Rule for Exponents

*My video summary*   Use the quotient rule to simplify each expression.

a. $\dfrac{t^9}{t^5}$     b. $\dfrac{7^5}{7^3}$     c. $\dfrac{y^{24}}{y^{15}}$     d. $\dfrac{(-4)^{14}}{(-4)^{11}}$

**Solutions**

a. The two exponential expressions $t^9$ and $t^5$ have the same base $t$, so we subtract the exponents:

Subtract the exponents.

$$\frac{t^9}{t^5} = t^{9-5} = t^4$$

Keep the common base.

b. $\dfrac{7^5}{7^3} = 7^{5-3} = 7^2 = 49$

**c.–d.** Try to simplify these expressions on your own. View the **answers**, or watch this **video** for detailed solutions to all four parts of this example.

**You Try It**   **Work through this You Try It problem.**

**Work Exercises 9–12 in this textbook or in the *MyMathLab*® Study Plan.**

When the **exponential expressions** being divided involve more than one **base**, we group **like bases** into individual **quotients**. Then we apply the **quotient rule** to each quotient and simplify.

### Example 4  Using the Quotient Rule for Exponents

*My video summary*   Simplify using the quotient rule.

a. $\dfrac{15x^6}{3x^2}$     b. $\dfrac{a^4b^9c^5}{a^2b^3c}$     c. $\dfrac{4m^6n^7}{12m^5n^2}$

## Solutions

**a.**   Begin with the original expression:   $\dfrac{15x^6}{3x^2}$

Group like bases into individual quotients:   $= \dfrac{15}{3} \cdot \dfrac{x^6}{x^2}$

Divide the constants; apply the quotient rule:   $= 5 \cdot x^{6-2}$

Simplify:   $= 5x^4$

**b.**   Begin with the original expression:   $\dfrac{a^4 b^9 c^5}{a^2 b^3 c}$

Group like bases into individual quotients:   $= \dfrac{a^4}{a^2} \cdot \dfrac{b^9}{b^3} \cdot \dfrac{c^5}{c^1}$ ← $\boxed{\text{Understood exponent of 1}}$

Apply the quotient rule to each quotient:   $= a^{4-2} \cdot b^{9-3} \cdot c^{5-1}$

Simplify:   $= a^2 b^6 c^4$

**c.**   Try to simplify this expression on your own. View the **answer**, or watch this **video** for detailed solutions to all three parts of this example.

**You Try It**   **Work through this You Try It problem.**

**Work Exercises 13–16 in this textbook or in the *MyMathLab* Study Plan.**

OBJECTIVE 3   USE THE ZERO-POWER RULE

If we multiply $2^3$ by $2^0$, we can use the **product rule** and write

$$2^3 \cdot 2^0 = 2^{3+0} = 2^3.$$

From the **multiplicative identity property** we know that $2^3 \cdot 1 = 2^3$. Therefore, it makes sense that $2^0 = 1$ because $2^3 \cdot 2^0 = 2^3 = 2^3 \cdot 1$. As long as the **base** is not 0, this result is true in general.

> **The Zero-Power Rule**
>
> A non-zero base raised to the 0 power equals 1.
>
> $$a^0 = 1 \quad (a \neq 0)$$

The zero-power rule can also be derived from the **quotient rule**. View this **popup box** to see how.

### Example 5   Using the Zero-Power Rule

Simplify using the zero-power rule.

**a.** $6^0$      **b.** $(-3)^0$      **c.** $-3^0$      **d.** $(2x)^0$      **e.** $2x^0$

### Solutions

**a.**   The base 6 is non-zero, so $6^0 = 1$.

**b.**   The parentheses indicate that the base is $-3$, which is non-zero, so $(-3)^0 = 1$.

**c.**   There are no parentheses, so the base is 3, not $-3$. We need to find the "opposite of $3^0$." So, $-3^0 = -(3^0) = -1$.

**d.** The parentheses indicate that the base is $2x$, which is non-zero provided that $x \neq 0$, so $(2x)^0 = 1$.

**e.** There are no parentheses, so the base is $x$, not $2x$. We have "2 times $x^0$." So, assuming $x \neq 0$, $2x^0 = 2 \cdot 1 = 2$.

**You Try It**   **Work through this You Try It problem.**

**Work Exercises 17–22 in this textbook or in the *MyMathLab*® Study Plan.**

OBJECTIVE 4   USE THE POWER-TO-POWER RULE

An **exponential expression** itself can be raised to a power, such as $(x^2)^3$. We can simplify this expression by expanding and then using the **product rule**.

$$(x^2)^3 = \underbrace{x^2 \cdot x^2 \cdot x^2}_{\text{3 factors of } x^2} = \underbrace{x^{2+2+2}}_{\text{Product rule}} = x^6$$

Notice the final exponent 6 is the result of multiplying the two original exponents 2 and 3, so $(x^2)^3 = x^{2 \cdot 3} = x^6$. This result is true in general.

---

**The Power-to-Power Rule**

When an exponential expression is raised to a power, multiply the exponents.

$$(a^m)^n = a^{m \cdot n}$$

---

**Example 6  Using the Power-to-Power Rule**

*My video summary*  Simplify using the **power-to-power rule**.

**a.** $(y^5)^6$          **b.** $[(-2)^3]^5$

**Solutions**

**a.** Because we are raising the exponential expression, $y^5$, to the sixth power, we multiply the exponents:

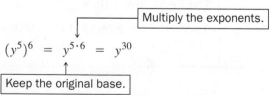

$$(y^5)^6 = y^{5 \cdot 6} = y^{30}$$

Keep the original base.

**b.** Try to simplify this expression on your own. View the **answer**, or watch this **video** for detailed solutions to both parts of this example.

**You Try It**   **Work through this You Try It problem.**

**Work Exercises 23–26 in this textbook or in the *MyMathLab*® Study Plan.**

OBJECTIVE 5   USE THE PRODUCT-TO-POWER RULE

When the **base** of an **exponential expression** is a **product**, such as $(xy)^4$, we can expand the expression and use the **commutative** and **associative** properties of multiplication to regroup like factors.

$$\text{Expand:} \quad (xy)^4 = (xy)\cdot(xy)\cdot(xy)\cdot(xy) \leftarrow 4 \text{ factors of } xy$$

$$\text{Group common factors:} \quad = \underbrace{x\cdot x\cdot x\cdot x}_{4 \text{ factors of } x} \cdot \underbrace{y\cdot y\cdot y\cdot y}_{4 \text{ factors of } y}$$

$$\text{Rewrite with exponents:} \quad = x^4 y^4$$

The final result is that each factor of the original base is raised to the exponent: $(xy)^4 = x^4y^4$. This result is true in general.

---

**The Product-to-Power Rule**

When raising a product to a power, raise each factor of the base to the common exponent.

$$(ab)^n = a^n b^n$$

---

### Example 7 Using the Product-to-Power Rule

Simplify using the **product-to-power rule**.

**a.** $(mn)^8$      **b.** $(x^2 y)^5$      **c.** $(3y)^4$      **d.** $(-4p^5 q^3)^2$

**Solutions**

**a.** The base is the product $mn$.

$$\text{Apply the product-to-power rule:} \quad (mn)^8 = m^8 n^8$$

**b.** The base is the product $x^2 y$.

$$\text{Apply the product-to-power rule:} \quad (x^2 y)^5 = (x^2)^5 \cdot y^5$$
$$\text{Apply the power-to-power rule:} \quad = x^{2\cdot 5} \cdot y^5$$
$$\text{Simplify:} \quad = x^{10} y^5$$

  **c.–d.** Try to simplify these expressions on your own. View the answers, or watch this video for detailed solutions to parts c and d.

**You Try It**     Work through this You Try It problem.

Work Exercises 27–32 in this textbook or In the **MyMathLab** Study Plan.

---

OBJECTIVE 6     USE THE QUOTIENT-TO-POWER RULE

When the **base** of an **exponential expression** is a **quotient**, such as $\left(\dfrac{x}{y}\right)^4$ with $y \neq 0$, we can expand the expression and multiply the resulting fractions.

$$\text{Expand:} \quad \left(\frac{x}{y}\right)^4 = \frac{x}{y}\cdot\frac{x}{y}\cdot\frac{x}{y}\cdot\frac{x}{y} \leftarrow 4 \text{ factors of } \frac{x}{y}$$

$$\text{Multiply the fractions:} \quad = \frac{x\cdot x\cdot x\cdot x}{y\cdot y\cdot y\cdot y} \begin{array}{l} \leftarrow 4 \text{ factors of } x \text{ in numerator} \\ \leftarrow 4 \text{ factors of } y \text{ in denominator} \end{array}$$

$$\text{Rewrite with exponents:} \quad = \frac{x^4}{y^4}$$

The final result is that the **numerator** and **denominator** are each raised to the **exponent**: $\left(\dfrac{x}{y}\right)^4 = \dfrac{x^4}{y^4}$. This result is true in general.

---

**The Quotient-to-Power Rule**

When raising a quotient to a power, raise both the numerator and denominator to the common exponent.

$$\left(\frac{a}{b}\right)^n = \frac{a^n}{b^n} \quad (b \neq 0)$$

---

**Example 8  Using the Quotient-to-Power Rule**

Simplify using the **quotient-to-power rule**.

a. $\left(\dfrac{m}{n}\right)^9$    b. $\left(\dfrac{x^2}{y^5}\right)^4$    c. $\left(\dfrac{x}{2}\right)^5$    d. $\left(\dfrac{3x^2}{5y^4}\right)^3$

**Solutions**

a.  The **base** is the **quotient** $\dfrac{m}{n}$.

Apply the quotient-to-power rule:    $\left(\dfrac{m}{n}\right)^9 = \dfrac{m^9}{n^9}$

b.  The base is the quotient $\dfrac{x^2}{y^5}$.

Apply the quotient-to-power rule:    $\left(\dfrac{x^2}{y^5}\right)^4 = \dfrac{(x^2)^4}{(y^5)^4}$

Apply the power-to-power rule:    $= \dfrac{x^{2\cdot4}}{y^{5\cdot4}}$

Simplify:    $= \dfrac{x^8}{y^{20}}$

*My video summary*   **c.–d.** Try to simplify these expressions on your own. View the **answers**, or watch this video for detailed solutions to parts c and d.

**You Try It**    Work through this You Try It problem.

Work Exercises 33–38 in this textbook or in the *MyMathLab*® Study Plan.

OBJECTIVE 7    SIMPLIFY EXPONENTIAL EXPRESSIONS USING A COMBINATION OF RULES

In some of the earlier examples, more than one rule for exponents was used to simplify the **exponential expressions**. To be considered **simplified**, an exponential expression must meet the following conditions:

· No parentheses or other **grouping symbols** are present.

· No zero **exponents** are present.

· No **powers** are raised to powers.

· Each **base** occurs only once.

Following is a summary of the exponent rules we have seen so far. We can combine these rules to simplify exponential expressions.

**Rules for Exponents**

| | |
|---|---|
| Product Rule | $a^m \cdot a^n = a^{m+n}$ |
| Quotient Rule | $\dfrac{a^m}{a^n} = a^{m-n} \quad (a \neq 0)$ |
| Zero-Power Rule | $a^0 = 1 \quad (a \neq 0)$ |
| Power-to-Power Rule | $(a^m)^n = a^{m \cdot n}$ |
| Product-to-Power Rule | $(ab)^n = a^n b^n$ |
| Quotient-to-Power Rule | $\left(\dfrac{a}{b}\right)^n = \dfrac{a^n}{b^n} \quad (b \neq 0)$ |

### Example 9  Using a Combination of Exponent Rules

*My interactive video summary*

 Simplify using the **rules for exponents**.

a. $(c^3)^5(c^2)^6$

b. $\left(\dfrac{15x^8y^5}{3x^6y}\right)^2$

c. $(-2w^3z^2)(-2wz^2)^4$

d. $\dfrac{(4m^2n^0)(2n^3)^2}{8mn^5}$

### Solutions

a. Apply the power-to-power rule: $\quad (c^3)^5(c^2)^6 = c^{3 \cdot 5} \cdot c^{2 \cdot 6}$

$\qquad\qquad\qquad$ Multiply: $\qquad\qquad\qquad = c^{15} \cdot c^{12}$

$\qquad\qquad$ Apply the product rule: $\qquad\quad = c^{15+12}$

$\qquad\qquad\qquad\qquad$ Add: $\qquad\qquad\qquad = c^{27}$

b. Within the parentheses, group like bases: $\left(\dfrac{15x^8y^5}{3x^6y}\right)^2 = \left(\dfrac{15}{3} \cdot \dfrac{x^8}{x^6} \cdot \dfrac{y^5}{y^1}\right)^2$

$\qquad$ Divide constants; apply the quotient rule: $\qquad = (5 \cdot x^{8-6} \cdot y^{5-1})^2$

$\qquad\qquad\qquad\qquad\qquad$ Simplify: $\qquad = (5x^2y^4)^2$

$\qquad\qquad$ Apply the product-to-power rule: $\qquad = 5^2(x^2)^2(y^4)^2$

$\qquad\qquad$ Apply the power-to-power rule: $\qquad = 5^2 \cdot x^{2 \cdot 2} \cdot y^{4 \cdot 2}$

$\qquad\qquad\qquad\qquad\qquad$ Simplify: $\qquad = 25x^4y^8$

c.–d. Try to simplify these expressions on your own. View the **answers**, or work through this **interactive video** for complete solutions to all four parts.

**You Try It**    Work through this You Try It problem.

Work Exercises 39–44 in this textbook or in the *MyMathLab*® Study Plan.

**Note:** There is often more than one way to simplify an exponential expression. Do not feel that you need to always work problems using the exact same steps as shown in the examples. Just make sure that your solutions meet the **conditions** for simplified exponential expressions.

# 5.1 Exercises

In Exercises 1–8, simplify using the product rule.

**You Try It**

**1.** $7^4 \cdot 7^2$

**2.** $x^8 \cdot x^5$

**3.** $w^5 \cdot w$

**4.** $t^6 \cdot t^2 \cdot t^5$

**You Try It**

**5.** $(6a^3)(5a^7)$

**6.** $(x^2y^7)(x^4y^3)$

**7.** $(-2m^9n)(-9m^4n^5)$

**8.** $(-6p^2)(5p^4)(2p^3)$

In Exercises 9–16, simplify using the quotient rule.

**You Try It**

**9.** $\dfrac{b^{12}}{b^8}$

**10.** $\dfrac{5^9}{5^7}$

**11.** $\dfrac{(-10)^7}{(-10)^4}$

**12.** $\dfrac{y^{17}}{y^{16}}$

**You Try It 13.** $\dfrac{24q^{15}}{6q^3}$

**14.** $\dfrac{60a^4b^{13}}{12ab^9}$

**15.** $\dfrac{x^7y^9z^8}{x^5y^3z^7}$

**16.** $\dfrac{10m^{21}n^9}{15m^{13}n^2}$

In Exercises 17–22, simplify using the zero-power rule.

**You Try It 17.** $4^0$

**18.** $(-8)^0$

**19.** $-12^0$

**20.** $6m^0$

**21.** $(3x)^0$

**22.** $3p^0 + 2q^0$

In Exercises 23–26, simplify using the power-to-power rule.

**You Try It 23.** $(x^3)^8$

**24.** $(y^9)^4$

**25.** $[(-2)^4]^3$

**26.** $[(m^4)^5]^3$

In Exercises 27–32, simplify using the product-to-power rule.

**You Try It 27.** $(pq)^6$

**28.** $(2m)^5$

**29.** $(a^3b)^7$

**30.** $(-5t^6)^4$

**31.** $(-7r^3s^9)^2$

**32.** $(2x^3y^2z)^4$

In Exercises 33–38, simplify using the quotient-to-power rule.

**You Try It 33.** $\left(\dfrac{a}{b}\right)^8$

**34.** $\left(\dfrac{3}{5}\right)^4$

**35.** $\left(\dfrac{t}{4}\right)^3$

**36.** $\left(\dfrac{m^4}{n^3}\right)^5$

**37.** $\left(\dfrac{u^5}{2}\right)^7$

**38.** $\left(\dfrac{3p^4}{4q^5}\right)^3$

In Exercises 39–44, simplify using the rules for exponents.

**You Try It 39.** $(x^2)^4(x^3)^6$

**40.** $(m^3n^2)^3(m^2n^5)^2$

**41.** $\left(\dfrac{16p^9q^5}{8p^3q^5}\right)^2$

**42.** $\dfrac{(18c^0d^{10})(4c^5d^2)}{12cd^5}$

**43.** $\left(\dfrac{x^2}{y}\right)^3\left(\dfrac{x^3}{y^2}\right)^2$

**44.** $\left(\dfrac{-2ac^2}{b^3}\right)^2\left(\dfrac{9a^5b^8}{c^3}\right)$

# 5.2 Introduction to Polynomials

## THINGS TO KNOW

Before working through this section, be sure you are familiar with the following concepts:

VIDEO    ANIMATION    INTERACTIVE

 **You Try It**
1. Evaluate Algebraic Expressions (Section 1.5, **Objective 1**)

 **You Try It**
2. Identify Terms, Coefficients, and Like Terms of an Algebraic Expression (Section 1.6, **Objective 1**)

 **You Try It**
3. Simplify Algebraic Expressions (Section 1.6, **Objective 2**)

 **You Try It**
4. Simplify Exponential Expressions Using a Combination of Rules (Section 5.1, **Objective 7**)

## OBJECTIVES

1 Classify Polynomials

2 Determine the Degree and Coefficient of a Monomial

3 Determine the Degree and Leading Coefficient of a Polynomial

4 Evaluate a Polynomial for a Given Value

5 Simplify Polynomials by Combining Like Terms

................................................................................

OBJECTIVE 1    CLASSIFY POLYNOMIALS

Recall that the **terms** of an algebraic expression are the quantities being added. A term can be a **constant**, a **variable**, or the product of a constant and one or more variables raised to **powers**. For example, the terms of the expression $3x^5 - 2x^2 + 7$ are $3x^5$, $-2x^2$, and 7.

If a term contains a single numeric **factor** and if none of the **variable factors** can be combined using the **rules for exponents**, then it is called a **simplified term**. Examples of simplified terms include

$$5x^3, -7y^2z, \frac{2}{3}b^2, \text{ and } 8.$$

The terms $-4m^2m^5$ and $\dfrac{12y^3}{20}$ are not simplified. Do you see **why**?

Some **simplified terms** are also *monomials*.

---

**Definition   Monomial**

A **monomial** is a simplified term in which all variables are raised to nonnegative integer powers and no variables appear in any **denominator**.

---

*My interactive video summary*

Note that a monomial can be a constant, such as 7. Some examples of monomials are $7x^4$, $-\frac{3}{4}x^3y^2$, and 9. The terms $2y^{-3}$, $6x^{1/4}$, and $\frac{2}{x}$ are not monomials. Do you see **why**? Watch this **interactive video** to determine if a given term is a monomial.

We studied **algebraic expressions** in **Section 1.5** and **Section 1.6**. We now look at special kinds of algebraic expressions called *polynomials*.

---

**Definition   Polynomial**

A **polynomial** in $x$ is a **monomial** or a **finite** sum of monomials of the form $ax^n$, where $a$ is any real number and $n$ is any whole number.

---

For example,

$$10x^7 + 6x^5 - 3x^2 - 5x + 4$$

is a polynomial in $x$. However, the expressions

$$5x^2 + 3x - 7x^{-4} \quad \text{and} \quad -3x^5 + \frac{2}{x^2} - 6x$$

are not polynomials. Do you see **why**?

The monomials that make up a polynomial are called the **terms of the polynomial**. A polynomial is a **simplified polynomial** if all of its **terms** are simplified and none of its terms are **like terms**. Polynomials can be defined using variables other than $x$ and may have terms with more than one variable. For now, we will consider **polynomials in one variable**. In **Section 5.8**, we visit polynomials in several variables.

We can classify simplified polynomials by the number of terms in the polynomial.

There are special names for **polynomials** with 1, 2, or 3 **terms**. If there are more than three terms, we use the general name, *polynomial*.

| Polynomial | Number of terms | Name | Hint |
|---|---|---|---|
| $4x$ | 1 | Monomial | ← "Mono-" means "one" |
| $3x^4 + 2$ | 2 | Binomial | ← "Bi-" means "two" |
| $-2x^4 - 3x + 1$ | 3 | Trinomial | ← "Tri-" means "three" |
| $2x^3 - 8x^2 + 5x - 9$ | 4 (or more) | Polynomial | ← "Poly-" means "many" |

Table 1

## Example 1  Classifying Polynomials

Classify each polynomial as a monomial, binomial, trinomial, or none of these.

**a.** $5x - 7$    **b.** $\frac{1}{3}x^2$    **c.** $5x^3 - 7x^2 + 4x + 1$    **d.** $-2x^3 - 5x^2 + 8x$

## Solutions

**a.–d.** Count the number of terms, then use Table 1 to help you classify the expressions on your own. View the **answers**.

**You Try It**    Work through this You Try It problem.

Work Exercises 1–6 in this textbook or in the *MyMathLab*® Study Plan.

OBJECTIVE 2   DETERMINE THE DEGREE AND COEFFICIENT OF A MONOMIAL

Every **monomial** has both a *degree* and a *coefficient*.

> **Definition   Degree of a Monomial**
>
> The **degree of a monomial** is the *sum* of the **exponents** on the **variables**.

For example, the degree of $-5x^2$ is 2, and the degree of $\frac{5}{8}x^3y^4$ is $3 + 4 = 7$.

 The degree of any constant term is 0 because there is no **variable factor**.

> **Definition   Coefficient of a Monomial**
>
> The **coefficient of a monomial** is the **constant** factor.

For example, the coefficient of $-5x^2$ is $-5$, and the coefficient of $\frac{5}{8}x^3y^4$ is $\frac{5}{8}$.

### Example 2  Determining the Coefficient and Degree of a Monomial

Determine the coefficient and degree of each monomial.

**a.** $4.6x^3$     **b.** $7x$     **c.** $x^2y^4$     **d.** $12$     **e.** $\frac{3}{4}x^2yz^3$     **f.** $-2xyz^7$

### Solutions

**a.** The **coefficient** of $4.6x^3$ is the **constant** factor 4.6. Because the **exponent** is 3, the **degree** is 3.

**b.** The coefficient of $7x$ is 7. The exponent of $x$ is **understood to be** 1, so the degree is 1.

**c.** Because $x^2y^4 = 1x^2y^4$, the coefficient is 1. The degree is the sum of the exponents on the variables: $2 + 4 = 6$.

  **d.–f.** Try answering these questions on your own, then view the **answers** to check. Watch this **video** for the solutions to all three parts.

**You Try It**   Work through this You Try It problem.

Work Exercises **7–12** in this textbook or in the *MyMathLab*® Study Plan.

OBJECTIVE 3   DETERMINE THE DEGREE AND LEADING COEFFICIENT OF A POLYNOMIAL

Because a **polynomial** is made up of **monomials**, every polynomial also has a degree.

> **Definition   Degree of a Polynomial**
>
> The **degree of a polynomial** is the largest degree of its terms.

For example, $4x + 5x^3 - x^2 + 6$ has a degree of 3, $-3x^5 - 2x^3 + 5x^2 - 7$ has a degree of 5, and 15 has a degree of 0 (think $15 = 15(1) = 15x^0$).

The polynomial $-3x^5 - 2x^3 + 5x^2 - 7$ is an example of a polynomial written in **descending order**. This means that the terms are listed so that the first term has the largest degree and the **exponents** on the variable decrease from left to right. This is called the **standard form** for polynomials. When a polynomial is in standard form, we can find its *leading coefficient*.

---

**Definition   Leading Coefficient of a Polynomial in One Variable**

When a polynomial in one variable is written in **standard form**, the **coefficient** of the first term (the term with the highest degree) is called the **leading coefficient**.

---

### Example 3  Determining the Degree and Leading Coefficient of a Polynomial

Write each polynomial in **standard form**. Then find its **degree** and **leading coefficient**.

**a.**  $4.2m - 3m^2 + 1.8 - 7m^3$ 　　　　　　**b.**  $\dfrac{2}{3}x^3 - 3x^2 + 5 - x^4 + \dfrac{1}{4}x$

### Solutions

**a.**  Start by determining the degree of each term.

| Term | Degree | |
|------|--------|--|
| $4.2m$ | 1 | ← $m = m^1$ |
| $-3m^2$ | 2 | |
| $1.8$ | 0 | ← $1.8 = 1.8m^0$ |
| $-7m^3$ | 3 | |

Now write the polynomial so that the terms descend in order from the largest degree, 3, to the smallest degree, 0 (the constant term): $-7m^3 - 3m^2 + 4.2m + 1.8$. The degree of the polynomial is 3, the largest degree. The leading coefficient is $-7$, the coefficient of the first term when the polynomial is written in standard form.

 *My video summary*    **b.** Try working this problem on your own. View the **answer**, or watch this **video** for a detailed solution.

**You Try It**   Work through this You Try It problem.

Work Exercises 13–18 in this textbook or in the *MyMathLab*® Study Plan.

---

OBJECTIVE 4   EVALUATE A POLYNOMIAL FOR A GIVEN VALUE

We **evaluate** polynomials exactly the same way we evaluated **algebraic expressions** in **Section 1.5**. We substitute the given value for the **variable** and simplify.

### Example 4  Evaluating Polynomials

Evaluate the **polynomial** $x^3 + 3x^2 + 4x - 5$ for the given values of $x$.

**a.**  $x = -2$ 　　　**b.** $x = 0$ 　　　**c.** $x = 2$ 　　　**d.** $x = \dfrac{5}{2}$

**Solutions**

a. Begin with the polynomial: $x^3 + 3x^2 + 4x - 5$

Substitute $-2$ for $x$: $(-2)^3 + 3(-2)^2 + 4(-2) - 5$ ← Put values in parentheses when substituting

Simplify the exponents: $= -8 + 3(4) + 4(-2) - 5$

Simplify the multiplication: $= -8 + 12 - 8 - 5$

Add and subtract: $= -9$

The value of the polynomial is $-9$ when $x$ is $-2$.

b. Begin with the polynomial: $x^3 + 3x^2 + 4x - 5$

Substitute 0 for $x$: $(0)^3 + 3(0)^2 + 4(0) - 5$ ← Put values in parentheses when substituting

Simplify the exponents: $= 0 + 3(0) + 4(0) - 5$

Simplify the multiplication: $= 0 + 0 + 0 - 5$

Add and subtract: $= -5$

The value of the polynomial is $-5$ when $x$ is 0.

 *My video summary*  **c.–d.** Try to evaluate the **polynomial** for the remaining values on your own. Then view the **answers**, or watch this **video** to see the complete solutions.

**You Try It** Work through this You Try It problem.

Work Exercises 19–25 in this textbook or in the *MyMathLab*® Study Plan.

OBJECTIVE 5    SIMPLIFY POLYNOMIALS BY COMBINING LIKE TERMS

We learned how to simplify **algebraic expressions** in **Section 1.6**. This often involved **combining like terms**. Similarly, we can simplify a **polynomial** with **like terms** by combining the like terms. This will be a necessary step as we perform operations on polynomials in the next few sections.

### Example 5  Simplifying Polynomials

 *My interactive video summary*    Simplify each polynomial by combining like terms.

a. $3x^2 + 8x - 4x + 2$    b. $2.3x - 3 - 5x + 8.4$

c. $2x + 3x^2 - 6 + x^2 - 2x + 9$    d. $\dfrac{2}{3}x^2 + \dfrac{1}{5}x - \dfrac{1}{10}x - \dfrac{1}{6}x^2 + \dfrac{1}{4}$

e. $6x^3 + x^2 - 7$

**Solutions**

a. Begin with the original expression: $3x^2 + 8x - 4x + 2$

Reverse the **distributive property**: $= 3x^2 + (8 - 4)x + 2$

Subtract: $= 3x^2 + 4x + 2$

b. Begin with the original expression: $2.3x - 3 - 5x + 8.4$

Rearrange to **collect like terms**: $= 2.3x - 5x - 3 + 8.4$

Reverse the **distributive property**: $= (2.3 - 5)x + (-3 + 8.4)$

Add or subtract: $= -2.7x + 5.4$

**c.–e.** Try to simplify these polynomials on your own. Then view the **answers**, or watch this interactive video to see the complete solution for any of the five parts.

**You Try It**    Work through this You Try It problem.

Work Exercises 26–34 in this textbook or in the *MyMathLab*® Study Plan.

# 5.2 Exercises

In Exercises 1–6, classify each polynomial as a monomial, binomial, trinomial, or none of these.

**You Try It**

**1.** $3x^2 - 5x + 1$

**2.** $-\dfrac{3}{5}x^2$

**3.** $x^5 - 7x^3 + x^2 - 4$

**4.** $5x^3 + 2x$

**5.** $2.4x^3 - 0.5x^2 + 0.25x + 1.35$

**6.** $\dfrac{1}{2}x^4 - \dfrac{4}{5}x + 1$

In Exercises 7–12, determine the coefficient and degree of each monomial.

**You Try It**

**7.** $4x^3y$

**8.** $8.1$

**9.** $x^4y^3$

**10.** $-7xy^2z^2$

**11.** $-9x$

**12.** $\dfrac{2}{3}x^2y^4z$

In Exercises 13–18, write each polynomial in standard form. Then find its degree and leading coefficient.

**You Try It**

**13.** $4 + 5x$

**14.** $\dfrac{2}{3}y - 12y^2 - 5$

**15.** $9 - 3x$

**16.** $m - 8 - m^7$

**17.** $4.1x^3 - 6.7x + 3.8 + x^2$

**18.** $7a - a^5 + 4a^2 + 3 - 8a^2$

In Exercises 19–25, evaluate each polynomial for the given value.

**You Try It**

**19.** $y + 4$ when $y = 3$.

**20.** $2x - 5$ when $x = -6$.

**21.** $5x^2 - 9x - 7$ when $x = 0$.

**22.** $x^2 - 7x - 12$ when $x = -1$

**23.** $3.5x^2 + 2.4$ when $x = 1.2$.

**24.** $x^2 + \dfrac{2}{5}x - 2$ when $x = \dfrac{3}{2}$

**25.** $-2x^3 + 6x^2 - 5x - 11$ for $x = -2$.

In Exercises 26–34, simplify each polynomial by combining like terms.

**You Try It**

**26.** $6x - 3x$

**27.** $2x + 5 - 7x$

**28.** $5x^2 - 9x + 3$

**29.** $14x + 3x - x + 1$

**30.** $-4x^2 + 9x + 7x^2 - 7$

**31.** $0.2x^2 - 4.1 + 3.2x + 1.1x^2 - 5$

**32.** $\dfrac{1}{3}x - \dfrac{2}{5} + 2x^2 - \dfrac{3}{2}x - \dfrac{4}{5}$

**33.** $x^3 - 2x + 3 + 6x^3 + 2x - 11$

**34.** $3x^4 + x^2 - 8 - 7x^4 + 3x - 2x^2$

# 5.3 Adding and Subtracting Polynomials

## THINGS TO KNOW

Before working through this section, be sure you are familiar with the following concepts:

VIDEO      ANIMATION      INTERACTIVE

**You Try It**

**1.** Find the Opposite of a Real Number
(Section 1.1, **Objective 3**)

**You Try It**

**2.** Identify Terms, Coefficients, and Like Terms of
an Algebraic Expression (Section 1.6, **Objective 1**)

**You Try It**

**3.** Simplify Algebraic Expressions
(Section 1.6, **Objective 2**)

## OBJECTIVES

**1** Add Polynomials

**2** Find the Opposite of a Polynomial

**3** Subtract Polynomials

................................................................................................................

OBJECTIVE 1    ADD POLYNOMIALS

In **Section 1.6**, we learned how to simplify algebraic expressions. This process often involves combining like terms.

To add **polynomials**, we remove all **grouping symbols**, use the **commutative** and **associative** **properties** of addition to rearrange the terms so that **like terms** are grouped, and combine all like terms.

> **Adding Polynomials**
>
> To add polynomials, remove all grouping symbols and combine like terms.

**Example 1  Adding Polynomials**

Add $(2x + 8) + (7x - 3)$.

**Solution**

$$
\begin{aligned}
\text{Begin with the original expression:} \quad & (2x + 8) + (7x - 3) \\
\text{Remove the grouping symbols:} \quad & = 2x + 8 + 7x - 3 \\
\text{Rearrange to group like terms:} \quad & = 2x + 7x + 8 - 3 \\
\text{Combine like terms:} \quad & = 9x + 5
\end{aligned}
$$

So, $(2x + 8) + (7x - 3) = 9x + 5$.

We can also add **polynomials** vertically. To do this, line up **like terms** in columns and then combine like terms. We repeat **Example 1** to show this method:

$$
\begin{array}{r}
2x + 8 \\
+ \quad 7x - 3 \\
\hline
9x + 5
\end{array}
$$

**You Try It**    Work through this You Try It problem.

Work Exercises 1–2 in this textbook or in the *MyMathLab*® Study Plan.

 Remember, when a **term** has no **coefficient** shown, it is understood to be 1.

### Example 2  Adding Polynomials

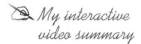 *My interactive video summary*

Add.

a. $(y^2 + 3y + 7) + (y^2 - 3y - 2)$
b. $(10p^3 + 7p - 13) + (5p^2 - 4p)$
c. $(3m^3 + m^2 - 8) + (2m^3 - 4m^2 + 3m) + (5m^2 + 4)$

**Solutions**  We follow the procedure for **adding polynomials**.

a. Begin with the original expression:   $(y^2 + 3y + 7) + (y^2 - 3y - 2)$
   Remove the grouping symbols:   $= y^2 + 3y + 7 + y^2 - 3y - 2$
   Rearrange to group like terms:   $= y^2 + y^2 + 3y - 3y + 7 - 2$
   Combine like terms:   $= 2y^2 + 5$

   Therefore, $(y^2 + 3y + 7) + (y^2 - 3y - 2) = 2y^2 + 5$.

b. Begin with the original expression:   $(10p^3 + 7p - 13) + (5p^2 - 4p)$
   Remove the grouping symbols:   $= 10p^3 + 7p - 13 + 5p^2 - 4p$
   Rearrange to group like terms:   $= 10p^3 + 5p^2 + 7p - 4p - 13$

   Finish this problem on your own. Check your **answer**, or watch this **interactive video** for the complete solutions to all three parts.

c. Try adding these polynomials on your own. Check your **answer**, or watch this **interactive video** to see the complete solutions to all three parts.

Now try reworking **Example 2** by adding vertically. See this **popup box** for the solutions. Which method do you prefer?

**You Try It**    Work through this You Try It problem.

Work Exercises 3–8 in this textbook or in the *MyMathLab*® Study Plan.

OBJECTIVE 2    FIND THE OPPOSITE OF A POLYNOMIAL

In **Section 1.1** we learned that a negative sign can be used to represent the "opposite" of a **real number**. For example, $-5$ is "the opposite of 5," and $-(-7)$ is "the opposite of negative 7." We find the opposite of a real number by changing its sign.

In the same way, $-(x^2 - 5x + 7)$ is "the opposite of the **polynomial** $x^2 - 5x + 7$," and we can find the *opposite of a polynomial* by changing the sign of each of its **terms**. So, the opposite of $x^2 - 5x + 7$, or $-(x^2 - 5x + 7)$, is $-x^2 + 5x - 7$.

---

**Opposite Polynomials**

To find the **opposite of a polynomial**, change the sign of each term.

---

**Note:** Finding the opposite of a polynomial, such as $-(x^2 - 5x + 7)$, can be thought of as **distributing** the negative sign through the polynomial.

$$-(x^2 - 5x + 7) = -x^2 - (-5x) + (-7) = -x^2 + 5x - 7$$

### Example 3  Finding Opposite Polynomials

Find the opposite of each polynomial.

**a.** $x^2 + 6x + 8$     **b.** $8y - 27$     **c.** $-m^3 - 5m^2 + m + 7$

**Solutions**  To find the opposite polynomial, change the sign of each **term**.

**a.** $-(x^2 + 6x + 8) = -x^2 - 6x - 8$

**b.** $-(8y - 27) = -8y + 27$

**c.** $-(-m^3 - 5m^2 + m + 7) = m^3 + 5m^2 - m - 7$

**You Try It**    Work through this You Try It problem.

Work Exercises 9–12 in this textbook or in the **MyMathLab**® Study Plan.

## OBJECTIVE 3   SUBTRACT POLYNOMIALS

In **Section 1.2**, we learned to subtract a real number by adding its opposite. In the same way, we **subtract** a polynomial by adding its **opposite polynomial**.

---

**Subtracting Polynomials**

To subtract a polynomial, add its opposite polynomial.

---

### Example 4  Subtracting Polynomials

Subtract.

**a.** $(9x + 13) - (6x - 4)$          **b.** $(3a^2 + 5a - 8) - (-2a^2 + a - 7)$

**Solutions**  We subtract by adding the opposite polynomial.

**a.** Begin with the original expression:    $(9x + 13) - (6x - 4)$

       Add the opposite polynomial:    $= (9x + 13) + (-6x + 4)$

       Remove the grouping symbols:    $= 9x + 13 - 6x + 4$

       Rearrange to group like terms:    $= 9x - 6x + 13 + 4$

         Combine like terms:    $= 3x + 17$

   So, $(9x + 13) - (6x - 4) = 3x + 17$.

**b.** Begin with the original expression:    $(3a^2 + 5a - 8) - (-2a^2 + a - 7)$

       Add the opposite polynomial:    $= (3a^2 + 5a - 8) + (+2a^2 - a + 7)$

Try finishing this problem on your own. Check your **answer**, or watch this **video** for detailed solutions to both parts.

*My video summary*

As with addition, polynomials can be subtracted vertically. We repeat **Example 4a** to show this method:

$$\begin{array}{r} 9x + 13 \\ -\ (6x - 4) \end{array} \xrightarrow{\substack{\text{Add the}\\ \text{opposite}\\ \text{polynomial}}} \begin{array}{r} 9x + 13 \\ +\ -6x + 4 \\ \hline 3x + 17 \end{array}$$

Now try reworking **Example 4b** by subtracting vertically. See this **popup box** for the solution.

**You Try It**    Work through this **You Try It** problem.

Work Exercises 13–20 in this textbook or in the **MyMathLab**® Study Plan.

# 5.3 Exercises

In Exercises 1–8, add.

**You Try It**

**1.** $(2x + 5) + (4x + 9)$

**2.** $(5y^2 - 3y) + (2y^2 - y)$

**3.** $(2m^2 - 5m - 13) + (m^2 + 5m + 23)$

**4.** $(7n^2 + 3n - 6) + (-2n^2 + 5)$

**You Try It**

**5.** $(12p^3 - 3p^2 - 7p - 5) + (-4p^3 + p^2 + 2p - 6)$

**6.** $(2a^2 - 11a + 9) + (8a^2 - 13) + (2a - 3)$

**7.** $(0.4w^2 - 5.6w + 3.1) + (1.3w^2 + 2.1w - 4.0)$

**8.** $\left(\dfrac{3}{5}x - \dfrac{1}{3}\right) + \left(\dfrac{1}{10}x + \dfrac{5}{6}\right)$

In Exercises 9–12, find the opposite of each polynomial.

**You Try It**

**9.** $7x^2 - 24$

**10.** $y^3 - 8y^2 - 17$

**11.** $-8z^3 + 3z^2 + 4z - 13$

**12.** $-\dfrac{7}{9}t^2 - \dfrac{4}{3}t + \dfrac{5}{6}$

In Exercises 13–20, subtract.

**You Try It**

**13.** $(5y + 8) - (2y - 14)$

**14.** $(7p^2 + 3p + 17) - (5p + 12)$

**15.** $(2t^2 + 13t - 17) - (-3t^2 + 11t - 13)$

**16.** $(-6z^2 + 3z - 14) - (8z^2 + z)$

**17.** $(-7x^3 + 6x^2 + x - 15) - (-2x^3 - 6x^2 + 3x - 15)$

**18.** $(m^2 - 9m + 1) - (3m^2 - 7) - (-2m + 3)$

**19.** $(5.1n^2 - 3.6n + 7.8) - (2.6n^2 + 1.9)$

**20.** $\left(\dfrac{7}{8}a - \dfrac{1}{3}\right) - \left(\dfrac{3}{8}a - \dfrac{1}{2}\right)$

# 5.4 Multiplying Polynomials

## THINGS TO KNOW

Before working through this section, be sure you are familiar with the following concepts:

VIDEO      ANIMATION      INTERACTIVE

**You Try It**

**1.** Simplify Algebraic Expressions
(Section 1.6, **Objective 2**)

**You Try It**

**2.** Simplify Exponential Expressions Using a
Combination of Rules (Section 5.1, **Objective 7**)

## OBJECTIVES

**1** Multiply Monomials

**2** Multiply a Polynomial by a Monomial

**3** Multiply Two Binomials

**4** Multiply Two or More Polynomials

--------

**OBJECTIVE 1**  MULTIPLY MONOMIALS

To multiply **monomials,** we use the **commutative** and **associative** properties to group **factors** with **like bases** and then apply the **product rule for exponents**. This is similar to the way we simplified some exponential expressions in **Section 5.1.**

> **Multiplying Monomials**
>
> Rearrange the factors to group the **coefficients** and to group like bases. Then, multiply the coefficients and apply the product rule for exponents.

### Example 1  Multiplying Monomials

Multiply.

**a.** $(6x^5)(7x^2)$      **b.** $\left(-\frac{3}{4}x^2\right)\left(-\frac{2}{9}x^8\right)$      **c.** $(3x^2)(-0.2x^3)$

### Solutions

**a.**

Begin with the original expression: $(6x^5)(7x^2)$

Rearrange factors to group coefficients and like bases: $= (6 \cdot 7)(x^5 x^2)$

Multiply coefficients; apply product rule for exponents: $= 42x^{5+2}$

Simplify: $= 42x^7$

**b.–c.** Try to work these problems on your own. View this **popup** for the complete solutions to both parts.

**You Try It**  **Work through this You Try It problem.**

**Work Exercises 1–6 in this textbook or in the *MyMathLab*® Study Plan.**

OBJECTIVE 2   MULTIPLY A POLYNOMIAL BY A MONOMIAL

To multiply a **monomial** and a **polynomial** with more than one **term**, we use the **distributive property**.

> **Multiplying Polynomials by Monomials**
>
> To multiply a polynomial by a monomial, use the distributive property to multiply each term of the polynomial by the monomial. Then simplify using the method for **multiplying monomials**.

If $m$ is a monomial and $p_1, p_2, p_3, \ldots, p_n$ are the terms of a polynomial, then

$$m \cdot (p_1 + p_2 + p_3 + \ldots + p_n) = m \cdot p_1 + m \cdot p_2 + m \cdot p_3 + \ldots + m \cdot p_n$$

### Example 2  Multiplying a Polynomial by a Monomial

Multiply.

**a.** $3x(4x - 5)$        **b.** $-4x^2(3x^2 + x - 7)$

### Solutions

**a.**  Begin with the original expression:   $3x(4x - 5)$

Distribute the monomial $3x$:   $= 3x \cdot 4x - 3x \cdot 5$

Rearrange the factors:   $= 3 \cdot 4 \cdot x \cdot x - 3 \cdot 5 \cdot x$

Multiply coefficients; apply product rule for exponents:   $= 12x^{1+1} - 15x$

Simplify:   $= 12x^2 - 15x$

**b.**  Begin with the original expression:   $-4x^2(3x^2 + x - 7)$

Distribute the monomial $-4x^2$:   $= -4x^2 \cdot 3x^2 + (-4x^2) \cdot x - (-4x^2) \cdot 7$

Try to finish this problem on your own. View this **popup** for the complete solution.

### Example 3  Multiplying a Polynomial by a Monomial

*My video summary*    Multiply.

**a.** $\dfrac{1}{2}x^2(4x^2 - 6x + 2)$        **b.** $0.25x^3(6x^3 - 10x^2 + 4x - 7)$

**Solutions**  Try to work these problems on your own. View the **answers**, or watch this **video** for the complete solutions to both parts.

**You Try It**   Work through this You Try It problem.

Work Exercises 7–12 in this textbook or in the *MyMathLab*® Study Plan.

## OBJECTIVE 3    MULTIPLY TWO BINOMIALS

To multiply two binomials, we can use the **distributive property** twice. We can distribute the first binomial to each **term** in the second binomial. We can then distribute each term from the second binomial through the first binomial (from the back).

$$(a + b)(c + d) = (a + b)c + (a + b)d = ac + bc + ad + bd$$

The end result is that each term in the first binomial gets multiplied by each term in the second binomial.

$$(a + b)(c + d) = ac + ad + bc + bd$$

---

**Multiplying Two Binomials**

To multiply two binomials, multiply each term of the first binomial by each term of the second binomial. To simplify, **combine like terms**, if any.

---

### Example 4  Multiplying Two Binomials

Multiply using the distributive property twice.

**a.** $(x + 3)(x + 2)$          **b.** $(x + 6)(x - 2)$          **c.** $(x - 4)(x - 5)$

### Solutions

**a.** Multiply each **term** in the first **binomial** by each term in the second binomial.

Begin with the original expression:  $(x + 3)(x + 2)$

Multiply $x$ by $x$ and 2;
multiply 3 by $x$ and 2:  $= x \cdot x + x \cdot 2 + 3 \cdot x + 3 \cdot 2$   ← Distributed

Simplify:  $= x^2 + 2x + 3x + 6$   ← $x \cdot x = x^2$

Combine like terms:  $= x^2 + 5x + 6$   ← $2x + 3x = 5x$

**b.** Multiply each term in the first binomial by each term in the second.

Begin with the original expression:  $(x + 6)(x - 2)$

Multiply $x$ by $x$ and $-2$;
multiply 6 by $x$ and $-2$:  $= x \cdot x + x \cdot (\ 2) + 6 \cdot x + 6 \cdot (-2)$  ← Distributed

Simplify:  $= x^2 - 2x + 6x - 12$  ← $x \cdot (-2) = -2x, 6 \cdot (-2) = -12$

Combine like terms:  $= x^2 + 4x - 12$  ← $-2x + 6x = 4x$

*My video summary*  **c.** Try to work this problem on your own. View the **answer**, or watch this **video** for the complete solution.

 Remember that a term contains the sign of its **coefficient**. For example, in the expression $3x^2 - 7x$ the terms are $3x^2$ and $-7x$.

The process used in Example 4 follows a specific order for multiplying **binomials**. The **acronym** FOIL summarizes the process and reminds us to multiply the two First terms, the

 *My animation summary*

 two Outside terms, the two Inside terms, and the two Last terms. Watch this **animation**, which illustrates how to use the **FOIL method** to multiply two binomials.

> **The FOIL Method**
>
> FOIL reminds us to multiply the two First terms, the two Outside terms, the two Inside terms, and the two Last terms.

**CAUTION** After using FOIL, be sure to simplify by combining any like terms.

### Example 5 Multiplying Two Binomials

 *My interactive video summary*

Multiply using the FOIL method.

**a.** $(x - 4)(2x + 3)$    **b.** $\left(\dfrac{1}{2}x - 6\right)(3x - 4)$    **c.** $(5x + 7)(4x + 3)$

### Solutions

**a.** The two first terms are $x$ and $2x$. The two outside terms are $x$ and $3$. The two inside terms are $-4$ and $2x$. The two last terms are $-4$ and $3$.

$$
\begin{aligned}
(x - 4)(2x + 3) &= \overset{\text{First}}{x \cdot 2x} + \overset{\text{Outside}}{x \cdot 3} + \overset{\text{Inside}}{(-4) \cdot 2x} + \overset{\text{Last}}{(-4) \cdot 3} \\
&= 2x^2 + 3x + (-8x) + (-12) \\
&= 2x^2 + 3x - 8x - 12 \\
&= 2x^2 - 5x - 12 \quad \leftarrow \boxed{3x - 8x = -5x}
\end{aligned}
$$

**b.** Apply the **FOIL method**.

$$
\left(\dfrac{1}{2}x - 6\right)(3x - 4) = \overset{\text{First}}{\dfrac{1}{2}x \cdot 3x} + \overset{\text{Outside}}{\dfrac{1}{2}x \cdot (-4)} + \overset{\text{Inside}}{(-6) \cdot 3x} + \overset{\text{Last}}{(-6) \cdot (-4)}
$$

Try finishing this problem on your own. View the **answer**, or watch this **interactive video** to see the complete solution.

**c.** Try working this problem on your own. View the **answer**, or watch this **interactive video** to see the complete solution.

**You Try It**    Work through this You Try It problem.

**Work Exercises 13–20 in this textbook or in the *MyMathLab*® Study Plan.**

 **CAUTION** The FOIL method can be used only when multiplying two binomials.

We can multiply two binomials vertically, just as we were able to add or subtract vertically. View this **popup** to see Example 5c worked in a vertical format.

OBJECTIVE 4  MULTIPLY TWO OR MORE POLYNOMIALS

*My animation summary*

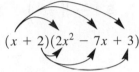 We can expand our work in Objective 3 to multiply any two **polynomials**. Watch this **animation** for an explanation.

> **Multiplying Two or More Polynomials**
>
> To multiply two polynomials, multiply each term of the first polynomial by each term of the second polynomial. To simplify, **combine like terms**, if any. If there are more than two polynomials, then we use this procedure to multiply two polynomials at a time.

### Example 6  Multiplying Two Polynomials

Multiply:  **a.** $(x + 2)(2x^2 - 7x + 3)$        **b.** $(y^2 + 2y - 9)(2y^2 - 4y + 7)$

**Solutions**

**a.** We multiply each term in the **binomial** by each term in the **trinomial**.

Begin with the original expression: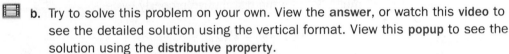

$$
\begin{aligned}
\text{Multiply } x \text{ by } 2x^2, -7x, \text{ and } 3; \\
\text{Multiply } 2 \text{ by } 2x^2, -7x, \text{ and } 3: \quad &= x \cdot 2x^2 + x \cdot (-7x) + x \cdot 3 + 2 \cdot 2x^2 + 2 \cdot (-7x) + 2 \cdot 3 \\
\text{Simplify:} \quad &= 2x^3 - 7x^2 + 3x + 4x^2 - 14x + 6 \\
\text{Combine like terms:} \quad &= 2x^3 - 3x^2 - 11x + 6
\end{aligned}
$$

View this **popup** to see the problem solved by multiplying vertically.

*My video summary*

 **b.** Try to solve this problem on your own. View the **answer**, or watch this **video** to see the detailed solution using the vertical format. View this **popup** to see the solution using the **distributive property**.

**You Try It**   **Work through this You Try It problem.**

**Work Exercises 21–27 in this textbook or in the *MyMathLab*®  Study Plan.**

To multiply three or more **polynomials**, we multiply two polynomials at a time.

### Example 7  Multiplying Three or More Polynomials

*My interactive video summary*

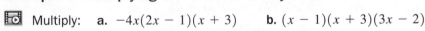 Multiply:  **a.** $-4x(2x - 1)(x + 3)$      **b.** $(x - 1)(x + 3)(3x - 2)$

**Solutions**

**a.** Because multiplication is **commutative**, we can perform the multiplications in any order we wish. For this problem, we will first multiply the two **binomials** and then distribute the **monomial** through the resulting **product**. Work through the steps below, or watch this **interactive video** to see the complete solutions to both parts of this example.

$$
\begin{aligned}
\text{Begin with the original expression:} \quad & -4x(2x - 1)(x + 3) \\
\text{Multiply } 2x - 1 \text{ and } x + 3: \quad &= -4x(2x \cdot x + 2x \cdot 3 - 1 \cdot x - 1 \cdot 3) \leftarrow \text{FOIL} \\
&= -4x(2x^2 + 6x - x - 3) \\
\text{Combine like terms:} \quad &= -4x(2x^2 + 5x - 3) \\
\text{Distribute } -4x: \quad &= (-4x) \cdot 2x^2 + (-4x) \cdot 5x + (-4x) \cdot (-3) \\
\text{Simplify:} \quad &= -8x^3 + (-20x^2) + 12x \\
&= -8x^3 - 20x^2 + 12x
\end{aligned}
$$

b. Work this problem by first multiplying two of the **binomials** and then **combining like terms**. Then multiply the resulting product by the remaining binomial. View the **answer**, or watch this **interactive video** for the complete solutions to both parts.

**You Try It**  Work through this You Try It problem.

Work Exercises 28–30 in this textbook or in the *MyMathLab* Study Plan.

# 5.4 Exercises

In Exercises 1–6, multiply each pair of monomials.

**You Try It**  **1.** $(3x^5)(2x^3)$    **2.** $(5x^4)(-x^5)$    **3.** $(-3y^5)(-2y^9)$

**4.** $(3.2w^3)(2.5w^2)$    **5.** $\left(\dfrac{2}{3}x\right)\left(\dfrac{6}{5}x^4\right)$    **6.** $(3x^2y^3)(5x^4y^2)$

In Exercises 7–12, multiply.

**You Try It**  **7.** $5x(3x + 2)$    **8.** $-8x(4x - 9)$    **9.** $3x(x^2 - x + 1)$

**10.** $-\dfrac{1}{2}x^2(2x^2 + 4x + 6)$    **11.** $-3x^3(x^3 + 2x - 7)$    **12.** $2x^2(3x^5 - 2x^3 + 7x - 4)$

In Exercises 13–16, multiply using the distributive property twice.

  **13.** $(x + 2)(x + 5)$    **14.** $(w - 3)(2w - 5)$

**15.** $\left(y + \dfrac{3}{5}\right)\left(y - \dfrac{4}{3}\right)$    **16.** $(2x^2 - 3)(3x + 2)$

In Exercises 17–20, multiply using the FOIL method.

**You Try It**  **17.** $(x - 2)(x + 9)$    **18.** $(2m + 1)(m + 3)$

**19.** $(7z + 6)(8z - 3)$    **20.** $(2x^2 - 3)(3x - 2)$

In Exercises 21–30, multiply the polynomials and simplify by combining like terms.

**You Try It**  **21.** $(x - 3)(x^2 + 3x + 9)$    **22.** $(m - 1)(3m^2 + 2m - 9)$

**23.** $(2x + 1)(x^2 - x + 1)$    **24.** $(3x + 2)(4x^3 - 2x + 11)$

**25.** $(x^2 + 3)(2x^3 + x - 3)$    **26.** $(x^2 + 3x + 1)(2x^2 + x - 3)$

**27.** $(x^3 - x + 1)(x^2 + x - 1)$    **28.** $(2x)(-3x)(x^2 + x + 1)$

  **29.** $(5x)(3x - 2)(2x + 3)$    **30.** $(x + 4)(2x - 1)(3x - 2)$

**You Try It**

# 5.5 Special Products

## THINGS TO KNOW

Before working through this section, be sure you are familiar with the following concepts:

| | VIDEO | ANIMATION | INTERACTIVE |
|---|---|---|---|

 **You Try It**   1. Simplify Algebraic Expressions (Section 1.6, **Objective 2**)

 **You Try It**   2. Simplify Exponential Expressions Using a Combination of Rules (Section 5.1, **Objective 7**)

 **You Try It**   3. Multiply a Polynomial by a Monomial (Section 5.4, **Objective 2**)

 **You Try It**   4. Multiply Two Binomials (Section 5.4, **Objective 3**)

## OBJECTIVES

**1** Square a Binomial Sum

**2** Square a Binomial Difference

**3** Multiply the Sum and Difference of Two Terms

---

OBJECTIVE 1   SQUARE A BINOMIAL SUM

In this section, we look at three special **products** involving **binomials**.

A **binomial sum** is a binomial in which the two **terms** are added such as $A + B$. We can use the **FOIL method** to find a general result for the square of a binomial sum, $(A + B)^2$.

$$(A + B)^2 = (A + B)(A + B) = \overbrace{A \cdot A}^{F} + \overbrace{A \cdot B}^{O} + \overbrace{B \cdot A}^{I} + \overbrace{B \cdot B}^{L}$$

$$= A^2 + AB + AB + B^2$$

$$= A^2 + 2AB + B^2$$

This result is our first special product rule.

---

**The Square of a Binomial Sum Rule**

$$\underbrace{(A + B)^2}_{\substack{\text{The square} \\ \text{of a bino-} \\ \text{mial sum}}} \underbrace{=}_{\text{equals}} \underbrace{A^2}_{\substack{\text{the square} \\ \text{of the} \\ \text{first term}}} \underbrace{+}_{\text{plus}} \underbrace{2AB}_{\substack{\text{2 times the} \\ \text{product of} \\ \text{the two} \\ \text{terms}}} \underbrace{+}_{\text{plus}} \underbrace{B^2}_{\substack{\text{the square} \\ \text{of the} \\ \text{second term.}}}$$

---

 Note that the square of a **binomial sum** is not the sum of the squares: $(A + B)^2 \neq A^2 + B^2$. See this **popup** for a geometric explanation of the square of a binomial sum.

**Example 1** Squaring a Binomial Sum

Multiply.

**a.** $(x + 7)^2$    **b.** $(0.2m + 1)^2$    **c.** $\left(z^2 + \dfrac{1}{4}\right)^2$    **d.** $\left(10y + \dfrac{2}{5}\right)^2$

**Solution** In each case, use the **square of a binomial sum rule**.

**a.** Square of a binomial sum rule:  $(A + B)^2 = A^2 + 2AB + B^2$

Substitute $x$ for $A$ and 7 for $B$:  $(x + 7)^2 = (x)^2 + 2(x)(7) + (7)^2$

Simplify:  $= x^2 + 14x + 49$ $\leftarrow 2(x)(7) = 2(7)(x) = 14x$

**b.** Square of a binomial sum rule:  $(A + B)^2 = A^2 + 2AB + B^2$

Substitute $0.2m$ for $A$ and 1 for $B$:  $(0.2m + 1)^2 = (0.2m)^2 + 2(0.2m)(1) + (1)^2$

Simplify:  $= 0.04m^2 + 0.4m + 1$ $\longleftarrow$ $\begin{array}{l}(0.2m)^2 = (0.2)^2 m^2 = 0.04m^2 \\ 2(0.2m)(1) = 2(1)(0.2)m = 0.4m\end{array}$

 *My video summary*   **c.–d.** Try working these parts on your own. View the **answers**, or watch this **video** for the complete solutions.

**You Try It**    Work through this You Try It problem.

Work Exercises 1–8 in this textbook or in the **MyMathLab**® Study Plan.

OBJECTIVE 2   SQUARE A BINOMIAL DIFFERENCE

A **binomial difference** is a binomial in which one **term** is subtracted from the other such as $A - B$. We can again use the **FOIL method** to find a general result for the square of a binomial difference, $(A - B)^2$.

$$
\overset{}{(A - B)^2} = (A - B)(A - B) = \overset{F}{\overbrace{A \cdot A}} + \overset{O}{\overbrace{A \cdot (-B)}} + \overset{I}{\overbrace{(-B) \cdot A}} + \overset{L}{\overbrace{(-B) \cdot (-B)}}
$$

$$
= A^2 + (-AB) + (-AB) + B^2
$$

$$
= A^2 - AB - AB + B^2
$$

$$
= A^2 - 2AB + B^2
$$

This result is our second special product rule.

---

**The Square of a Binomial Difference Rule**

$$(A - B)^2 = A^2 - 2AB + B^2$$

| The square of a binomial difference | equals | the square of the first term | minus | 2 times the product of the two terms | plus | the square of the second term. |

---

Note that the square of a **binomial difference** is not the difference of the squares: $(A - B)^2 \neq A^2 - B^2$.

### Example 2 Squaring a Binomial Difference

Multiply.

**a.** $(x - 3)^2$    **b.** $\left(2z - \dfrac{1}{6}\right)^2$    **c.** $(w^3 - 0.7)^2$    **d.** $(5p - 1.2)^2$

**Solution** In each case, use the **square of a binomial difference rule**.

**a.** Square of a binomial difference rule:   $(A - B)^2 = A^2 - 2AB + B^2$

     Substitute $x$ for $A$ and 3 for $B$:   $(x - 3)^2 = (x)^2 - 2(x)(3) + (3)^2$

                    Simplify:   $= x^2 - 6x + 9 \leftarrow 2(x)(3) = 2(3)(x) = 6x$

**b.** Square of a binomial difference rule:   $(A - B)^2 = A^2 - 2AB + B^2$

     Substitute $2z$ for $A$ and $\dfrac{1}{6}$ for $B$:   $\left(2z - \dfrac{1}{6}\right)^2 = (2z)^2 - 2(2z)\left(\dfrac{1}{6}\right) + \left(\dfrac{1}{6}\right)^2$

                    Simplify:   $= 4z^2 - \dfrac{2}{3}z + \dfrac{1}{36} \leftarrow \begin{array}{l} (2z)^2 = 2^2 z^2 = 4z^2 \\ 2(2z)\left(\frac{1}{6}\right) = 2(2)\left(\frac{1}{6}\right)z = \frac{4}{6}z = \frac{2}{3}z \end{array}$

 *My video summary*      **c.–d.** Try working these parts on your own. View the **answers**, or watch this **video** for the complete solutions.

**You Try It**   **Work through this You Try It problem.**

**Work Exercises 9–16 in this textbook or in the _MyMathLab_® Study Plan.**

Notice that the first two special product rules are very similar:

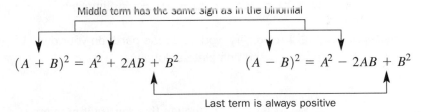

Middle term has the same sign as in the binomial

$$(A + B)^2 = A^2 + 2AB + B^2 \qquad (A - B)^2 = A^2 - 2AB + B^2$$

Last term is always positive

The **trinomials** $A^2 + 2AB + B^2$ and $A^2 - 2AB + B^2$ are called **perfect square trinomials** because they come from squaring a **binomial**. Recognizing perfect square trinomials will be helpful when we get to Chapter 6.

OBJECTIVE 3   MULTIPLY THE SUM AND DIFFERENCE OF TWO TERMS

When a **binomial sum** and a **binomial difference** are made from the same two terms, they are called **conjugates** of each other. For example, the binomial sum $x + 5$ and the binomial difference $x - 5$ are **conjugates** of each other.

We can use the **FOIL method** to find a general result for the **product** of conjugates, $(A + B)(A - B)$.

$$
\begin{aligned}
(A + B)(A - B) &= \overbrace{A \cdot A}^{F} + \overbrace{A \cdot (-B)}^{O} + \overbrace{B \cdot A}^{I} + \overbrace{B \cdot (-B)}^{L} \\
&= A^2 - AB + AB - B^2 \\
&= A^2 - B^2
\end{aligned}
$$

So, the product of the sum and difference of two terms (product of conjugates) equals the difference of the squares of the two terms.

This result is our third special product rule.

---

**The Sum and Difference of Two Terms Rule (Product of Conjugates Rule)**

$$\underbrace{(A + B)(A - B)}_{\substack{\text{The product of the} \\ \text{sum and difference} \\ \text{of two terms}}} = \underbrace{A^2}_{\substack{\text{equals the square} \\ \text{of the} \\ \text{first term}}} - \underbrace{B^2}_{\substack{\text{minus the square of} \\ \text{the second} \\ \text{term.}}}$$

---

### Example 3  Multiplying the Sum and Difference of Two Terms

Multiply.

**a.** $(x + 4)(x - 4)$

**b.** $\left(5y + \dfrac{1}{2}\right)\left(5y - \dfrac{1}{2}\right)$

**c.** $(8 - x)(8 + x)$

**d.** $(3z^2 + 0.5)(3z^2 - 0.5)$

**Solution**  In each case, use the **sum and difference of two terms rule** (product of conjugates).

**a.**  Sum and difference of two terms rule: $\quad (A + B)(A - B) = A^2 - B^2$

Substitute $x$ for $A$ and $4$ for $B$: $\quad (x + 4)(x - 4) = (x)^2 - (4)^2$

Simplify: $\quad = x^2 - 16$

**b.**  Sum and difference of two terms rule: $\quad (A + B)(A - B) = A^2 - B^2$

Substitute $5y$ for $A$ and $\dfrac{1}{2}$ for $B$: $\quad \left(5y + \dfrac{1}{2}\right)\left(5y - \dfrac{1}{2}\right) = (5y)^2 - \left(\dfrac{1}{2}\right)^2$

Simplify: $\quad = 25y^2 - \dfrac{1}{4}$

*My video summary*  **c.–d.**  Try working these parts on your own. View the **answers**, or watch this **video** for the complete solutions.

**You Try It**    **Work through this You Try It problem.**

**Work Exercises 17–24 in this textbook or in the *MyMathLab*® Study Plan.**

In summary, the results from Examples 1–3 give us the following **special product rules for binomials.**

---

**Special Product Rules for Binomials**

| | | |
|---|---|---|
| The square of a binomial sum | $(A + B)^2 = A^2 + 2AB + B^2$ | Perfect square trinomial |
| The square of a binomial difference | $(A - B)^2 = A^2 - 2AB + B^2$ | Perfect square trinomial |
| The product of the sum and difference of two terms (product of conjugates) | $(A + B)(A - B) = A^2 - B^2$ | Difference of two squares |

---

Recognizing these special binomial sums and differences will allow us to find their **products** more quickly by using the special product rules.

# 5.5 Exercises

Multiply.

**You Try It**

1. $(y + 2)^2$

2. $\left(n + \dfrac{4}{3}\right)^2$

3. $(x + 3.5)^2$

4. $(2x + 5)^2$

5. $\left(\dfrac{3}{4}z + 1\right)^2$

6. $(3.2m + 7)^2$

7. $\left(6x + \dfrac{3}{5}\right)^2$

8. $(y^3 + 0.8)^2$

**You Try It**

9. $(x - 6)^2$

10. $\left(y - \dfrac{3}{7}\right)^2$

11. $(z^2 - 4.2)^2$

12. $(4x - 1)^2$

13. $(8 - x)^2$

14. $(1.5 - 3m)^2$

15. $\left(\dfrac{5}{2}z - 4\right)^2$

16. $(1.2x - 2.5)^2$

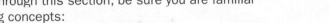
**You Try It**

17. $(x + 9)(x - 9)$

18. $(y - 2)(y + 2)$

19. $(3z + 4)(3z - 4)$

20. $\left(w + \dfrac{1}{5}\right)\left(w - \dfrac{1}{5}\right)$

21. $(2.1 + 3y)(2.1 - 3y)$

22. $\left(\dfrac{5}{3}v + 2\right)\left(\dfrac{5}{3}v - 2\right)$

23. $(4x^2 + 7)(4x^2 - 7)$

24. $(3.5 - x^2)(3.5 + x^2)$

# 5.6 Negative Exponents and Scientific Notation

## THINGS TO KNOW
Before working through this section, be sure you are familiar with the following concepts:

VIDEO    ANIMATION    INTERACTIVE

**You Try It**

1. Add Two Real Numbers with the Same Sign
(Section 1.2, **Objective 1**)

**You Try It**

2. Add Two Real Numbers with Different Signs
(Section 1.2, **Objective 2**)

**You Try It**

3. Subtract Real Numbers
(Section 1.2, **Objective 3**)

**You Try It**

4. Multiply Real Numbers
   (Section 1.3, **Objective 1**)

**You Try It**

5. Divide Real Numbers
   (Section 1.3, **Objective 2**)

**You Try It**

6. Simplify Exponential Expressions Using a
   Combination of Rules (Section 5.1, **Objective 7**)

## OBJECTIVES

**1** Use the Negative Power Rule

**2** Simplify Expressions Containing Negative Exponents Using a Combination of Rules

**3** Convert a Number from Standard Form to Scientific Notation

**4** Convert a Number from Scientific Notation to Standard Form

**5** Multiply and Divide with Scientific Notation

---

### OBJECTIVE 1   USE THE NEGATIVE POWER RULE

So far, we have encountered only **exponential expressions** with non-negative exponents. But what if an exponential expression contains a negative exponent, such as $x^{-2}$? We can use the **quotient rule** to understand the meaning of such an expression. Consider $\dfrac{x^2}{x^5}$, where $x \neq 0$. We can find this quotient using two different methods.

Method 1: Apply the **quotient rule for exponents**: $\dfrac{x^2}{x^5} = x^{2-5} = x^{-3}$

> Subtract the denominator exponent from the numerator exponent.

> Keep the common base.

Method 2: Expand the numerator and denominator: $\dfrac{x^2}{x^5} = \dfrac{x \cdot x}{x \cdot x \cdot x \cdot x \cdot x}$ ← 2 factors of $x$
← 5 factors of $x$

Divide out common factors: $= \dfrac{\cancel{x} \cdot \cancel{x}}{\cancel{x} \cdot \cancel{x} \cdot x \cdot x \cdot x}$

Simplify: $= \dfrac{1}{x \cdot x \cdot x} = \dfrac{1}{x^3}$

Because $\dfrac{x^2}{x^5}$ equals both $x^{-3}$ and $\dfrac{1}{x^3}$, this means $x^{-3} = \dfrac{1}{x^3}$. So, $x^{-3}$ is the **reciprocal** of $x^3$.

---

**Definition   Negative Exponent**

A negative exponent can be expressed with a positive exponent as follows:

$$a^{-n} = \dfrac{1}{a^n} \quad (a \neq 0)$$

---

## Example 1  Using the Definition of a Negative Exponent

Write each expression with positive exponents. Then simplify if possible.

**a.** $x^{-4}$    **b.** $2^{-3}$    **c.** $7x^{-3}$    **d.** $(-2)^{-4}$    **e.** $-3^{-2}$    **f.** $2^{-1} + 3^{-1}$

**Solutions** In each case, we use the definition of a negative exponent to rewrite the expression with all positive exponents. Then we simplify if necessary.

**a.**  $x^{-4} = \dfrac{1}{x^4}$

**b.**  $2^{-3} = \dfrac{1}{2^3} = \dfrac{1}{2 \cdot 2 \cdot 2} = \dfrac{1}{8}$

**c.** There are no parentheses, so only $x$ is the **base** of the exponential expression.

$$7x^{-3} = 7 \cdot \dfrac{1}{x^3} = \dfrac{7}{x^3}$$

 **d.–f.** Try to simplify these expressions on your own. View the **answers**, or watch this video for detailed solutions to parts d through f.

**You Try It**   Work through this You Try It problem.

**Work Exercises 1–8 in this textbook or in the *MyMathLab*® Study Plan.**

 A **negative exponent** does not mean the value of the expression is negative. For example, $2^{-1} = \dfrac{1}{2}$, which is positive.

Sometimes, a negative exponent appears in a **denominator**. Consider the following:

$$\frac{1}{x^{-3}} = 1 \div x^{-3} = 1 \div \frac{1}{x^3} = 1 \cdot \frac{x^3}{1} = x^3$$

As long as the **base** is not 0, this result is true in general.

---

**The Negative-Power Rule**

To remove a negative exponent, switch the location of the base (**numerator** or denominator) and change the exponent to be positive.

$$a^{-n} = \frac{1}{a^n} \quad \text{and} \quad \frac{1}{a^{-n}} = a^n \quad (a \neq 0)$$

---

## Example 2  Using the Negative Power Rule

Write each expression with positive exponents. Then simplify if possible.

**a.** $\dfrac{1}{y^{-5}}$    **b.** $\dfrac{1}{6^{-2}}$    **c.** $\dfrac{3}{4t^{-7}}$    **d.** $\dfrac{-8}{q^{-11}}$    **e.** $\dfrac{m^{-9}}{n^{-4}}$    **f.** $\dfrac{5^{-3}}{2^{-4}}$

**Solutions** In each case, we use the **negative power rule** to rewrite the expression with all positive exponents. Then we simplify if necessary.

**a.** The **base** $y$ is raised to the $-5$ power. Switch the location of $y$ from the **denominator** to the **numerator** and change its exponent from $-5$ to 5.

$$\frac{1}{y^{-5}} = y^5 \quad \begin{array}{l} \leftarrow \text{Exponent is now positive} \\ \\ \nwarrow \text{Base is now in the numerator} \end{array}$$

**b.** The base 6 is raised to the −2 power. Switch the location of 6 from the denominator to the numerator and change the exponent from −2 to 2. Then simplify.

$$\frac{1}{6^{-2}} = 6^2 = 36$$

**c.** There are no parentheses, so the **base** is only $t$, which is raised to the −7 power. Switch the location of $t$ from the **denominator** to the **numerator**, and change the exponent from −7 to 7.

$$\frac{3}{4t^{-7}} = \frac{3t^7}{4}$$

Note that the 3 and 4 do not switch locations because they are not affected by the negative exponent.

*My video summary*    **d.–f.** Try to simplify these expressions on your own. View the **answers**, or watch this **video** for detailed solutions to parts d through f.

**You Try It**    **Work through this You Try It problem.**

**Work Exercises 9–16 in this textbook or in the *MyMathLab*® Study Plan.**

OBJECTIVE 2    SIMPLIFY EXPRESSIONS CONTAINING NEGATIVE EXPONENTS USING A COMBINATION OF RULES

An **exponential expression** is not simplified if it contains negative exponents. Adding this to the conditions from **Section 5.1** gives the following complete list of requirements for **simplified exponential expressions**:

- No parentheses or other **grouping symbols** are present.
- No zero or **negative exponents** are present.
- No **powers** are raised to powers.
- Each **base** occurs only once.

The following table summarizes all of our rules for exponents.

| **Rules for Exponents** | |
|---|---|
| Product Rule | $a^m \cdot a^n = a^{m+n}$ |
| Quotient Rule | $\dfrac{a^m}{a^n} = a^{m-n} \quad (a \neq 0)$ |
| Zero-Power Rule | $a^0 = 1 \quad (a \neq 0)$ |
| Power-to-Power Rule | $(a^m)^n = a^{m \cdot n}$ |
| Product-to-Power Rule | $(ab)^n = a^n b^n$ |
| Quotient-to-Power Rule | $\left(\dfrac{a}{b}\right)^n = \dfrac{a^n}{b^n} \quad (b \neq 0)$ |
| Negative-Power Rule | $a^{-n} = \dfrac{1}{a^n}$ and $\dfrac{1}{a^{-n}} = a^n \quad (a \neq 0)$ |

We can combine the **rules for exponents** to simplify exponential expressions.

**CAUTION** Remember, there is typically more than one path that leads to a correct, simplified exponential expression. Just make sure your path correctly applies the rules for exponents.

### Example 3  Using a Combination of Exponent Rules

 *My interactive video summary*

Simplify.

a. $(9x^{-5})(7x^2)$    b. $(p^{-4})^2$    c. $\dfrac{52m^{-4}}{13m^{-10}}$    d. $(w^{-1}z^3)^{-4}$

### Solutions

a.  Begin with the original expression:  $(9x^{-5})(7x^2)$

Rearrange factors to group like bases:  $= 9 \cdot 7 \cdot x^{-5} \cdot x^2$

Multiply constants; apply **product rule**:  $= 63 \cdot x^{-5+2}$

Simplify:  $= 63x^{-3}$

Apply the **negative-power rule**:  $= \dfrac{63}{x^3}$ ← Switch location of base $x$ from numerator to denominator, and change exponent from $-3$ to 3.

b.  Begin with the original expression:  $(p^{-4})^2$

Apply the **power-to-power rule**:  $= p^{(-4)(2)}$

Simplify:  $= p^{-8}$

Apply the **negative-power rule**:  $= \dfrac{1}{p^8}$

c.–d. Try to simplify these expressions on your own. View the **answers**, or work through this **interactive video** for complete solutions to all four parts.

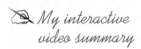

**You Try It**    Work through this You Try It problem.

Work Exercises 17–22 in this textbook or in the **MyMathLab**® Study Plan.

### Example 4  Using a Combination of Exponent Rules

*My interactive video summary*

Simplify.

a. $\dfrac{(3xz)^{-2}}{(2yz)^{-3}}$    b. $\left(\dfrac{10}{x}\right)^{-3}$    c. $\dfrac{(2a^5b^{-6})^3}{4a^{-1}b^5}$    d. $\left(\dfrac{-5xy^{-3}}{x^{-2}y^5}\right)^4$

### Solutions

a. Begin with the original expression:  $\dfrac{(3xz)^{-2}}{(2yz)^{-3}}$

Apply the **negative-power rule**:  $= \dfrac{(2yz)^3}{(3xz)^2}$

Apply the **product-to-power rule**:  $= \dfrac{8y^3z^3}{9x^2z^2}$ ← $\boxed{(2yz)^3 = 2^3y^3z^3 = 8y^3z^3}$
                                                    ← $\boxed{(3xz)^2 = 3^2x^2z^2 = 9x^2z^2}$

Apply the **quotient rule**:  $= \dfrac{8y^3z}{9x^2}$ ← $\boxed{\dfrac{z^3}{z^2} = z^{3-2} = z^1 = z}$

b. Begin with the original expression:  $\left(\dfrac{10}{x}\right)^{-3}$

Apply the **quotient-to-power rule**:  $= \dfrac{10^{-3}}{x^{-3}}$

Apply the **negative-power rule:** $= \dfrac{x^3}{10^3}$

Simplify: $= \dfrac{x^3}{1000}$

**c.–d.** Try to simplify these expressions on your own. View the **answers,** or work through this **interactive video** for complete solutions to all four parts.

**You Try It**   **Work through this You Try It problem.**

**Work Exercises 23–28 in this textbook or in the *MyMathLab*® Study Plan.**

OBJECTIVE 3   CONVERT A NUMBER FROM STANDARD FORM TO SCIENTIFIC NOTATION

Scientists frequently work with very large or very small numbers. For example, the distance from the Sun to the center of the Milky Way is about 19,200,000,000,000 miles, and the mass of a neutron is about 0.0000000000000000000000001675 grams.

Calculating with such numbers in **standard form** can be difficult. To make the numbers more manageable, we can use **scientific notation**.

---

**Scientific Notation**

A number is written in **scientific notation** if it has the form

$$a \times 10^n,$$

where $a$ is a **real number**, such that $1 \le |a| < 10$, and $n$ is an **integer**.

---

**Note:** When writing scientific notation, typically we use a times sign ($\times$) instead of a dot ($\cdot$) to indicate multiplication.

In scientific notation, we write the distance from the Sun to the center of the Milky Way as $1.92 \times 10^{13}$ miles and the mass of a neutron as $1.675 \times 10^{-24}$ grams. Notice that for the very large number, the **exponent** on 10 is positive, while for the very small number, the exponent on 10 is negative.

The following procedure can be used to convert a **real number** from **standard form** to **scientific notation**.

---

**Converting from Standard Form to Scientific Notation**

**Step 1.**   Write the real number factor $a$ by moving the **decimal point** so that $|a|$ is greater than or equal to 1 but less than 10. To do this, place the decimal point to the right of the first non-zero **digit**.

**Step 2.**   Multiply the number by $10^n$, where $|n|$ is the number of places that the decimal point moves. If the decimal point moves to the left, then $n > 0$ ($n$ is positive). If the decimal point moves to the right, then $n < 0$ ($n$ is negative). Remove any zeros lying to the right of the last non-zero digit or to the left of the first non-zero digit.

---

### Example 5  Converting from Standard Form to Scientific Notation

 *My video summary*  Write each number in scientific notation.

**a.** 56,800,000,000,000,000    **b.** 0.0000000467    **c.** 0.00009012    **d.** 200,000,000

**Solutions**

**a.** Move the **decimal point** to the right of the 5, the first non-zero **digit**.

$$56,800,000,000,000,000.$$

16 places

The decimal point has moved sixteen places to the left, so the exponent on the 10 will be 16. Remove all of the zeros behind the last non-zero digit 8.

$$56,800,000,000,000,000 = 5.68 \times 10^{16}$$

**b.** Move the **decimal point** to the right of the 4, the first non-zero **digit**.

$$0.0000000467$$

8 places

The decimal point has moved eight places to the right, so the exponent on the 10 will be $-8$. Remove all of the zeros in front of the 4.

$$0.0000000467 = 4.67 \times 10^{-8}$$

**c.–d.** Try to convert these numbers to scientific notation on your own. View the **answers**, or watch this **video** for detailed solutions to all four parts.

**You Try It**    Work through this You Try It problem.

**Work Exercises 29–36 in this textbook or in the *MyMathLab*® Study Plan.**

**OBJECTIVE 4**    CONVERT A NUMBER FROM SCIENTIFIC NOTATION TO STANDARD FORM

---

**Converting from Scientific Notation to Standard Form**

1. Remove the exponential factor, $10^n$.

2. Move the **decimal point** $|n|$ places, inserting zero placeholders as needed. If $n > 0$ ($n$ is positive), move the decimal point to the right. If $n < 0$ ($n$ is negative), move the decimal point to the left.

---

**CAUTION** Remember, if the **exponent** on 10 is positive, then the absolute value of the number is larger than 1. If the exponent on 10 is negative, then the absolute value of the number is smaller than 1.

### Example 6  Converting from Scientific Notation to Standard Form

 *My video summary*  Write each number in **standard form**.

**a.** $4.98 \times 10^{-5}$    **b.** $9.4 \times 10^7$    **c.** $-3.015 \times 10^9$    **d.** $1.203 \times 10^{-4}$

**Solutions**

**a.** The exponent is $-5$, so we move the decimal point five places to the left, filling in zero placeholders as needed.

$$00004.98$$

5 places

So, $4.98 \times 10^{-5} = 0.0000498.$

**b.** The exponent is 7, so we move the decimal point seven places to the right, filling in zero placeholders as needed.

$$9.\underset{\text{7 places}}{\underbrace{4000000}}$$

So, $9.4 \times 10^7 = 94,000,000$.

**c.–d.** Try to convert these numbers to standard form on your own. View the **answers**, or watch this **video** for detailed solutions to all four parts.

**You Try It**     Work through this You Try It problem.

**Work Exercises 37–44 in this textbook or in the *MyMathLab*® Study Plan.**

OBJECTIVE 5   MULTIPLY AND DIVIDE WITH SCIENTIFIC NOTATION

We use **rules for exponents** to multiply or divide numbers written in **scientific notation**.

### Example 7  Multiplying and Dividing with Scientific Notation

Perform the indicated operations. Write your results in scientific notation.

**a.** $(1.8 \times 10^5)(3 \times 10^8)$

**b.** $\dfrac{2.16 \times 10^{12}}{4.5 \times 10^3}$

**c.** $(-7.4 \times 10^9)(6.5 \times 10^{-4})$

**d.** $\dfrac{5.7 \times 10^{-3}}{7.5 \times 10^{-7}}$

### Solutions

**a.**

| | |
|---|---|
| Begin with original expression: | $(1.8 \times 10^5)(3 \times 10^8)$ |
| Regroup factors: | $= (1.8 \times 3) \times (10^5 \times 10^8)$ |
| Multiply numeric factors; apply **product rule**: | $= 5.4 \times 10^{5+8}$ |
| Simplify: | $= 5.4 \times 10^{13}$ ← This answer is in scientific notation. |

So, $(1.8 \times 10^5)(3 \times 10^8) = 5.4 \times 10^{13}$.

**b.**

| | |
|---|---|
| Begin with original expression: | $\dfrac{2.16 \times 10^{12}}{4.5 \times 10^3}$ |
| Regroup factors: | $= \left(\dfrac{2.16}{4.5}\right) \times \left(\dfrac{10^{12}}{10^3}\right)$ |
| Divide numeric factors; apply **quotient rule**: | $= 0.48 \times 10^9$ ← Not scientific notation because 0.48 is not between 1 and 10. |
| Write 0.48 in scientific notation: | $= (4.8 \times 10^{-1}) \times 10^9$ |
| Regroup factors: | $= 4.8 \times (10^{-1} \times 10^9)$ |
| Apply the **product rule**: | $= 4.8 \times 10^8$ |

So, $\dfrac{2.16 \times 10^{12}}{4.5 \times 10^3} = 4.8 \times 10^8$.

 *My video summary*

 **c.–d.** Try to work these problems on your own. View the **answers**, or watch this **video** for detailed solutions to parts c and d.

**You Try It**     Work through this You Try It problem.

**Work Exercises 45–48 in this textbook or in the *MyMathLab*® Study Plan.**

# 5.6 Exercises

In Exercises 1–16, write each expression with positive exponents. Then simplify if possible.

**You Try It**

**1.** $y^{-3}$      **2.** $5^{-3}$      **3.** $6w^{-5}$      **4.** $-3t^{-8}$

**5.** $(-8)^2$      **6.** $-10^{-4}$      **7.** $2^{-3} + 4^{-1}$      **8.** $3^{-1} \cdot 5^{-2}$

**You Try It**

**9.** $\dfrac{1}{z^{-5}}$      **10.** $\dfrac{1}{4^{-3}}$      **11.** $\dfrac{1}{(-3)^{-4}}$      **12.** $\dfrac{5}{7x^{-4}}$

**13.** $\dfrac{-12}{t^{-3}}$      **14.** $\dfrac{a^{-8}}{b^{-15}}$      **15.** $\dfrac{9^{-2}}{2^{-5}}$      **16.** $\dfrac{x^{-2}y^7}{z^{-5}}$

In Exercises 17–28, simplify each exponential expression using the rules for exponents.

**You Try It**

**17.** $(8x^{-7})(4x^3)$      **18.** $(q^{-6})^4$      **19.** $\dfrac{45t^{-2}}{9t^{-8}}$      **20.** $\dfrac{-24x^{-7}y^5}{4x^{-3}y^2}$

**21.** $(6x^{-2}y^5)^{-3}$      **22.** $\dfrac{5r^5s^0t^4}{-3r^3s^{-4}t^9}$      **23.** $\dfrac{(5ac)^{-4}}{(3bc)^{-2}}$      **24.** $\left(\dfrac{x}{2}\right)^{-5}$

**You Try It**

**25.** $\left(\dfrac{p^2}{q^7}\right)^{-8}$      **26.** $(2x^4)^{-3}(-3y^{-7})$      **27.** $\dfrac{(3a^5b^{-6})^4}{9a^{-7}b^{-1}}$      **28.** $\left(\dfrac{8m^{-4}n^5}{m^3n^{-1}}\right)^3$

In Exercises 29–36, write each number in scientific notation.

**You Try It**

**29.** 89,000,000,000               **30.** 1,025,000,000,000,000

**31.** 0.000000406                 **32.** −0.000005

**33. Microns** A micron is a metric unit for measuring very small lengths or distances. There are 25,400 microns in 1 inch. Write 25,400 in scientific notation.

**34. X-rays** An X-ray has a wavelength of 0.000000724 millimeters. Write 0.000000724 in scientific notation.

**35. Response Time** A computer's RAM response time is 0.000000005 second. Write 0.000000005 in scientific notation.

**36. National Debt** On November 21, 2010, the U.S. national debt was approximately $13,800,000,000,000. Write 13,800,000,000,000 in scientific notation. (*Source*: usdebtclock.org)

In Exercises 37–44, write each number in standard form.

**You Try It**
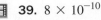

**37.** $2.97 \times 10^{-6}$            **38.** $9.63 \times 10^{13}$

**39.** $8 \times 10^{-10}$             **40.** $-6.045 \times 10^8$

**41. Light-Year** Astronomers measure large distances in light-years. One light-year is the distance light can travel in one year, which is approximately $5.878 \times 10^{12}$ miles. Write $5.878 \times 10^{12}$ in standard form.

**42. Chickenpox** A chickenpox virus measures $1.6 \times 10^{-4}$ millimeters in diameter. Write $1.6 \times 10^{-4}$ in standard form.

**43. Dust Mites** A dust mite weighs $5.3 \times 10^{-5}$ grams. Write $5.3 \times 10^{-5}$ in standard form.

**44. Google Search** In February 2010, an average of $1.21 \times 10^{8}$ searches per hour were conducted using the Google search engine. Write $1.21 \times 10^{8}$ in standard form. (*Source*: searchengineland.com)

**You Try It**

In Exercises 45–48, perform the indicated operations. Write your results in scientific notation.

**45.** $(4.6 \times 10^6)(1.5 \times 10^9)$        **46.** $(9.2 \times 10^{-5})(6.8 \times 10^7)$

**47.** $\dfrac{7.2 \times 10^9}{2.4 \times 10^4}$        **48.** $\dfrac{5.2 \times 10^5}{6.5 \times 10^{-3}}$

# 5.7 Dividing Polynomials

## THINGS TO KNOW

Before working through this section, be sure you are familiar with the following concepts:

|  |  | VIDEO | ANIMATION | INTERACTIVE |
|---|---|---|---|---|

 **You Try It**  **1.** Simplify Exponential Expressions Using the Quotient Rule (Section 5.1, **Objective 2**)

 **You Try It**  **2.** Add Polynomials (Section 5.3, **Objective 1**)

 **You Try It**  **3.** Subtract Polynomials (Section 5.3, **Objective 3**)

 **You Try It**  **4.** Multiply Monomials (Section 5.4, **Objective 1**)

 **You Try It**  **5.** Multiply a Polynomial by a Monomial (Section 5.4, **Objective 2**)

 **You Try It**  **6.** Multiply Two Binomials (Section 5.4, **Objective 3**)

 **You Try It**  **7.** Multiply Two or More Polynomials (Section 5.4, **Objective 4**)

## OBJECTIVES

1 Divide Monomials

2 Divide a Polynomial by a Monomial

3 Divide Polynomials Using Long Division

## OBJECTIVE 1   DIVIDE MONOMIALS

In **Section 5.1**, we used the **quotient rule for exponents** to simplify **exponential expressions**. We grouped **like bases** into individual **quotients** and then applied the quotient rule to simplify. Dividing monomials is a similar process.

---

**Dividing Monomials**

Group the **coefficients** and group like bases into individual quotients. Then divide the coefficients and apply the quotient rule for exponents.

---

### Example 1  Dividing Monomials

Divide.

**a.** $\dfrac{32x^7}{4x^3}$    **b.** $\dfrac{9y^4}{45y^4}$    **c.** $\dfrac{60y}{5y^4}$

### Solutions

**a.**

Begin with the original expression: $\dfrac{32x^7}{4x^3}$

Group like bases into individual quotients: $= \dfrac{32}{4} \cdot \dfrac{x^7}{x^3}$

Divide the coefficients; apply the quotient rule: $= 8 \cdot x^{7-3}$

Simplify: $= 8x^4$

**b.**

Begin with the original expression: $\dfrac{9y^4}{45y^4}$

Group like bases into individual quotients: $= \dfrac{9}{45} \cdot \dfrac{y^4}{y^4}$

Divide the coefficients; apply the quotient rule: $= \dfrac{1}{5} \cdot y^0$ ← $\boxed{4-4}$

Apply the zero-power rule: $= \dfrac{1}{5}$ ← $\boxed{\frac{1}{5} \cdot y^0 = \frac{1}{5} \cdot 1 = \frac{1}{5}}$

**c.**

Begin with the original expression: $\dfrac{60y}{5y^4}$

Group like bases into individual quotients: $= \dfrac{60}{5} \cdot \dfrac{y^1}{y^4}$ ← $\boxed{y = y^1}$

Divide the coefficients; apply the quotient rule: $= 12 \cdot y^{-3}$ ← $\boxed{1-4}$

Apply the negative-power rule: $= \dfrac{12}{y^3}$

**You Try It**   Work through this You Try It problem.

Work Exercises 1–4 in this textbook or in the *MyMathLab*® Study Plan.

OBJECTIVE 2   DIVIDE A POLYNOMIAL BY A MONOMIAL

From **Section 5.4**, we learned how to multiply a **polynomial** by a **monomial**. Using the **distributive property**, we multiplied each **term** of the polynomial by the monomial and then simplified the expression. Dividing a polynomial by a monomial is a similar process. We divide each term of the polynomial by the monomial.

> **Dividing Polynomials by Monomials**
>
> To divide a polynomial by a monomial, divide each term of the polynomial by the monomial and simplify using the method for **dividing monomials**.

We can check our division answers by multiplying. For example, $\dfrac{10}{2} = 5$ checks because $2 \cdot 5 = 10$. In general, if $\dfrac{\text{dividend}}{\text{divisor}} = \text{quotient}$, then $(\text{divisor})(\text{quotient}) = \text{dividend}$.

### Example 2  Dividing a Polynomial by a Monomial

Divide.

**a.** $\dfrac{12x^3 - 28x^2}{4x^2}$        **b.** $(9m^5 - 15m^4 + 18m^3) \div 3m^3$

### Solutions

**a.**                    Begin with the original expression:   $\dfrac{12x^3 - 28x^2}{4x^2}$

Divide each term of the **polynomial** by the **monomial**:   $= \dfrac{12x^3}{4x^2} - \dfrac{28x^2}{4x^2}$

Use the method for **dividing monomials**:   $= \dfrac{12}{4}x^{3-2} - \dfrac{28}{4}x^{2-2}$

Simplify:   $= 3x^1 - 7x^0$

$= 3x - 7$

**Check:** If $\dfrac{12x^3 - 28x^2}{4x^2} = 3x - 7$, then $4x^2(3x - 7)$ must equal $12x^3 - 28x^2$. We multiply to check.

$$4x^2(3x - 7) = 4x^2 \cdot 3x - 4x^2 \cdot 7 = 12x^3 - 28x^2$$

So, our **quotient** $3x - 7$ checks.

*My video summary*      **b.** Try to work this problem on your own. View the **answer**, or watch this **video** for a complete solution to part b.

**You Try It**   Work through this You Try It problem.

Work Exercises 5–8 in this textbook or in the *MyMathLab*® Study Plan.

### Example 3  Dividing a Polynomial by a Monomial

Divide $\dfrac{54t^3 - 12t^2 - 24t}{6t^2}$.

**Solution**

Begin with the original expression: $\dfrac{54t^3 - 12t^2 - 24t}{6t^2}$

Divide each term of the **polynomial** by the **monomial**: $= \dfrac{54t^3}{6t^2} - \dfrac{12t^2}{6t^2} - \dfrac{24t}{6t^2}$

Use the method for **dividing monomials**: $= \dfrac{54}{6}t^{3-2} - \dfrac{12}{6}t^{2-2} - \dfrac{24}{6}t^{1-2}$

Simplify: $= 9t^1 - 2t^0 - 4t^{-1}$

Apply the **negative-power rule**: $= 9t - 2 - \dfrac{4}{t}$

**Check:** $6t^2\left(9t - 2 - \dfrac{4}{t}\right) = 6t^2 \cdot 9t - 6t^2 \cdot 2 - 6t^2 \cdot \dfrac{4}{t}$

$$= 54t^3 - 12t^2 - 24t$$

Take note that $9t - 2 - \dfrac{4}{t}$ is not a polynomial because of the term $-\dfrac{4}{t}$. This illustrates that the **quotient** of two polynomials is not always a polynomial.

**You Try It**    **Work through this You Try It problem.**

**Work Exercises 9–10 in this textbook or in the *MyMathLab*® Study Plan.**

---

OBJECTIVE 3    DIVIDE POLYNOMIALS USING LONG DIVISION

When dividing a **polynomial** by a polynomial, we use **long division**. **Polynomial long division** follows the same approach as long division for **real numbers**. View this **popup box** to review an example using long division to divide real numbers.

We can think of long division as a four-step process: **divide, multiply, subtract,** and **drop**.

 Before performing polynomial long division, write the **dividend** and **divisor** in **descending order**. If any **powers** are missing, then insert them with a **coefficient** of 0 as a placeholder.

> **Process for Polynomial Long Division**
>
> 1. **Divide** the first term of the dividend by the first term of the divisor.
> 2. **Multiply** the result of the division by the divisor. Write this result under the dividend, making sure to line up like terms.
> 3. **Subtract** the result of the multiplication from the dividend.
> 4. **Drop** down the next term from the original dividend to form a reduced polynomial.
>
> Repeat these four steps using the reduced polynomial. Continue repeating until the **remainder** can no longer be divided, which occurs when the **degree** of the remainder is less than the degree of the divisor.

The final result can be written in the form

$$\frac{\text{dividend}}{\text{divisor}} = \text{quotient} + \frac{\text{remainder}}{\text{divisor}}.$$

We can check this result by noting that

$$\text{dividend} = (\text{divisor})(\text{quotient}) + \text{remainder}.$$

We demonstrate **polynomial long division** in the following examples.

### Example 4 Dividing Polynomials Using Long Division

Divide $(2x^2 + x - 15) \div (x + 3)$.

**Solution** The **dividend** is $2x^2 + x - 15$ and the **divisor** is $x + 3$. Both are in **descending order**. We apply the long division process of divide, multiply, subtract, and drop until the **remainder** can no longer be divided.

**Divide** the first **term** of the dividend, $2x^2$, by the first term of the divisor, $x$.

Because $\dfrac{2x^2}{x} = 2x$, $2x$ is the first term in the **quotient**.

$$x + 3 \overline{)2x^2 + x - 15} \quad \overset{2x}{\phantom{)}}$$

**Multiply** the result of the division, $2x$, by the divisor, $x + 3$, to get $2x(x + 3) = 2x^2 + 6x$. Write this result under the dividend, making sure to line up **like terms**.

$$\begin{array}{r} 2x \phantom{-15} \\ x + 3 \overline{)2x^2 + \phantom{6}x - 15} \\ 2x^2 + 6x \phantom{-15} \end{array}$$

**Subtract** the result of the multiplication, $2x^2 + 6x$, from the first two terms of the dividend. To do the subtraction, we change the sign of each term in $(2x^2 + 6x)$ and add.

$$\begin{array}{r} 2x \phantom{-15} \\ x + 3 \overline{)2x^2 + \phantom{6}x - 15} \\ -(2x^2 + 6x) \phantom{} \end{array} \quad \rightarrow \quad \begin{array}{r} 2x \phantom{-15} \\ x + 3 \overline{)2x^2 + \phantom{6}x - 15} \\ \text{Add} \rightarrow \underline{-2x^2 - 6x \phantom{}} \\ -5x \phantom{} \end{array}$$

**Drop** down the next term, $-15$, and repeat the process.

$$\begin{array}{r} 2x \phantom{-15} \\ x + 3 \overline{)2x^2 + \phantom{6}x - 15} \\ \underline{-2x^2 - 6x \phantom{-15}} \\ -5x - 15 \end{array}$$

**Divide** $-5x$ by $x$. Because $\dfrac{-5x}{x} = -5$, the next term in the **quotient** is $-5$.

$$\begin{array}{r} 2x - 5 \phantom{} \\ x + 3 \overline{)2x^2 + \phantom{6}x - 15} \\ \underline{-2x^2 - 6x \phantom{-15}} \\ -5x - 15 \end{array}$$

**Multiply** $-5$ by $x + 3$ to get $-5(x + 3) = -5x - 15$.

$$\begin{array}{r} 2x - 5 \phantom{} \\ x + 3 \overline{)2x^2 + \phantom{6}x - 15} \\ \underline{-2x^2 - 6x \phantom{-15}} \\ -5x - 15 \\ -5x - 15 \end{array}$$

**Subtract** $-5x - 15$ from $-5x - 15$. We do this by changing the sign of each term in $(-5x - 15)$ and adding.

$$
\begin{array}{r}
2x - 5 \\
x + 3\overline{)2x^2 + \phantom{2}x - 15} \\
\underline{-2x^2 - 6x} \\
-5x - 15 \\
\underline{-(-5x - 15)}
\end{array}
\qquad \rightarrow \qquad
\begin{array}{r}
2x - 5 \\
x + 3\overline{)2x^2 + \phantom{2}x - 15} \\
\underline{-2x^2 - 6x} \\
-5x - 15 \\
\text{Add} \rightarrow \underline{5x + 15} \\
0 \leftarrow \text{Remainder}
\end{array}
$$

The remainder is 0. So, $(2x^2 + x - 15) \div (x + 3) = 2x - 5$.

**Check** We must show that dividend = (divisor)(quotient) + remainder.

$$
\underbrace{(x + 3)}_{\text{Divisor}}\underbrace{(2x - 5)}_{\text{Quotient}} + \underbrace{(0)}_{\text{Remainder}} = 2x^2 - 5x + 6x - 15 + 0 = \underbrace{2x^2 + x - 15}_{\text{Dividend}}
$$

Our result checks, so the division is correct.

**You Try It**   **Work through this You Try It problem.**

**Work Exercises 11–14 in this textbook or in the *MyMathLab*® Study Plan.**

### Example 5  Dividing Polynomials Using Long Division

Divide $\dfrac{x^2 + 26x - 6x^3 - 12}{2x - 3}$.

**Solution**  First, we rewrite the dividend in **descending order:**   $6x^3 + x^2 + 26x - 12$. The divisor is $2x - 3$. We apply the **long division process** of divide, multiply, subtract, and drop until the **remainder** can no longer be divided.

**Divide** the first **term** of the dividend, $-6x^3$, by the first term of the divisor, $2x$. Because $\dfrac{-6x^3}{2x} = -3x^2$, $-3x^2$ is the first term in the **quotient**.

$$
\begin{array}{r}
-3x^2 \phantom{+ 26x - 12} \\
2x - 3\overline{)-6x^3 + x^2 + 26x - 12}
\end{array}
$$

**Multiply** the result of the division, $-3x^2$, by the divisor, $2x - 3$, to get $-3x^2(2x - 3) = -6x^3 + 9x^2$. Write this result under the dividend, making sure to line up **like terms.**

$$
\begin{array}{r}
-3x^2 \phantom{+ 26x - 12} \\
2x - 3\overline{)-6x^3 + \phantom{2}x^2 + 26x - 12} \\
-6x^3 + 9x^2 \phantom{+ 26x - 12}
\end{array}
$$

**Subtract** the result of the multiplication, $-6x^3 + 9x^2$, from the first two terms of the dividend. To do the subtraction, we change the sign of each term in $(-6x^3 + 9x^2)$ and add.

$$
\begin{array}{r}
-3x^2 \phantom{+ 26x - 12} \\
2x - 3\overline{)-6x^3 + \phantom{2}x^2 + 26x - 12} \\
\underline{-(-6x^3 + 9x^2)}
\end{array}
\qquad \rightarrow \qquad
\begin{array}{r}
-3x^2 \phantom{+ 26x - 12} \\
2x - 3\overline{)-6x^3 + \phantom{2}x^2 + 26x - 12} \\
\text{Add} \rightarrow \underline{+6x^3 - 9x^2} \phantom{+ 26x - 12} \\
-8x^2 \phantom{+ 26x - 12}
\end{array}
$$

**Drop** down the next term, $26x$, and repeat the process.

$$2x - 3\overline{)-6x^3 + \phantom{x}x^2 + 26x - 12}$$
$$\underline{+6x^3 - 9x^2}$$
$$-8x^2 + 26x$$

Try finishing the problem on your own. See the **answer**, or watch this **video** for the complete solution.

**Check** Show that dividend = (divisor)(quotient) + remainder. View the **check**.

**You Try It**    **Work through this You Try It problem.**

**Work Exercises 15–18 in this textbook or in the *MyMathLab*® Study Plan.**

### Example 6  Dividing Polynomials Using Long Division

 Divide $\dfrac{3t^3 - 11t - 12}{t + 4}$.

**Solution** The **dividend** is $3t^3 - 11t - 12$, and the **divisor** is $t + 4$. There is no $t^2$-term in the dividend, so we insert one with a **coefficient** of 0 as a placeholder.

$$t + 4\overline{)3t^3 + 0t^2 - 11t - 12}$$

Try to complete this problem on your own. See the **answer**, or watch this **video** for a detailed solution.

**You Try It**    **Work through this You Try It problem.**

**Work Exercises 19–20 in this textbook or in the *MyMathLab*® Study Plan.**

# 5.7  Exercises

In Exercises 1–4, divide the monomials.

**You Try It**

**1.** $\dfrac{54x^{11}}{9x^3}$

**2.** $\dfrac{-108z^{36}}{27z^{12}}$

**3.** $\dfrac{5m^3}{30m^3}$

**4.** $\dfrac{72t^4}{8t^9}$

In Exercises 5–10, divide the polynomial by the monomial.

**You Try It**

**5.** $\dfrac{14x^4 - 35x^3}{7x^3}$

**6.** $\dfrac{12m^4 - 16m^3 + 32m^2}{4m}$

**7.** $(48y^7 - 24y^5 + 6y^2) \div 6y^2$

**8.** $\dfrac{-10p^5 + 75p^4}{-5p^4}$

**You Try It**

**9.** $\dfrac{18a^3 - 63a^2 - 27a}{9a^2}$

**10.** $\dfrac{8w^3 - 14w + 6}{-2w^2}$

In Exercises 11–20, divide the polynomials using polynomial long division.

**You Try It** **11.** $\dfrac{x^2 + 18x + 72}{x + 6}$

**12.** $\dfrac{6t^2 - t - 43}{2t + 5}$

**13.** $\dfrac{x^3 - 2x^2 - 8x + 21}{x + 3}$

**14.** $\dfrac{6a^3 - 5a^2 - 11a + 13}{3a - 1}$

**You Try It** **15.** $\dfrac{2x^2 + x^3 - 9x - 4}{x + 5}$

**16.** $\dfrac{20 - 11x - 3x^2}{4 - 3x}$

**17.** $\dfrac{30 + 5x - 6x^2}{5 - 2x}$

**18.** $\dfrac{x^4 - 2x^3 - 13x^2 + 24x - 36}{x - 6}$

**You Try It** **19.** $\dfrac{4m^3 + 5m - 7}{m + 2}$

**20.** $\dfrac{9x^4 - 2x^2 + 2x + 9}{x^2 + 5}$

# 5.8  Polynomials in Several Variables

## THINGS TO KNOW

Before working through this section, be sure you are familiar with the following concepts:

| | | VIDEO | ANIMATION | INTERACTIVE |
|---|---|---|---|---|
 **You Try It** | 1. Evaluate Algebraic Expressions (Section 1.5, **Objective 1**) |  | | |
 **You Try It** | 2. Determine the Degree and Leading Coefficient of a Polynomial (Section 5.2, **Objective 3**) |  | | |
 **You Try It** | 3. Add Polynomials (Section 5.3, **Objective 1**) | | |  |
 **You Try It** | 4. Subtract Polynomials (Section 5.3, **Objective 3**) | ▤ | | |
 **You Try It** | 5. Multiply a Polynomial by a Monomial (Section 5.4, **Objective 2**) | ▤ | | |
 **You Try It** | 6. Multiply Two Binomials (Section 5.4, **Objective 3**) | ▤ |  | ▣ |
 **You Try It** | 7. Multiply Two or More Polynomials (Section 5.4, **Objective 4**) | ▤ | ↻ | |

## OBJECTIVES

**1** Determine the Degree of a Polynomial in Several Variables

**2** Evaluate Polynomials in Several Variables

**3** Add or Subtract Polynomials in Several Variables

**4** Multiply Polynomials in Several Variables

OBJECTIVE 1   DETERMINE THE DEGREE OF A POLYNOMIAL IN SEVERAL VARIABLES

So far in this chapter, we have focused on **polynomials in one variable**. A polynomial containing two or more variables, such as the following example, is called a **polynomial in several variables**.

$$-2x^2y^2z^4 + 7xyz^3 + 5xz^2$$

---
**Definition   Polynomial in Several Variables**

A **polynomial in several variables** is a polynomial containing two or more variables.

---

Our discussion of polynomials in one variable extends to polynomials in several variables as well.

Recall that the degree of a monomial is the sum of the exponents on the variables, and that the degree of a polynomial is the largest degree of any of its terms.

### Example 1  Determining the Degree of a Polynomial in Several Variables

Determine the **coefficient** and **degree** of each **term**, then find the **degree of the polynomial**.

**a.** $2x^3y - 7x^2y^3 + xy^2$          **b.** $3x^2yz^3 - 4xy^3z + xy^2z^4$

**Solution**

**a.**

| Term | Coefficient | Degree |
|------|------------|--------|
| $2x^3y$ | 2 | $3 + 1 = 4$  $\leftarrow x^3y = x^3y^1$ |
| $-7x^2y^3$ | $-7$ | $2 + 3 = 5$ |
| $xy^2$ | 1 | $1 + 2 = 3$  $\leftarrow xy^2 = x^1y^2$ |

The degree of the polynomial is 5.

**b.** Try working this problem on your own. View this **popup** to check your answer.

**You Try It**    Work through this You Try It problem.

Work Exercises 1–3 in this textbook or in the *MyMathLab*® Study Plan.

OBJECTIVE 2   EVALUATE POLYNOMIALS IN SEVERAL VARIABLES

We **evaluate** polynomials in several variables the same way we evaluated **algebraic expressions** in **Section 1.5**. We substitute the given values for the **variables** and simplify.

### Example 2  Evaluating Polynomials in Several Variables

**a.** Evaluate $3x^2y - 2xy^3 + 5$ for $x = -2$ and $y = 3$.

**b.** Evaluate $-a^3bc^2 + 5a^2b^2c - 2ab$ for $a = 2$, $b = -1$ and $c = 4$.

**Solution**

a.  Begin with the polynomial:    $3x^2y - 2xy^3 + 5$

     Substitute $-2$ for $x$ and 3 for $y$:    $3(-2)^2(3) - 2(-2)(3)^3 + 5$ ← Put values in parentheses when substituting

     Simplify the exponents:    $= 3(4)(3) - 2(-2)(27) + 5$

     Simplify the multiplication:    $= 36 - (-108) + 5$

                                $= 36 + 108 + 5$ ← $-(-108) = +108$

               Add:    $= 149$

The value of the polynomial is 149 when $x$ is $-2$ and $y = 3$.

 **My video summary**       **b.** Try to evaluate this **polynomial** for the given values on your own. Then view the **answer**, or watch this **video** to see the complete solution.

**You Try It**    Work through this You Try It problem.

Work Exercises 4–6 in this textbook or in the **MyMathLab** Study Plan.

OBJECTIVE 3   ADD OR SUBTRACT POLYNOMIALS IN SEVERAL VARIABLES

We add **polynomials in several variables** by removing **grouping symbols** and **combining like terms**. This is the same way we added polynomials in one variable in **Section 5.3**.

We subtract polynomials in several variables just like we subtract polynomials in one variable—by adding its **opposite**.

**Example 3** Adding or Subtracting Polynomials in Several Variables

Add or Subtract as indicated.

**a.** $(2x^2 + 3xy - 7y^2) + (4x^2 - xy + 11y^2)$

**b.** $(4a^2 - 3ab + 2b^2) - (6a^2 - 5ab + 7b^2)$

**c.** $(7x^4 + 3x^3y^3 - 2xy^3 + 5) + (2x^4 - x^3y^3 + 8xy^3 - 10)$

**d.** $(10x^3y + 2x^2y^2 - 5xy^3 - 8) - (6x^3y + x^2y^2 - 3xy^3)$

**Solution**

**a.** Write the original expression:   $(2x^2 + 3xy - 7y^2) + (4x^2 - xy + 11y^2)$

     Remove grouping symbols:   $= 2x^2 + 3xy - 7y^2 + 4x^2 - xy + 11y^2$

     Rearrange terms:   $= 2x^2 + 4x^2 + 3xy - xy - 7y^2 + 11y^2$

     Combine like terms:   $= 6x^2 + 2xy + 4y^2$

**b.** Write the original expression:   $(4a^2 - 3ab + 2b^2) - (6a^2 - 5ab + 7b^2)$

     Change to add the opposite:   $= (4a^2 - 3ab + 2b^2) + (-6a^2 + 5ab - 7b^2)$

     Remove grouping symbols:   $= 4a^2 - 3ab + 2b^2 - 6a^2 + 5ab - 7b^2$

     Rearrange terms:   $= 4a^2 - 6a^2 - 3ab + 5ab + 2b^2 - 7b^2$

     Combine like terms:   $= -2a^2 + 2ab - 5b^2$

 **My video summary**       **c.–d.** Try working these parts on your own. View the **answers**, or watch this **video** for the complete solutions to both parts.

**You Try It**    Work through this You Try It problem.

Work Exercises 7–16 in this textbook or in the **MyMathLab** Study Plan.

## OBJECTIVE 4   MULTIPLY POLYNOMIALS IN SEVERAL VARIABLES

As with addition and subtraction, we multiply polynomials in several variables just like we multiply polynomials in one variable. The **FOIL method** and the **special product rules** from Section 5.5 can also be used to simplify the multiplication of these polynomials.

### Example 4  Multiplying a Polynomial by a Monomial

Multiply $5xy^2(4x^2 - 3xy + 2y^2)$.

### Solution

| | |
|---|---|
| Begin with the original expression: | $5xy^2(4x^2 - 3xy + 2y^2)$ |
| Distribute the monomial $5xy^2$: | $= 5xy^2 \cdot 4x^2 - 5xy^2 \cdot 3xy + 5xy^2 \cdot 2y^2$ |
| Rearrange the factors: | $= 5 \cdot 4 \cdot x \cdot x^2 \cdot y^2 - 5 \cdot 3 \cdot x \cdot x \cdot y^2 \cdot y + 5 \cdot 2 \cdot x \cdot y^2 \cdot y^2$ |
| Multiply the coefficients and apply the product rule for exponents: | $= 20x^{1+2}y^2 - 15x^{1+1}y^{2+1} + 10xy^{2+2}$ |
| Simiplify: | $= 20x^3y^2 - 15x^2y^3 + 10xy^4$ |

**You Try It**   Work through this You Try It problem.

**Work Exercises 17–20 in this textbook or in the *MyMathLab*® Study Plan.**

### Example 5  Multiplying Two Binomials in Several Variables

Multiply $(3x - 2y)(4x + 3y)$.

### Solution  Apply the **FOIL method**.

$$(3x - 2y)(4x + 3y) = \overbrace{3x \cdot 4x}^{\text{First}} + \overbrace{3x \cdot 3y}^{\text{Outside}} + \overbrace{(-2y) \cdot 4x}^{\text{Inside}} + \overbrace{(-2y) \cdot 3y}^{\text{Last}}$$
$$= 3 \cdot 4 \cdot x \cdot x + 3 \cdot 3 \cdot x \cdot y - 2 \cdot 4 \cdot x \cdot y - 2 \cdot 3 \cdot y \cdot y$$
$$= 12x^{1+1} + 9xy - 8xy - 6y^{1+1}$$
$$= 12x^2 + 9xy - 8xy - 6y^2$$
$$= 12x^2 + xy - 6y^2 \leftarrow \text{Combine like terms}$$

**You Try It**   Work through this You Try It problem.

**Work Exercises 21–24 in this textbook or in the *MyMathLab*® Study Plan.**

### Example 6  Using Special Product Rules with Polynomials in Several Variables

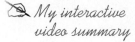

Multiply.

a. $(6x^2 + 5y)^2$    b. $(4x^3 - 9y^2)^2$    c. $(2x^2y - 7)(2x^2y + 7)$

### Solution

| | |
|---|---|
| a.   Square of a binomial sum rule: | $(A + B)^2 = A^2 + 2AB + B^2$ |
| Substitute $6x^2$ for $A$ and $5y$ for $B$: | $(6x^2 + 5y)^2 = (6x^2)^2 + 2(6x^2)(5y) + (5y)^2$ |
| Simplify: | $= 6^2(x^2)^2 + 2(6 \cdot 5)(x^2y) + 5^2y^2$ |
| | $= 36x^4 + 60x^2y + 25y^2$ |

**b.–c.** Try working these problems on your own by applying a **special product rule for binomials**. View the **answers**, or watch this **interactive video** to see solutions for all three parts.

**You Try It**    Work through this You Try It problem.

Work Exercises 25–30 in this textbook or in the *MyMathLab*® Study Plan.

### Example 7  Multiplying Two Polynomials in Several Variables

Multiply $(x + 2y)(x^2 - 4xy + y^2)$.

**Solution**  We multiply each term in the **binomial** by each term in the **trinomial**.

Begin with the original expression:  $(x + 2y)(x^2 - 4xy + y^2)$

Multiply $x$ by $x^2$, $-4xy$, and $y^2$;
Multiply $2y$ by $x^2$,  $4xy$, and $y^2$:  $-x \cdot x^2 + x \cdot (-4xy) + x \cdot y^2 + 2y \cdot x^2 + 2y \cdot (-4xy) + 2y \cdot y^2$

Simplify:  $= x^3 - 4x^2y + xy^2 + 2x^2y - 8xy^2 + 2y^3$

Combine like terms:  $= x^3 - 2x^2y - 7xy^2 + 2y^3$

**You Try It**    Work through this You Try It problem.

Work Exercises 31–34 in this textbook or in the *MyMathLab*® Study Plan.

Remember that multiplication is **commutative** when identifying like terms. For example, $-4x^2y$ and $2yx^2$ are like terms because $2yx^2 = 2x^2y$ so the terms have the same variable factor, $x^2y$.

# 5.8  Exercises

In Exercises 1–3, determine the coefficient and degree of each term, then find the degree of the polynomial.

**You Try It**  **1.** $3x^2y^5 - 4xy^3 + 12y^2 - 5$

**2.** $-\dfrac{1}{2}xy^3 + xy + 3x^4y$

**3.** $6x^2yz^2 - 2.5xy^2z^2 + 7y^3z^5 - 4.3$

In Exercises 4–6, evaluate the polynomial for the given values of the variables.

**You Try It**  **4.** Evaluate $x^2 - 3xy + 2y^2$ for $x = 5$ and $y = 4$.

**5.** Evaluate $xy^3 + y^3 - 2xy + 1$ for $x = -4$ and $y = 2$.

**6.** Evaluate $2x^2y + 3xy - y^2$ for $x = 3$ and $y = -2$.

   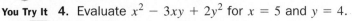
In Exercises 7–16, add or subtract as indicated.

**You Try It**

**7.** $(-12a + 11b) - (6a - 7b)$        **8.** $(3x + 2y) + (8x - 7y)$

9. $(2m^2 + n) - (4m^2 - 16mn - 5n)$

10. $(a^3 + b^3) + (2a^3 - 3a^2b + 9b^3)$

11. $(x - y) - (4x + 2xy - 3y)$

12. $(x^2 + 2xy - 5y^2) + (3x^2 - 4xy + 6y^2)$

13. $(5x^3 + xy - 7) + (-2x^3 - 4xy + 3)$

14. $(3a^4b^2 + 7a^2b^2 - 5ab) - (2a^4b^2 - 8a^2b^2 + ab)$

15. $(mn^2 + n) - (m + 3n) + (2m - 7mn^2)$

16. $(5a + 7b - c) + (2b - 13c) - (a - b + 9c)$

In Exercises 17–34, multiply. Simplify by combining like terms.

**You Try It** 17. $3x(x^2y - 3xy)$

18. $\frac{1}{2}y^2(3x^2 - y^2)$

19. $3x^2y(x + y)$

20. $-4y^3(4xy + 15y^2)$

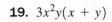

**You Try It** 21. $(5x - 3y)(3x + 5y)$

22. $(x + 7y)(3x + 8y)$

23. $(0.2a^2 + 3b)(3a - 0.4b^2)$

24. $\left(\frac{1}{3}x - 3y\right)\left(\frac{2}{3}x + 2y\right)$

**You Try It** 25. $(a + 3b)^2$

26. $(3a^2 - b^2)(3a^2 + b^2)$

27. $(4m - n)^2$

28. $\left(\frac{2}{5}xy + 3\right)^2$

29. $(5x - 2y)^2$

30. $(x^2y^3 - 4xy^2)(x^2y^3 + 4xy^2)$

31. $(2x - y)(x^2 + 3xy - y^2)$

32. $(x + 5y)(x^2 - 5xy + 25y^2)$

**You Try It** 33. $-2x(x - y)(3x^2 + y)$

34. $(3a - b)(9a^2 - 6ab + b^2)$

# Factoring Polynomials

## CHAPTER SIX CONTENTS

## 6.1 Greatest Common Factor and Factoring by Grouping

### THINGS TO KNOW

Before working through this section, be sure you are familiar with the following concepts:

|  |  | VIDEO | ANIMATION | INTERACTIVE |
|---|---|---|---|---|
| You Try It | 1. Use the Distributive Property (Section 1.5, **Objective 3**) | 🎞 |  |  |
| You Try It | 2. Identify Terms, Coefficients, and Like Terms of an Algebraic Expression (Section 1.6, **Objective 1**) | 🎞 |  |  |
| You Try It | 3. Multiply a Polynomial by a Monomial (Section 5.4, **Objective 2**) | 🎞 |  |  |
| You Try It | 4. Multiply Two Binomials (Section 5.4, **Objective 3**) | 🎞 | 🔄 | 📺 |
| You Try It | 5. Divide a Polynomial by a Monomial (Section 5.7, **Objective 2**) | 🎞 |  |  |

## OBJECTIVES

1 Find the Greatest Common Factor of a Group of Integers

2 Find the Greatest Common Factor of a Group of Monomials

3 Factor Out the Greatest Common Factor from a Polynomial

4 Factor by Grouping

---

### OBJECTIVE 1   FIND THE GREATEST COMMON FACTOR OF A GROUP OF INTEGERS

**Factoring** a **polynomial** is the reverse process of **multiplying polynomials**. When a polynomial is written as an **equivalent expression** that is a **product** of polynomials, we say that the polynomial has been **factored** or written in **factored form**. In **Example 5a of Section 5.4**, we multiplied the **binomials** $x - 4$ and $2x + 3$ to result in the **trinomial** $2x^2 - 5x - 12$. Reversing the process, we factor $2x^2 - 5x - 12$ into the product $(x - 4)(2x + 3)$.

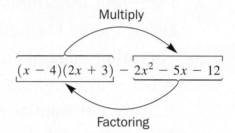

Multiply

$(x - 4)(2x + 3) - 2x^2 - 5x - 12$

Factoring

Therefore, $x - 4$ and $2x + 3$ are **factors** of $2x^2 - 5x - 12$.

 The word *factor* can be used as a noun or verb. As a noun, a *factor* is one of the numbers or expressions being multiplied to form a product. As a verb, *factor* means to rewrite the polynomial as a product.

When factoring a **polynomial**, first we find the *greatest common factor* or GCF of its **terms**. In this section, we always **factor over the integers**, which means that all **coefficients** in both the original polynomial and its factors will be **integers**.

The **greatest common factor (GCF) of a group of integers** is the largest integer that divides evenly into each integer in the group. For example, the GCF of 12 and 18 is 6 because 6 is the largest integer that divides both 12 and 18 evenly ($12 \div 6 = 2$ and $18 \div 6 = 3$). When the GCF is not easy to find, we use the following process:

---

**Finding the GCF of a Group of Integers**

**Step 1.**   Write out the **prime factorization** for each integer in the group.

**Step 2.**   Write down the prime factors common to all the integers in the group.

**Step 3.**   Multiply the **common prime factors** listed in Step 2 to find the GCF. If there are no common prime factors in Step 2, then the GCF is 1.

---

#### Example 1  Finding the GCF of a Group of Integers

Find the GCF of each group of integers.

**a.** 36 and 60     **b.** 28 and 45     **c.** 75, 90, and 105

**Solutions** We follow the steps for finding the GCF of a group of integers.

a. **Step 1.** $36 = \underbrace{2 \cdot 2 \cdot 3} \cdot 3$   and   $60 = \underbrace{2 \cdot 2 \cdot 3} \cdot 5$

Common Prime Factors

**Step 2.** There are three **common prime factors**: $2, 2,$ and $3.$

**Step 3.** The GCF is $2 \cdot 2 \cdot 3 = 12.$

b. **Step 1.** $28 = 2 \cdot 2 \cdot 7$   and   $45 = 3 \cdot 3 \cdot 5$

**Step 2.** There are no common prime factors.

**Step 3.** The GCF is $1.$

*My video summary*  c. Try to find this GCF on your own. View the **answer,** or watch this **video** for a detailed solution to part c.

**You Try It** **Work through this You Try It problem.**

**Work Exercises 1–4 in this textbook or in the *MyMathLab*®️ Study Plan.**

OBJECTIVE 2   FIND THE GREATEST COMMON FACTOR OF A GROUP OF MONOMIALS

The **greatest common factor (GCF) of a group of monomials** is the **monomial** with the largest **coefficient** and highest **degree** that divides each monomial evenly.

Consider the **exponential expressions** $x^3, x^4,$ and $x^5$, each of which is a monomial with a coefficient of $1.$

$$x^3 = \underbrace{x \cdot x \cdot x}, \qquad x^4 = \underbrace{x \cdot x \cdot x} \cdot x, \quad \text{and} \quad x^5 = \underbrace{x \cdot x \cdot x} \cdot x \cdot x$$

Common Factors      Common Factors

The expressions have three factors of the **base** $x$ in common, so the greatest common *factor* of the expressions is $x \cdot x \cdot x = x^3$. Notice that this GCF is the expression with the lowest **power.** This is true in general.

---

**Common Variable Factors for a GCF**

If a variable is a **common factor** of a group of monomials, then the lowest power on that variable in the group will be a factor of the GCF.

---

**Example 2  Finding the GCF of a Group of Exponential Expressions**

Find the GCF of each group of exponential expressions.

a. $x^4$ and $x^7$      b. $y^3, y^6,$ and $y^9$      c. $w^6 z^2, w^3 z^5,$ and $w^5 z^4$

**Solutions**

a. The variable $x$ is a common factor of both expressions. The lowest power is 4, so the GCF is $x^4$.

b. The variable $y$ is a **common factor** of all three expressions. The smallest **power** is 3, so the GCF is $y^3$.

c. The variables $w$ and $z$ are common factors of all three expressions. The lowest power of $w$ is 3 and the lowest power of $z$ is 2, so the GCF is $w^3 z^2$.

**You Try It**     Work through this You Try It problem.

Work Exercises 5–8 in this textbook or in the *MyMathLab*® Study Plan.

We can use the following process to find the **GCF of a group of monomials**.

---

**Finding the GCF of a Group of Monomials**

**Step 1.**  Find the GCF of the coefficients.

**Step 2.**  Find the lowest power for each **common variable factor**.

**Step 3.**  The GCF is the **monomial** with the coefficient from Step 1 and the variable factor(s) from Step 2.

---

**Example 3  Finding the GCF of a Group of Monomials**

Find the GCF of each group of monomials.

**a.** $14x^6$ and $21x^8$
**b.** $6a^2, 10ab$, and $14b^2$
**c.** $40x^5y^6, -48x^9y$, and $24x^2y^4$
**d.** $14m^3n^2, 6m^5n$, and $9m^4$

**Solutions**

**a.**  We follow the steps for **finding the GCF of a group of monomials**.

  **Step 1.**  The **prime factorizations** of the **coefficients** 14 and 21 are

$$14 = 2 \cdot 7 \text{ and } 21 = 3 \cdot 7.$$

  So, the **GCF** for the coefficients is 7.

  **Step 2.**  The variable $x$ is a **common factor** of both expressions, and the lowest power is 6. So, $x^6$ is a factor of the GCF.

  **Step 3.**  Combining Steps 1 and 2, the GCF of the group of monomials is $7x^6$.

**b.**  **Step 1.**  The prime factorizations of the coefficients 6, 10, and 14 are

$$6 = 2 \cdot 3, 10 = 2 \cdot 5, \text{ and } 14 = 2 \cdot 7.$$

  The GCF for the coefficients is 2.

  **Step 2.**  There is no common variable factor for all three monomials. (Note: $6a^2$ does not have a factor of $b$, and $14b^2$ does not have a factor of $a$.) So, no variables are included in the GCF.

  **Step 3.**  The GCF for the group of monomials is 2.

 **c.–d.**  Try finding these GCFs on your own, then check your **answers**. Watch this video for complete solutions to parts c and d.

**You Try It**     Work through this You Try It problem.

Work Exercises 9–14 in this textbook or in the *MyMathLab*® Study Plan.

---

OBJECTIVE 3     FACTOR OUT THE GREATEST COMMON FACTOR FROM A POLYNOMIAL

The **greatest common factor (GCF) of a polynomial** is the **expression** with the largest **coefficient** and highest **degree** that divides each term of the **polynomial** evenly.

Consider the **binomial** $6x^2 + 10x$. The **GCF** of the two **monomials** $6x^2$ and $10x$ is $2x$, so $2x$ is the GCF of $6x^2 + 10x$. Once we know the GCF, we can **factor** it out of the binomial. We write each term of the binomial as a **product** that includes the **factor** $2x$.

$$6x^2 + 10x = 2x \cdot 3x + 2x \cdot 5$$

Reversing the **distributive property**, we factor $2x$ out of each term.

$$2x \cdot 3x + 2x \cdot 5 = 2x(3x + 5)$$

So, $6x^2 + 10x = 2x(3x + 5)$.

Notice that the final result is the GCF times a **binomial factor**. This is expected because the original expression is a binomial. When the GCF is factored from a polynomial, the **polynomial factor** of the resulting product will have the same number of terms as the original polynomial.

Generalizing our process for this example gives the following steps for factoring out the GCF from a polynomial.

---

**Factoring Out the GCF from a Polynomial**

**Step 1.** Find the **GCF** of all **terms** in the **polynomial**.

**Step 2.** Write each term as a **product** that includes the GCF.

**Step 3.** Use the **distributive property** in reverse to factor out the GCF.

**Step 4.** Check the answer. The terms of the **polynomial factor** should have no more **common factors**. Multiplying the answer back out should give the original polynomial.

---

### Example 4  Factoring Out the GCF from a Binomial

Factor out the GCF from each binomial.

**a.** $6x + 12$ **b.** $w^5 + w^4$ **c.** $8y^3 - 12y^2$

**Solutions**  We follow our four-step process.

a. **Step 1.** The two terms are $6x$ and $12$. The largest **integer** that divides evenly into the two **coefficients** $6$ and $12$ is $6$. There is no common variable factor, so the GCF is $6$.

   **Step 2.** Write each term as the product of $6$ and another factor.

   $$6x + 12 = 6 \cdot x + 6 \cdot 2$$

   **Step 3.** Factor out $6$ using the distributive property in reverse.

   $$6 \cdot x + 6 \cdot 2 = 6(x + 2)$$

   **Step 4.** The **terms** of the **binomial factor** $x + 2$ have no **common factors**, so our **GCF** should be correct. Multiplying the final answer,

   $$6(x + 2) = 6 \cdot x + 6 \cdot 2 = 6x + 12.$$

   This is the original binomial, so our answer checks.

   So, $6x + 12 = 6(x + 2)$.

*My video summary*  **b. Step 1.** The two terms are $w^5$ and $w^4$. Both **coefficients** are understood to be 1, so the coefficient of the GCF is 1. The variable $w$ is a common factor of both terms, and the lowest **power** is 4. So the GCF is $w^4$.

**Step 2.** Write each term as the product of $w^4$ and another factor.

$$w^5 + \underbrace{w^4} = w^4 \cdot w + \underbrace{w^4 \cdot 1}$$

Because this term is the GCF, $w^4$, write it as $w^4 \cdot 1$.

Try to complete Steps 3 and 4 on your own to finish the problem. View the **answer**, or watch this **video** for fully worked solutions to parts b and c.

**c.** Try factoring out the GCF on your own. View the **answer**, or watch this **video** for complete solutions to parts b and c.

When checking answers after factoring out the **GCF**, multiplying is not enough. We must also make sure that the **terms** of the **polynomial factor** have no more **common factors**. For example, consider $6x + 12 = 2(3x + 6)$. The GCF has not been factored out of the binomial correctly because the terms $3x$ and 6 still have a common factor of 3. However, multiplying the right side results in the left side, so the problem appears to check. From Example 4a, we know the correct answer is $6x + 12 = 6(x + 2)$.

**You Try It**    **Work through this You Try It problem.**

**Work Exercises 15–18 in this textbook or in the *MyMathLab*® Study Plan.**

## Example 5  Factoring Out the GCF from a Polynomial

Factor out the GCF from each **polynomial**.

**a.** $9p^5 + 18p^4 + 54p^3$          **b.** $10a^4b^6 - 15a^3b^7 + 35a^2b^8$

**Solutions**  We follow our **four-step process**.

**a. Step 1.** The three terms are $9p^5, 18p^4$, and $54p^3$. The largest **integer** that divides evenly into the three **coefficients** 9, 18, and 54 is 9. The variable $p$ is in all three terms and $p^3$ has the lowest **power**. So, the GCF is $9p^3$.

**Step 2.** We write each term as the product of $9p^3$ and another factor.

$$9p^5 + 18p^4 + 54p^3 = 9p^3 \cdot p^2 + 9p^3 \cdot 2p + 9p^3 \cdot 6$$

**Step 3.** Factor out $9p^3$ using the **distributive property** in reverse.

$$9p^3 \cdot p^2 + 9p^3 \cdot 2p + 9p^3 \cdot 6 = 9p^3(p^2 + 2p + 6)$$

**Step 4.** View the **check**.

The answer checks, so $9p^5 + 18p^4 + 54p^3 = 9p^3(p^2 + 2p + 6)$.

*My video summary*  **b.** Try factoring out the GCF on your own. View the **answer**, or watch this **video** for a complete solution to part b.

**You Try It**    **Work through this You Try It problem.**

**Work Exercises 19–22 in this textbook or in the *MyMathLab*® Study Plan.**

When the **leading coefficient** of a polynomial is negative, we may wish to factor out the negative sign with the GCF.

## Example 6  Factoring Out a Negative Sign with the GCF

Factor out the negative sign with the GCF. $-8x^3 + 28x^2 - 20x$

**Solution**  We again follow our **four-step process**.

**Step 1.** The three terms are $-8x^3, 28x^2$, and $-20x$. The largest **integer** that divides evenly into $-8, 28$, and $-20$ is 4. The variable $x$ is in all three terms, and the lowest **power** is 1. So, $x$ is a factor of the GCF. Because the leading coefficient is negative, we include a negative sign as part of the GCF. Therefore, we use $-4x$ as the GCF.

**Step 2.** We write each term as the product of $-4x$ and another factor. Because the second term $28x^2$ has a positive **coefficient**, both factors must be negative.

$$-8x^3 + 28x^2 - 20x = (-4x) \cdot 2x^2 + (-4x) \cdot (-7x) + (-4x) \cdot 5$$

**Step 3.** Factor out $-4x$ using the **distributive property** in reverse.

$$(-4x) \cdot 2x^2 + (-4x) \cdot (-7x) + (-4x) \cdot 5 = -4x(2x^2 - 7x + 5)$$

**Step 4.** View the **check**.

The answer checks, so $-8x^3 + 28x^2 - 20x = -4x(2x^2 - 7x + 5)$.

**You Try It**  Work through this You Try It problem.

**Work Exercises 23–26 in this textbook or in the *MyMathLab*®️ Study Plan.**

In each example so far, the GCF has been a **monomial**. In the next example, we factor out a **binomial** as the GCF.

## Example 7  Factoring Out a Binomial as the GCF

Factor out the common **binomial factor** as the GCF.

**a.** $4x(y + 5) + 11(y + 5)$          **b.** $7x(x + y) - (x + y)$

## Solutions

**a.** Treating $4x(y + 5)$ and $11(y + 5)$ as **terms**, the only **common factor** is the binomial factor $y + 5$. So, the GCF is $y + 5$. Each term is already written as a product of $y + 5$, so we factor it out:

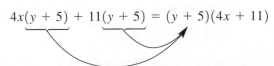

So, $4x(y + 5) + 11(y + 5) = (y + 5)(4x + 11)$.

*My video summary*  **b.** Try factoring out the common binomial factor as the GCF on your own. View the **answer**, or watch this **video** for a complete solution.

**You Try It**  Work through this You Try It problem.

**Work Exercises 27–30 in this textbook or in the *MyMathLab*®️ Study Plan.**

## OBJECTIVE 4  FACTOR BY GROUPING

Suppose the **polynomial** $4x(y + 5) + 11(y + 5)$ from **Example 7a** had been simplified so that its **terms** did not contain the common **binomial factor** $y + 5$. Can we **factor** the simplified polynomial $4xy + 20x + 11y + 55$? The answer is yes. To do this, we use a method called **factoring by grouping**. Let's examine this method using the polynomial

$$4xy + 20x + 11y + 55.$$

Looking at all four terms in the polynomial, there are no **common factors** other than 1. However, if we group the first two terms together and consider them by themselves, they have a common factor of $4x$. **Factoring out** $4x$ from these two terms results in

$$4xy + 20x = 4x(y + 5).$$

Similarly, if we group the last two terms, they have a common factor of 11. Factoring out the 11 gives us

$$11y + 55 = 11(y + 5).$$

Using the grouped pairs, we arrive at the original polynomial in **Example 7a**. We can then factor out the common binomial factor to complete the process.

$$4xy + 20x + 11y + 55 = (4xy + 20x) + (11y + 55)$$
$$= 4x(y + 5) + 11(y + 5)$$
$$= (y + 5)(4x + 11)$$

We summarize how to *factor by grouping* with the following steps:

---

**Factoring a Polynomial by Grouping**

**Step 1.**  Group **terms** with a **common factor**. It may be necessary to rearrange the terms.

**Step 2.**  For each group, **factor** out the **greatest common factor**.

**Step 3.**  Factor out the common **polynomial factor**, if there is one.

**Step 4.**  Check your answer by multiplying out the factors.

---

CAUTION  If no arrangement of terms leads to a common polynomial factor for Step 3, then the polynomial cannot be factored by grouping.

### Example 8  Factoring by Grouping

*My interactive video summary*

 Factor by grouping.

a.  $2x^2 - 6x + xy - 3y$  b.  $5xy + 6 + 5x + 6y$

c.  $3m^2 + 3m - 2mn - 2n$  d.  $4w^3 - 14w^2 - 10w + 35$

### Solutions

a.  The first two terms have a common factor of $2x$, and the last two terms have a common factor of $y$. We follow our process for factoring by grouping.

Begin with the original polynomial expression:  $2x^2 - 6x + xy - 3y$

Group the first two terms and last two terms:  $= (2x^2 - 6x) + (xy - 3y)$

Factor out the GCF from each group:  $= 2x(x - 3) + y(x - 3)$

Factor out the common binomial factor:  $= (x - 3)(2x + y)$

This answer **checks**, so $2x^2 - 6x + xy - 3y = (x - 3)(2x + y)$.

**b.** For $5xy + 6 + 5x + 6y$, the first two terms and the last two terms have no common factors other than 1, so we must rearrange the terms to group terms with common factors. Notice that $5xy$ and $5x$ have a common factor of $5x$ and that $6y$ and 6 have a common factor of 6.

| | |
|---|---|
| Begin with the original polynomial expression: | $5xy + 6 + 5x + 6y$ |
| Rearrange to group terms with common factors: | $= (5xy + 5x) + (6y + 6)$ |
| Factor out the GCF from each group: | $= 5x(y + 1) + 6(y + 1)$ |
| Factor out the common binomial factor: | $= (y + 1)(5x + 6)$ |

This answer **checks**, so $5xy + 6 + 5x + 6y = (y + 1)(5x + 6)$.

**c.–d.** Try factoring these polynomials on your own. View the **answers**, or watch this **interactive video** for complete solutions to all four parts.

 If minus signs are involved when factoring by grouping, pay close attention to the placement of grouping symbols. For example, consider $wx - wy - xz + yz$. It is incorrect to write the grouping as $(wx - wy) - (xz + yz)$ because $-(xz + yz) = -xz - yz$, not $-xz + yz$. It is correct to write the grouping as $(wx - wy) + (-xz + yz)$ or $(wx - wy) - (xz - yz)$.

**You Try It** Work through this You Try It problem.

Work Exercises 31–40 in this textbook or in the *MyMathLab*® Study Plan.

# 6.1 Exercises

In Exercises 1–4, find the GCF of each group of integers.

 **You Try It**

**1.** 21 and 35

**2.** 48 and 80

**3.** 45, 63, and 90

**4.** 70, −105, and 175

In Exercises 5–14, find the GCF of each group of monomials.

 **You Try It**

**5.** $x^6$ and $x^2$

**6.** $w^4, w^7,$ and $w^{10}$

**7.** $a^3b^9$ and $a^5b$

**8.** $x^4y^7, x^8y^6,$ and $x^{12}y^5$

 **You Try It** **9.** $12x^8$ and $30x^5$

**10.** $32y^2$ and $27y^5$

**11.** $10p^2, 25pq,$ and $15q^2$

**12.** $8x^3, 16x^5,$ and $4x^7$

**13.** $54m^3n^4, -36mn^5,$ and $72m^8n^6$

**14.** $24a^2b^4, 15ab^4,$ and $25b^5$

In Exercises 15–22, factor out the GCF from each polynomial.

 **You Try It** **15.** $30x + 12$

**16.** $y^7 + y^5$

 **17.** $20w^4 - 28w^3$

**18.** $27x^2y + 63xy$

 **You Try It** **19.** $4y^5 - 16y^4 + 28y^3$

**20.** $x^2y^6 - 5x^3y^5 + 9x^4y^4$

**21.** $25ab^5 - 10a^2b^4 + 5a^3b^3$

**22.** $6m^4 - 10m^3 + 12m^2 - 14m$

In Exercises 23–26, factor out the negative sign with the GCF.

**You Try It** **23.** $-8x^2 + 24x$                             **24.** $-10m^2n - 35mn^2$

**25.** $-9x^2 - 27x + 12$                      **26.** $-3x^3 + 15x^2 - 21x$

In Exercises 27–30, factor out the common binomial factor as the GCF.

**27.** $5x(y - 4) + 8(y - 4)$               **28.** $2x(x + 8) - (x + 8)$

**You Try It** **29.** $m(m - 2n) - 3n(m - 2n)$        **30.** $z^2(w^2 + 4) + 7(w^2 + 4)$

In Exercises 31–40, factor by grouping.

**31.** $5xy + 15x + 6y + 18$               **32.** $wz + 10w - 2z - 20$

**You Try It** **33.** $9ab - 12a - 6b + 8$            **34.** $4x^2 + 5xy + 8x + 10y$

**35.** $6x^2 - 11xy - 6x + 11y$         **36.** $5x^2 - 3y + 5xy - 3x$

**37.** $x^3 + 6x^2 + 7x + 42$              **38.** $x^3 - x^2 + 13x - 13$

**39.** $x^3 - 5x^2 + x - 5$                **40.** $m^2 + 7n + mn + 7m$

# 6.2   Factoring Trinomials of the Form $x^2 + bx + c$

## THINGS TO KNOW

Before working through this section, be sure you are familiar with the following concepts:

|  |  | VIDEO | ANIMATION | INTERACTIVE |
|---|---|---|---|---|

   **1.** Identify Terms, Coefficients, and Like Terms of an Algebraic Expression (Section 1.6, **Objective 1**)    VIDEO

**You Try It** **2.** Determine the Degree and Leading Coefficient of a Polynomial (Section 5.2, **Objective 3**)

**You Try It** **3.** Multiply Two Binomials (Section 5.4, **Objective 3**)    VIDEO     ANIMATION     INTERACTIVE

**You Try It** **4.** Factor Out the Greatest Common Factor from a Polynomial (Section 6.1, **Objective 3**)

## OBJECTIVES

**1** Factor Trinomials of the Form $x^2 + bx + c$

**2** Factor Trinomials of the Form $x^2 + bxy + cy^2$

**3** Factor Trinomials of the Form $x^2 + bx + c$ after Factoring Out the GCF

OBJECTIVE 1    FACTOR TRINOMIALS OF THE FORM $x^2 + bx + c$

In this section, we continue learning how to **factor polynomials** by focusing on **trinomials**. From **Section 5.4**, we know the **product** of two **binomials** is often a trinomial. Since factoring is the reverse of multiplication, trinomials typically factor into the product of two binomials.

*My animation summary*

Let's look at trinomials with a **leading coefficient** of 1 and a **degree** of 2. These trinomials have the form, $x^2 + bx + c$. Begin by working through this **animation** about factoring trinomials.

Using **FOIL**, pay close attention to the multiplication process of two binomials. For example, look at $(x + 3)(x + 4)$.

$$\overbrace{\phantom{xx}}^{F}\ \overbrace{\phantom{xx}}^{O}\ \overbrace{\phantom{xx}}^{I}\ \overbrace{\phantom{xx}}^{L}$$
$$(x + 3)(x + 4) = x^2 + 4x + 3x + 12 = x^2 + 7x + 12$$

The product is a trinomial of the form $x^2 + bx + c$. Study the **terms** of the trinomial and the terms of each **binomial factor**, and then answer the following questions:

- Where does the term $x^2$ in the trinomial come from? View the **answer**.
- Where does the term $12$ in the trinomial come from? View the **answer**.
- Where does the term $7x$ in the trinomial come from? View the **answer**.

Reversing our multiplication, the **trinomial** $x^2 + 7x + 12$ factors into the **product** $(x + 3)(x + 4)$. The numbers 3 and 4 are the "Last" **terms** in the **binomial factors** because 3 and 4 multiply to 12 and add to 7.

$$\begin{array}{ccccc} x^2 & + & 7x & + & 12 = (x + 3)(x + 4) \\ \uparrow & & \uparrow & & \uparrow \\ x \cdot x & & 3x + 4x & & 3 \cdot 4 \end{array}$$

The relationship between the terms of the trinomial and the terms of the binomial factors leads us to the following steps for *factoring trinomials of the form* $x^2 + bx + c$. Remember that we will be **factoring over integers**.

---

**Factoring Trinomials of the Form $x^2 + bx + c$**

**Step 1.**  Find two **integers**, $n_1$ and $n_2$, whose product is the **constant** term $c$ and whose sum is the **coefficient** $b$. So, $n_1 \cdot n_2 = c$ and $n_1 + n_2 = b$.

**Step 2.**  Write the trinomial in the **factored form** $(x + n_1)(x + n_2)$.

**Step 3.**  Check the answer by multiplying out the factored form.

---

**Example 1  Factoring Trinomials of the Form $x^2 + bx + c$**

Factor each trinomial.

**a.** $x^2 + 11x + 18$          **b.** $x^2 + 13x + 30$

**Solutions**  Follow the **three-step** process.

**a. Step 1.**  We need to find two **integers** whose **product** is 18 and whose **sum** is 11. Because 18 and 11 are both positive, we begin by listing the pairs of **positive factors** for 18:

| Positive Factors of 18 | Sum of Factors |
|---|---|
| 1, 18 | $1 + 18 = 19$ |
| 2, 9 | $2 + 9 = 11$ ← This is the pair. |
| 3, 6 | $3 + 6 = 9$ |

**Step 2.** Because $2 \cdot 9 = 18$ and $2 + 9 = 11$, we write

$$x^2 + 11x + 18 = (x + 2)(x + 9).$$

**Step 3.** Using **FOIL**, we check our answer by multiplying:

$$\overbrace{}^{F} \quad \overbrace{}^{O} \quad \overbrace{}^{I} \quad \overbrace{}^{L}$$
$$(x + 2)(x + 9) = x^2 + 9x + 2x + 18$$
$$= x^2 + 11x + 18 \leftarrow \text{Original trinomial}$$

Our result checks, so $x^2 + 11x + 18 = (x + 2)(x + 9)$.

**Note:** Because multiplication is **commutative**, the order in which we list the factors in the product does not matter. For example, we could also write this answer as
$x^2 + 11x + 18 = (x + 9)(x + 2)$.

*My video summary*  **b.** We need to find two **integers** whose **product** is 30 and whose **sum** is 13. Try to find this pair of integers and finish factoring the trinomial on your own. Remember to check your answer by multiplying out the **factors**. View the **answer**, or watch this **video** for a fully worked solution to part b.

**You Try It** **Work through this You Try It problem.**

**Work Exercises 1–3 in this textbook or in the *MyMathLab*®  Study Plan.**

Recall that a **prime number** is a **whole number** greater than 1 whose only whole number factors are 1 and itself. For example, the first ten prime numbers are $2, 3, 5, 7, 11, 13, 17, 19, 23,$ and $29$. As with numbers, there are also *prime polynomials*.

---

**Definition** **Prime Polynomial**

A polynomial is a **prime polynomial** if its only **factors over the integers** are 1 and itself.

---

For example, the **binomial** $4x + 3$ is a prime polynomial because it cannot be written as the product of any two factors other than 1 and itself. The binomial $4x + 8$ is not prime because it can be factored into $4(x + 2)$.

 When deciding if a polynomial is prime, consider only factors over the integers. For example, even though $4x + 3 = 4(x + 0.75)$, the binomial $4x + 3$ is prime. Because 0.75 is not an integer, $4(x + 0.75)$ is not factored over the integers.

## Example 2  Recognizing a Prime Trinomial of the Form $x^2 + bx + c$

Factor $x^2 + 14x + 20$.

## Solution

**Step 1.** We need to find two **integers** whose **product** is 20 and whose **sum** is 14. Because 20 and 14 are both positive, we list the pairs of **positive factors** for 20:

| Positive Factors of 20 | Sum of Factors |
|---|---|
| 1, 20 | $1 + 20 = 21$ |
| 2, 10 | $2 + 10 = 12$ |
| 4, 5 | $4 + 5 = 9$ |

**Step 2.** Because none of the three possible pairs of factors have a sum of 14, we cannot factor this **trinomial** into the form $(x + n_1)(x + n_2)$. Therefore, $x^2 + 14x + 20$ is a **prime polynomial**.

**You Try It**  Work through this You Try It problem.

**Work Exercise 4 in this textbook or in the MyMathLab® Study Plan.**

In all of the examples so far involving $x^2 + bx + c$, the **coefficients** $b$ and $c$ have both been positive integers, but these coefficients can also be negative. When this happens, we must consider the sign of the **factors** of $c$ as we look for the pair of factors whose sum is $b$.

### Example 3  Factoring Trinomials of the Form $x^2 + bx + c$

Factor.

**a.** $x^2 - 13x + 40$       **b.** $m^2 - 5m - 36$       **c.** $w^2 + 7w - 60$

**Solutions**  Follow the **three-step process**.

**a. Step 1.**  We need to find two **integers** whose **product** is 40 and whose **sum** is $-13$. The only way that the product can be positive and the sum can be negative is if both integers are negative. We begin by listing the pairs of **negative factors** for 40:

| Negative Factors of 40 | Sum of Factors |
|---|---|
| $-1, -40$ | $(-1) + (-40) = -41$ |
| $-2, -20$ | $(-2) + (-20) = -22$ |
| $-4, -10$ | $(-4) + (-10) = -14$ |
| $-5, -8$ | $(-5) + (-8) = -13$  ← This is the pair. |

**Step 2.**  Because $(-5)(-8) = 40$ and $(-5) + (-8) = -13$, we write

$$x^2 - 13x + 40 = (x - 5)(x - 8).$$

**Step 3.**  Multiply to check the answer. View the **check**.

Our result checks, so $x^2 - 13x + 40 = (x - 5)(x - 8)$.

**b.**  We need to find two **integers** whose **product** is $-36$ and whose **sum** is $-5$. For the product to be negative, one of the integers must be positive and the other must be negative. For the sum to be negative, the integer with the larger **absolute value** must be negative. We list such pairs of **factors** for $-36$:

| Factors of $-36$ | Sum of Factors |
|---|---|
| $1, -36$ | $(1) + (-36) = -35$ |
| $2, -18$ | $(2) + (-18) = -16$ |
| $3, -12$ | $(3) + (-12) = -9$ |
| $4, -9$ | $(4) + (-9) = -5$ |
| $6, -6$ | $(6) + (-6) = 0$ |

Use this information to finish factoring the **trinomial** on your own. View the **answer**, or work through this **interactive video** for the complete solution.

**c.**  Try factoring this trinomial on your own. View the **answer**, or work through this **interactive video** for the complete solution.

**You Try It**  Work through this You Try It problem.

**Work Exercise 5–16 in this textbook or in the MyMathLab® Study Plan.**

OBJECTIVE 2    FACTOR TRINOMIALS OF THE FORM $x^2 + bxy + cy^2$

We can factor a **trinomial** in two variables using the same approach as that for factoring a trinomial in one variable. For example, we factor a trinomial of the form $x^2 + bxy + cy^2$ just as we would factor $x^2 + bx + c$. The difference is that the second **term** of each **binomial factor** must contain the $y$ variable. So, we find two **integers**, $n_1$ and $n_2$, whose **product** is the **coefficient** $c$ and whose **sum** is the coefficient $b$. Then we write

$$x^2 + bxy + cy^2 = (x + n_1 y)(x + n_2 y)$$

### Example 4   Factoring a Trinomial of the Form $x^2 + bxy + cy^2$

Factor.

**a.** $x^2 + 10xy + 24y^2$        **b.** $m^2 + 22mn - 48n^2$

**Solutions**

**a.** We need to find two integers whose product is 24 and whose sum is 10. Both 24 and 10 are positive, so we consider the pairs of **positive factors** for 24.

| Positive Factors of 24 | Sum of Factors |
|---|---|
| 1, 24 | $1 + 24 = 25$ |
| 2, 12 | $2 + 12 = 14$ |
| 3, 8 | $3 + 8 = 11$ |
| 4, 6 | $4 + 6 = 10 \leftarrow$ This is the pair. |

Because $4 \cdot 6 = 24$ and $4 + 6 = 10$, we write

$$x^2 + 10xy + 24y^2 = (x + 4y)(x + 6y)$$

Check the answer by multiplying. View the **check**.

*My video summary*    **b.** Try factoring this **trinomial** on your own. View the **answer**, or watch this **video** for a detailed solution.

**You Try It**    Work through this You Try It problem.

**Work Exercise 17–20 in this textbook or in the *MyMathLab*®️ Study Plan.**

OBJECTIVE 3    FACTOR TRINOMIALS OF THE FORM $x^2 + bx + c$ AFTER FACTORING OUT THE GCF

A **polynomial** is **factored completely** if it is written as the **product** of all **prime polynomials**. For example, we factored each **trinomial** in **Example 3** as a product of prime **binomials**, so we factored each trinomial completely.

In **Section 6.1**, we learned how to factor out the **greatest common factor** from a polynomial. If a trinomial has a **common factor** other than 1, we factor out the GCF first.

### Example 5   Factoring Trinomials with a Common Factor

Factor completely.

**a.** $4x^2 - 28x - 32$        **b.** $2y^3 - 36y^2 + 64y$

**Solutions**

a. The three **terms** have a GCF of 4. First, we factor out the 4, then we factor the remaining trinomial.

$$\text{Begin with the original trinomial:} \quad 4x^2 - 28x - 32$$
$$\text{Write each term as a product of the GCF 4:} \quad = 4 \cdot x^2 - 4 \cdot 7x - 4 \cdot 8$$
$$\text{Factor out the 4:} \quad = 4(x^2 - 7x - 8)$$

The trinomial $x^2 - 7x - 8$ is of the form $x^2 + bx + c$, so we use the **three-step process.**

We need to find two **integers** whose **product** is $-8$ and whose **sum** is $-7$. For the product to be negative, one of the integers must be positive and the other must be negative. For the sum to be negative, the integer with the larger **absolute value** must be negative. We list such pairs of **factors** for $-8$:

| Factors of $-8$ | Sum of Factors |
|---|---|
| $1, -8$ | $1 + (-8) = -7$ ‹ This is the pair. |
| $2, -4$ | $2 + (-4) = -2$ |

Because $1(-8) = -8$ and $1 + (-8) = -7$, we write

$$x^2 - 7x - 8 = (x + 1)(x - 8)$$

Therefore, $4x^2 - 28x - 32 = 4(x + 1)(x - 8)$.
Check the answer by multiplying. View the **check.**

✎ *My video summary*  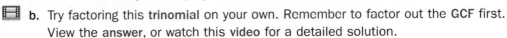 b. Try factoring this **trinomial** on your own. Remember to factor out the **GCF** first. View the **answer**, or watch this **video** for a detailed solution.

**You Try It**   Work through this You Try It problem.

**Work Exercise 21–26 in this textbook or in the *MyMathLab*® Study Plan.**

A trinomial will sometimes have a **leading coefficient** of $-1$, such as $-x^2 + 7x + 60$. Because it is much easier to factor a trinomial when the leading coefficient is 1, we often begin by factoring out $-1$.

## Example 6 Factoring a Trinomial with a Leading Coefficient of $-1$

✎ *My video summary*  Factor $-x^2 + 3x + 10$.

**Solution**  The leading coefficient is $-1$, so let's begin by factoring it out.

$$\text{Begin with the original trinomial:} \quad -x^2 + 3x + 10$$
$$\text{Write each term as a product of } -1: \quad = (-1)x^2 + (-1)(-3x) + (-1)(-10)$$
$$\text{Factor out the } -1: \quad = -1(x^2 - 3x - 10)$$
$$= -(x^2 - 3x - 10)$$

The trinomial $x^2 - 3x - 10$ is of the form $x^2 + bx + c$. Try to finish factoring this trinomial on your own. View the **answer**, or watch this **video** for a detailed solution.

**You Try It**   Work through this You Try It problem.

**Work Exercise 27–30 in this textbook or in the *MyMathLab*® Study Plan.**

## 6.2 Exercises

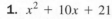

In Exercises 1–20, factor each trinomial, or state that the trinomial is prime.

 **You Try It**

1. $x^2 + 10x + 21$

2. $y^2 + 16y + 28$

3. $w^2 + 17w + 72$

4. $z^2 + 12z + 16$

 **You Try It**

5. $n^2 - 9n + 14$

6. $k^2 - 4k - 5$

7. $p^2 + 3p - 54$

8. $x^2 - 24x + 44$

9. $y^2 + 10y - 36$

10. $m^2 - 3m - 70$

11. $x^2 - 24x + 108$

12. $n^2 + 18n - 40$

13. $x^2 - 31x - 66$

14. $t^2 - 10t + 25$

15. $w^2 + 4w - 96$

16. $q^2 + 20q + 36$

 **You Try It**

17. $x^2 + 6xy + 8y^2$

18. $p^2 + 3pq - 54q^2$

19. $a^2 - 11ab + 24b^2$

20. $m^2 - mn - 90n^2$

 **You Try It**

In Exercises 21–30, factor completely.

21. $5x^2 + 50x + 80$

22. $t^4 - 17t^3 + 60t^2$

23. $2x^3 - 12x^2 - 270x$

24. $8y^2z + 16yz - 280z$

25. $20m^2 - 520m + 500$

26. $2x^3 + 18x^2y + 40xy^2$

 **You Try It**

27. $-x^2 + 2x + 3$

28. $-y^2 - 10y + 144$

29. $-x^2 + 7xy - 10y^2$

30. $-4t^3 + 32t^2 - 60t$

## 6.3 Factoring Trinomials of the Form $ax^2 + bx + c$ Using Trial and Error

### THINGS TO KNOW

Before working through this section, be sure you are familiar with the following concepts:

VIDEO   ANIMATION   INTERACTIVE

 **You Try It**

1. Identify Terms, Coefficients, and Like Terms of an Algebraic Expression (Section 1.6, **Objective 1**)

 **You Try It**

2. Determine the Degree and Leading Coefficient of a Polynomial (Section 5.2, **Objective 3**)

**You Try It**    3.  Multiply Two Binomials
(Section 5.4, **Objective 3**)

**You Try It**    4.  Factor Out the Greatest Common Factor from a
Polynomial (Section 6.1, **Objective 3**)

## OBJECTIVES

**1** Factor Trinomials of the Form $ax^2 + bx + c$ Using Trial and Error

**2** Factor Trinomials of the Form $ax^2 + bxy + cy^2$ Using Trial and Error

OBJECTIVE 1    FACTOR TRINOMIALS OF THE FORM $ax^2 + bx + c$ USING TRIAL AND ERROR

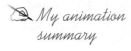
*My animation summary*

If the **leading coefficient** of a **trinomial** is not 1, we must consider the coefficient as we look for the **binomial factors**. Work through this **animation**.

Once again, we can use **FOIL** to see how to factor this type of trinomial. For example, look at the **product** $(2x + 3)(2x + 5)$.

$$\overset{\text{F}}{\overbrace{\phantom{}}}\quad\overset{\text{O}}{\overbrace{\phantom{}}}\quad\overset{\text{I}}{\overbrace{\phantom{}}}\quad\overset{\text{L}}{\overbrace{\phantom{}}}$$
$$(2x + 3)(2x + 5) = 2x \cdot 2x + 2x \cdot 5 + 3 \cdot 2x + 3 \cdot 5 = 4x^2 + 16x + 15$$

The product results in a trinomial of the form $ax^2 + bx + c$ with $a \neq 1$. Study the **terms** of the trinomial and the terms of each binomial factor, and then answer the following questions:

- Where does the first term, $4x^2$, in the trinomial come from? View the **answer**.
- Where does the last term, 15, in the trinomial come from? View the **answer**.
- Where does the middle term, $16x$, in the trinomial come from? View the **answer**.

Reversing the multiplication, the **trinomial** $4x^2 + 16x + 15$ factors into the product $(2x + 3)(2x + 5)$. Using FOIL, the "first" **terms** in our **factors** must multiply to $4x^2$. The "last" terms must multiply to 15. And, the "outside" and "inside" **products** must add to $16x$.

$$\begin{array}{ccccc} 4x^2 & + & 16x & + & 15 = (2x + 3)(2x + 5) \\ \uparrow & & \uparrow & & \uparrow \\ 2x \cdot 2x & & 2x \cdot 5 + 3 \cdot 2x & & 3 \cdot 5 \end{array}$$

If a trinomial of the form $ax^2 + bx + c$ can be factored, it will factor to the form $(m_1x + n_1)(m_2x + n_2)$. We can find the **integers** $m_1$, $m_2$, $n_1$, and $n_2$ by using the following *trial and error strategy*.

---

**Trial-and-Error Strategy for Factoring Trinomials of the Form $ax^2 + bx + c$**

**Step 1.**    Find all pairs of factors for the **leading coefficient** $a$.

**Step 2.**    Find all pairs of factors for the **constant term** $c$.

**Step 3.**    By trial and error, check different combinations of factors from Step 1 and factors from Step 2 in the form $(\Box x + \Box)(\Box x + \Box)$ until the correct middle term $bx$ is found by adding the "outside" and "inside" products. If no such combination of factors exists, the trinomial is **prime**.

**Step 4.**    Check your answer by multiplying out the **factored form**.

---

### Example 1  Factoring Trinomials of the Form $ax^2 + bx + c$ Using Trial and Error

Factor $3x^2 + 7x + 2$.

**Solution**  We follow the **trial-and-error strategy**. Because all of the **coefficients** are positive, we need to check only combinations of **positive factors**.

**Step 1.**  The **leading coefficient** $a$ is 3. The only pair of positive factors is $3 \cdot 1$.

**Step 2.**  The **constant** term $c$ is 2. The only pair of positive factors is $2 \cdot 1$.

**Step 3.**  Using the form $(\Box x + \Box)(\Box x + \Box)$, we try different combinations of the possible factors until we find one that gives the middle term $7x$.

**Trial 1.**  $(3x + 2)(1x + 1)$

$$\begin{array}{r} 2x \\ +3x \\ \hline 5x \end{array} \;\text{◀Not } 7x$$

The "outside" **product** is $3x \cdot 1 = 3x$ and the "inside" product is $2 \cdot 1x = 2x$, giving a **sum** of $3x + 2x = 5x$. This is not the correct combination.

**Trial 2.**  $(3x + 1)(1x + 2)$

$$\begin{array}{r} x \\ +6x \\ \hline 7x \end{array} \;\text{◀Correct}$$

The "outside" product is $3x \cdot 2 = 6x$ and the "inside" product is $1 \cdot 1x = x$, giving a sum of $6x + x = 7x$. This is the correct combination.

**Step 4.**  Check the answer by multiplying out $(3x + 1)(x + 2)$. View the **check**.

The answer checks, so $3x^2 + 7x + 2 = (3x + 1)(x + 2)$.

⬦ CAUTION  Remember, because multiplication is **commutative**, the order in which the **binomial factors** are listed does not matter. So, we could also write the answer to Example 1 as $3x^2 + 7x + 2 = (x + 2)(3x + 1)$.

In Example 1, there were only two possible combinations of factors to check. Often there are many more combinations to consider.

### Example 2  Factoring Trinomials of the Form $ax^2 + bx + c$ Using Trial and Error

*My video summary*  Factor $5x^2 + 17x + 6$.

**Solution**  We follow the **trial-and-error strategy**. Because all of the **coefficients** are positive, we need to check only combinations of **positive factors**.

**Step 1.**  The **leading coefficient** $a$ is 5. The only pair of positive factors is $5 \cdot 1$.

**Step 2.**  The constant term $c$ is 6. The pairs of positive factors are $1 \cdot 6$ and $2 \cdot 3$.

**Step 3.**  We use the form $(\Box x + \Box)(\Box x + \Box)$ with different combinations of the factors from Step 2 until we find one that gives the middle term $17x$.

**Trial 1.**  $(5x + 6)(1x + 1)$

$$\begin{array}{r} 6x \\ +5x \\ \hline 11x \end{array} \;\text{◀Not } 17x$$

The "outside" **product** is $5x \cdot 1 = 5x$ and the "inside" product is $6 \cdot 1x = 6x$, giving a **sum** of $5x + 6x = 11x$. This is not the correct combination.

**Trial 2.**  $(5x + 1)(1x + 6)$

$$\begin{array}{r} x \\ +30x \\ \hline 31x \end{array} \;\text{◀Not } 17x$$

The "outside" product is $5x \cdot 6 = 30x$ and the "inside" product is $1 \cdot 1x = x$, giving a sum of $30x + x = 31x$. This is not the correct combination.

**Trial 3.** $(5x + 3)(1x + 2)$

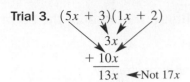

$3x$
$+ \ 10x$
_____
$13x$ ◄Not $17x$

The "outside" product is $5x \cdot 2 = 10x$ and the "inside" product is $3 \cdot 1x = 3x$, giving a sum of $10x + 3x = 13x$. This is not the correct combination.

**Trial 4.** $(5x + 2)(1x + 3)$

$2x$
$+ \ 15x$
_____
$17x$ ◄Correct

The "outside" product is $5x \cdot 3 = 15x$ and the "inside" product is $2 \cdot 1x = 2x$, giving a sum of $15x + 2x = 17x$. This is the correct combination.

**Step 4.** Check the answer by multiplying out $(5x + 2)(x + 3)$. View the **check**, or watch this **video** for a complete solution.

The answer checks, so $5x^2 + 17x + 6 = (5x + 2)(x + 3)$.

**You Try It**   Work through this You Try It problem.

Work Exercises 1–4 in this textbook or in the ***MyMathLab***® Study Plan.

Now let's try factoring **trinomials** with negative **coefficients**.

### Example 3  Factoring Trinomials of the Form $ax^2 + bx + c$ Using Trial and Error

  Factor.

**a.** $4x^2 - 5x - 6$          **b.** $12n^2 - 16n + 5$

**Solutions**

**a.** We follow the **trial-and-error strategy.**

  **Step 1.** The leading coefficient $a$ is 4. For convenience, we will consider only the **positive factors** of 4, which are $4 \cdot 1$ and $2 \cdot 2$.

  **Step 2.** The constant term $c$ is $-6$. In this case, we must consider the negative sign. The pairs of factors are $1 \cdot (-6)$, $-1 \cdot 6$, $2 \cdot (-3)$, and $-2 \cdot 3$.

  **Step 3.** We use the form $(\Box x + \Box)(\Box x + \Box)$ with different combinations of the factors from Step 2 until we find one that gives the middle term $-5x$. Try to complete this problem on your own. Compare your **answer**, or work through this **interactive video** for complete solutions to parts a and b.

**b.** Try factoring this trinomial on your own, then compare your **answer**. Work through this **interactive video** for complete solutions to parts a and b.

**You Try It**   Work through this You Try It problem.

Work Exercises 5–8 in this textbook or in the ***MyMathLab***® Study Plan.

If the **terms** of a **polynomial** have no **common factor** (other than 1), then the terms of its **polynomial factors** will have no common factors either. For example, the terms of the trinomial $4x^2 - 5x - 6$ have no common factors, which means the terms of its binomial factors will have no common factors either. Therefore, some potential **binomial factors**, such as $4x + 6$ and $2x - 2$, can be disregarded because they contain a common factor.

When a trinomial is **prime**, no combination of binomial factors results in the correct middle term.

### Example 4  Recognizing a Prime Trinomial of the Form $ax^2 + bx + c$

Factor $2y^2 - 19y + 15$.

**Solution**  We follow the **trial-and-error strategy**.

**Step 1.**  The **leading coefficient** $a$ is 2. The **positive factors** are $2 \cdot 1$.

**Step 2.**  The **constant term** $c$ is 15. Because the middle **coefficient** $b$ is negative and comes from adding the "outside" and "inside" **products**, we need to consider only the pairs of **negative factors** for 15, which are $-1 \cdot (-15)$ and $-3 \cdot (-5)$.

**Step 3.**  Using the form $(\Box x + \Box)(\Box x + \Box)$, we try different combinations of the factors until we find one that gives the middle term $-19y$.

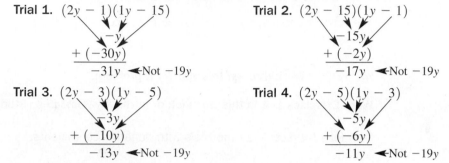

**Trial 1.** $(2y - 1)(1y - 15)$
$-y$
$+ (-30y)$
$-31y$ ◄Not $-19y$

**Trial 2.** $(2y - 15)(1y - 1)$
$-15y$
$+ (-2y)$
$-17y$ ◄Not $-19y$

**Trial 3.** $(2y - 3)(1y - 5)$
$-3y$
$+ (-10y)$
$-13y$ ◄Not $-19y$

**Trial 4.** $(2y - 5)(1y - 3)$
$-5y$
$+ (-6y)$
$-11y$ ◄Not $-19y$

We have checked all the possible combinations, and none result in the middle term $-19y$. This means $2y^2 - 19y + 15$ is **prime**.

**You Try It**    Work through this You Try It problem.

**Work Exercises 9–26 in this textbook or in the *MyMathLab*® Study Plan.**

OBJECTIVE 2    FACTOR TRINOMIALS OF THE FORM $ax^2 + bxy + cy^2$ USING TRIAL AND ERROR

We can factor a **trinomial** of the form $ax^2 + bxy + cy^2$ just as we would factor $ax^2 + bx + c$, except that the second term of each **binomial factor** must contain the variable $y$. So, we use **trial and error** to check different combinations of factors for $a$ and factors for $c$ in the form $(\Box x + \Box y)(\Box x + \Box y)$ until the correct middle term $bxy$ is found.

### Example 5  Factoring a Trinomial of the Form $ax^2 + bxy + cy^2$

Factor.

**a.** $6x^2 + 17xy - 3y^2$          **b.** $2m^2 + 11mn + 12n^2$

**Solutions**

**a.** The pairs of **positive factors** of the **leading coefficient** 6 are $6 \cdot 1$ and $3 \cdot 2$.

The **coefficient** of the last **term** is $-3$, and its **factors** are $1 \cdot (-3)$ and $-1 \cdot 3$.

Using the form $(\Box x + \Box y)(\Box x + \Box y)$, we try different combinations of the factors until we find one that gives the middle term $17xy$.

**Trial 1.** $(6x + 1y)(1x - 3y)$
$xy$
$+ (-18xy)$
$-17xy$ ◄Wrong sign

The "outside" **product** is $6x \cdot (-3y) = -18x$ and the "inside" product is $1y \cdot 1x = xy$, giving a **sum** of $-18xy + xy = -17xy$. This is not the correct combination. Only the sign is wrong, so switch the signs in the binomial factors.

**Trial 2.** $(6x - 1y)(1x + 3y)$

$-xy$
$+18xy$
$17xy$ ◄─Correct

The "outside" **product** is $6x \cdot 3y = 18xy$ and the "inside" product is $-1y \cdot 1x = -xy$, giving a **sum** of $18xy - xy = 17xy$. This is the correct combination.

Because we have found the correct factored form, we do not need to test any more combinations. Check the answer by multiplying out $(6x - y)(x + 3y)$. View the **check**.

The answer does check, so $6x^2 + 17xy - 3y^2 = (6x - y)(x + 3y)$.

*My video summary* ▣ **b.** Try factoring this trinomial on your own, then compare your **answer**. Work through this **video** for a complete solution.

**You Try It** Work through this You Try It problem.

Work Exercises 27–30 in this textbook or in the *MyMathLab* Study Plan.

# 6.3 Exercises

In Exercises 1–30, factor each trinomial, or state that the trinomial is prime.

**1.** $3x^2 + 16x + 5$

**2.** $2y^2 + 7y + 6$

**You Try It** ▣ **3.** $5x^2 + 22x + 8$

**4.** $4w^2 + 20w + 21$

**5.** $6n^2 - 17n + 7$

**6.** $5m^2 + 13m - 6$

**You Try It** ▣ **7.** $12t^2 + 5t - 7$

**8.** $8x^2 - 6x - 9$

**9.** $3x^2 - 4x + 8$

**10.** $10m^2 - 23m + 12$

**You Try It** **11.** $20p^2 - 8p - 1$

**12.** $9y^2 - 24y + 16$

**13.** $2w^2 + 23w - 39$

**14.** $8x^2 + 15x + 9$

**15.** $15t^2 - 49t + 24$

**16.** $14p^2 + 28p - 36$

**17.** $4z^2 - 21z - 18$

**18.** $5p^2 + 8p - 21$

**19.** $12x^2 + 16x - 3$

**20.** $2q^2 + 15q + 7$

**21.** $7y^2 - y - 8$

**22.** $6x^2 - 25x + 4$

**23.** $2w^2 - 15w + 27$

**24.** $22y^2 + 43y - 30$

**25.** $24m^2 - 26m - 15$

**26.** $16x^2 + 32x + 15$

**27.** $9x^2 - 21xy - 8y^2$

**28.** $10p^2 + 7pq - 3q^2$

**You Try It** **29.** $2a^2 - 27ab + 70b^2$

**30.** $3m^2 + 17mn + 20n^2$

# 6.4 Factoring Trinomials of the Form $ax^2 + bx + c$ Using the $ac$ Method

## THINGS TO KNOW

Before working through this section, be sure you are familiar with the following concepts:

## OBJECTIVES

1 Factor Trinomials of the Form $ax^2 + bx + c$ Using the $ac$ Method

2 Factor Trinomials of the Form $ax^2 + bxy + cy^2$ Using the $ac$ Method

3 Factor Trinomials of the Form $ax^2 + bx + c$ after Factoring Out the GCF

---

OBJECTIVE 1    FACTOR TRINOMIALS OF THE FORM $ax^2 + bx + c$ USING THE $ac$ METHOD

The **trial-and-error method** for factoring **trinomials** of the form $ax^2 + bx + c$ can be time consuming. In this section, we learn a second method, known as the **$ac$ method**, for factoring these trinomials. We continue to **factor over the integers**.

> **The $ac$ Method for Factoring Trinomials of the Form $ax^2 + bx + c$**
>
> **Step 1.** Multiply $a \cdot c$.
>
> **Step 2.** Find two **integers**, $n_1$ and $n_2$, whose **product** is $ac$ and whose **sum** is $b$. So, $n_1 \cdot n_2 = ac$ and $n_1 + n_2 = b$. If no such pair of integers exists, the trinomial is **prime**.
>
> **Step 3.** Rewrite the middle term as the sum of two terms using the integers found in Step 2. So, $ax^2 + bx + c = ax^2 + n_1x + n_2x + c$.
>
> **Step 4.** **Factor by grouping.**
>
> **Step 5.** Check your answer by multiplying out the **factored form**.

*My animation video summary*

 Watch this **animation** to see how the *ac* method works.

The *ac* method is also known as the **grouping method** or the **expansion method**. It is favored by some over the trial-and-error method because its approach is more systematic, but it can still be time consuming. We recommend that you practice using both methods and then choose the one that you like best.

**Example 1  Factoring Trinomials of the Form $ax^2 + bx + c$ Using the *ac* Method**

*My video summary*

Factor $3x^2 + 14x + 8$ using the *ac* method.

**Solution**  For $3x^2 + 14x + 8$, we have $a = 3$, $b = 14$, and $c = 8$.

**Step 1.**  $a \cdot c = 3 \cdot 8 = 24$.

**Step 2.**  We must find two **integers** whose **product** is $ac = 24$ and whose **sum** is $b = 14$. Because the product and sum are both positive, we consider only the **positive factors** of 24.

| Positive Factors of 24 | Sum of Factors |
|---|---|
| 1, 24 | $1 + 24 = 25$ |
| 2, 12 | $2 + 12 = 14$ ← This is the pair. |
| 3, 8 | $3 + 8 = 11$ |
| 4, 6 | $4 + 6 = 10$ |

From the list above, 2 and 12 are the integers we need.

**Step 3.**  $3x^2 + 14x + 8 = 3x^2 + 2x + 12x + 8$

**Step 4.**  Begin with the new polynomial from Step 3:  $3x^2 + 2x + 12x + 8$

Group the first two terms and last two terms:  $= (3x^2 + 2x) + (12x + 8)$

Factor out the GCF from each group:  $= x(3x + 2) + 4(3x + 2)$

Factor out the common binomial factor:  $= (3x + 2)(x + 4)$

**Step 5.**  Check the answer by multiplying out $(3x + 2)(x + 4)$. View the **check**, or watch this **video** for a complete solution.

The answer checks, so $3x^2 + 14x + 8 = (3x + 2)(x + 4)$.

**You Try It**    Work through this You Try It problem.

**Work Exercises 1 and 2 in this textbook or in the *MyMathLab*® Study Plan.**

**Example 2  Factoring Trinomials of the Form $ax^2 + bx + c$ Using the *ac* Method**

*My video summary*

Factor $2x^2 - 3x - 20$ using the *ac* method.

**Solution**  For $2x^2 - 3x - 20$, we have $a = 2$, $b = -3$, and $c - -20$.

**Step 1.**  $a \cdot c = 2 \cdot (-20) = -40$.

**Step 2.**  We must find two **integers** whose **product** is $ac = -40$ and whose **sum** is $b = 3$. Because the product is negative, one of the integers must be positive and the other must be negative. For the sum to be negative, the integer with the larger **absolute value** must be negative. We list such pairs of factors for $-40$.

| Factors of $-40$ | Sum of Factors |
|---|---|
| $1, -40$ | $1 + (-40) = -39$ |
| $2, -20$ | $2 + (-20) = -18$ |
| $4, -10$ | $4 + (-10) = -6$ |
| $5, -8$ | $5 + (-8) = -3$ ← This is the pair. |

From the list above, 5 and $-8$ are the integers we need.

**Step 3.** $2x^2 - 3x - 20 = 2x^2 + 5x - 8x - 20$

**Step 4.** **Factor by grouping** to complete this problem on your own. Remember to check your answer by multiplying out your final result. Compare your **answer**, or watch this **video** for a complete solution.

**You Try It**   Work through this You Try It problem.

Work Exercises 3–6 in this textbook or in the *MyMathLab* ® Study Plan.

### Example 3 Factoring Trinomials of the Form $ax^2 + bx + c$ Using the $ac$ Method

*My interactive video summary*

Factor each **trinomial** using the $ac$ **method**. If the trinomial is **prime**, state this as your answer.

**a.** $2x^2 + 9x - 18$    **b.** $6x^2 - 23x + 20$    **c.** $5x^2 + x + 6$

**Solutions**  Try factoring each of these trinomials on your own, and then compare your **answers**. Work through this **interactive video** for detailed solutions to all three parts.

**You Try It**   Work through this You Try It problem.

Work Exercises 7–20 in this textbook or in the *MyMathLab* ® Study Plan.

OBJECTIVE 2   FACTOR TRINOMIALS OF THE FORM $ax^2 + bxy + cy^2$ USING THE $ac$ METHOD

In **Section 6.3**, we used **trial and error** to factor **trinomials** of the form $ax^2 + bxy + cy^2$. We can also use the $ac$ **method** to factor such trinomials, but we must be careful to include the second variable throughout the process.

### Example 4 Factoring a Trinomial of the Form $ax^2 + bxy + cy^2$

*My video summary*

Factor $2p^2 + 7pq - 15q^2$ using the $ac$ method.

**Solution**  For $2p^2 + 7pq - 15q^2$, we have $a = 2, b = 7$, and $c = -15$.

**Step 1.** $a \cdot c = 2 \cdot (-15) = -30$.

**Step 2.** We must find two **integers** whose **product** is $ac = -30$ and whose **sum** is $b = 7$. The product is negative, so one integer must be positive and the other must be negative. The sum is positive, so the integer with the larger **absolute value** must be positive. We list such pairs of factors for $-30$.

| Factors of $-30$ | Sum of Factors |
|---|---|
| $-1, 30$ | $-1 + 30 = 29$ |
| $-2, 15$ | $-2 + 15 = 13$ |
| $-3, 10$ | $-3 + 10 = 7$  ← This is the pair. |
| $-5, 6$ | $-5 + 6 = 1$ |

From the list above, $-3$ and $10$ are the integers we need.

**Step 3.** $2p^2 + 7pq - 15q^2 = 2p^2 - 3pq + 10pq - 15q^2$

**Step 4.** **Factor by grouping** to complete this problem on your own. Remember to check your answer by multiplying out your final result. Compare your **answer**, or watch this **video** for a complete solution.

**You Try It**   Work through this You Try It problem.

Work Exercises 21–24 in this textbook or in the **MyMathLab**® Study Plan.

OBJECTIVE 3   FACTOR TRINOMIALS OF THE FORM $ax^2 + bx + c$ AFTER FACTORING OUT THE GCF

When a **trinomial** has a **common factor** other than 1, we factor out the **GCF** first.

**Example 5  Factoring Trinomials with a Common Factor**

Factor completely:  $24t^5 - 52t^4 - 20t^3$

**Solution**  The terms in this trinomial have a GCF of $4t^3$. We first factor out $4t^3$ and then factor the remaining trinomial.

$$\begin{aligned}
\text{Begin with the original trinomial:} \quad & 24t^5 - 52t^4 - 20t^3 \\
\text{Write each term as a product of the GCF } 4t^3\text{:} \quad & = 4t^3 \cdot 6t^2 - 4t^3 \cdot 13t - 4t^3 \cdot 5 \\
\text{Factor out the } 4t^3\text{:} \quad & = 4t^3(6t^2 - 13t - 5)
\end{aligned}$$

Try to finish the problem on your own by factoring $6t^2 - 13t - 5$. You can use your preferred method, the **ac method** or **trial and error**. Compare your **answer**, or watch this **interactive video** for the complete solution.

**You Try It**   Work through this You Try It problem.

Work Exercises 25 and 26 in this textbook or in the **MyMathLab**® Study Plan.

Consider the trinomial $-2x^2 + 9x + 35$. Notice that the three **terms** do not have a **common factor**. However, when a trinomial has a negative number for a **leading coefficient**, we typically begin by factoring out a $-1$ as we did in **Example 6 of Section 6.2**. We do this because it is much easier to factor a trinomial when the leading coefficient is positive.

**Example 6  Factoring Trinomials with a Common Factor**

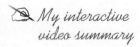

Factor completely. $-2x^2 + 9x + 35$

**Solution**  Because the leading coefficient is negative, we factor out a $-1$.

$$-2x^2 + 9x + 35 = -1(2x^2 - 9x - 35) = -(2x^2 - 9x - 35)$$

*(eText Screens 6.4-1–6.4-12)*

Finish the problem on your own by factoring $2x^2 - 9x - 35$. You can use either the *ac* method or trial-and-error. View the **answer**, or watch this **interactive video** for the complete solution.

**You Try It** Work through this You Try It problem.

Work Exercises 27–30 in this textbook or in the *MyMathLab*® Study Plan.

# 6.4 Exercises

In Exercises 1–30, factor each trinomial completely, or state that the trinomial is prime.

**You Try It**

1. $5x^2 + 17x + 6$

2. $2y^2 + 27y + 13$

3. $4t^2 + 7t - 15$

4. $7m^2 - 8m - 12$

**You Try It**

5. $3p^2 - 26p + 16$

6. $12n^2 + 28n - 5$

7. $2w^2 - 17w + 33$

8. $4x^2 + 3x + 2$

**You Try It**

9. $10y^2 + 47y - 15$

10. $16a^2 + 32a - 9$

11. $18q^2 - 13q + 2$

12. $2m^2 - m - 21$

13. $20x^2 + 29x + 5$

14. $15n^2 - 38n + 24$

15. $8t^2 + 6t - 5$

16. $5y^2 + 11y - 12$

17. $6z^2 - 19z - 20$

18. $21t^2 + 8t - 5$

19. $8n^2 - 51n - 56$

20. $22x^2 + 81x + 14$

**You Try It**

21. $4x^2 + 11xy + 6y^2$

22. $3m^2 - 23mn + 40n^2$

23. $14a^2 + 15ab - 9b^2$

24. $3p^2 - 11pq - 20q^2$

**You Try It**

25. $10x^4 + 44x^3 - 30x^2$

26. $40y^2 - 285y + 35$

27. $-3m^2 - 10m + 8$

28. $-20n^2 - 26n - 8$

**You Try It**

29. $-36a^2b + 147ab - 12ab^2$

30. $24x^2 - 4xy - 60y^2$

# 6.5 Factoring Special Forms

## THINGS TO KNOW

Before working through this section, be sure you are familiar with the following concepts:

|  |  | VIDEO | ANIMATION | INTERACTIVE |

**You Try It**   1.  Square a Binomial Sum (Section 5.5, **Objective 1**)

**You Try It**   2.  Square a Binomial Difference (Section 5.5, **Objective 2**)

**You Try It**   3.  Multiply the Sum and Difference of Two Terms (Section 5.5, **Objective 3**)

**You Try It**   4.  Factor Out the Greatest Common Factor from a Polynomial (Section 6.1, **Objective 3**)

## OBJECTIVES

**1** Factor the Difference of Two Squares

**2** Factor Perfect Square Trinomials

**3** Factor the Sum or Difference of Two Cubes

................................................................................................................................

OBJECTIVE 1   FACTOR THE DIFFERENCE OF TWO SQUARES

In this section, we focus on the *form*. For example, the **standard form** for a **linear equation in two variables** is $Ax + By = C$, where $A$ and $B$ are not both zero. The equations $y = -3x$ and $4 + 5y = 8x$ may look different and have different **coefficients**, but they are both linear equations in two variables because they can be written in the same standard form, $Ax + By = C$.

| Equation | Standard Form | Coefficients |
|---|---|---|
| $y = -3x$ | $3x + y = 0$ | $A = 3, B = 1, C = 0$ |
| $4 + 5y = 8x$ | $8x - 5y = 4$ | $A = 8, B = -5, C = 4$ |

Focusing on *form* is particularly helpful with **factoring**. In **Section 5.5**, we used **special product rules** to multiply two **binomials**. In that section, recognizing how two binomial factors fit a special form helped us to quickly find the product. Working in reverse, we can use the results of special binomial products to help us factor certain expressions.

Recall from **Section 5.5** that the product of a binomial and its **conjugate** (the sum and difference of two terms) is the difference of the **squares** of the **terms** in the binomial. So, $(A + B)(A - B) = A^2 - B^2$. We can use this result to form a rule for factoring the **difference of two squares**.

---

**Factoring the Difference of Two Squares**

If $A$ and $B$ are **real numbers**, **variables**, or **algebraic expressions**, then the difference of their squares can be factored into the product of the sum and difference of the two quantities.

$$A^2 - B^2 = (A + B)(A - B)$$

---

**Note:** Because multiplication is **commutative**, we could also write $A^2 - B^2 = (A - B)(A + B)$.

 Remember to focus on *form*. The quantities $A$ and $B$ could be numbers, variables, or algebraic expressions.

## Example 1 Factoring the Difference of Two Squares

Factor each expression completely.

**a.** $x^2 - 9$          **b.** $16 - y^2$

## Solutions

**a.** Start by rewriting the expression as the difference of two squares.

$$x^2 - 9 = x^2 - (3)^2$$

Here, we have $A = x$ (a **variable**) and $B = 3$ (a **real number**). Applying the **difference of two squares rule**, we get

$$x^2 - 9 = \underbrace{x^2 - 3^2}_{A^2 - B^2} = \underbrace{(x + 3)(x - 3)}_{(A + B)(A - B)}.$$

**b.** Start by rewriting the expression as the **difference of two squares**.

$$16 - y^2 = (4)^2 - y^2$$

Here we have $A = 4$ (a real number) and $B = y$ (a variable). Applying the difference of two squares rule, we get

$$16 - y^2 = \underbrace{(4)^2 - y^2}_{A^2 - B^2} = \underbrace{(4 + y)(4 - y)}_{(A + B)(A - B)}.$$

**You Try It**     Work through this You Try It problem.

**Work Exercises 1–5 in this textbook or in the *MyMathLab*® Study Plan.**

To use the **difference of two squares rule**, both terms must be **perfect squares**. An integer is a perfect square if it is the square of another integer. View this **popup** to review some **perfect square** integers. A rational number is a perfect square if it can be written as the square of another rational number. For example, $\dfrac{4}{9}$ is a perfect square because we can write $\dfrac{4}{9} = \left(\dfrac{2}{3}\right)^2$.

A **monomial** with a perfect square **coefficient** and variables raised to even **powers** is also a perfect square. For example, $9x^2$ and $4x^2y^6$ are perfect squares.

An **algebraic expression** is a perfect square if it is raised to an even power. For example, $(5n)^2$ is a perfect square.

## Example 2  Factoring the Difference of Two Squares

Factor each expression completely.

**a.** $z^2 - \dfrac{25}{16}$    **b.** $36x^2 - 25$    **c.** $4 - 49n^6$    **d.** $81m^2 - n^2$

### Solutions

**a.** Start by rewriting the expression as the **difference of two squares**.

$$z^2 - \frac{25}{16} = z^2 - \left(\frac{5}{4}\right)^2$$

Here, we have $A = z$ (a variable) and $B = \dfrac{5}{4}$ (a real number). Applying the **difference of two squares rule**, we get

$$z^2 - \frac{25}{16} = \underbrace{z^2 - \left(\frac{5}{4}\right)^2}_{A^2 - B^2} = \underbrace{\left(z + \frac{5}{4}\right)}_{(A + B)}\underbrace{\left(z - \frac{5}{4}\right)}_{(A - B)}.$$

**b.** Start by rewriting the expression as the difference of two squares.

$$36x^2 - 25 = 6^2x^2 - 5^2 = (6x)^2 - 5^2$$

Here we have $A = 6x$ (an **algebraic expression**) and $B = 5$ (a real number). Apply the difference of two squares rule, then view the **answer** to see if you are correct.

  **c.–d.** Try factoring these two expressions on your own using the **difference of two squares rule**. View the **answers**, or watch this **video** for detailed solutions to parts c and d.

**You Try It**    Work through this You Try It problem.

Work Exercises **6–10** in this textbook or in the **MyMathLab**® Study Plan.

## Example 3  Factoring the Difference of Two Squares with a Greatest Common Factor

Factor each expression completely.

**a.** $3x^2 - 75$    **b.** $36x^3 - 64x$

### Solutions

**a.** Start by factoring out the **greatest common factor**, 3.

$$3x^2 - 75 = 3 \cdot x^2 - 3 \cdot 25$$
$$= 3(x^2 - 25)$$

Next, rewrite the expression in parentheses as the **difference of two squares**.

$$3(x^2 - 25) = 3(x^2 - 5^2)$$

Here, we have $A = x$ and $B = 5$. Applying the **difference of two squares rule**, we get

$$3(x^2 - \overbrace{5^2}^{A^2 - B^2}) = 3\underbrace{(x + 5)}_{(A + B)}\underbrace{(x - 5)}_{(A - B)}.$$

So, $3x^2 - 75 = 3(x + 5)(x - 5)$

 *My video summary*    **b.** Try factoring this expression on your own. Begin by looking for a **greatest common factor**, then use the difference of two squares rule. View the **answer**, or watch this **video** for a detailed solution to part b.

**You Try It**   **Work through this You Try It problem.**

**Work Exercises 11–16 in this textbook or in the *MyMathLab*® Study Plan.**

⬥ CAUTION   The *sum* of two **perfect squares** cannot be factored using real numbers other than to factor out the greatest common factor.

When factoring, we want to make sure we always factor *completely*. This means we should always check each **factor** to see if it can be factored further.

### Example 4 Factoring the Difference of Two Squares More than Once

 *My video summary*    Factor completely.

$$16x^4 - 81$$

**Solution**   Work through the following, or watch this **video** solution.
Start by rewriting the expression as the **difference of two squares**.

$$16x^4 - 81 = (4x^2)^2 - (9)^2$$

Here, we have $A = 4x^2$ and $B = 9$. Applying the **difference of two squares rule**, we get

$$16x^4 - 81 = \underbrace{(4x^2)^2 - (9)^2}_{A^2 - B^2} = \underbrace{(4x^2 + 9)}_{(A + B)}\underbrace{(4x^2 - 9)}_{(A - B)}.$$

Notice that the factor $(4x^2 - 9)$ can be written as $(2x)^2 - 3^2$, so it too is the **difference of two squares** and can be factored further. The factor $(4x^2 + 9)$ is the sum of two squares, so it cannot be factored further.

$$4x^2 - 9 = \underbrace{(2x)^2 - (3)^2}_{A^2 - B^2} = \underbrace{(2x + 3)}_{(A + B)}\underbrace{(2x - 3)}_{(A - B)}$$

We write

$$16x^4 - 81 = (4x^2 + 9)\overbrace{(2x + 3)(2x - 3)}^{4x^2 - 9}.$$

This polynomial is now factored completely.

**You Try It**   **Work through this You Try It problem.**

**Work Exercises 17–22 in this textbook or in the *MyMathLab*® Study Plan.**

OBJECTIVE 2   FACTOR PERFECT SQUARE TRINOMIALS

In **Section 5.5**, we had two **special product rules** for squaring the sum or difference of two terms. The results of these products were called **perfect square trinomials**. We can use these special product rules to factor perfect square trinomials by reversing the process.

---

**Factoring Perfect Square Trinomials**

If $A$ and $B$ are real numbers, variables, or algebraic expressions, then

$$A^2 - 2AB + B^2 = (A - B)(A - B) = (A - B)^2 \quad \text{and}$$
$$A^2 + 2AB + B^2 = (A + B)(A + B) = (A + B)^2.$$

---

Again, we focus on the form of the **trinomial**. To be a perfect square trinomial, the first and last **terms** must be **perfect squares**, and the middle term must be twice the product of the two quantities being squared or the opposite of the product. To **factor** a perfect square trinomial, we first identify the quantities being squared in the first and last terms and then apply the appropriate rule.

### Example 5  Factoring Perfect Square Trinomials

Factor each expression completely.

**a.** $x^2 + 6x + 9$         **b.** $y^2 - 10y + 25$

### Solutions

**a.** The first **term** is a **perfect square**, $x^2 = (x)^2$, and the last term is also a perfect square, $9 = 3^2$. Because $6x = 2(x)(3)$, we have a **perfect square trinomial** with a positive middle term.

$$x^2 + 6x + 9 = \underbrace{(x)^2 + 2(x)(3) + 3^2}_{A^2 + 2 \cdot A \cdot B + B^2} = \underbrace{(x + 3)^2}_{(A + B)^2}$$

**b.** The first term is a perfect square, $y^2 = (y)^2$, and the last term is also a perfect square, $25 = 5^2$. Because $10y = 2(y)(5)$, we have a perfect square trinomial with a negative middle term.

$$y^2 - 10y + 25 = \underbrace{(y)^2 - 2(y)(5) + 5^2}_{A^2 - 2 \cdot A \cdot B + B^2} = \underbrace{(y - 5)^2}_{(A - B)^2}$$

**You Try It**     Work through this You Try It problem.

Work Exercises 23–24 in this textbook or in the **MyMathLab**® Study Plan.

### Example 6  Factoring Perfect Square Trinomials

Factor each expression completely.

**a.** $4x^2 + 12x + 9$         **b.** $25y^2 - 60y + 36$

### Solutions

**a.** We write $4x^2 = 2^2x^2 = (2x)^2$ and $9 = 3^2$. Because $12x = 2(2x)(3)$, we have a **perfect square trinomial** with a positive middle term.

$$4x^2 + 12x + 9 = \underbrace{(2x)^2 + 2(2x)(3) + 3^2}_{A^2 + 2 \cdot A \cdot B + B^2} = \underbrace{(2x + 3)^2}_{(A + B)^2}$$

 *My video summary*    **b.** Try to do this factoring on your own. When finished, view the **answer**, or watch this **video** for a detailed solution to part b.

**You Try It**    **Work through this You Try It problem.**

**Work Exercises 25–27 in this textbook or in the *MyMathLab*®  Study Plan.**

Perfect square trinomials may contain several variables as shown in Example 7a.

### Example 7   Factoring Perfect Square Trinomials

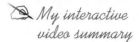

 Factor each expression completely.

a. $16x^2 + 24xy + 9y^2$      b. $m^4 - 12m^2 + 36$

### Solutions

a. Write the first and last **terms** as **perfect squares** to determine $A$ and $B$. Now complete the factorization on your own. View the **answer**, or watch this **interactive video** for a detailed solution.

b. Using **rules for exponents**, we can write $m^4 = (m^2)^2$. Now complete the factorization on your own. View the **answer**, or watch this **interactive video** for a detailed solution.

**You Try It**    **Work through this You Try It problem.**

**Work Exercises 28–34 in this textbook or in the *MyMathLab*®  Study Plan.**

OBJECTIVE 3    FACTOR THE SUM OR DIFFERENCE OF TWO CUBES

Like the difference of two **squares**, the difference of two **cubes** is the result of a special product rule. Consider the following example:

$$\overbrace{(A - B)(A^2 + AB + B^2)}$$
$$(x - 3)(x^2 + 3x + 9)$$

| | |
|---|---|
| Distribute: | $= x \cdot x^2 + x \cdot 3x + x \cdot 9 - 3 \cdot x^2 - 3 \cdot 3x - 3 \cdot 9$ |
| Simplify: | $= x^3 + 3x^2 + 9x - 3x^2 - 9x - 27$ |
| Collect like terms: | $= x^3 + 3x^2 - 3x^2 + 9x - 9x - 27$ |
| Combine like terms: | $= x^3 - 27$ |
| | $= \underbrace{x^3}_{A^3} - \underbrace{3^3}_{B^3}$   ← Difference of two cubes |

However, unlike the sum of two squares, the sum of two cubes *can* be factored. These two new factor rules are given next.

---

**Factoring the Sum and Difference of Two Cubes**

If $A$ and $B$ are **real numbers**, **variables**, or **algebraic expressions**, then the sum or difference of their cubes can be factored as follows:

$$A^3 + B^3 = (A + B)(A^2 - AB + B^2)$$
$$A^3 - B^3 = (A - B)(A^2 + AB + B^2)$$

---

We can check these rules using two special products. View this **popup** to see the steps.

In order to use these rules, both terms in the expression must be **perfect cubes**. An integer is a perfect cube if it is the cube of another **integer**. View this **popup** to review some **perfect cube** integers. A **monomial** with a perfect cube coefficient and variables raised to powers that are multiples of 3 is also considered a perfect cube (Example: $27x^3$). An algebraic expression is a perfect cube if it is raised to a power that is a multiple of 3 (Example: $(2m - 1)^3$).

Notice that the form of the results can help us remember how to factor the sum or difference of two cubes. See the **details**.

To factor the sum or difference of two cubes, first we identify the quantities being cubed and then apply the appropriate rule.

### Example 8   Factoring the Sum or Difference of Two Cubes

Factor each expression completely.

**a.** $x^3 + 64$        **b.** $z^3 - 8$

### Solutions

**a.** The first **term** is a **perfect cube**, $x^3 = (x)^3$, and the second term is also a perfect cube, $64 = 4^3$. We have the **sum of two cubes**, so we apply the rule for $A^3 + B^3$.

$$A^3 + B^3 = (A + B)(A^2 - AB + B^2)$$
$$x^3 + 64 = (x)^3 + (4)^3 = (x + 4)(x^2 - 4x + 4^2)$$
$$= (x + 4)(x^2 - 4x + 16)$$

**b.** The first term is a perfect cube, $z^3 = (z)^3$, and the second term is also a perfect cube, $8 = 2^3$. We have the **difference of two cubes**, so we apply the rule for $A^3 - B^3$.

$$A^3 - B^3 = (A - B)(A^2 + AB + B^2)$$
$$z^3 - 8 = (z)^3 - (2)^3 = (z - 2)(z^2 + 2z + 2^2)$$
$$= (z - 2)(z^2 + 2z + 4)$$

**You Try It**    **Work through this You Try It problem.**

**Work Exercises 35–36 in this textbook or in the *MyMathLab*® Study Plan.**

### Example 9   Factoring the Sum or Difference of Two Cubes

Factor each expression completely.

**a.** $125y^3 - 1$        **b.** $128z^3 + 54y^3$        **c.** $8x^3y^2 + y^5$

### Solutions

**a.** Because $125y^3 = (5y)^3$ and $1 = 1^3$, we have the **difference of two cubes**. Apply the rule for $A^3 - B^3$.

$$A^3 - B^3 = (A - B)(A^2 + AB + B^2)$$
$$125y^3 - 1 = (5y)^3 - (1)^3 = (5y - 1)((5y)^2 + (5y)(1) + (1)^2)$$
$$= (5y - 1)(25y^2 + 5y + 1)$$

**b.** First, factor out the **GCF**, 2. Write the remaining two **terms** as **perfect cubes** to determine the quantities $A = 4z$ and $B = 3y$. Try to complete the factorization on your own. When finished, view the rest of the **solution**.

*My video summary*  c. First, factor out the GCF. Using **rules for exponents**, we write $y^5 = y^2 \cdot y^3$. Now complete this factorization on your own. Check your **answer**, or watch this **video** for a detailed solution.

**You Try It**   Work through this You Try It problem.

Work Exercises 37–43 in this textbook or in the *MyMathLab* Study Plan.

# 6.5  Exercises

In Exercises 1–43, factor each expression completely. If a polynomial is prime, then state this as your answer.

**You Try It**

**1.** $y^2 - 16$   **2.** $49 - z^2$   **3.** $x^2 - y^2$   **4.** $-p^2 + q^2$

**You Try It**
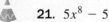

**5.** $x^{10} - 4$   **6.** $x^2 - \dfrac{49}{9}$   **7.** $9y^2 - 1$   **8.** $x^2y^2 - 25$

**9.** $16 - 25m^2$   **10.** $4x^2y^4 - \dfrac{9}{4}z^8$   **11.** $12x^2 - 27$   **12.** $50 - 8y^2$

**You Try It**

**13.** $20x^2 + 45$   **14.** $8a^2 - 18b^2$   **15.** $3x^2y - 12y$   **16.** $a^3b - ab^3$

**You Try It**

**17.** $x^4 - 1$   **18.** $81x^4 - 16$   **19.** $x^4 - y^4$   **20.** $m^8 - 16n^4$

**You Try It**

**21.** $5x^8 - 5$   **22.** $y^{16} - z^{16}$   **23.** $x^2 - 2x + 1$   **24.** $m^2 + 16m + 64$

**You Try It**

**25.** $4x^2 + 28x + 49$   **26.** $9h^2 - 30h + 25$   **27.** $\dfrac{1}{4}x^2 + x + 4$   **28.** $x^2 + 2xy + y^2$

**You Try It**

**29.** $m^2 - 14mn + 49n^2$   **30.** $4a^2 - 20ab + 25b^2$   **31.** $25xy^2 + 70xy + 49x$   **32.** $y^4 - 2y^2 + 1$

**You Try It**

**33.** $z^6 + 10z^3 + 25$   **34.** $16a^4 - 72a^2b^2 + 81b^4$   **35.** $y^3 - 125$   **36.** $z^3 + 27$

**You Try It**

**37.** $8x^3 - 1$   **38.** $5t^3 + 40$   **39.** $64p^3 + 121q^3$   **40.** $3z^5 - 24z^2$

**41.** $x^3y^3 - 27$   **42.** $a^3b^3 + 8c^3$   **43.** $x^9 - 1$

# 6.6 A General Factoring Strategy

## THINGS TO KNOW

Before working through this section, be sure you are familiar with the following concepts:

|  | VIDEO | ANIMATION | INTERACTIVE |

**You Try It**
1. Factor Out the Greatest Common Factor from a Polynomial (Section 6.1, **Objective 3**)

**You Try It**
2. Factor by Grouping (Section 6.1, **Objective 4**)

**You Try It**
3. Factor Trinomials of the Form $x^2 + bx + c$ (Section 6.2, **Objective 1**)

**You Try It**
4. Factor Trinomials of the Form $ax^2 + bx + c$ Using Trial and Error (Section 6.3, **Objective 1**)

**You Try It**
5. Factor Trinomials of the Form $ax^2 + bx + c$ Using the $ac$ Method (Section 6.4, **Objective 1**)

## OBJECTIVE

**1** Factor Polynomials Completely

........................................................................................................................

OBJECTIVE 1   FACTOR POLYNOMIALS COMPLETELY

Now we know several techniques and rules for factoring **polynomials**. We can use this knowledge to create a general strategy for **factoring polynomials completely**. A polynomial is factored completely if all its polynomial **factors**, other than **monomials**, are **prime**.

---

**General Strategy for Factoring Polynomials Completely**

**Step 1.** If necessary, factor out the **greatest common factor**. If the **leading coefficient** is negative, factor out a **common factor** with a negative coefficient.

**Step 2.** Select a strategy based on the number of **terms**.

　　a. If there are two terms, try to use one of the following special factor rules:

　　　　Difference of two squares: $A^2 - B^2 = (A + B)(A - B)$

　　　　Sum of two cubes: $A^3 + B^3 = (A + B)(A^2 - AB + B^2)$

　　　　Difference of two cubes: $A^3 - B^3 = (A - B)(A^2 + AB + B^2)$

　　b. If there are three terms, see if the **trinomial** is a **perfect square trinomial**. If so, factor using one of these special factor rules:

$$A^2 + 2AB + B^2 = (A + B)^2$$
$$A^2 - 2AB + B^2 = (A - B)^2$$

　　　If the trinomial is not a perfect square, factor using **trial and error** or the *ac* **method**.

　　c. If there are four or more terms, try **factoring by grouping**.

*continued*

---

> **Step 3.** Check to see if any factors can be factored further. Check each factor, other than **monomial factors**, to make sure they are **prime**.
>
> **Step 4.** Check your answer by multiplying. Multiply out your result to see if it equals the original expression.

### Example 1 Factoring a Polynomial Completely

*My interactive video summary*

Factor each expression completely.

**a.** $w^2 - w - 20$   **b.** $4y^4 - 32y$   **c.** $x^2 - 14x + 49$   **d.** $3z^3 - 15z^2 - 42z$

### Solutions

**a.** Follow the **general factoring strategy**.

**Step 1.** Looking at each **term**, we see there is no common factor.

**Step 2.** The expression has three terms. The first and last terms are not perfect squares, so this is not one of our special forms.

Because the **leading coefficient** of the trinomial factor is 1, we need two **factors** whose product is the **constant**, $-20$, and whose sum is the middle **coefficient** $-1$. Because the constant is negative, the two factors have opposite signs. And because the middle term is negative, the factor with the larger absolute value must be negative.

| Factor 1 | Factor 2 | Sum |
|----------|----------|-----|
| 1 | $-20$ | $-19$ |
| 2 | $-10$ | $-8$ |
| 4 | $-5$ | $-1$ ← This is what we need. |

The required factors are 4 and $-5$. We can factor the trinomial as $w^2 - w - 20 = (w + 4)(w - 5)$.

**Step 3.** All the factors are **prime**, so our factorization is complete.

**Step 4.** Check by multiplying.

$$(w + 4)(w - 5) = w \cdot w + w(-5) + 4 \cdot w + 4(-5)$$
$$= w^2 - 5w + 4w - 20$$
$$= w^2 - w - 20 \ \checkmark$$

**b.** Follow the **general factoring strategy**.

**Step 1.** Looking at each **term**, we can factor out a **common factor** of $4y$.

$$4y^4 - 32y = 4y \cdot y^3 - 4y \cdot 8 = 4y(y^3 - 8)$$

**Step 2.** The expression in parentheses, $y^3 - 8$, has two terms that are both cubes $(8 = 2^3)$, so this is one of our special forms. We factor this expression using the **difference of two cubes** rule with $A = y$ and $B = 2$.

$$y^3 - 8 = y^3 - 2^3$$
$$= (y - 2)(y^2 + 2 \cdot y + 2^2)$$
$$= (y - 2)(y^2 + 2y + 4)$$

Thus, we have $4y^4 - 32y = 4y(y - 2)(y^2 + 2y + 4)$.

**Step 3.** All the factors, other than the **monomial factor**, are **prime**, so our factorization is complete.

**Step 4.** Check by multiplying.

$$4y(y - 2)(y^2 + 2y + 4) = (4y \cdot y - 4y \cdot 2)(y^2 + 2y + 4)$$
$$= (4y^2 - 8y)(y^2 + 2y + 4)$$

Distribute $\rightarrow = 4y^2 \cdot y^2 + 4y^2 \cdot 2y + 4y^2 \cdot 4 - 8y \cdot y^2 - 8y \cdot 2y - 8y \cdot 4$

Simplify $\rightarrow = 4y^4 + 8y^3 + 16y^2 - 8y^3 - 16y^2 - 32y$

Combine like terms $\rightarrow = 4y^4 - 32y$ ✓

**c.–d.** Follow the **general factoring strategy** to factor these trinomials on your own. Once finished, view the **answers**. If you need help, watch this **interactive video** for complete solutions to all four parts.

**You Try It**     Work through this **You Try It** problem.

Work Exercises 1–8 in this textbook or in the *MyMathLab* ® **Study Plan.**

## Example 2  Factoring a Polynomial Completely

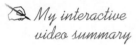

▣ Factor each expression completely.

**a.** $2x^3 - 5x^2 - 8x + 20$        **b.** $3a^2 - 10a - 8$        **c.** $3z^2 + z - 1$

### Solutions

**a.** Use the **general factoring strategy**. Other than 1 or $-1$, there is no **common factor**. There are four **terms**, so we consider **factoring by grouping**. We try grouping the first two terms and the last two terms.

$$\underbrace{2x^3 \quad 5x^2} \quad \underbrace{8x \; | \; 20}$$

From the first two terms, we can factor out $x^2$. From the last two terms, we can factor out $-4$.

$$2x^3 - 5x^2 - 8x + 20 = (2x^3 - 5x^2) + (-8x + 20)$$
$$= x^2(2x - 5) - 4(2x - 5)$$

Try to complete this factorization on your own. View the **answer**, or watch this **interactive video** for the full solution.

**b.–c.** Use the **general factoring strategy**. Other than 1 or $-1$, there is no **common factor**. There are three **terms** in both cases, but neither is a **perfect square trinomial**. Try factoring on your own using **trial and error** or the *ac* **method**. View the **answers**, or watch this **interactive video** for complete solutions to all three parts.

**You Try It**     Work through this **You Try It** problem.

Work Exercises 9–16 in this textbook or in the *MyMathLab* ® **Study Plan.**

## Example 3  Factoring a Polynomial Completely

▣ Factor each expression completely.

**a.** $10x^2 + 11xy - 6y^2$        **b.** $2p^2 - 32pq + 128q^2$

**c.** $7x^2z - 14x$        **d.** $-3y^4z - 24yz^4$

**Solutions**

**a.** Other than 1 or −1, there is no common factor. We then try to factor the expression using trial and error, or the *ac* method. We will use the *ac* method.

$$ac = (10)(-6) = -60$$

We need to find two **integers** whose **product** is −60 and whose **sum** is 11. For the product to be negative, one of the integers must be positive and the other must be negative. For the sum to be positive, the integer with the larger **absolute value** must be positive. We list such pairs of **factors** for −60:

| Factor 1 | Factor 2 | Sum |
|:---:|:---:|:---:|
| −1 | 60 | 59 |
| −2 | 30 | 28 |
| −3 | 20 | 17 |
| −4 | 15 | 11 ← This is what we need |
| −5 | 12 | 7 |
| −6 | 10 | 4 |

Because −4 and 15 multiply to −60 and add to 11, the integers we need are −4 and 15. We use these values to split up the middle term.

$$10x^2 + 11xy - 6y^2 = 10x^2 - 4xy + 15xy - 6y^2$$

We now **factor by grouping**.

Original expression: $10x^2 + 11xy - 6y^2$

Split up middle term: $= 10x^2 - 4xy + 15xy - 6y^2$

Group first two terms and last two: $= (10x^2 - 4xy) + (15xy - 6y^2)$

Factor $2x$ from first two and $3y$ from last two: $= 2x(5x - 2y) + 3y(5x - 2y)$

Factor out $(5x - 2y)$: $= (5x - 2y)(2x + 3y)$

Check the answer by multiplying. View the **check**.

**b.–d.** Try factoring on your own using the **general factoring strategy**. View the **answers**, or watch this **interactive video** for complete solutions to all four parts.

**You Try It**     Work through this You Try It problem.

Work Exercises 17–24 in this textbook or in the *MyMathLab*® Study Plan.

# 6.6 Exercises

In Exercises 1–30, factor completely. If a polynomial is prime, then state this as your answer.

**You Try It**

**1.** $3x^2 - 12$

**2.** $2y^3 - 10y^2 - 28y$

**3.** $a^3 + 8b^3$

**4.** $y^2 + 3y + 3$

**5.** $x^2 + x - 6$

**6.** $w^2 + 12w + 20$

**7.** $5z^2 - 30z + 45$

**8.** $-4m^4 + 4m$

**You Try It**

**9.** $4x^2 - 20x + 25$

**11.** $5x^3 + 14x^2 - 3x$

**13.** $3w^2 - 5w + 4$

**15.** $9x^2 + 12x + 4$

**17.** $-6p^2 + 9pq + 6q^2$

**You Try It**  **19.** $9y + 18y^2$

**21.** $25m^2 + 40mn + 16n^2$

**23.** $b^3 + 2ab^2 - a^2b - 2a^3$

**25.** $x^2 - x - 20$

**27.** $48y^3 + 6y^2 - 9y$

**29.** $p^2 - 10p + 9$

**10.** $2x^3 + 5x^2 - 8x - 20$

**12.** $6x^2 - x - 7$

**14.** $9t^2 - 15t - 36$

**16.** $3x^4 + 3x^3 - x^2 - x$

**18.** $5y^2z^3 - 20z$

**20.** $6xz^2 - 14xz - 40x$

**22.** $6x^2 - 5xy - 6y^2$

**24.** $-6b^2 + 3ab + a - 2b$

**26.** $1 - a^4$

**28.** $r^3(t^3 - 8) + 8(t^3 - 8)$

**30.** $y^5 - y^3 + y^2 - 1$

# 6.7 Solving Polynomial Equations by Factoring

## THINGS TO KNOW

Before working through this section, be sure that you are familiar with the following concepts:

VIDEO       ANIMATION       INTERACTIVE

**You Try It**

1. Solve Linear Equations Using Both Properties of Equality (Section 2.1, **Objective 5**)

**You Try It**

2. Factor Polynomials Completely (Section 6.6, **Objective 1**)

## OBJECTIVES

**1** Solve Quadratic Equations by Factoring

**2** Solve Polynomial Equations by Factoring

---

OBJECTIVE 1   SOLVE QUADRATIC EQUATIONS BY FACTORING

A **polynomial equation** results when we set two **polynomials** equal to each other. Some examples of polynomial equations are

$$2x - 7 = 4, \quad 3x^2 + 5x = x - 2, \quad \text{and} \quad 2x^3 + 7 = 3x^2 - x.$$

A polynomial equation is in **standard form** if one side equals zero and the other side is a simplified polynomial written in **descending order**.

The **standard forms** of the above polynomial equations are

$$2x - 11 = 0, \quad 3x^2 + 4x + 2 = 0, \quad \text{and} \quad 2x^3 - 3x^2 + x + 7 = 0.$$

The **degree of a polynomial equation** in standard form is the same as the highest **degree** of any of its terms. Notice that a polynomial of degree 1, such as $2x - 7 = 4$, is a **linear equation**. We learned how to solve these types of equations in **Section 2.1** and **Section 2.2**.

We now look at how to solve polynomial equations of degree 2. These types of equations are called **quadratic equations**. A quadratic equation in one variable is any equation that can be written in the standard form, $ax^2 + bx + c = 0$.

---

**Definition   Quadratic Equation**

A **quadratic equation** in standard form is written as

$$ax^2 + bx + c = 0,$$

where $a$, $b$, and $c$ are real numbers and $a \neq 0$.

---

Some examples of quadratic equations are

$$5x^2 - 7x + 4 = 0, \quad 3x^2 + 5x = x - 2, \quad 2x^2 + 4 = 9, \quad \text{and} \quad 7x^2 = x.$$

The first equation above is written in standard form. The **standard forms** of the remaining equations are

$$3x^2 + 4x + 2 = 0, \quad 2x^2 - 5 = 0, \quad \text{and} \quad 7x^2 - x = 0.$$

To solve quadratic equations, we can use the **factoring techniques** discussed in this chapter together with the **zero product property**.

---

**Zero Product Property**

If $A$ and $B$ are **real numbers** or **algebraic expressions** and $A \cdot B = 0$, then $A = 0$ or $B = 0$.

---

Recall that **factors** are quantities that are multiplied together to form a **product**. The **zero product property** tells us that if the product of two factors equals zero, then at least one of the factors must equal zero. This property extends to any number of factors.

 Note that the zero product property only works when the product equals zero. For instance, if we have $A \cdot B = 18$, it is *not* true that one or both of the factors must be equal to 18. In fact, we can multiply $2 \cdot 9$ to get 18 (or even $3 \cdot 6$), and neither factor is equal to 18. View this **popup** to see another example.

### Example 1  Solving a Quadratic Equation by Factoring

Solve each equation.

**a.** $(x + 10)(x - 3) = 0$          **b.** $x(3x + 5) = 0$

**Solutions**

**a.**  We have a product of two **factors** that is equal to zero. Using the **zero product property**, we set each factor equal to zero and solve the resulting equations.

$$(x + 10)(x - 3) = 0$$
$$x + 10 = 0 \quad \text{or} \quad x - 3 = 0$$
$$x = -10 \qquad\qquad x = 3$$

To check that these values are **solutions**, substitute them back into the original equation to see if a true statement results.

Check $x = -10$:

$$(x + 10)(x - 3) = 0$$
$$(-10 + 10)(-10 - 3) \overset{?}{=} 0$$
$$(0)(-13) \overset{?}{=} 0$$
$$0 = 0 \quad \text{True}$$

Check $x = 3$:

$$(x + 10)(x - 3) = 0$$
$$(3 + 10)(3 - 3) \overset{?}{=} 0$$
$$(13)(0) \overset{?}{=} 0$$
$$0 = 0 \quad \text{True}$$

Both values check, so the **solution set** is $\{-10, 3\}$.

  **b.** Set each factor equal to zero and solve the resulting equations. View the **answer**, or watch this **video** for the complete solution.

**You Try It**  Work through this You Try It problem.

Work Exercises 1–4 in this textbook or in the *MyMathLab*® Study Plan.

In Example 1, we solved equations in which the quadratic expression was given in factored form. This will not always be the case. In general, we can use the following steps to solve a quadratic equation, or other polynomial equations, by factoring.

---

**Solving Polynomial Equations by Factoring**

**Step 1.** Write the equation in **standard form** so that one side is zero and the other side is a simplified polynomial written in **descending order**.

**Step 2.** Factor the polynomial **completely**.

**Step 3.** Set each **distinct factor** with a variable equal to zero, and solve the resulting equations.

**Step 4.** Check each **solution** in the original equation.

---

## Example 2  Solving Quadratic Equations by Factoring

Solve each equation by factoring.

**a.** $z^2 + 4z - 12 = 0$          **b.** $-4x^2 + 28x - 40 = 0$

**Solutions**  We follow the **four-step process** for solving polynomial equations by factoring.

**a.  Step 1.**  The equation is already in **standard form**, so we move on to Step 2.

**Step 2.**  Using the **general strategy for factoring**, we need two numbers whose product is the constant $-12$ and whose sum is the middle coefficient 4. Because the product is negative, the two numbers have opposite signs. Because the sum is positive, the number with the larger **absolute value** is positive.

| Factor 1 | Factor 2 | Sum |
|:---:|:---:|:---:|
| $-1$ | 12 | 11 |
| $-2$ | 6 | 4  ← This is the desired sum |
| $-3$ | 4 | 1 |

The required numbers are $-2$ and 6, which can be used to factor the polynomial.

$$z^2 + 4z - 12 = 0$$
$$(z - 2)(z + 6) = 0$$

**Step 3.** Set each factor equal to zero and solve the equations.

$$z - 2 = 0 \quad \text{or} \quad z + 6 = 0$$
$$z = 2 \qquad\qquad z = -6$$

**Step 4.** Check these values on your own to confirm that the **solution set** is $\{-6, 2\}$.

 *My video summary*  **b.** The equation is in standard form. We factor the expression on the left by first factoring out the **greatest common factor**, $-4$.

$$-4x^2 + 28x - 40 = 0$$
$$-4(x^2 - 7x + 10) = 0$$

Now **factor** $x^2 - 7x + 10$. Set each variable factor equal to 0 and solve the resulting equations to form the solution set. View the **answer**, or watch this **video** for the complete solution.

**You Try It**  **Work through this You Try It problem.**

Work Exercises 5–10 in this textbook or in the *MyMathLab*® Study Plan.

**CAUTION** Note in Example 2b that the factor $-4$ is a constant and will never equal 0. When solving equations by factoring, we set only factors that contain a **variable** equal to 0.

## Example 3  Solving Quadratic Equations by Factoring

Solve each equation by factoring.

**a.** $9w^2 + 64 = 48w$   **b.** $4m^2 = 49$   **c.** $3x(x - 2) = 2 - x$

**Solutions** We follow the **four-step process** for solving polynomial equations by factoring.

**a. Step 1.** Write the equation in **standard form** by subtracting $48w$ from both sides.

Original equation: $\qquad 9w^2 + 64 = 48w$

Subtract $48w$ from both sides: $\quad 9w^2 - 48w + 64 = 0$

**Step 2.** The **quadratic trinomial** on the left is a **perfect square trinomial** with a negative middle term. Factoring gives us the following:

$$9w^2 - 48w + 64 = 0$$
$$(3w)^2 - 2(3w)(8) + 8^2 = 0 \leftarrow A^2 - 2AB + B^2$$
$$(3w - 8)^2 = 0 \leftarrow (A - B)^2$$
$$\text{or}$$
$$(3w - 8)(3w - 8) = 0$$

**Step 3.** Set each **distinct factor** equal to zero and solve the equations. The factor $(3w - 8)$ occurs twice, but we only set it equal to 0 once.

$$3w - 8 = 0$$
$$3w = 8$$
$$w = \frac{8}{3}$$

**Step 4.** Check this value on your own to confirm that the **solution set** is $\left\{\dfrac{8}{3}\right\}$.

*My video summary*

⬦ CAUTION When writing a solution set, we include only distinct solutions. This is why in Step 3 we set only each *distinct* variable factor equal to zero.

🎞 **b.** Write the equation in **standard form** by subtracting 49 from both sides.

Begin with the original equation: $\qquad 4m^2 = 49$

Subtract 49 from both sides: $\quad 4m^2 - 49 = 0$

The left side is now the **difference of two squares**, $4m^2 - 49 = (2m)^2 - 7^2$. Factor this special form and set each variable factor equal to 0. Solve the resulting equations to form the solution set. View the **answer**, or watch this **video** for the complete solution.

 *My video summary*

🎞 **c.** To write this equation in **standard form**, we first simplify the left side of the equation, then move all terms to one side so the other side is 0.

Begin with the original equation: $\qquad 3x(x - 2) = 2 - x$

Distribute $3x$ on the left: $\qquad 3x^2 - 6x = 2 - x$

Add $x$ to both sides: $\qquad 3x^2 - 5x = 2$

Subtract 2 from both sides: $\quad 3x^2 - 5x - 2 = 0$

Now factor the left side. Set each **variable factor** equal to 0 and solve the resulting equations to form the solution set. View the **answer**, or watch this **video** for the complete solution.

**You Try It** Work through this You Try It problem.

Work Exercises 11–18 in this textbook or in the ***MyMathLab***® Study Plan.

**Example 4 Solving Quadratic Equations by Factoring**

Solve each equation by factoring.

 *My interactive video summary*

🎞 **a.** $(x + 2)(x - 5) = 18$ $\qquad$ **b.** $(x + 3)(3x - 5) = 5(x + 1) - 10$

**Solutions** We follow the **four-step process** for solving polynomial equations by factoring.

**a.** Begin by simplifying both sides; then write the equation in **standard form**.

Begin with the original equation: $\qquad (x + 2)(x - 5) = 18$

Expand the left side: $\quad x^2 - 5x + 2x - 10 = 18$

Simplify: $\qquad x^2 - 3x - 10 = 18$

Subtract 18 from both sides: $\qquad x^2 - 3x - 28 = 0$

Solve the equation on your own. View the **answer**, or watch this **interactive video** for detailed solutions to both parts.

**b.** Begin by simplifying both sides; then write the equation in standard form. Solve the equation on your own, then view the **answer**. Or watch this **interactive video** for detailed solutions to both parts.

**You Try It** Work through this You Try It problem.

Work Exercises 19–24 in this textbook or in the ***MyMathLab***® Study Plan.

⬦ CAUTION Not every **quadratic equation** can be solved by factoring. We will look at additional methods for solving quadratic equations in Chapter 9.

OBJECTIVE 2   SOLVE POLYNOMIAL EQUATIONS BY FACTORING

The **zero product property** can be extended to any number of factors. This allows us to solve some **polynomial equations** of degree larger than 2 by factoring.

### Example 5  Solving Polynomial Equations by Factoring

Solve each equation by factoring.

**a.** $(x + 7)(2x - 1)(5x + 4) = 0$     **b.** $24x^3 + 8x^2 = 100x^2 - 28x$

**c.** $z^3 + z^2 = z + 1$

### Solutions

**a.** The equation has 0 on the right and the polynomial on the left is given in factored form. Therefore, we can apply the zero product property directly. We set each **variable factor** equal to 0 and solve the resulting equations.

$$(x + 7)(2x - 1)(5x + 4) = 0$$

$$x + 7 = 0 \quad \text{or} \quad 2x - 1 = 0 \quad \text{or} \quad 5x + 4 = 0$$

$$x = -7 \qquad\qquad 2x = 1 \qquad\qquad 5x = -4$$

$$x = \frac{1}{2} \qquad\qquad x = -\frac{4}{5}$$

Check these values on your own to confirm that the **solution set** is $\left\{ -7, -\frac{4}{5}, \frac{1}{2} \right\}$.

*My video summary*    **b.** Write the equation in **standard form**.

Begin with the original equation:     $24x^3 + 8x^2 = 100x^2 - 28x$

Subtract $100x^2$ from both sides:     $24x^3 - 92x^2 = -28x$

Add $28x$ to both sides:    $24x^3 - 92x^2 + 28x = 0$

Now factor the left side. Begin by factoring out the GCF, $4x$.

$$24x^3 - 92x^2 + 28x = 4x(6x^2 - 23x + 7)$$

Complete the factoring, then apply the **zero product property** and finish solving the equation. View the **answer**, or watch this **video** for the complete solution to part b.

*My video summary*    **c.** Write the equation in standard form by subtracting $z + 1$ from both sides.

Begin with the original equation:     $z^3 + z^2 = z + 1$

Subtract $z + 1$ from both sides:    $z^3 + z^2 - (z + 1) = 0$

Take the opposite of $(z + 1)$:    $z^3 + z^2 - z - 1 = 0$

To factor the left side, notice that there are four terms so try **factoring by grouping**. Once factored, set each **variable factor** equal to zero, and solve the resulting equations on your own. View the **answer**, or watch this **video** for a detailed solution to part c.

**You Try It**    Work through this **You Try It** problem.

Work Exercises **25–34** in this textbook or in the **MyMathLab**® Study Plan.

### Example 6  Solving Polynomial Equations by Factoring

Solve by factoring:

$$(2x - 9)(3x^2 - 16x - 12) = 0$$

**Solution**

Begin with the original equation:   $(2x - 9)(3x^2 - 16x - 12) = 0$

Factor the trinomial:   $(2x - 9)(x - 6)(3x + 2) = 0$ ← See the details

Set each **variable factor** equal to 0 and solve the resulting equations.

$$2x - 9 = 0 \quad \text{or} \quad x - 6 = 0 \quad \text{or} \quad 3x + 2 = 0$$
$$2x = 9 \qquad\qquad x = 6 \qquad\qquad 3x = -2$$
$$x = \frac{9}{2} \qquad\qquad\qquad\qquad x = -\frac{2}{3}$$

Check these values on your own to confirm that the **solution set** is $\left\{ -\frac{2}{3}, \frac{9}{2}, 6 \right\}$.

**You Try It**    Work through this You Try It problem.

**Work Exercises 35–40 in this textbook or in the *MyMathLab*® Study Plan.**

Note that in Example 6, we did not begin by multiplying the factors on the left together as we did in Example 4a. This was because the right side of the equation was already 0 as required by the zero product property.

# 6.7 **Exercises**

In Exercises 1–4, solve each quadratic equation using the zero product property.

**You Try It**

**1.** $(x - 1)(x + 2) = 0$

**2.** $2x(2x - 5) = 0$

**3.** $(7x + 1)(3x - 1) = 0$

**4.** $5(7 - x)(x + 9) = 0$

In Exercises 5–24, use factoring techniques to solve each quadratic equation.

**You Try It**

**5.** $x^2 - 5x + 6 = 0$

**6.** $y^2 + 3y - 10 = 0$

**7.** $2x^2 + 10x + 12 = 0$

**8.** $w^2 - w - 6 = 0$

**9.** $z^2 + 9z + 14 = 0$

**10.** $5x^2 + 10x = 0$

**11.** $3z^2 + 5z + 1 = 3$

**12.** $4x^2 + 1 = 4x$

**13.** $2x^2 = 50$

**14.** $3q^2 + 5q - 2 = 2q - 2$

**You Try It**

**15.** $9x(x + 3) = 3x - 16$

**16.** $x(x + 3) = 4x + 2$

**17.** $2x(x + 1) = 3 - 3x$

**18.** $7x(x + 3) = 2x + 6$

**You Try It** **19.** $(x - 4)(x + 2) = 7$

**20.** $(x + 3)(x - 2) = x + 3$

**21.** $(x - 4)(x + 6) = 7x$

**22.** $(3x - 2)(x + 2) = -4$

**23.** $(8x + 5)(2x - 5) = 10x - 50$

**24.** $(x + 10)(x - 1) = 2(x - 5) - 12$

In Exercises 25–28, solve each polynomial equation using the zero product property.

**25.** $(x - 1)(x + 12)(x - 3) = 0$

**26.** $x(4x + 1)(x - 2) = 0$

**27.** $(x + 2)(x - 8)(7 - x) = 0$

**28.** $(x - 4)(3x + 5)(x + 6)(x - 10) = 0$

In Exercises 29–40, use factoring techniques to solve each polynomial equation.

 **29.** $x^3 + x^2 = 2x$

**30.** $6y^3 + 7y^2 = 8y^2 + 5y$

**31.** $x^2(7x + 3) - 4(7x + 3) = 0$

**32.** $z^3 + 4 = z^2 + 4z$

 **You Try It** **33.** $12x^3 - 75x = 25 - 4x^2$

**34.** $8h^3 + 41 = 14$

 **35.** $(x - 2)(x^2 - 4x + 3) = 0$

**36.** $(2x + 1)(3x^2 + 8x - 3) = 0$

**You Try It** **37.** $(x^2 - 3x - 10)(3x + 5) = 0$

**38.** $(x^2 - 9)(x^2 - 25) = 0$

**39.** $(x^2 - 1)(x^2 - 4x + 4) = 0$

**40.** $x^2 + 4 = 5x^2$

# 6.8 **Applications of Quadratic Equations**

## THINGS TO KNOW

Before working through this section, be sure that you are familiar with the following concepts:

| | | VIDEO | ANIMATION | INTERACTIVE |
|---|---|---|---|---|

**You Try It**    1. Solve Linear Equations Using Both Properties of Equality (Section 2.1, **Objective 5**)   

**You Try It**    2. Use Linear Equations to Solve Application Problems (Section 2.2, **Objective 5**)   

**You Try It**    3. Factor Polynomials Completely (Section 6.6, **Objective 1**)   

**You Try It**    4. Solve Quadratic Equations by Factoring (Section 6.7, **Objective 1**)   

## OBJECTIVES

**1** Solve Application Problems Involving Consecutive Numbers

**2** Solve Application Problems Involving Geometric Figures

**3** Solve Application Problems Using the Pythagorean Theorem

**4** Solve Application Problems Involving Quadratic Models

---

OBJECTIVE 1    SOLVE APPLICATION PROBLEMS INVOLVING CONSECUTIVE NUMBERS

Solving some real-world applications will require us to solve a quadratic equation. To solve these types of applications, we follow the same **problem-solving strategy** used for **linear equations**.

 Recall that we need to pay attention to **feasible solutions** when working with applications. Based on the problem's context, not every **solution** to an equation will be a solution to the problem. We need to discard any solutions that do not make sense.

### Example 1  House Numbers

The house numbers on the west side of a street are consecutive positive odd **integers**. The product of the house numbers for two next-door neighbors on the west side of the street is 575. Find the house numbers.

### Solution

**Step 1. Define the Problem.**   We know the house numbers on the west side of a street are consecutive positive odd numbers and that the product of the house numbers for two next-door neighbors on that side of the street is 575. We want to find the house numbers.

**Step 2. Assign Variables.**   Let $x$ = the first odd house number. Then $x + 2$ is the next consecutive odd house number.

**Step 3. Translate into an Equation.**   To write an equation, we write the product of the two house numbers and set it equal to 575.

$$x(x + 2) = 575$$

**Step 4. Solve the Equation.**

| | |
|---|---|
| Begin with the original equation: | $x(x + 2) = 575$ |
| Distribute on the left: | $x^2 + 2x = 575$ |
| Subtract 575 from both sides: | $x^2 + 2x - 575 = 0$ |
| Factor the left side: | $(x - 23)(x + 25) = 0$  ← See the details |

Set each **variable factor** equal to zero and solve the resulting equations.

$$x - 23 = 0 \quad \text{or} \quad x + 25 = 0$$
$$x = 23 \qquad\qquad x = -25$$

**Step 5. Check the Reasonableness of Your Answer.**   Because the house numbers are positive odd integers, we discard the negative solution. The only reasonable solution is $x = 23$.

**Step 6. Answer the Question.**   The house numbers are 23 and $23 + 2 = 25$.

**You Try It**   **Work through this You Try It problem.**

**Work Exercises 1–5 in this textbook or in the *MyMathLab*® Study Plan.**

OBJECTIVE 2   SOLVE APPLICATION PROBLEMS INVOLVING GEOMETRIC FIGURES

Let's look at an application that involves a quadratic equation and a geometric figure.

### Example 2  Swimming Pool Border

*My video summary*   A swimming pool is 20 feet wide and 30 feet long. A sidewalk border around the pool has uniform width and an **area** that is equal to the area of the pool. See Figure 1. Find the width of the border.

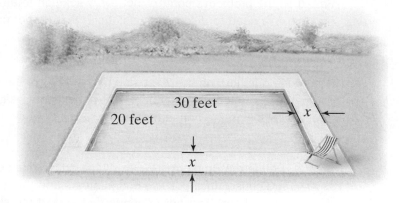

**Figure 1**

**Solution**   Follow the **problem-solving strategy.**

**Step 1.** The area of the pool is $20(30) = 600$ sq ft., which is also the area of the sidewalk border. We want to find the width of the border.

**Step 2.** From the figure, the width of the sidewalk border is $x$. The combined width of the pool and border is $20 + 2x$ because we are adding one border width to each side of the pool. The combined length is $30 + 2x$ for the same reason. The area of the pool and the area of the border are both 600 ft², so the combined area is 1200 ft².

**Step 3.** We can use the formula for the **area of a rectangle** to write an equation for the total combined area.

Begin with the rectangle area formula:     $A = lw$

Substitute:   $1200 = (20 + 2x)(30 + 2x)$

Finish the problem on your own by solving the equation to find the width of the sidewalk border, $x$. View the **answer**, or watch this **video** for a detailed solution.

**You Try It**   **Work through this You Try It problem.**

**Work Exercises 6–10 in this textbook or in the *MyMathLab*® Study Plan.**

OBJECTIVE 3  SOLVE APPLICATION PROBLEMS USING THE PYTHAGOREAN THEOREM

In **Section 2.4**, we explored **common formulas** for geometric figures such as triangles. **Right triangles** are triangles with a 90° angle, or **right angle**. The **hypotenuse** of a right triangle is the side opposite the right angle and is the longest of the three sides. The other two sides are called the **legs** of the triangle.

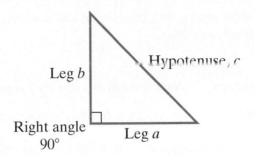

Leg $b$

Hypotenuse, $c$

Right angle
90°

Leg $a$

**Figure 2**

---

**Pythagorean Theorem**

For **right triangles**, the sum of the **squares** of the lengths of the **legs** of the triangle equals the square of the length of the **hypotenuse**.

$$a^2 + b^2 = c^2$$

---

**Example 3  Support Wire**

*My video summary*

A wire is attached to a cell phone tower for support. See Figure 3. The length of the wire is 40 meters less than twice the height of the tower. The wire is fixed to the ground at a distance that is 40 meters more than the height of the tower. Find the length of the wire.

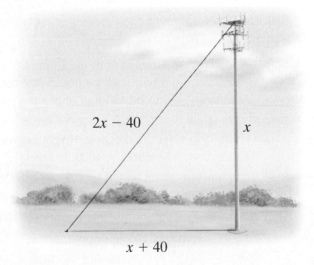

$2x - 40$

$x$

$x + 40$

**Figure 3**

**Solution**  Follow the **problem-solving strategy**.

Step 1.  Figure 3 shows us that we can use a **right triangle** to describe this situation. One **leg** is the height of the tower, and the other leg is the distance from the tower where the support wire is fixed to the ground. The length of the wire is the **hypotenuse**.

Step 2.  The figure is labeled with the height of the tower as $x$ and the distance from the tower to where the wire is fixed to the ground as $x + 40$. The length of the wire is represented by $2x - 40$.

**Step 3.** We can use the **Pythagorean Theorem** to find the height of the tower.

$$\text{Pythagorean Theorem:} \qquad a^2 + b^2 = c^2$$

$$\text{Substitute expressions for the lengths:} \qquad (x + 40)^2 + (x)^2 = (2x - 40)^2$$

**Step 4.** Expand using special products:  $x^2 + 80x + 1600 + x^2 = 4x^2 - 160x + 1600$

Simplify the left side:  $2x^2 + 80x + 1600 = 4x^2 - 160x + 1600$

Finish solving this problem on your own. View the **answer**, or watch this **video** for a detailed solution.

**You Try It**  **Work through this You Try It problem.**

**Work Exercises 11–14 in this textbook or in the *MyMathLab*®  Study Plan.**

 Be sure to pay attention to the question being asked. When solving application problems, solving for a variable is not always enough. Sometimes, as in Example 3, we will need to do some computation using the value of the variable in order to answer the question.

OBJECTIVE 4    SOLVE APPLICATION PROBLEMS INVOLVING QUADRATIC MODELS

Now we can solve application problems that are **modeled** by quadratic equations.

### Example 4  Falling Object

 The Grand Canyon Skywalk sits 4000 ft above the Colorado River. If an object is dropped from the observation deck, its height $h$, in feet after $t$ seconds, is given by

$$h = -16t^2 + 4000.$$

How long will it take for the object to be 400 feet above the Colorado River?

**Solution**  Work through the following solution, or watch this **video** to see the solution worked out.

**Step 1.** An object is dropped from the Grand Canyon Skywalk. Given the equation that describes the height of the object after some time, we must find the time it takes for the object to be 400 feet above the river.

**Step 2.** We know that $h$ is the height of the object in feet and $t$ is the time in seconds after the object is dropped. Height is given in terms of time through the quadratic equation, $h = -16t^2 + 4000$.

**Step 3.** When the object is 400 feet above the Colorado River, $h = 400$. Substituting this value into the equation, we write $400 = -16t^2 + 4000$.

**Step 4.**    Begin with the original equation:  $400 = -16t^2 + 4000$

Subtract 400 from both sides:  $0 = -16t^2 + 3600$

Factor out the **GCF**:  $0 = -16(t^2 - 225)$

Factor the **difference of two squares**:  $0 = -16(t + 15)(t - 15)$

Set each **variable factor** equal to zero and solve the resulting equations.

$$t + 15 = 0 \qquad \text{or} \quad t - 15 = 0$$

$$t = -15 \qquad\qquad t = 15$$

**Step 5.** Because the time for the object to fall cannot be negative, we discard the negative solution. The only reasonable solution is $t = 15$ seconds.

**Step 6.** The object will be 400 feet above the Colorado River 15 seconds after it is dropped.

**You Try It**   Work through this You Try It problem.

Work Exercises 15–17 in this textbook or in the *MyMathLab*® Study Plan.

### Example 5   Broadband Access

For household incomes under $100,000, the relationship between the percentage of households with home broadband access and the annual household income can be approximated by the model,

$$y = -0.01x^2 + 1.7x + 9.5.$$

Here $x$ is the annual household income (in $1000s) and $y$ is the percentage of households with home broadband access. Use the model to estimate the annual household income if 75.5 percent of such households have home broadband access.

(*Source:* Pew Research Center's Internet & American Life Project, Aug. 9–Sept. 13, 2010)

**Solution**   Follow the **problem-solving strategy**. We are given the relationship between household income and percent with broadband access. We want to determine the household income when 75.5 percent have broadband access. Substitute 75.5 for $y$ and solve the resulting equation for $x$. Keep in mind that the model is for household incomes *under* $100,000. View the **answer**, or watch this **video** for the complete solution.

**You Try It**   Work through this You Try It problem.

Work Exercises 18–21 in this textbook or in the *MyMathLab*® Study Plan.

# 6.8 Exercises

In Exercises 1–21, solve each application problem.

**1. Room Numbers**   The room numbers on one side of a school corridor are consecutive positive even integers. The product of the room numbers for two adjoining rooms on that side is 6560. Find the room numbers.

**2. Flooring Repair**   To repair his kitchen subfloor, John cut a rectangular piece of plywood such that the width and length were two consecutive odd integers. If the product of the length and width is 323, what were the dimensions of the piece of plywood?

**3. Consecutive Integers**   Find three consecutive integers whose product is 161 larger than the cube of the smallest integer.

**4. Multiples**   Find two consecutive positive multiples of 3 such that the sum of their squares is 225.

**5. Mobile Access**   According to the 2010 Digital Community College survey, the percent of community colleges that provide access to grades through mobile devices (such as smart phones) and the percent that allow students to register for classes through such devices are consecutive integers. If more allowed students to register than access to grades, and the product of the two percentages is 380, find the percentages.

6. **Garden Path** Aliesha has a rectangular plot of land where she can plant a garden. She wants to have a path around the garden with a uniform width of $x$. If the plot of land measures 30 feet by 45 feet and Aliesha has a usable garden area of 700 square feet, what is the width of the path?

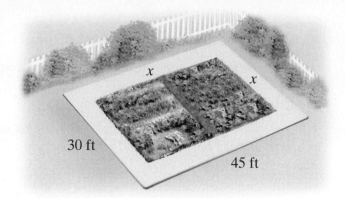

7. **Picture Frame** A portrait has dimensions 20 inches by 30 inches and is surrounded by a frame of uniform width. If the total area (portrait plus frame) is 1064 square inches, find the width of the frame.

8. **Rectangular Table Top** A rectangular table top has a perimeter of 38 inches and an area of 84 square inches. Find its dimensions.

9. **Triangular Sail** The height of a triangular sail is 4 yards less than twice the length of its base. If the area of the sail is 48 square yards, find the dimensions of the sail.

10. **Banner Size** The length of a rectangular banner is 5 feet longer than its width. If the area is 36 square feet, find the dimensions.

**You Try It**

11. **Dining Canopy** A dining canopy has corner posts that are anchored by support lines. The length of each support line is 5 feet longer than the height of the post, and the support line is anchored at a distance that is 5 feet less than twice the height of the post. Determine the height of the corner post.

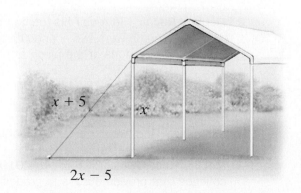

12. **Modern Architecture**  An art studio purchased a triangular plot of land in downtown San Diego and constructed a new building whose base is in the shape of a right triangle. The two legs have lengths of $x - 2$ feet and $x + 5$ feet. The hypotenuse is $2x - 27$ feet. Find the lengths of the three sides.

13. **Desktop Organizer**  A dorm room desktop organizer is roughly in the shape of a right triangle. The height and base are the solutions to the equation $x^2 + 200 = 28x + 8$. Find the hypotenuse, $h$.

14. **Bike Ride**  At 8 AM, Todd rides his bike due north from campus at 10 mph. One hour later, Chad rides his bike due east from the same point, also at 10 mph. When will they be 50 miles apart?

**You Try It**

15. **Falling Object**  The Infinity Room at Wisconsin's House on the Rock is an observation deck that extends 200 feet out above a scenic valley. The deck sits 150 feet above the valley floor. If an object is dropped from the observation deck, its height $h$ in feet, after $t$ seconds, is given by

$$h = -16t^2 + 150.$$

How long will it take for the object to be 6 feet above the valley floor?

16. **Projectile Motion**  A cannonball is fired across a field. Its height above the ground after $t$ seconds is given by $-16t^2 + 40t$. How long does it take for the ball to reach its maximum height of 25 feet?

17. **Projectile Motion**  Cierra stands on the edge of her dorm roof and throws a ball into the air. The ball's height above the ground after $t$ seconds is given by

$$h = -16t^2 + 32t + 48.$$

When will the ball hit the ground?

18. **Facebook Growth**  The number of Facebook users worldwide can be approximated by the model $N = 24t^2 - 240t + 600$, where $N$ is the number of users in millions and $t$ is the number of years after 2000. Based on this model, predict when the number of Facebook users worldwide will reach 2400 million (i.e. 2.4 billion). (*Source:* internetworldstats.com)

19. **Population**  A councilwoman conducts a study and finds the population of her city can be estimated by the equation $p = x^2 - 5x + 1$, where $p$ is thousands of people and $x$ is the number of years after 2000. In what year did the city have a population of 7000?

20. **Wrench Cost**  Eduardo notices that the number of oil filter wrenches $n$ that he sells at his hardware store each week is related to the price $p$ (in dollars) by the model,

$$n = -p^2 - 2p + 263.$$

What should Eduardo charge for the oil filter wrench if he wants to sell 200 per week?

21. **Rental Cost**  The manager of a 50-unit apartment complex is trying to decide what rent to charge. He knows that all units will be rented if he charges $600 per month, but for every increase of $25 over the $600, there will be a unit vacant. Find the number of occupied units if his total monthly rental income is $33,600 and the rent per unit is less than $1000.

# CHAPTER SEVEN
# Rational Expressions and Equations

## CHAPTER SEVEN CONTENTS

# 7.1 Simplifying Rational Expressions

## THINGS TO KNOW

Before working through this section, be sure you are familiar with the following concepts:

|  | | VIDEO | ANIMATION | INTERACTIVE |
|---|---|---|---|---|

You Try It

**1.** Write Fractions in Simplest Form (Section R.1, **Objective 2**)

You Try It

**2.** Evaluate Algebraic Expressions (Section 1.5, **Objective 1**)

You Try It

**3.** Factor Polynomials Completely (Section 6.6, **Objective 1**)

## OBJECTIVES

**1** Evaluate Rational Expressions

**2** Find Restricted Values for Rational Expressions

**3** Simplify Rational Expressions

### OBJECTIVE 1  EVALUATE RATIONAL EXPRESSIONS

Recall from **Section 1.1** that a number is a **rational number** if it can be written as a fraction $\frac{p}{q}$, where $p$ and $q$ are **integers** and $q \neq 0$. A *rational expression* has a similar definition.

---

**Definition   Rational Expression**

A **rational expression** is an **expression** that can be written as the quotient $\dfrac{P}{Q}$ of two polynomials $P$ and $Q$ as long as $Q \neq 0$.

---

Some examples of rational expressions are

$$\frac{x-2}{5}, \quad \frac{x+4}{x-7}, \quad \frac{x^2+2x-8}{x^2-x-20}, \quad \text{and} \quad \frac{x+y}{x-y}.$$

The first three expressions are *rational expressions in one variable*, $x$. The last expression is a *rational expression in two variables*, $x$ and $y$.

Given values for the variables, we can **evaluate** rational expressions. Using the same process for evaluating expressions in **Section 1.5**, we substitute the given value for the **variable** and then **simplify**.

## Example 1  Evaluating a Rational Expression

Evaluate $\dfrac{x+8}{x-2}$ for the given value of $x$.

**a.** $x = 4$                      **b.** $x = -6$

**Solutions**  Substitute the given value for the **variable** and simplify.

**a.**  Substitute 4 for $x$ in the expression:    $\dfrac{x+8}{x-2} = \dfrac{4+8}{4-2}$

Add and subtract:            $= \dfrac{12}{2}$

Simplify the fraction:       $= 6$

The value of $\dfrac{x+8}{x-2}$ is 6 when $x = 4$.

**b.**  Substitute $-6$ for $x$ in the expression:    $\dfrac{x+8}{x-2} = \dfrac{-6+8}{-6-2}$

Add and subtract:            $= \dfrac{2}{-8}$

Simplify the fraction:       $= -\dfrac{1}{4}$   ←   $\dfrac{2}{-8} = \dfrac{1 \cdot 2}{-4 \cdot 2} = \dfrac{1 \cdot \cancel{2}}{-4 \cdot \cancel{2}} = -\dfrac{1}{4}$

The value of $\dfrac{x+8}{x-2}$ is $-\dfrac{1}{4}$ when $x = -6$.

**You Try It**   Work through this You Try It problem.

Work Exercises 1–6 in this textbook or in the **_MyMathLab_**® Study Plan.

**Example 2  Evaluating a Rational Expression**

*My video summary*   Evaluate $\dfrac{x^2 - y}{9x + 5y}$ for $x = 3$ and $y = -1$.

**Solution**  Substitute $x = 3$ and $y = -1$:  $\dfrac{x^2 - y}{9x + 5y} = \dfrac{(3)^2 - (-1)}{9(3) + 5(-1)}$

Try to finish simplifying this expression on your own by using the correct order of operations. Check your **answer**, or watch this **video** for a complete solution.

**You Try It**   Work through this You Try It problem.

Work Exercises 7–10 in this textbook or in the *MyMathLab*® Study Plan.

OBJECTIVE 2   FIND RESTRICTED VALUES FOR RATIONAL EXPRESSIONS

In the definition of a **rational expression**, the statement $Q \neq 0$ means that the **polynomial** $Q$ in the **denominator** cannot equal zero. For example, if we **evaluate** $\dfrac{x + 8}{x - 2}$ from **Example 1** for $x = 2$, the result is $\dfrac{2 + 8}{2 - 2} = \dfrac{10}{0}$, which is **undefined**. So, we must *restrict* $\dfrac{x + 8}{x - 2}$ to values of $x$ such that $x \neq 2$. We call $x = 2$ a *restricted value* for $\dfrac{x + 8}{x - 2}$.

---

**Definition  Restricted Value**

A **restricted value** for an **algebraic expression** is a value that makes the expression undefined.

---

When working with rational expressions in one variable, values of the variable that cause the denominator to equal zero are restricted values.

---

**Finding Restricted Values for Rational Expressions in One Variable**

To find the restricted values for a rational expression in one variable, set the denominator equal to zero. Then, solve the resulting **equation** and restrict the **solutions**. If the equation has no real-number solution, then the expression has no restricted values.

---

**Example 3  Finding Restricted Values for Rational Expressions**

Find any **restricted values** for each **rational expression**.

a. $\dfrac{3x + 5}{3x - 2}$  b. $\dfrac{x^2 + 2x - 35}{x^2 + x - 30}$

**Solutions**  For each rational expression, we **set** the **denominator** equal to zero and solve the resulting **equation**. The **solutions** are the restricted values.

a. Set the denominator equal to 0:  $3x - 2 = 0$

Add 2 to both sides:  $3x = 2$

Divide both sides by 3:  $x = \dfrac{2}{3}$

The restricted value for $\dfrac{3x + 5}{3x - 2}$ is $\dfrac{2}{3}$. This means that the rational expression is undefined when $x = \dfrac{2}{3}$.

*My video summary*  **b.** Set the denominator equal to 0:   $x^2 + x - 30 = 0$

Try to solve this equation on your own by factoring. The solutions will be the restricted values for the rational expression. See the **answer**, or watch this **video** for a complete solution.

**You Try It**   Work through this You Try It problem.

Work Exercises 11–18 in this textbook or in the **MyMathLab**® Study Plan.

### Example 4  Finding Restricted Values for Rational Expressions

Find any **restricted values** for each **rational expression**.

**a.** $\dfrac{2x + 9}{4}$          **b.** $\dfrac{2x}{x^2 + 1}$

### Solutions

**a.** The **denominator** is a **constant**, 4, which is not equal to 0. This means there are no restricted values for the rational expression $\dfrac{2x + 9}{4}$.

**b.** Any restricted value would be a solution to the equation $x^2 + 1 = 0$, but $x^2 + 1$ is a **prime polynomial** and cannot be factored. For any real number that is substituted for $x$, the expression $x^2$ will have a non-negative value. Adding 1 will then produce a positive value, which means $x^2 + 1$ cannot equal 0. This means there are no restricted values for $\dfrac{2x}{x^2 + 1}$.

**You Try It**   Work through this You Try It problem.

Work Exercises 19 and 20 in this textbook or in the **MyMathLab**® Study Plan.

OBJECTIVE 3   SIMPLIFY RATIONAL EXPRESSIONS

A fraction is written in **lowest terms**, or **simplest form**, if the **numerator** and **denominator** have no **common factors** other than 1. For example, the fraction $\dfrac{5}{8}$ is in lowest terms because 5 and 8 have no common factors (except 1). However, the fraction $\dfrac{6}{10}$ is not in lowest terms because 6 and 10 have a common factor of 2.

Recall that multiplying or dividing both the numerator and denominator of a fraction by the same nonzero value results in an **equivalent fraction**. Therefore, we can **simplify a fraction** by dividing both the numerator and the denominator by their **greatest common factor**. To do this, we can write both the numerator and denominator in terms of their **prime factorizations**. We then **divide out** each common factor by dividing both the numerator and denominator by the common factor. This process is shown by placing slash marks through each common factor and then replacing the factor with 1, the result of dividing the common factor by itself.

Separately, in the numerator and denominator, we multiply the 1's by any remaining factors to result in the simplified fraction.

$$\frac{6}{10} = \frac{2 \cdot 3}{2 \cdot 5} = \frac{\overset{1}{\cancel{2}} \cdot 3}{\underset{1}{\cancel{2}} \cdot 5} = \frac{1 \cdot 3}{1 \cdot 5} = \frac{3}{5}.$$

We simplify **rational expressions** in the same way that we **simplify fractions**. We **divide out** the **common factors**.

---

**Simplification Principle for Rational Expressions**

If $P$, $Q$, and $R$ are **polynomials**, then $\dfrac{P \cdot R}{Q \cdot R} = \dfrac{P \cdot \overset{1}{\cancel{R}}}{Q \cdot \underset{1}{\cancel{R}}} = \dfrac{P \cdot 1}{Q \cdot 1} = \dfrac{P}{Q}$ for $Q \neq 0$ and $R \neq 0$.

---

We can use the following steps to simplify rational expressions.

---

**Simplifying Rational Expressions**

**Step 1.** **Factor** the **numerator** and **denominator** completely.

**Step 2.** Divide out each common factor of the numerator and denominator.

**Step 3.** Separately, in the numerator and denominator, multiply any factors that were not divided out to obtain the simplified rational expression. If all **factors** in the numerator divide out, the numerator will be 1.

---

In most cases, we will not multiply out the polynomial factors in a simplified rational expression. Instead, we will leave them in **factored form**.

### Example 5 Simplifying a Rational Expression

Simplify $\dfrac{2x^2 - 6x}{7x - 21}$.

**Solution** We follow the **three-step process**.

Factor the numerator and denominator: $\dfrac{2x^2 - 6x}{7x - 21} = \dfrac{2x(x - 3)}{7(x - 3)}$

Divide out the common factor $x - 3$: $= \dfrac{2x\overset{1}{\cancel{(x - 3)}}}{7\underset{1}{\cancel{(x - 3)}}}$

Write the simplified rational expression: $= \dfrac{2x}{7} \quad \text{for} \quad x \neq 3$

The original expression $\dfrac{2x^2 - 6x}{7x - 21}$ and the simplified expression $\dfrac{2x}{7}$ are equal only when $x \neq 3$. This is why we write "for $x \neq 3$." Do you see **why**? Typically, we will not list such restricted values when simplifying rational expressions.

**You Try It**    Work through this You Try It problem.

Work Exercises 21 and 22 in this textbook or in the *MyMathLab*® Study Plan.

## Example 6  Simplifying a Rational Expression

Simplify $\dfrac{5x}{x^2 + 5x}$.

**Solution**

Factor the denominator:  $\dfrac{5x}{x^2 + 5x} = \dfrac{5x}{x(x + 5)}$

**Divide** out the common factor $x$:  $= \dfrac{5\overset{1}{\cancel{x}}}{\cancel{x}(x + 5)}$
$\phantom{= \dfrac{5}{1}}\underset{1}{}$

Write the simplified rational expression:  $= \dfrac{5}{x + 5}$

**You Try It**  Work through this You Try It problem.

**Work Exercises 23 and 24 in this textbook or in the *MyMathLab* Study Plan.**

 Only **common factors** of an expression can be divided out. It is incorrect to divide out **terms**. In Example 6, $5x$ is a term of the original denominator, and $5$ is a term of the simplified denominator. Neither can be divided out.

|  |  |
|:---:|:---:|
| **Incorrect** | **Incorrect** |
| $\dfrac{5x}{x^2 + 5x} = \dfrac{5\overset{1}{\cancel{x}}}{x^2 + 5\underset{1}{\cancel{x}}} = \dfrac{1}{x^2 + 1}$ | $\dfrac{5}{x + 5} = \dfrac{\overset{1}{\cancel{5}}}{x + \underset{1}{\cancel{5}}} = \dfrac{1}{x + 1}$ |

 When simplifying a **rational expression**, replacing each divided-out **factor** with a 1 can cause unnecessary clutter. Therefore, it is fine to leave out the 1's. However, if every factor in the numerator divides out, be sure to include a 1 in the numerator of the simplified rational expression.

## Example 7  Simplifying a Rational Expression

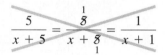 Simplify $\dfrac{y^2 + 2y - 24}{y^2 + 4y - 32}$.

**Solution**  Follow the **three-step process**.

Begin with the original expression:  $\dfrac{y^2 + 2y - 24}{y^2 + 4y - 32}$

Factor the numerator and denominator:  $= \dfrac{(y - 4)(y + 6)}{(y - 4)(y + 8)}$

Try to finish simplifying this rational expression on your own. View the **answer**, or watch this **video** for a complete solution.

**You Try It**  Work through this You Try It problem.

**Work Exercises 25–28 in this textbook or in the *MyMathLab* Study Plan.**

### Example 8  Simplifying a Rational Expression

 *My video summary*

Simplify $\dfrac{2m^2 + m - 15}{2m^3 - 5m^2 - 18m + 45}$.

**Solution** Follow the **three-step process**, and try to simplify this **rational expression** on your own. View the **answer**, or watch this **video** for a complete solution.

**You Try It**   Work through this You Try It problem.

**Work Exercises 29 and 30 in this textbook or in the *MyMathLab*® Study Plan.**

### Example 9  Simplifying a Rational Expression in Two Variables

 *My video summary*

Simplify $\dfrac{x^2 - xy - 12y^2}{2x^2 + 7xy + 3y^2}$.

**Solution**

Begin with the original expression: $\dfrac{x^2 - xy - 12y^2}{2x^2 + 7xy + 3y^2}$

Factor the numerator and denominator: $= \dfrac{(x - 4y)(x + 3y)}{(2x + y)(x + 3y)}$

Try to finish this problem on your own. View the **answer**, or watch this **video** for a complete solution.

**You Try It**   Work through this You Try It problem.

**Work Exercises 31 and 32 in this textbook or in the *MyMathLab*® Study Plan.**

The **commutative property of addition** states that a sum is not affected by the order of the terms. So, $a + b = b + a$. Sometimes, when factoring the numerator and denominator of a **rational expression**, the terms of **common factors** may be arranged in a different order. We can still **divide out** these common factors.

### Example 10  Simplifying a Rational Expression

Simplify $\dfrac{w^2 - y^2}{2xy + 2xw}$.

**Solution**

Begin with the original expression: $\dfrac{w^2 - y^2}{2xy + 2xw}$

Factor the numerator and denominator: $= \dfrac{(w + y)(w - y)}{2x(y + w)}$

Divide out the common factors $w + y = y + w$: $= \dfrac{\cancel{(w + y)}(w - y)}{2x\cancel{(y + w)}}$

Write the simplified rational expression: $= \dfrac{w - y}{2x}$

**You Try It**   Work through this You Try It problem.

**Work Exercises 33 and 34 in this textbook or in the *MyMathLab*® Study Plan.**

If the numerator and denominator of a **rational expression** have factors that are **opposite polynomials**, we can simplify by first factoring $-1$ from one of them.

### Example 11  Simplifying a Rational Expression Involving Opposites

Simplify $\dfrac{3x - 10}{10 - 3x}$.

**Solution**  We recognize that $3x - 10$ and $10 - 3x$ are opposite polynomials, so we factor $-1$ from $10 - 3x$:

$$10 - 3x = -1(-10 + 3x) = -1(3x - 10)$$

This gives

$$\frac{3x - 10}{10 - 3x} = \frac{3x - 10}{-1(3x - 10)} = \frac{3x - 10}{-1(3x - 10)} = \frac{1}{-1} = -1.$$

**You Try It**  Work through this You Try It problem.

Work Exercises 35 and 36 in this textbook or in the *MyMathLab*® Study Plan.

**Example 11** shows us that **opposite polynomials** in the numerator and denominator of a rational expression will **divide out**, leaving a factor of $-1$.

### Example 12  Simplifying a Rational Expression Involving Opposites

 Simplify $\dfrac{2x^2 - 27x + 70}{49 - 4x^2}$.

**Solution**  Begin with the original expression: $\dfrac{2x^2 - 27x + 70}{49 - 4x^2}$

Factor the numerator and denominator: $= \dfrac{(2x - 7)(x - 10)}{(7 - 2x)(7 + 2x)}$

Notice that $2x - 7$ and $7 - 2x$ are opposite polynomials. Finish simplifying the **rational expression** on your own. View the **answer**, or watch this **video** for a complete solution.

**You Try It**  Work through this You Try It problem.

Work Exercises 37–40 in this textbook or in the *MyMathLab*® Study Plan.

## 7.1  Exercises

In Exercises 1–10, evaluate each rational expression for the given value(s) of the variable(s).

**You Try It**

1. $\dfrac{x + 5}{x - 1}$ for $x = 3$

2. $\dfrac{2p + 1}{3p - 1}$ for $p = 4$

3. $\dfrac{w + 6}{w + 8}$ for $w = -4$

4. $\dfrac{y^2 - 1}{y^3 - 1}$ for $y = 3$

**5.** $\dfrac{x^2 - 5x - 14}{5 - x}$ for $x = 3$

**6.** $\dfrac{2t^3 - 7}{t^2 + 5}$ for $t = -1$

**7.** $\dfrac{y^2 - x}{5x - 2y}$ for $x = 4$ and $y = -6$

**8.** $\dfrac{w + 2z}{w - 2z}$ for $w = -10$ and $z = 3$

**You Try It**

**9.** $\dfrac{m^2 - n^2}{m^3 + n^3}$ for $m = 5$ and $n = 3$

**10.** $\dfrac{x^2 + xy}{5 - y}$ for $x = -3$ and $y = 8$

In Exercises 11–20, find any restricted values for the given rational expression.

**11.** $\dfrac{3}{x - 8}$

**12.** $\dfrac{y - 4}{y + 7}$

**You Try It**

**13.** $\dfrac{3t + 1}{7t - 4}$

**14.** $\dfrac{2m - 3}{5m}$

**15.** $\dfrac{p + 6}{(p - 3)(p + 2)}$

**16.** $\dfrac{x^2 - 16}{x^2 - x - 20}$

**17.** $\dfrac{2y^2 - 11y - 6}{3y^2 - 13y - 30}$

**18.** $\dfrac{n^3 - 3n^2 - 4n}{n^3 + 5n^2 - n - 5}$

**19.** $\dfrac{2t - 7}{3}$

**20.** $\dfrac{5x - 1}{x^2 + 3}$

**You Try It**

In Exercises 21–40, simplify each rational expression. If the expression is already in simplest form, state this in your answer.

**21.** $\dfrac{3m - 9}{5m - 15}$

**22.** $\dfrac{5x^2 - 10x}{13x - 26}$

**You Try It**

**23.** $\dfrac{6y}{y^2 + 6y}$

**24.** $\dfrac{4n + 24}{9n^3 + 54n^2}$

**You Try It**

**25.** $\dfrac{z^2 - 2z - 15}{z^2 + 10z + 21}$

**26.** $\dfrac{m^2 + 4m - 117}{m^2 + 18m + 80}$

**You Try It**

**27.** $\dfrac{t^2 + 5t - 24}{2t^2 - 11t + 15}$

**28.** $\dfrac{6w^2 + 5w - 4}{3w^2 + 13w + 12}$

**29.** $\dfrac{2x^2 + 11x + 12}{2x^3 + 3x^2 - 32x - 48}$

**30.** $\dfrac{12a^3 + 4a^2 + 9a + 3}{3a^2 - 11a - 4}$

**You Try It**

**31.** $\dfrac{x^2 - 7xy - 8y^2}{x^2 - 3xy - 4y^2}$

**32.** $\dfrac{4p^2 + 5pq - 6q^2}{3p^2 + 5pq - 2q^2}$

**You Try It**

**33.** $\dfrac{12 + 5x}{5x^2 + 22x + 24}$    **34.** $\dfrac{a^2 - 49b^2}{56bc + 8ac}$
You Try It

 **35.** $\dfrac{2m - 9}{9 - 2m}$    **36.** $\dfrac{3x + 8}{8 - 3x}$
You Try It

 **37.** $\dfrac{w^2 - 6w - 16}{64 - w^2}$    **38.** $\dfrac{2 - z}{z^2 + z - 6}$
You Try It

**39.** $\dfrac{n^3 - 5n^2 + 7n - 35}{n^2 - 10n + 25}$    **40.** $\dfrac{y^2 - 81}{-y - 9}$

# 7.2 Multiplying and Dividing Rational Expressions

## THINGS TO KNOW

Before working through this section, be sure you are familiar with the following concepts:

|  | VIDEO | ANIMATION | INTERACTIVE |

 **1.** Simplify Exponential Expressions Using the Quotient Rule (Section 5.1, **Objective 2**)
You Try It

**2.** Factor Polynomials Completely (Section 6.6, **Objective 1**)
You Try It

**3.** Simplify Rational Expressions (Section 7.1, **Objective 3**)
You Try It

## OBJECTIVES

**1** Multiply Rational Expressions

**2** Divide Rational Expressions

................................................................................................................

OBJECTIVE 1    MULTIPLY RATIONAL EXPRESSIONS

Recall that to multiply **rational numbers** (fractions), we multiply straight across the **numerators**, multiply straight across the **denominators**, and **simplify**. If $\dfrac{p}{q}$ and $\dfrac{r}{s}$ are rational numbers, then

$$\frac{p}{q} \cdot \frac{r}{s} = \frac{p \cdot r}{q \cdot s} = \frac{pr}{qs}.$$

For example,

$$\frac{3}{14} \cdot \frac{8}{9} = \frac{3 \cdot 8}{14 \cdot 9} = \frac{24}{126}.$$

You may wish to review multiplying fractions in **Section R.1**. Recall that we write fractions in simplified form by **dividing out** common factors from the numerator and denominator.

$$\frac{3}{14} \cdot \frac{8}{9} = \frac{24}{126} = \frac{4 \cdot 6}{21 \cdot 6} = \frac{4 \cdot 6^1}{21 \cdot 6_1} = \frac{4}{21}$$

It is often easier to simplify fractions by first writing the numerator and the denominator as a product of **prime factors**, then divide out any common factors.

$$\frac{3}{14} \cdot \frac{10}{9} = \underbrace{\frac{3}{2 \cdot 7} \cdot \frac{2 \cdot 5}{3 \cdot 3}}_{\substack{\text{Factor into} \\ \text{prime} \\ \text{factors}}} = \underbrace{\frac{3^1}{2_1 \cdot 7} \cdot \frac{2^1 \cdot 5}{3 \cdot 3_1}}_{\substack{\text{Divide out} \\ \text{common} \\ \text{factors}}} = \underbrace{\frac{5}{7 \cdot 3}}_{\substack{\text{Multiply} \\ \text{remaining} \\ \text{factors}}} = \frac{5}{21}$$

We follow the same approach when multiplying **rational expressions**. We factor all numerators and denominators **completely**, **divide out** any **common factors**, multiply remaining factors in the numerators, and multiply remaining factors in the denominators. So, if $\dfrac{P}{Q}$ and $\dfrac{R}{S}$ are rational expressions, then

$$\frac{P}{Q} \cdot \frac{R}{S} = \frac{PR}{QS}.$$

We keep the result in **factored form** just as we did when simplifying rational expressions.

---

**Multiplying Rational Expressions**

**Step 1.** Factor each numerator and denominator completely into **prime factors**.

**Step 2.** Divide out common factors.

**Step 3.** Multiply remaining factors in the numerators and multiply remaining factors in the denominators.

---

 Factoring is a critical step in the multiplication process. If necessary, review the **factoring techniques** discussed in Chapter 6.

### Example 1  Multiplying Rational Expressions

Multiply $\dfrac{5x^2}{2y} \cdot \dfrac{6y^2}{25x^3}$.

**Solution**  We follow the **three-step process** for multiplying **rational expressions**.

Begin with the original expression: $\dfrac{5x^2}{2y} \cdot \dfrac{6y^2}{25x^3}$

Factor numerators and denominators: $= \dfrac{5 \cdot x \cdot x}{2 \cdot y} \cdot \dfrac{2 \cdot 3 \cdot y \cdot y}{5 \cdot 5 \cdot x \cdot x \cdot x}$

Divide out common factors: $= \dfrac{5 \cdot x \cdot x}{2 \cdot y} \cdot \dfrac{2 \cdot 3 \cdot y \cdot y}{5 \cdot 5 \cdot x \cdot x \cdot x}$

Multiply remaining factors: $= \dfrac{3 \cdot y}{5 \cdot x} = \dfrac{3y}{5x}$

**You Try It**    **Work through this You Try It problem.**

**Work Exercises 1–4 in this textbook or in the *MyMathLab*® Study Plan.**

Notice that we could also have worked Example 1 by using the **rules for exponents** (view the **details**).

### Example 2  Multiplying Rational Expressions

Multiply $\dfrac{3x - 6}{2x} \cdot \dfrac{8}{5x - 10}$.

**Solution**  We follow the **three-step process** for multiplying **rational expressions**.

$$\text{Begin with the original expression:} \quad \frac{3x - 6}{2x} \cdot \frac{8}{5x - 10}$$

$$\text{Factor numerators and denominators:} \quad = \frac{3(x - 2)}{2 \cdot x} \cdot \frac{2 \cdot 4}{5(x - 2)}$$

$$\text{Divide out common factors:} \quad = \frac{3\cancel{(x - 2)}}{2 \cdot x} \cdot \frac{2 \cdot 4}{5\cancel{(x - 2)}}$$

$$\text{Multiply remaining factors:} \quad = \frac{3 \cdot 4}{x \cdot 5} = \frac{12}{5x}$$

**You Try It**   Work through this You Try It problem.

Work Exercises 5–8 in this textbook or in the *MyMathLab*® Study Plan.

### Example 3  Multiplying Rational Expressions

*My video summary*   Multiply $\dfrac{x^2 - 4}{x^2 + 2x - 35} \cdot \dfrac{x^2 - 25}{x + 2}$.

**Solution**  Follow the **three-step process** for multiplying **rational expressions**.

$$\text{Begin with the original expression:} \quad \frac{x^2 - 4}{x^2 + 2x - 35} \cdot \frac{x^2 - 25}{x + 2}$$

$$\text{Factor numerators and denominators:} \quad = \frac{(x + 2)(x - 2)}{(x + 7)(x - 5)} \cdot \frac{(x + 5)(x - 5)}{(x + 2)}$$

Continue by dividing out **common factors**, multiplying remaining factors in the numerator, and multiplying remaining factors in the denominator. Remember that even though we are multiplying numerators and denominators, we typically will leave them in **factored form**. View the **answer**, or watch this **video** for a complete solution.

**You Try It**   Work through this You Try It problem.

Work Exercises 9–14 in this textbook or in the *MyMathLab*® Study Plan.

### Example 4  Multiplying Rational Expressions

*My interactive video summary*   Multiply $\dfrac{2x^2 + 3x - 2}{3x^2 - 2x - 1} \cdot \dfrac{3x^2 + 4x + 1}{2x^2 + x - 1}$.

**Solution**  Work through this **interactive video** to complete the solution by factoring the expressions and **dividing out** common factors.

**You Try It**    Work through this You Try It problem.

**Work Exercises 15–18 in this textbook or in the *MyMathLab*® Study Plan.**

**Common factors** are not limited to constants or **binomial factors**. It is possible to have other common **polynomial factors**, as shown in the next example.

### Example 5 Multiplying Rational Expressions

Multiply $\dfrac{3x^2 + 9x + 27}{x - 1} \cdot \dfrac{x + 3}{x^3 - 27}$.

**Solution**    Follow the **three-step process** for multiplying **rational expressions**.

Begin with the original expression:    $\dfrac{3x^2 + 9x + 27}{x - 1} \cdot \dfrac{x + 3}{x^3 - 27}$

Factor numerators and denominators:    $= \dfrac{3(x^2 + 3x + 9)}{(x - 1)} \cdot \dfrac{(x + 3)}{(x - 3)(x^2 + 3x + 9)}$

Notice there is a common **trinomial factor** in a numerator and denominator. This trinomial factor happens to be **prime**. So, these factors **divide out** just like any other common factor.

Divide out common factors:    $\dfrac{3\cancel{(x^2 + 3x + 9)}}{(x - 1)} \cdot \dfrac{(x + 3)}{(x - 3)\cancel{(x^2 + 3x + 9)}}$

Multiply remaining factors:    $= \dfrac{3(x + 3)}{(x - 1)(x - 3)}$

**You Try It**    Work through this You Try It problem.

**Work Exercises 19–22 in this textbook or in the *MyMathLab*® Study Plan.**

Sometimes **factors** in the numerator and denominator are **opposites (or additive inverses)**. The **quotient** of opposites equals $-1$. This can be seen by factoring out $-1$ from either the numerator or denominator and dividing out the **common factors**. For example,

$$\frac{4x - 7}{7 - 4x} = \frac{4x - 7}{-1(-7 + 4x)} = \frac{4x - 7}{-1(4x - 7)} = \frac{\cancel{4x - 7}}{-1\cancel{(4x - 7)}} = \frac{1}{-1} = -1.$$

Usually, we want the factors to have positive **leading coefficients**. If a leading coefficient is negative, we can factor out a **negative constant**.

### Example 6 Multiplying Rational Expressions

    Multiply $\dfrac{3x^2 + 10x - 8}{2x - 3x^2} \cdot \dfrac{4x + 1}{x + 4}$.

**Solution**    Multiply the expressions on your own. Remember to factor out a negative constant if the leading coefficient is negative. View the **answer**, or watch this **video** for a complete solution.

**You Try It**    Work through this You Try It problem.

**Work Exercises 23–26 in this textbook or in the *MyMathLab*® Study Plan.**

As we saw when factoring in **Section 6.2**, rational expressions may contain more than one **variable**.

### Example 7 Multiplying Rational Expressions

*My video summary*

 Multiply $\dfrac{x^2 + xy}{3x + y} \cdot \dfrac{3x^2 + 7xy + 2y^2}{x^2 - y^2}$.

**Solution** Remember to factor out the **GCF** first. For **polynomials** of the form $ax^2 + bxy + cy^2$, remember that the second **term** in each **binomial factor** must contain the variable $y$.

Begin with the original expression: $\dfrac{x^2 + xy}{3x + y} \cdot \dfrac{3x^2 + 7xy + 2y^2}{x^2 - y^2}$

Factor the numerators and denominators: $= \dfrac{x(x + y)}{3x + y} \cdot \dfrac{(3x + y)(x + 2y)}{(x + y)(x - y)}$

Finish the multiplication on your own. View the **answer**, or watch this **video** for a complete solution.

**You Try It** **Work through this You Try It problem.**

**Work Exercises 27–30 in this textbook or in the *MyMathLab*® Study Plan.**

OBJECTIVE 2 DIVIDE RATIONAL EXPRESSIONS

To divide **rational numbers**, recall that if $\dfrac{p}{q}$ and $\dfrac{r}{s}$ are rational numbers such that $q$, $r$, and $s$ are not zero, then

$$\frac{p}{q} \div \frac{r}{s} = \frac{p}{q} \cdot \frac{s}{r} = \frac{ps}{qr}$$

The numbers $\dfrac{r}{s}$ and $\dfrac{s}{r}$ are **reciprocals** of each other.

For example,

$$\frac{4}{15} \div \frac{6}{25} = \underbrace{\frac{4}{15} \cdot \frac{25}{6}}_{\substack{\text{Change to}\\\text{multiplication}\\\text{by the}\\\text{reciprocal}}} = \underbrace{\frac{4 \cdot 25}{15 \cdot 6}}_{\substack{\text{Multiply}\\\text{numerators}\\\text{and multiply}\\\text{denominators}}} = \underbrace{\frac{2 \cdot 2 \cdot 5 \cdot 5}{3 \cdot 5 \cdot 2 \cdot 3}}_{\substack{\text{Factor into}\\\text{prime}\\\text{factors}}} = \underbrace{\frac{2 \cdot 2 \cdot 5 \cdot 5}{3 \cdot 5 \cdot 2 \cdot 3}}_{\substack{\text{Divide out}\\\text{common}\\\text{factors}}} = \underbrace{\frac{2 \cdot 5}{3 \cdot 3}}_{\substack{\text{Multiply}\\\text{remaining}\\\text{factors}}} = \frac{10}{9}.$$

Refer to **Section R.1** to review dividing rational numbers in fraction form.

We follow the same approach when dividing **rational expressions**. To find the **quotient** of two rational expressions, we find the **product** of the first rational expression and the reciprocal of the second rational expression. So, if $\dfrac{P}{Q}$ and $\dfrac{R}{S}$ are rational expressions, then

$$\frac{P}{Q} \div \frac{R}{S} = \frac{P}{Q} \cdot \frac{S}{R} = \frac{PS}{QR}.$$

> **Dividing Rational Expressions**
>
> **Step 1.**  Change the division to multiplication and replace the **divisor** by its **reciprocal**.
>
> **Step 2.**  Multiply the expressions.

### Example 8  Dividing Rational Expressions

Divide each rational expression.

**a.** $\dfrac{6x^5}{9y^3} \div \dfrac{5x^4}{3y^2}$

**b.** $\dfrac{(x+2)(x-1)}{(3x-5)} \div \dfrac{(x-1)(x+4)}{(2x+3)}$

### Solutions

**a.** We follow the **two-step process** for dividing **rational expressions**.

Begin with the original expression: $\dfrac{6x^5}{9y^3} \div \dfrac{5x^4}{3y^2}$

Change to multiplication by
the reciprocal: $= \dfrac{6x^5}{9y^3} \cdot \dfrac{3y^2}{5x^4}$   ← $\dfrac{3y^2}{5x^4}$ is the reciprocal of $\dfrac{5x^4}{3y^2}$

Factor: $= \dfrac{2 \cdot 3 \cdot x \cdot x \cdot x \cdot x \cdot x}{3 \cdot 3 \cdot y \cdot y \cdot y} \cdot \dfrac{3 \cdot y \cdot y}{5 \cdot x \cdot x \cdot x \cdot x}$

Divide out common factors: $= \dfrac{2 \cdot \cancel{3} \cdot \cancel{x \cdot x \cdot x \cdot x} \cdot x}{\cancel{3} \cdot 3 \cdot \cancel{y \cdot y} \cdot y} \cdot \dfrac{\cancel{3 \cdot y \cdot y}}{5 \cdot \cancel{x \cdot x \cdot x \cdot x}}$

Multiply remaining factors: $= \dfrac{2 \cdot x}{y \cdot 5} = \dfrac{2x}{5y}$

**b.** Begin with the original expression: $\dfrac{(x+2)(x-1)}{(3x-5)} \div \dfrac{(x-1)(x+4)}{(2x+3)}$

Change to multiplication by
the reciprocal: $= \dfrac{(x+2)(x-1)}{(3x-5)} \cdot \dfrac{(2x+3)}{(x-1)(x+4)}$

Divide out common factors: $= \dfrac{(x+2)\cancel{(x-1)}}{(3x-5)} \cdot \dfrac{(2x+3)}{\cancel{(x-1)}(x+4)}$

Multiply remaining factors: $= \dfrac{(x+2)(2x+3)}{(3x-5)(x+4)}$

**You Try It**    Work through this **You Try It** problem.

Work Exercises 31–34 in this textbook or in the ***MyMathLab***® Study Plan.

### Example 9  Dividing Rational Expressions

Divide $\dfrac{9y^2 - 81}{4y^2} \div \dfrac{y+3}{8}$.

**Solution**    Begin with the original expression: $\dfrac{9y^2 - 81}{4y^2} \div \dfrac{y+3}{8}$

Change to multiplication by the reciprocal: $= \dfrac{9y^2 - 81}{4y^2} \cdot \dfrac{8}{y+3}$

$$\text{Factor out the GCF:} \quad = \frac{9(y^2 - 9)}{4y^2} \cdot \frac{8}{y + 3}$$

$$\text{Factor difference of two squares:} \quad = \frac{9(y - 3)(y + 3)}{4y^2} \cdot \frac{8}{y + 3}$$

**Divide out** common factors and finish the problem on your own. View the **answer**, or watch this **video** for the complete solution.

**You Try It**    Work through this You Try It problem.

**Work Exercises 35–40** in this textbook or in the *MyMathLab*® Study Plan.

### Example 10  Dividing Rational Expressions

*My video summary*    Divide $\dfrac{2x^2 + 21x + 40}{3x^2 + 23x - 8} \div \dfrac{4x^2 + 16x + 15}{x + 2}$.

**Solution**    Begin with the original expression: $\dfrac{2x^2 + 21x + 40}{3x^2 + 23x - 8} \div \dfrac{4x^2 + 16x + 15}{x + 2}$

$$\text{Change to multiplication by the reciprocal:} \quad = \frac{2x^2 + 21x + 40}{3x^2 + 23x - 8} \cdot \frac{x + 2}{4x^2 + 16x + 15}$$

**Factor** the numerators and denominators and then multiply. View the **answer**, or watch this **video** for a complete solution.

**You Try It**    Work through this You Try It problem.

**Work Exercises 41–43** in this textbook or in the *MyMathLab*® Study Plan.

### Example 11  Dividing Rational Expressions

*My interactive video summary*    Divide $\dfrac{x^3 - 8}{2x^2 - x - 6} \div \dfrac{x^2 + 2x + 4}{6x^2 + 11x + 3}$.

**Solution**    Work through this **interactive video** to complete the solution, or view the **answer**.

**You Try It**    Work through this You Try It problem.

**Work Exercises 44–46** in this textbook or in the *MyMathLab*® Study Plan.

As with multiplication, when dividing **rational expressions**, we may have more than one variable as shown in the next example.

### Example 12  Dividing Rational Expressions

*My video summary*    Divide $\dfrac{x^3 - 8y^3}{3x + y} \div \dfrac{4x - 8y}{6x^2 + 17xy + 5y^2}$.

## Solution

Begin with the original expression:
$$\frac{x^3 - 8y^3}{3x + y} \div \frac{4x - 8y}{6x^2 + 17xy + 5y^2}$$

Change to multiplication:
$$= \frac{x^3 - 8y^3}{3x + y} \cdot \frac{6x^2 + 17xy + 5y^2}{4x - 8y}$$

Factor:
$$= \frac{(x - 2y)(x^2 + 2xy + 4y^2)}{3x + y} \cdot \frac{(3x + y)(2x + 5y)}{4(x - 2y)}$$

Finish the problem on your own. View the **answer**, or watch the **video** for a complete solution.

**You Try It**   Work through this **You Try It** problem.

**Work Exercises 47 and 48 in this textbook or in the _MyMathLab_® Study Plan.**

If we need to multiply or divide more than two rational expressions, first we change any divisions to multiplication using the appropriate **reciprocal**. Then we follow the **three-step process** for multiplying rational expressions.

### Example 13  Multiplying and Dividing Rational Expressions

*My video summary*   Perform the indicated operations.

$$\frac{x^2 + 2x - 15}{x^2 + 2x - 8} \cdot \frac{x^2 + 3x + 2}{x^2 + 4x - 21} \div \frac{x + 2}{x^2 + 9x + 14}$$

**Solution**   Try this problem on your own. Remember to first change all divisions to multiplication using the **reciprocal** of the **divisor**. View the **answer**, or watch the **video** to see a complete solution.

**You Try It**   Work through this **You Try It** problem.

**Work Exercises 49–54 in this textbook or in the _MyMathLab_® Study Plan.**

# 7.2  **Exercises**

In Exercises 1–30, multiply the rational expressions.

**You Try It**

**1.** $\dfrac{6a^3}{2b^2} \cdot \dfrac{5b^3}{3a^4}$

**2.** $\dfrac{14x^2y}{3y^2x} \cdot \dfrac{9x^3y^2}{7x^2y^3}$

**3.** $\dfrac{2xyz}{5y^2z^2} \cdot \dfrac{15xz^2}{6x^2y}$

**4.** $\dfrac{-3x^2z}{2y^3z^2} \cdot \dfrac{4xy^2}{z^3}$

**You Try It**

**5.** $\dfrac{5x}{3x + 6} \cdot \dfrac{x + 2}{7}$

**6.** $\dfrac{3m + 9}{10m} \cdot \dfrac{5m^2}{2m + 6}$

**7.** $\dfrac{6x^2}{2x - 2} \cdot \dfrac{x^2 - x}{5x^3}$

**8.** $\dfrac{2xy + y}{3y^2z} \cdot \dfrac{9yz^2}{2xz + z}$

**You Try It**

9. $\dfrac{x+1}{x^2-4} \cdot \dfrac{x-2}{x^2-1}$

10. $\dfrac{x-6}{x^2+3x-10} \cdot \dfrac{2x-4}{x^2-36}$

11. $\dfrac{3r^2+12r}{r-1} \cdot \dfrac{r^2-5r+4}{r+2}$

12. $\dfrac{x^2-3x}{x^2+x} \cdot \dfrac{x^2+3x+2}{x^2-2x-3}$

13. $\dfrac{x^2+4x+3}{x^2+5x+6} \cdot \dfrac{x+2}{x-1}$

14. $\dfrac{x^2+12x+36}{x+1} \cdot \dfrac{4x+3}{x^2+2x-24}$

**You Try It**

15. $\dfrac{x^2-1}{3x^2+11x+6} \cdot \dfrac{3x^2+7x-6}{x^2-2x+1}$

16. $\dfrac{x^2+3x}{x^2-4x-5} \cdot \dfrac{x^2-25}{2x^2+x}$

17. $\dfrac{3b^2+4b-4}{b^2+b-2} \cdot \dfrac{b^2-1}{3b^2+b-2}$

18. $\dfrac{4x^2+20x-24}{6x^2+7x-3} \cdot \dfrac{4x^2+20x+21}{9x^2+51x-18}$

**You Try It**

19. $\dfrac{x^3-8}{x^2+4x+3} \cdot \dfrac{x^2+3x}{x^3+2x^2+4x}$

20. $\dfrac{x+10}{2x^2-4x+8} \cdot \dfrac{3x^3+24}{x-5}$

21. $\dfrac{5x^2-5x+5}{10x^2+20x} \cdot \dfrac{x^2-2x}{x^3+1}$

22. $\dfrac{2n^3+2n^2+n+1}{n^2+5n+6} \cdot \dfrac{n^2-2n}{n^3+1}$

**You Try It**

23. $\dfrac{4-x^2}{x^2-x-6} \cdot \dfrac{5x^2-14x-3}{2x^2-3x-2}$

24. $\dfrac{12-2x-2x^2}{2y^2} \cdot \dfrac{x+3y}{-x^2+xy-y^2}$

25. $\dfrac{p^2-p}{3p-2p^2} \cdot \dfrac{2p^2-p-3}{1-p^2}$

26. $\dfrac{3-8x-3x^2}{9-x^2} \cdot \dfrac{x^2-x-6}{9x^2-1}$

**You Try It**

27. $\dfrac{x^2+2xy-3y^2}{3x^2+2xy-y^2} \cdot \dfrac{x^2+3xy+2y^2}{x^2+xy-2y^2}$

28. $\dfrac{a^3+a^3b+a+b}{2a^3+2a} \cdot \dfrac{18a^2}{6a^2-6b^2}$

29. $\dfrac{x^2-y^2}{2x^2+xy-y^2} \cdot \dfrac{2xy-y^2}{x^3-y^3}$

30. $\dfrac{2x^2+xy-y^2}{2x^2-3xy+y^2} \cdot \dfrac{x^2-y^2}{x^2+2xy+y^2}$

In Exercises 31–48, divide the rational expressions.

**You Try It**

31. $\dfrac{4y^3}{3x^2} \div \dfrac{8y^2}{15x^3}$

32. $\dfrac{5x}{6} \div \dfrac{10x+15}{6x+18}$

33. $\dfrac{3x+2}{(x-2)(x+1)} \div \dfrac{2x+3}{(x-2)(x+3)}$

34. $\dfrac{x(x+1)}{(x-1)(x+2)} \div \dfrac{2x+2}{(x-1)(x+3)}$

**You Try It**

35. $\dfrac{2x^2}{3x^2-3} \div \dfrac{14}{x+1}$

36. $\dfrac{3m}{4m-12} \div \dfrac{m^2+m}{6-2m}$

37. $\dfrac{x-2}{3x-9} \div \dfrac{x^2-2x}{x-3}$

38. $\dfrac{6y^2}{3y^2+5y+2} \div \dfrac{10y^2-6y}{2y^2-3y-5}$

39. $\dfrac{x^2-4}{x^3-27} \div \dfrac{x^3-8}{x^2-9}$

40. $\dfrac{5x-x^2}{x^3-125} \div \dfrac{x}{x^2+4x+8}$

**41.**  $\dfrac{x^2 + x - 6}{x + 5} \div \dfrac{x - 2}{x^2 + 4x - 5}$

**You Try It**

**42.** $\dfrac{x^2 - 12x + 36}{x^2 - x - 42} \div \dfrac{x^2 - 36}{4}$

**43.**  $\dfrac{2x^2 + x - 3}{x^2 + 3x + 2} \div \dfrac{x^2 + 4x - 5}{3x^2 + 5x + 2}$

**44.** $\dfrac{x^2 - 2x - 3}{2x^2 + x - 1} \div \dfrac{3x^2 - 7x - 6}{6x^2 + x - 2}$

**You Try It 45.**  $\dfrac{3h^2 - 3h - 60}{7h^2 - 37h + 10} \div \dfrac{12h^2 + 20h + 3}{2h^2 - h - 6}$

**46.** $\dfrac{10x^2 + 3x - 1}{3x + 2} \div \dfrac{2x + 1}{9x^2 + 3x - 2}$

**47.** $\dfrac{x^2 - y^2}{4x + 2y} \div \dfrac{x^2 - 3xy - 4y^2}{2x^2 - 7xy - 4y^2}$

**You Try It**

**48.** $\dfrac{x^3 - y^3}{x^2 - xy + y^2} \div \dfrac{x^2 - y^2}{x^3 + y^3}$

In Exercises 49–54, perform the indicated operations.

**49.**  $\dfrac{5xy}{x^3} \cdot \dfrac{xy^3}{8y} \div \dfrac{2y^3}{3x^4y}$

**You Try It**

**50.** $\dfrac{7}{x^2} \div \dfrac{9x^2y}{x^3} \cdot \dfrac{4x}{x^8}$

**51.** $\dfrac{x^2 - 2x - 3}{x - 1} \cdot \dfrac{x + 4}{x^2 + 2x - 15} \cdot \dfrac{x^2 + 4x - 5}{x^2 - x - 2}$

**52.** $\dfrac{x^2 + 6x + 8}{x^2 + 6x + 5} \div \dfrac{x^2 - 5x - 14}{x^2 + 3x - 10} \cdot \dfrac{x^2 - 6x - 7}{x^2 + x - 6}$

**53.** $\dfrac{x^2 + 5x + 4}{x - 3} \cdot \dfrac{x^2 - 4x + 3}{x^2 + 6x + 5} \div \dfrac{x^2 - 3x + 2}{x + 5}$

**54.** $\dfrac{x^2 + x - 2}{x^2 - 5x + 6} \div \dfrac{x^2 + 4x + 3}{x^2 + 2x - 15} \div \dfrac{(x + 2)^2}{x^2 + 3x - 10}$

# 7.3 Least Common Denominators

## THINGS TO KNOW

Before working through this section, be sure you are familiar with the following concepts:

|  | VIDEO | ANIMATION | INTERACTIVE |

**1.** Factor Polynomials Completely
(Section 6.6, **Objective 1**)

**You Try It**

**2.** Simplify Rational Expressions
(Section 7.1, **Objective 3**)

**You Try It**

## OBJECTIVES

**1** Find the Least Common Denominator of Rational Expressions

**2** Write Equivalent Rational Expressions

OBJECTIVE 1   FIND THE LEAST COMMON DENOMINATOR OF RATIONAL EXPRESSIONS

Recall that when adding or subtracting **fractions**, we need a **common denominator**. Typically, we use the least common denominator (**LCD**). The same is true when adding or subtracting rational expressions, which we will visit in the next section.

To find the LCD of rational expressions, first we **factor** each denominator. Then we include each **unique factor** the largest number of times that it occurs in any denominator. We can illustrate this idea by considering **rational numbers**.

To find the LCD of $\frac{4}{15}$, $\frac{7}{18}$, and $\frac{3}{8}$, first factor each denominator into its **prime factors**.

$$15 = 3 \cdot 5$$
$$18 = 2 \cdot 3 \cdot 3$$
$$8 = 2 \cdot 2 \cdot 2$$

Next, form the LCD by including each unique factor the largest number of times that it occurs in any denominator. The factor 2 occurs at most three times in any denominator, the factor 3 occurs at most two times, and the factor 5 occurs at most once. So, our LCD is

$$\text{LCD} = 2 \cdot 2 \cdot 2 \cdot 3 \cdot 3 \cdot 5 = 2^3 \cdot 3^2 \cdot 5 = 360.$$

View this **popup** for a summary of the steps to find the LCD of rational numbers.

We follow a similar process to find the **LCD** of **rational expressions**.

---

**Finding the Least Common Denominator (LCD) of Rational Expressions**

**Step 1.** Factor each **denominator** completely.

**Step 2.** List each **unique factor** from any denominator.

**Step 3.** The least common denominator is the product of the unique factors, each raised to a **power** equivalent to the largest number of times that the factor occurs in any denominator.

---

 Don't forget numerical factors when forming the LCD.

**Example 1  Finding the Least Common Denominator of Rational Expressions**

Find the LCD of the rational expressions.

a. $\dfrac{7}{10x^3}, \dfrac{3}{5x^2}$      b. $\dfrac{x+2}{3x}, \dfrac{x-1}{2x^2+6x}$

**Solutions**

a. Follow the **three-step process**.

  **Step 1.**  $10x^3 = 2 \cdot 5 \cdot x \cdot x \cdot x$
      $5x^2 = 5 \cdot x \cdot x$

  **Step 2.** We have **unique factors** of 2, 5, and $x$. Make sure to include each factor the largest number of times that it occurs in any **denominator**.

  **Step 3.** The factors 2 and 5 occur at most once, and the factor $x$ occurs at most three times.

      $$\text{LCD} = 2 \cdot 5 \cdot x \cdot x \cdot x = 10x^3$$

**b.** Follow the three-step process.

**Step 1.** $3x = 3 \cdot x$

$2x^2 + 6x = 2 \cdot x(x + 3)$

**Step 2.** We have unique factors of 3, $x$, 2, and $(x + 3)$. Notice that we must list both $x$ and $x + 3$ as factors because $x$ is not a factor of $x + 3$.

**Step 3.** The factors 3, $x$, 2, and $(x + 3)$ each occur at most once.

$\text{LCD} = 3 \cdot 2 \cdot x \cdot (x + 3) = 6x(x + 3)$

**You Try It**   Work through this You Try It problem.

Work Exercises 1–6 in this textbook or in the *MyMathLab*® Study Plan.

## Example 2  Finding the Least Common Denominator of Rational Expressions

Find the LCD of the rational expressions.

**a.** $\dfrac{z^2}{6 - z}, \dfrac{9}{2z - 12}$    **b.** $\dfrac{y + 2}{y^2 + 2y - 3}, \dfrac{2y}{y^2 + 5y + 6}$

### Solutions

**a.** Follow the **three-step process**.

**Step 1.** $6 - z = -1 \cdot (z - 6)$

$2z - 12 = 2 \cdot (z - 6)$

**Step 2.** We have unique factors of $-1$, $(z - 6)$, and 2. Notice that we factored out a $-1$ in the first denominator. We will do this when the **leading coefficient** is negative to avoid including **opposite factors** in the **LCD**. In this case, we include $-1$ and $z - 6$ instead of $z - 6$ and $6 - z$.

Determine the largest number of times that each unique factor occurs and form the LCD. Then view the **answer**.

*My video summary*   **b.** Follow the three-step process.

**Step 1.** $y^2 + 2y - 3 = (y + 3)(y - 1)$

$y^2 + 5y + 6 = (y + 3)(y + 2)$

Determine the unique factors and the largest number of times that each occurs. Then form the LCD. View the **answer**, or watch this **video** for a complete solution.

**You Try It**   Work through this You Try It problem.

Work Exercises 7–12 in this textbook or in the *MyMathLab*® Study Plan.

## Example 3  Finding the Least Common Denominator of Rational Expressions

*My interactive video summary*   Find the **LCD** of the rational expressions.

**a.** $\dfrac{4x}{10x^2 - 7x - 12}, \dfrac{2x - 3}{5x^2 - 11x - 12}$

**b.** $\dfrac{10 - x}{6x^2 + 5x + 1}, \dfrac{-4}{9x^2 + 6x + 1}, \dfrac{x^2 - 7x}{10x^2 - x - 3}$

**Solutions** Try to find the LCDs on your own. View the **answers**, or watch this **interactive video** for complete solutions to both parts.

**You Try It**    **Work through this You Try It problem.**

**Work Exercises 13–17 in this textbook or in the *MyMathLab*®️ Study Plan.**

OBJECTIVE 2    WRITE EQUIVALENT RATIONAL EXPRESSIONS

Let's look at how to write an **equivalent rational expression** using the **LCD**. The key to doing this lies with our previous work on dividing out **common factors** in **Section 7.1**. Recall that

$$\frac{\text{common factor}}{\text{common factor}} = 1.$$

We can therefore **divide out** common factors that occur in the **numerator** and denominator because the quotient of the common factors equals 1.

When writing **equivalent fractions**, we are actually working the other way. We multiply the original expression by 1, but in such a way as to obtain the desired LCD. To do this, we first ask the question,

*"What should we multiply the original denominator by to get the LCD?"*

Whatever the answer is, we multiply both the numerator and denominator by this answer.

For example, to write $\frac{3}{5}$ as an equivalent fraction with a denominator of 20, we ask, "What do we multiply 5 by to get 20?" The answer is 4, so we multiply both the numerator and denominator by 4 to get $\frac{3}{5} = \frac{3}{5} \cdot 1 = \frac{3}{5} \cdot \frac{4}{4} = \frac{12}{20}$. Notice that $\frac{4}{4} = 1$, so we have not changed the value, just the way it looks.

To work with **rational expressions**, we ask the same question: *"What do we multiply the original denominator by to get the LCD?"* Then we multiply both the **numerator** and **denominator** by the answer.

---

**Writing Equivalent Rational Expressions**

To write an equivalent rational expression with a desired denominator:

**Step 1.**    Factor the given denominator and the desired denominator.

**Step 2.**    Determine the missing factor(s) from the given denominator that must be multiplied to get the desired denominator.

**Step 3.**    Multiply the numerator and denominator of the given rational expression by the missing factor(s) from Step 2.

---

From the commutative property of multiplication, the order of the factors in either the numerator or denominator does not matter. For example,

$$\frac{(x + 1)(x - 4)}{(x + 3)(2x - 1)} = \frac{(x - 4)(x + 1)}{(2x - 1)(x + 3)}.$$

## Example 4  Writing Equivalent Rational Expressions

Write each rational expression as an equivalent rational expression with the desired denominator.

a. $\dfrac{3}{2x} = \dfrac{}{10x^3}$

b. $\dfrac{x+2}{3x+15} = \dfrac{}{3(x-1)(x+5)}$

**Solutions**  We follow the three-step process.

a. **Step 1.** $2x = 2 \cdot x$

$10x^3 = 2 \cdot 5 \cdot x \cdot x \cdot x$

**Step 2.** The given denominator is missing $5 \cdot x \cdot x$. What do we multiply $2x$ by to get $10x^3$? The answer is $5 \cdot x \cdot x = 5x^2$.

**Step 3.** $\dfrac{3}{2x} = \dfrac{3}{2x} \cdot \underbrace{\dfrac{5x^2}{5x^2}}_{1} = \dfrac{15x^2}{10x^3}$

So, $\dfrac{3}{2x} = \dfrac{15x^2}{10x^3}$.

b. **Step 1.** $3x + 15 = 3 \cdot (x+5)$

$3(x-1)(x+5)$ is already factored.

**Step 2.** The given denominator is missing $(x-1)$. What do we multiply $3x + 15$ by to get $3(x-1)(x+5)$? The answer is $(x-1)$.

**Step 3.** $\dfrac{x+2}{3x+15} = \dfrac{x+2}{3(x+5)} \cdot \underbrace{\dfrac{(x-1)}{(x-1)}}_{1} = \dfrac{(x+2)(x-1)}{3(x+5)(x-1)}$  or  $\dfrac{(x+2)(x-1)}{3(x-1)(x+5)}$

Commutative property of multiplication

So, $\dfrac{x+2}{3x+15} = \dfrac{(x+2)(x-1)}{3(x-1)(x+5)}$.

**You Try It**  Work through this You Try It problem.

**Work Exercises 18–23** in this textbook or in the *MyMathLab* Study Plan.

## Example 5  Writing Equivalent Rational Expressions

Write each rational expression as an equivalent rational expression with the desired denominator.

a. $\dfrac{-7}{1-4y} = \dfrac{}{8y^2 - 2y}$

b. $\dfrac{5z}{z^2 + z - 6} = \dfrac{}{(z-4)(z-2)(z+3)}$

**Solutions**

a. **Step 1.** $1 - 4y = -1 \cdot (4y - 1)$

$8y^2 - 2y = 2 \cdot y \cdot (4y - 1)$

**Step 2.** What do we multiply $1 - 4y$ by to get $8y^2 - 2y$? The answer is $-2 \cdot y = -2y$. Notice that we need to use $-2$ so that $-1 \cdot -2 = 2$.

Multiply the numerator and denominator of the given rational expression by $-2y$ on your own. Then view the answer.

✎ *My video summary*  **b. Step 1.** $z^2 + z - 6 = (z + 3)(z - 2)$

$(z - 4)(z - 2)(z + 3)$ is already factored.

Try finishing this problem on your own. Determine what is missing from the given denominator, then multiply both the numerator and denominator of the given rational expression by this quantity. View the **answer**, or watch this **video** for a complete solution.

**You Try It**   Work through this You Try It problem.

Work Exercises 24–28 in this textbook or in the *MyMathLab* ® Study Plan.

# 7.3  Exercises

In Exercises 1–17, find the LCD for the given rational expressions.

**You Try It**

**1.** $\dfrac{5}{6x^5}, \dfrac{4}{9x^2}$

**2.** $\dfrac{6}{5a^3b^4}, \dfrac{7}{15a^2b^8}$

**3.** $\dfrac{3}{7a^3c^3}, \dfrac{5}{14b^5c^2}$

**4.** $\dfrac{x + 1}{4x^2}, \dfrac{x - 2}{3x^2 + 9x}$

**5.** $\dfrac{3x - 5}{6x^2 - 3x}, \dfrac{2x + 1}{12x^2 + 4x}$

**6.** $\dfrac{5}{3x^3 + 9x^2 + 6x}, \dfrac{7x}{2x^3 - 2x}$

**You Try It**

**7.** $\dfrac{x}{3 - x}, \dfrac{2x + 1}{x^2 - x - 6}$

**8.** $\dfrac{3x + 1}{x^2 - 4x - 5}, \dfrac{x + 7}{2 + x - x^2}$

**9.** $\dfrac{2x}{3 + 2x - x^2}, \dfrac{5x - 1}{9 - x^2}$

**10.** $\dfrac{y - 5}{y^2 - y - 2}, \dfrac{y + 5}{y^2 + y - 6}$

**11.** $\dfrac{2y - 1}{y^2 - 9}, \dfrac{1 - 5y}{y^2 + 6y + 9}$

**12.** $\dfrac{2x}{4 - x^2}, \dfrac{1 - x}{6 + x - x^2}$

**You Try It**

**13.** $\dfrac{7 - x}{2x^2 - 5x - 3}, \dfrac{2x + 3}{3x^2 - 11x + 6}$

**14.** $\dfrac{3x^2 + 2x + 1}{10x^2 - 15x}, \dfrac{2x^2 - 7x}{6x^2 - 13x + 6}$

**15.** $\dfrac{5x + 1}{6x^3 + 26x^2 - 20x}, \dfrac{5x - 1}{6x^4 + 33x^3 + 15x^2}$

**16.** $\dfrac{7 + x}{4 - x^2}, \dfrac{6 + x}{x^2 + x - 2}, \dfrac{5 + x}{x^2 - x - 2}$

**17.** $\dfrac{x + 1}{2x^2 + x - 3}, \dfrac{x + 2}{3x^2 + 13x - 10}, \dfrac{x + 3}{6x^2 + 5x - 6}$

In Exercises 18–28, write each rational expression as an equivalent rational expression with the desired denominator.

**You Try It**

**18.** $\dfrac{2}{3x} = \dfrac{}{15x^4}$

**19.** $\dfrac{3}{5x^2y^3} = \dfrac{}{15x^4y^5}$

**20.** $\dfrac{2b}{3a^2c^3} = \dfrac{}{21a^5b^3c^3}$

**21.** $\dfrac{x + 1}{2x + 6} = \dfrac{}{2(x - 1)(x + 3)}$

**22.** $\dfrac{7}{x^2 - x - 6} = \dfrac{}{(x + 2)(x - 2)(x + 3)(x - 3)}$

**23.** $\dfrac{3x - 5}{2x + 3} = \dfrac{}{(2x + 3)(x + 2)}$

**You Try It**

**24.** $\dfrac{5}{2 - x}, \dfrac{}{(x^2 - 4)(x^2 + x)}$

**25.** $\dfrac{x + 1}{9x - 3x^2} = \dfrac{}{3x(x - 1)(x - 2)(x - 3)}$

**26.** $\dfrac{2x + 1}{x^2 + x - 2} = \dfrac{}{(x - 1)(x + 1)(x + 2)}$

**27.** $\dfrac{3x - 5}{2x^2 - 4x - 6} = \dfrac{}{10(x - 3)(x - 2)(x + 1)}$

**28.** $\dfrac{6 - z}{z^2 + 8z + 15} = \dfrac{}{(z^2 + 7z + 12)(z^2 + 7z + 10)}$

# 7.4 Adding and Subtracting Rational Expressions

## THINGS TO KNOW

Before working through this section, be sure you are familiar with the following concepts:

| | VIDEO | ANIMATION | INTERACTIVE |
|---|---|---|---|

**You Try It**

**1.** Add Polynomials
(Section 5.3, **Objective 1**)                                           

**You Try It**

**2.** Subtract Polynomials
(Section 5.3, **Objective 3**)          

**You Try It**

**3.** Factor Polynomials Completely
(Section 6.6, **Objective 1**)                                       

**You Try It**

**4.** Simplify Rational Expressions
(Section 7.1, **Objective 3**)         

**You Try It**

## OBJECTIVES

**1** Add and Subtract Rational Expressions with Common Denominators

**2** Add and Subtract Rational Expressions with Unlike Denominators

---

OBJECTIVE 1    ADD AND SUBTRACT RATIONAL EXPRESSIONS WITH COMMON DENOMINATORS

Recall that when adding or subtracting **fractions**, we first check for a **common denominator**. Then we add or subtract the **numerators** and keep the common denominator. To add or subtract **rational numbers** with common **denominators**, we follow this general approach:

$$\frac{p}{q} + \frac{r}{q} = \frac{p + r}{q} \quad \text{or} \quad \frac{p}{q} - \frac{r}{q} = \frac{p - r}{q}.$$

For example,

$$\frac{4}{5} + \frac{3}{5} = \frac{4+3}{5} = \frac{7}{5} \quad \text{or} \quad \frac{11}{3} - \frac{7}{3} = \frac{11-7}{3} = \frac{4}{3}.$$

The same is true when working with **rational expressions**. Once we have a common denominator, we add or subtract the numerators and keep the common denominator.

---

**Adding and Subtracting Rational Expressions with Common Denominators**

If $\dfrac{P}{Q}$ and $\dfrac{R}{Q}$ are rational expressions, then

$$\frac{P}{Q} + \frac{R}{Q} = \frac{P+R}{Q} \quad \text{and} \quad \frac{P}{Q} - \frac{R}{Q} = \frac{P-R}{Q}.$$

---

### Example 1  Adding and Subtracting Rational Expressions with Common Denominators

Add or subtract.

a. $\dfrac{4z}{3} + \dfrac{5z}{3}$   b. $\dfrac{3r}{7s^2} - \dfrac{2r}{7s^2}$

### Solutions

a. The **rational expressions** have **common denominators**, so we write the sum of the **numerators** over the common denominator and **simplify**.

Begin with the original expression: $\dfrac{4z}{3} + \dfrac{5z}{3}$

Add the numerators: $= \dfrac{4z + 5z}{3}$

Combine like terms: $= \dfrac{9z}{3}$

Divide out common factor of 3: $= \dfrac{\cancel{9}^{3}z}{\cancel{3}_{1}}$

$= 3z$

b. The **rational expressions** have a **common denominator**, so we write the difference of the **numerators** over the common denominator and **simplify**.

Begin with the original expression: $\dfrac{3r}{7s^2} - \dfrac{2r}{7s^2}$

Subtract the numerators: $= \dfrac{3r - 2r}{7s^2}$

Simplify: $= \dfrac{r}{7s^2}$

**You Try It**   **Work through this You Try It problem.**

**Work Exercises 1–5 in this textbook or in the *MyMathLab*® Study Plan.**

When subtracting rational expressions, it is a good idea to use **grouping symbols** around each numerator that involves more than one **term**. This will help remind us to subtract each term in the second numerator. We see an illustration of this point in the next example.

### Example 2 Adding and Subtracting Rational Expressions with Common Denominators

Add or subtract.

a. $\dfrac{9x}{x-4} + \dfrac{7x-2}{x-4}$

b. $\dfrac{5y+1}{y-2} - \dfrac{2y+3}{y-2}$

**Solutions**

a. The **rational expressions** have **common denominators**, so we write the sum of the **numerators** over the common denominator and **simplify**.

Begin with the original expression: $\dfrac{9x}{x-4} + \dfrac{7x-2}{x-4}$

Add the numerators: $= \dfrac{9x+7x-2}{x-4}$

Finish simplifying. Check the resulting rational expression to see if there are any **common factors** that can be **divided out**. Then, view the **answer**.

b. The **rational expressions** have **common denominators**, so we write the difference of the **numerators** over the common denominator and **simplify**.

Begin with the original expression: $\dfrac{5y+1}{y-2} - \dfrac{2y+3}{y-2}$

Subtract the numerators: $= \dfrac{(5y+1)-(2y+3)}{y-2}$

Use the distributive property: $= \dfrac{5y+1-2y-3}{y-2}$

Finish simplifying. Check the resulting rational expression to see if there are any **common factors** that can be **divided out**. Then, view the **answer**.

**You Try It**  Work through this You Try It problem.

Work Exercises 6–10 in this textbook or in the *MyMathLab* Study Plan.

### Example 3 Adding and Subtracting Rational Expressions with Common Denominators

*My interactive video summary*

Add or subtract.

a. $\dfrac{4}{x^2+2x-8} + \dfrac{x}{x^2+2x-8}$

b. $\dfrac{x}{x+2} - \dfrac{x-3}{x+2}$

c. $\dfrac{x^2-2}{x-5} - \dfrac{4x+3}{x-5}$

**Solutions**  Try to perform the operations on your own. Remember to check your result for any **common factors**. View the **answers**, or watch this **interactive video** for complete solutions to all three parts.

**You Try It**    **Work through this You Try It problem.**

Work Exercises 11–15 in this textbook or in the *MyMathLab*® Study Plan.

OBJECTIVE 2    ADD AND SUBTRACT RATIONAL EXPRESSIONS WITH UNLIKE DENOMINATORS

To add or subtract **rational expressions** with unlike **denominators**, first we write each fraction as an **equivalent expression** using the **LCD**. Then we can add or subtract the numerators, keeping the common denominator.

---

**Adding and Subtracting Rational Expressions with Unlike Denominators**

**Step 1.**    Find the **LCD** for all expressions being added or subtracted.

**Step 2.**    Write **equivalent expressions** for each term using the LCD as the denominator.

**Step 3.**    Add/subtract the numerators, but keep the denominator the same (the LCD).

**Step 4.**    Simplify if possible.

---

We focus on the LCD because it is the easiest **common denominator** to use. Any common denominator will allow us to add or subtract rational expressions.

**Example 4  Adding and Subtracting Rational Expressions**

Perform the indicated operations and simplify.

**a.** $\dfrac{7}{6x} + \dfrac{3}{2x^3}$        **b.** $\dfrac{3x}{x-3} - \dfrac{x-2}{x+3}$

**Solutions**  We follow the **four-step process**.

**a. Step 1.**
$$6x = 2 \cdot 3 \cdot x$$
$$2x^3 = 2 \cdot x \cdot x \cdot x$$

We have **unique factors** of 2, 3, and $x$. Make sure to include each factor the largest number of times that it occurs in any **denominator**.

$$LCD = 2 \cdot 3 \cdot x \cdot x \cdot x = 6x^3$$

**Step 2.**  In the first expression, we multiply the denominator by $x^2$ to get the **LCD**, so we obtain an **equivalent expression** by multiplying both the numerator and denominator by $x^2$. In the second expression, we multiply the denominator by 3 to get the LCD, so we obtain an equivalent expression by multiplying both the numerator and denominator by 3.

The overall expression becomes

$$\frac{7}{6x} + \frac{3}{2x^3} = \frac{7}{6x} \cdot \frac{x^2}{x^2} + \frac{3}{2x^3} \cdot \frac{3}{3}$$

$$= \frac{7x^2}{6x^3} + \frac{9}{6x^3}$$

**Step 3.** With a **common denominator**, we can now add:

$$\frac{7x^2}{6x^3} + \frac{9}{6x^3} = \frac{7x^2 + 9}{6x^3}$$

**Step 4.** The expression cannot be simplified further, so $\dfrac{7}{6x} + \dfrac{3}{2x^3} = \dfrac{7x^2 + 9}{6x^3}$.

*My video summary*  **b. Step 1.** The **denominators** cannot be factored further.

We have **unique factors** of $x - 3$ and $x + 3$. Make sure to include each factor the largest number of times that it occurs in any denominator.

$$LCD = (x - 3)(x + 3)$$

**Step 2.** In the first expression, we multiply the denominator by $x + 3$ to get the **LCD** and obtain an **equivalent expression** by multiplying both the **numerator** and denominator by $x + 3$. In the second expression, we multiply the denominator by $x - 3$ to get the LCD and obtain an equivalent expression by multiplying both the numerator and denominator by $x - 3$.

So, we write

$$\frac{3x}{x - 3} - \frac{x - 2}{x + 3} = \frac{3x}{x - 3} \cdot \frac{x + 3}{x + 3} - \frac{x - 2}{x + 3} \cdot \frac{x - 3}{x - 3}$$

$$= \frac{3x(x + 3)}{(x - 3)(x + 3)} - \frac{(x - 2)(x - 3)}{(x - 3)(x + 3)}.$$

Carry out the subtraction on your own and **simplify** if necessary. View the **answer**, or watch this **video** for a complete solution.

**You Try It** Work through this You Try It problem.

Work Exercises 16–20 in this textbook or in the *MyMathLab*® Study Plan.

## Example 5 Adding and Subtracting Rational Expressions

Perform the indicated operations and simplify.

**a.** $\dfrac{z + 2}{3z} - \dfrac{5}{3z + 12}$

**b.** $\dfrac{5}{4m - 12} + \dfrac{3}{2m}$

### Solutions

**a. Step 1.**

$$3z = 3 \cdot z$$

$$3z + 12 = 3(z + 4)$$

We have **unique factors** of $3, z,$ and $(z + 4)$. Make sure to include each factor the largest number of times that it occurs in any **denominator**.

$$LCD = 3 \cdot z \cdot (z + 4) = 3z(z + 4).$$

**Step 2.** In the first expression, we multiply the denominator by $z + 4$ to get the LCD and obtain an **equivalent expression** by multiplying both the numerator and denominator by $z + 4$. In the second expression, we multiply the denominator by $z$ to get the LCD, so we obtain an equivalent expression by multiplying both the numerator and denominator by $z$.

The overall expression becomes

$$\frac{z + 2}{3z} - \frac{5}{3z + 12} = \frac{z + 2}{3z} \cdot \frac{z + 4}{z + 4} - \frac{5}{3(z + 4)} \cdot \frac{z}{z}$$

$$= \frac{z^2 + 6z + 8}{3z(z + 4)} - \frac{5z}{3z(z + 4)}$$

**Step 3.** With a **common denominator**, we can now subtract:

$$\text{Subtract:} \quad = \frac{z^2 + 6z + 8 - 5z}{3z(z + 4)}$$

$$\text{Simplify:} \quad = \frac{z^2 + z + 8}{3z(z + 4)}$$

**Step 4.** The expression cannot be simplified further, so

$$\frac{z + 2}{3z} - \frac{5}{3z + 12} = \frac{z^2 + z + 8}{3z(z + 4)}.$$

*My video summary*  **b.** Write equivalent expressions using the LCD and carry out the addition on your own. **Simply** if necessary, then view the **answer** or watch this **video** for a complete solution.

**You Try It** Work through this You Try It problem.

Work Exercises 21–25 in this textbook or in the *MyMathLab*® Study Plan.

## Example 6 Adding and Subtracting Rational Expressions

Perform the indicated operations and simplify.

**a.** $2 + \dfrac{4}{x - 5}$        **b.** $\dfrac{x^2 - 2}{x^2 + 6x + 8} - \dfrac{x - 3}{x + 4}$

**Solutions** Follow the **four-step process**.

**a. Step 1.** Note that the first term is a **rational expression** since we can write it as $2 = \dfrac{2}{1}$. The **LCD** is $x - 5$.

**Step 2.** In the first expression, we multiply the denominator by $x - 5$ to get the LCD and obtain an **equivalent expression** by multiplying both the numerator and denominator by $x - 5$.

We write

$$2 + \frac{4}{x - 5} = \frac{2}{1} + \frac{4}{x - 5}$$

$$= \frac{2}{1} \cdot \frac{x - 5}{x - 5} + \frac{4}{x - 5}$$

$$= \frac{2(x - 5)}{x - 5} + \frac{4}{x - 5}.$$

**Step 3.** With a **common denominator**, we can now add:

$$\text{Add:} \quad = \frac{2(x - 5) + 4}{x - 5}$$

$$\text{Distribute:} \quad = \frac{2x - 10 + 4}{x - 5}$$

$$\text{Simplify:} \quad = \frac{2x - 6}{x - 5}$$

$$= \frac{2(x - 3)}{x - 5}$$

**Step 4.** The expression cannot be simplified further, so

$$2 + \frac{4}{x - 5} = \frac{2(x - 3)}{x - 5}.$$

 ✎ *My video summary*   **b. Step 1.** Factor the denominators and determine the **LCD**. The second denominator is **prime**, but the first denominator can be factored as
$$x^2 + 6x + 8 = (x + 4)(x + 2)$$

The LCD is $(x + 4)(x + 2)$.

**Step 2.** In the first expression, we already have the **LCD**. In the second expression, we multiply the denominator by $x + 2$ to get the LCD and obtain an **equivalent expression** by multiplying both the **numerator** and **denominator** by $x + 2$.

Doing this gives

$$\frac{x^2 - 2}{x^2 + 6x + 8} - \frac{x - 3}{x + 4} = \frac{x^2 - 2}{(x + 4)(x + 2)} - \frac{x - 3}{x + 4} \cdot \frac{x + 2}{x + 2}$$

$$= \frac{x^2 - 2}{(x + 4)(x + 2)} - \frac{(x - 3)(x + 2)}{(x + 4)(x + 2)}.$$

Carry out the subtraction on your own and **simplify** if necessary. View the **answer**, or watch this **video** for a complete solution.

**You Try It**   Work through this You Try It problem.

Work Exercises 26–30 in this textbook or in the **MyMathLab**® Study Plan.

**Example 7**   Adding and Subtracting Rational Expressions with Unlike Denominators

✎ *My interactive video summary*   Perform the indicated operations and simplify.

**a.** $\dfrac{x + 7}{x^2 - 9} + \dfrac{3}{x + 3}$      **b.** $\dfrac{x + 1}{2x^2 + 5x - 3} - \dfrac{x}{2x^2 + 3x - 2}$

**Solutions** Perform the operations on your own and **simplify**. View the **answers**, or watch this **interactive video** to see complete solutions.

**You Try It**   Work through this You Try It problem.

Work Exercises 31–38 in this textbook or in the **MyMathLab**® Study Plan.

If we are adding or subtracting **rational expressions** whose denominators are **opposites**, we can find a **common denominator** by multiplying the numerator and denominator of either rational expression (but not both) by $-1$.

## Example 8 Adding and Subtracting Rational Expressions with Unlike Denominators

Perform the indicated operation and **simplify**.

$$\frac{2y}{y - 5} + \frac{y - 1}{5 - y}$$

**Solution**

The denominators are opposites:  $\dfrac{2y}{y - 5} + \dfrac{y - 1}{5 - y}$

Multiply the numerator and denominator of the second rational expression by $-1$:  $= \dfrac{2y}{y - 5} + \dfrac{(-1)}{(-1)} \cdot \dfrac{y - 1}{5 - y}$

Distribute:  $= \dfrac{2y}{y - 5} + \dfrac{-y + 1}{-5 + y}$

Rewrite the denominator in second expression:  $= \dfrac{2y}{y - 5} + \dfrac{-y + 1}{y - 5}$

Add:  $= \dfrac{2y - y + 1}{y - 5}$

Simplify:  $= \dfrac{y + 1}{y - 5}$

**You Try It** **Work through this You Try It problem.**

**Work Exercises 39–40 in this textbook or in the *MyMathLab*® Study Plan.**

## Example 9 Adding and Subtracting Rational Expressions with Unlike Denominators

Perform the indicated operations and **simplify**.

$$\frac{x + 1}{x^2 - 6x + 9} + \frac{3}{x - 3} - \frac{6}{x^2 - 9}$$

**Solution** Following the **four-step process**, we start by finding the **LCD** for all the terms. $x^2 - 6x + 9 = (x - 3)(x - 3)$; $x - 3$ is prime; $x^2 - 9 = (x + 3)(x - 3)$

The LCD is $(x - 3)(x - 3)(x + 3)$.

Next, we rewrite each term as an **equivalent expression** using the LCD.

$$\frac{x + 1}{x^2 - 6x + 9} + \frac{3}{x - 3} - \frac{6}{x^2 - 9}$$

$$= \frac{x + 1}{(x - 3)(x - 3)} \cdot \frac{x + 3}{x + 3} + \frac{3}{x - 3} \cdot \frac{(x - 3)(x + 3)}{(x - 3)(x + 3)} - \frac{6}{(x + 3)(x - 3)} \cdot \frac{x - 3}{x - 3}$$

$$= \frac{(x + 1)(x + 3)}{(x - 3)(x - 3)(x + 3)} + \frac{3(x - 3)(x + 3)}{(x - 3)(x - 3)(x + 3)} - \frac{6(x - 3)}{(x - 3)(x - 3)(x + 3)}$$

Now we combine the three **numerators** and keep the **common denominator**.

$$= \frac{(x + 1)(x + 3) + 3(x - 3)(x + 3) - 6(x - 3)}{(x - 3)(x - 3)(x + 3)}$$

Finish **simplifying** the expression on your own. View the **answer**, or watch this **video** for a complete solution.

**You Try It**   Work through this You Try It problem.

Work **Exercises 41–47** in this textbook or in the *MyMathLab* ® Study Plan.

# 7.4  Exercises

In Exercises 1–47, add or subtract as indicated.

**1.** $\dfrac{4x}{7} + \dfrac{2x}{7}$

**2.** $\dfrac{8}{7y^3} + \dfrac{6}{7y^3}$

**3.** $\dfrac{a^2 + 1}{5c^2} + \dfrac{b^2 + 2}{5c^2}$

**4.** $\dfrac{2x}{8y^3 + 2} - \dfrac{x}{8y^3 + 2}$

**5.** $\dfrac{5x + 1}{9a^3} - \dfrac{2x + 1}{9a^3}$

**6.** $\dfrac{x}{x + 5} + \dfrac{6}{x + 5}$

**7.** $\dfrac{x + 1}{2x + 3} + \dfrac{2x + 1}{2x + 3}$

**8.** $\dfrac{5x + 1}{3x + 2} + \dfrac{x + 3}{3x + 2}$

**9.** $\dfrac{2x - 3}{3x} - \dfrac{2x + 9}{3x}$

**10.** $\dfrac{7x + 3}{5x - 1} - \dfrac{2x + 5}{5x - 1}$

**11.** $\dfrac{2}{x^2 + 4x + 3} + \dfrac{2x}{x^2 + 4x + 3}$

**12.** $\dfrac{x^2}{2x + 1} + \dfrac{x^2 + x}{2x + 1}$

**13.** $\dfrac{x^2 + 9}{x^2 - 9} + \dfrac{6x}{x^2 - 9}$

**14.** $\dfrac{x}{x^2 - 4} - \dfrac{2}{x^2 - 4}$

**15.** $\dfrac{x^2 + x + 6}{x^2 - x - 6} - \dfrac{3x + 9}{x^2 - x - 6}$

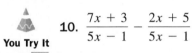

**16.** $\dfrac{x + 5}{5x^2} + \dfrac{3}{10x}$

**17.** $\dfrac{2}{3x} + \dfrac{3}{2x}$

**18.** $\dfrac{5x}{x + 2} + \dfrac{7}{2x + 1}$

**19.** $\dfrac{x + 2}{x - 2} - \dfrac{x - 2}{x + 2}$

**20.** $\dfrac{2x + 1}{x} - \dfrac{2x}{x + 1}$

**21.** $\dfrac{x + 1}{2x} + \dfrac{3x}{5x - 10}$

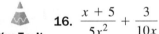

**22.** $\dfrac{2m}{3m + 3} + \dfrac{m + 1}{2m}$

**23.** $\dfrac{x + 2}{2x} - \dfrac{3}{2x + 10}$

**24.** $\dfrac{3}{2x + 2} - \dfrac{2}{3x + 3}$

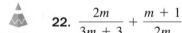

**25.** $\dfrac{x}{x^2 - 1} - \dfrac{1}{x}$

**26.** $\dfrac{3x}{x + 1} + 2$

**27.** $\dfrac{2(x + 11)}{2x^2 - x - 10} + \dfrac{x + 4}{x + 2}$

**28.** $\dfrac{3x}{x + 1} - 2$

**29.** $\dfrac{x^2 - 3}{x^2 - 5x + 6} - \dfrac{x - 3}{x - 2}$

**30.** $\dfrac{x}{2} - \dfrac{2}{x}$

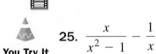

**31.** $\dfrac{x - 1}{x^2 - 4} + \dfrac{5}{x - 2}$

**32.** $\dfrac{4}{x^2 - 4} + \dfrac{2x}{x + 2}$

**33.** $\dfrac{1 - 12x}{2x^2 - 5x - 3} + \dfrac{x + 2}{x - 3}$

**34.** $\dfrac{3x - 4}{x + 3} - \dfrac{6 - 24x}{x^2 - 9}$

**35.** $\dfrac{2x}{x^2 - x - 2} - \dfrac{x}{x^2 - 1}$

**36.** $\dfrac{2x - 1}{x + 1} + \dfrac{3(3x + 1)}{x^2 - 1}$

**37.** $\dfrac{8}{x - 2} - \dfrac{3x + 26}{x^2 - 4}$

**38.** $\dfrac{6x}{2x^2 - 8xy + 4y^2} - \dfrac{3y}{x^2 - 3xy + 2y^2}$

**You Try It**

**39.** $\dfrac{3x}{x-1} + \dfrac{2x+3}{1-x}$

**40.** $\dfrac{x^2}{x-3} + \dfrac{9}{3-x}$

**41.** $\dfrac{1}{x} + \dfrac{1}{x+1} - \dfrac{1}{x-1}$

**You Try It**

**42.** $\dfrac{1}{x^2+x} + \dfrac{1}{x^2-x} + \dfrac{1}{x^2-1}$

**43.** $\dfrac{x-3}{x^2+12x+36} + \dfrac{1}{x+6} - \dfrac{2x+3}{2x^2+3x-54}$

**44.** $\dfrac{1}{x+1} + \dfrac{2}{x+2} - \dfrac{3}{x+3}$

**45.** $\dfrac{1}{x+2} + \dfrac{1}{x-2} - \dfrac{4}{x^2-4}$

**46.** $\dfrac{5}{y} - \dfrac{5}{y-3} + \dfrac{16}{(y-3)^2}$

**47.** $\left(\dfrac{1}{2} + \dfrac{3}{x}\right) - \left(\dfrac{1}{2} - \dfrac{1}{x}\right)$

# 7.5 Complex Rational Expressions

## THINGS TO KNOW

Before working through this section, be sure you are familiar with the following concepts:

| | | VIDEO | ANIMATION | INTERACTIVE |
|---|---|---|---|---|

**You Try It**
1. Use the Negative Power Rule (Section 5.6, **Objective 1**)   

**You Try It**
2. Simplify Rational Expressions (Section 7.1, **Objective 3**)   

**You Try It**
3. Multiply Rational Expressions (Section 7.2, **Objective 1**)       

**You Try It**
4. Divide Rational Expressions (Section 7.2, **Objective 2**)   

**You Try It**
5. Find the Least Common Denominator of Rational Expressions (Section 7.3, **Objective 1**)

**You Try It**
6. Add and Subtract Rational Expressions with Unlike Denominators (Section 7.4, **Objective 2**)

## OBJECTIVES

**1** Simplify Complex Rational Expressions by First Simplifying the Numerator and Denominator

**2** Simplify Complex Rational Expressions by Multiplying by a Common Denominator

OBJECTIVE 1   **SIMPLIFY COMPLEX RATIONAL EXPRESSIONS BY FIRST SIMPLIFYING THE NUMERATOR AND DENOMINATOR**

Sometimes the **numerator** and/or **denominator** of a **rational expression** will contain one or more rational expressions. In this section, we learn to simplify such *complex rational expressions*.

---

**Definition   Complex Rational Expression**

A **complex rational expression**, or **complex fraction**, is a rational expression in which the numerator and/or denominator contain(s) rational expressions.

---

Some examples of **complex rational expressions** are

$$\frac{\dfrac{2}{9x}}{\dfrac{5}{6xy}}, \quad \frac{\dfrac{x+2}{x-4}}{\dfrac{x^2-4}{x+1}}, \quad \frac{\dfrac{1}{y}+\dfrac{y+1}{y-1}}{\dfrac{1}{y}-\dfrac{y-1}{y+1}}, \quad \frac{\dfrac{1}{x}+\dfrac{1}{y}}{z}, \quad \text{and} \quad \frac{z^{-2}+z^{-1}}{z^{-1}-z^{-2}}.$$

The rational expressions within the numerator and denominator are called **minor rational expressions** or **minor fractions**. The numerator and denominator of the complex rational expression are separated by the **main fraction bar**.

$$\text{Minor rational expressions}\underset{\searrow}{\overset{\nearrow}{}}\quad \frac{\left.\dfrac{x+2}{x-4}\right\}}{\left.\dfrac{x^2-4}{x+1}\right\}}$$

← Numerator of the complex rational expression
← Main fraction bar
← Denominator of the complex rational expression

To **simplify a complex rational expression**, we rewrite it in the form $\dfrac{P}{Q}$, $Q \neq 0$, where $P$ and $Q$ are **polynomials** with no **common factors**. In this section, we show two methods that can be used to **simplify complex rational expressions**. We call these *Method I* and *Method II*. Method I results from recognizing that the **main fraction bar** is a division symbol. So, we divide the **numerator** by the **denominator**.

### Example 1  Simplifying a Complex Rational Expression

Simplify $\dfrac{\dfrac{2}{9x}}{\dfrac{5}{6xy}}$.

**Solution**  We divide the **minor rational expression** from the **numerator** by the minor rational expression from the **denominator**. To do this, we multiply the numerator by the **reciprocal** of the denominator.

Begin with the original expression:
$$\frac{\dfrac{2}{9x}}{\dfrac{5}{6xy}} = \frac{2}{9x} \div \frac{5}{6xy}$$

Change to multiplication by the reciprocal:
$$= \frac{2}{9x} \cdot \frac{6xy}{5}$$

Factor:
$$= \frac{2}{3\cdot 3\cdot x} \cdot \frac{2\cdot 3\cdot x\cdot y}{5}$$

Divide out common factors:
$$= \frac{2}{\cancel{3}\cdot 3\cdot \cancel{x}} \cdot \frac{2\cdot \cancel{3}\cdot \cancel{x}\cdot y}{5}$$

Multiply the remaining factors:
$$= \frac{4y}{15}$$

**You Try It**  Work through this You Try It problem.

Work Exercises 1–3 in this textbook or in the *MyMathLab* Study Plan.

In addition to acting as a division symbol, the **main fraction bar** of a **complex fraction** also serves as a **grouping symbol**. If the **numerator** and **denominator** are not each written as single **rational expressions**, then they must be simplified as such before dividing.

---

**Method I for Simplifying Complex Rational Expressions**

**Step 1.**  Simplify the expression in the numerator into a single rational expression.

**Step 2.**  Simplify the expression in the denominator into a single rational expression.

**Step 3.**  Divide the expression in the numerator by the expression in the denominator. To do this, multiply the **minor rational expression** in the numerator by the **reciprocal** of the minor rational expression in the denominator. Simplify if possible.

---

## Example 2  Simplifying a Complex Rational Expression Using Method I

Use **Method I** to simplify each complex rational expression.

a. $\dfrac{\dfrac{1}{3} - \dfrac{1}{x}}{\dfrac{1}{9} - \dfrac{1}{x^2}}$

b. $\dfrac{4 - \dfrac{5}{x-1}}{\dfrac{6}{x-1} - 7}$

## Solutions

a. **Step 1.**  The **LCD** for the **minor fractions** $\dfrac{1}{3}$ and $-\dfrac{1}{x}$ in the **numerator** is $3x$.

$$\frac{1}{3} - \frac{1}{x} = \frac{1}{3} \cdot \frac{x}{x} - \frac{1}{x} \cdot \frac{3}{3} = \frac{x}{3x} - \frac{3}{3x} = \frac{x-3}{3x}$$

**Step 2.**  The LCD for the minor fractions $\dfrac{1}{9}$ and $-\dfrac{1}{x^2}$ in the **denominator** is $9x^2$.

$$\frac{1}{9} - \frac{1}{x^2} = \frac{1}{9} \cdot \frac{x^2}{x^2} - \frac{1}{x^2} \cdot \frac{9}{9} = \frac{x^2}{9x^2} - \frac{9}{9x^2} = \frac{x^2 - 9}{9x^2}$$

**Step 3.**  Substitute the simplified **rational expressions** from Steps 1 and 2 for the numerator and denominator of the **complex fraction**, and then divide.

$$\frac{\dfrac{1}{3} - \dfrac{1}{x}}{\dfrac{1}{9} - \dfrac{1}{x^2}} = \frac{\dfrac{x-3}{3x}}{\dfrac{x^2-9}{9x^2}} = \frac{x-3}{3x} \div \frac{x^2-9}{9x^2}$$

Change to multiplication by the reciprocal:  $= \dfrac{x-3}{3x} \cdot \dfrac{9x^2}{x^2-9}$

Factor:  $= \dfrac{x-3}{3x} \cdot \dfrac{3 \cdot 3 \cdot x \cdot x}{(x+3)(x-3)}$

Divide out common factors: $= \dfrac{\cancel{x-3}}{3\cancel{x}} \cdot \dfrac{3 \cdot 3 \cdot \cancel{x} \cdot x}{(x+3)\cancel{(x-3)}}$

Multiply remaining factors: $= \dfrac{3x}{x+3}$

*My video summary*  **b. Step 1.** The **LCD** for the **numerator** is $x - 1$.

$$4 - \frac{5}{x-1} = 4 \cdot \frac{x-1}{x-1} - \frac{5}{x-1} = \frac{4(x-1)-5}{x-1} = \frac{4x-4-5}{x-1} = \frac{4x-9}{x-1}$$

**Step 2.** The LCD for the **denominator** is $x - 1$.

$$\frac{6}{x-1} - 7 = \frac{6}{x-1} - 7 \cdot \frac{x-1}{x-1} = \frac{6 - 7(x-1)}{x-1} = \frac{6 - 7x + 7}{x-1} = \frac{13 - 7x}{x-1}$$

**Step 3.** Try to finish simplifying the complex rational expression on your own. View the answer, or watch this video for the complete solution to part b.

**You Try It** **Work through this You Try It problem.**

**Work Exercises 4–6 in this textbook or in the *MyMathLab*® Study Plan.**

OBJECTIVE 2 SIMPLIFY COMPLEX RATIONAL EXPRESSIONS BY MULTIPLYING BY A COMMON DENOMINATOR

In **Section 7.3**, we saw that multiplying the **numerator** and **denominator** of a **rational expression** by the same nonzero **expression** resulted in an equivalent expression. For example, $\dfrac{3}{4} = \dfrac{3}{4} \cdot \dfrac{7}{7} = \dfrac{21}{28}$ because $\dfrac{7}{7} = 1$.

The second method for simplifying **complex rational expressions** is based on this same concept. Let's take another look at the complex fraction from **Example 1**.

### Example 3 Simplifying a Complex Rational Expression

Simplify $\dfrac{\dfrac{2}{9x}}{\dfrac{5}{6xy}}$.

**Solution** Let's multiply the numerator and denominator of the complex rational expression by $18xy$, the **LCD** of $\dfrac{2}{9x}$ and $\dfrac{5}{6xy}$.

Begin with the original expression: $\dfrac{\dfrac{2}{9x}}{\dfrac{5}{6xy}}$

Multiply the numerator and denominator by $18xy$: $= \dfrac{\dfrac{2}{9x} \cdot 18xy}{\dfrac{5}{6xy} \cdot 18xy}$

Divide out common factors: $\quad = \dfrac{\dfrac{2}{\cancel{9x}}\cdot \overset{2}{\cancel{18xy}}}{\dfrac{5}{\cancel{6xy}}\cdot \overset{3}{\cancel{18xy}}}$

Rewrite the remaining factors: $\quad = \dfrac{2\cdot 2\cdot y}{5\cdot 3}$

Multiply: $\quad = \dfrac{4y}{15}$

**You Try It**  **Work through this You Try It problem.**

**Work Exercises 7–9 in this textbook or in the *MyMathLab*® Study Plan.**

In **Example 3**, we multiplied the **numerator** and **denominator** of the **complex rational expression** by the **LCD** of its **minor rational expressions**. Multiplying by the LCD is the foundation of Method II for simplifying complex rational expression.

---

**Method II for Simplifying Complex Rational Expressions**

**Step 1.**  Determine the LCD of all the minor rational expressions within the complex rational expression.

**Step 2.**  Multiply the numerator and denominator of the complex rational expression by the LCD from Step 1.

**Step 3.**  Simplify.

---

For comparison, let's revisit the complex rational expressions from **Example 2**.

### Example 4  Simplifying a Complex Rational Expression Using Method II

Use **Method II** to simplify each complex rational expression.

a. $\dfrac{\dfrac{1}{3}-\dfrac{1}{x}}{\dfrac{1}{9}-\dfrac{1}{x^2}}$

b. $\dfrac{4-\dfrac{5}{x-1}}{\dfrac{6}{x-1}-7}$

### Solutions

a. **Step 1.**  The **denominators** of all the **minor fractions** are $3, x, 9,$ and $x^2$, so the **LCD** is $9x^2$.

**Step 2.**  Multiply the **numerator** and denominator by $9x^2$.

$$\dfrac{\dfrac{1}{3}-\dfrac{1}{x}}{\dfrac{1}{9}-\dfrac{1}{x^2}}\cdot \dfrac{9x^2}{9x^2}$$

**Step 3.**  Use the distributive property: $\quad = \dfrac{\dfrac{1}{3}\cdot 9x^2-\dfrac{1}{x}\cdot 9x^2}{\dfrac{1}{9}\cdot 9x^2-\dfrac{1}{x^2}\cdot 9x^2}$

Divide out common factors:   $= \dfrac{\dfrac{1}{\cancel{3}} \cdot \overset{3}{\cancel{9}}x^2 - \dfrac{1}{\cancel{x}} \cdot 9x^{\overset{x}{\cancel{2}}}}{\dfrac{1}{\cancel{9}} \cdot 9x^2 - \dfrac{1}{\cancel{x^2}} \cdot 9\cancel{x^2}}$

Simplify:   $= \dfrac{3x^2 - 9x}{x^2 - 9}$

Factor each polynomial:   $= \dfrac{3x(x-3)}{(x+3)(x-3)}$

Divide out common factors:   $= \dfrac{3x\cancel{(x-3)}}{(x+3)\cancel{(x-3)}}$

Simplify:   $= \dfrac{3x}{x+3}$

 *My video summary*    **b. Step 1.** The **LCD** for all the **minor fractions** is $x - 1$.

**Step 2.** Multiply the **numerator** and **denominator** by $x - 1$.

$$\dfrac{4 - \dfrac{5}{x-1}}{\dfrac{6}{x-1} - 7} \cdot \dfrac{x-1}{x-1}$$

**Step 3.** Try to finish simplifying the complex rational expression on your own. View the **answer**, or watch this **video** for the complete solution.

**You Try It**   **Work through this You Try It problem.**

**Work Exercises 10–12 in this textbook or in the** ***MyMathLab*** **Study Plan.**

As you have seen in Examples 1–4, **Method I** and **Method II** both give the same simplification. Have you developed a preference for one method over the other? Whether you have or not, we suggest that you practice using both methods for a while to discover when one method might be better suited than the other.

### Example 5  Simplifying a Complex Rational Expression

*My interactive video summary*   Simplify the **complex rational expression** using Method I or Method II.

$$\dfrac{\dfrac{5}{n-2} - \dfrac{3}{n}}{\dfrac{6}{n^2 - 2n} + \dfrac{2}{n}}$$

**Solution**  Try simplifying on your own. View the **answer**, or watch this **interactive video** to see the complete solution.

**You Try It**   **Work through this You Try It problem.**

**Work Exercises 13–24 in this textbook or in the** ***MyMathLab*** **Study Plan.**

**Complex rational expressions** can be written using **negative exponents**. For example, the

expression $\dfrac{3^{-1} - x^{-1}}{3^{-2} - x^{-2}}$ is equivalent to the complex fraction $\dfrac{\dfrac{1}{3} - \dfrac{1}{x}}{\dfrac{1}{9} - \dfrac{1}{x^2}}$ from **Example 2**.

When a complex rational expression contains negative exponents, we can rewrite it as an equivalent expression with positive exponents by using the **negative power rule**. Then we simplify it by using Method I or Method II.

### Example 6  Simplifying a Complex Rational Expression Containing Negative Exponents

Simplify the complex rational expression.

$$\frac{1 - 9y^{-1} + 14y^{-2}}{1 + 3y^{-1} - 10y^{-2}}$$

**Solution**  Use the **negative power rule** to rewrite the expression with positive **exponents**.

$$\frac{1 - 9y^{-1} + 14y^{-2}}{1 + 3y^{-1} - 10y^{-2}} = \frac{1 - \dfrac{9}{y} + \dfrac{14}{y^2}}{1 + \dfrac{3}{y} - \dfrac{10}{y^2}}$$

Try to finish simplifying the **complex rational expression** on your own using **Method I** or **Method II**. View the **answer**, or watch this **interactive video** for the complete solution.

**You Try It**   **Work through this You Try It problem.**

**Work Exercises 25 and 26 in this textbook or in the *MyMathLab*®  Study Plan.**

## 7.5  Exercises

In Exercises 1–6, use Method I to simplify each complex rational expression.

**1.** $\dfrac{\dfrac{14}{5x}}{\dfrac{7}{15x}}$

**2.** $\dfrac{\dfrac{x + 7}{2}}{\dfrac{3x - 1}{10}}$

**3.** $\dfrac{\dfrac{5x + 5}{x^2 - 9}}{\dfrac{x^2 - 1}{x + 3}}$

**4.** $\dfrac{\dfrac{5}{6} + \dfrac{3}{4}}{\dfrac{8}{3} - \dfrac{5}{9}}$

**5.** $\dfrac{\dfrac{3}{x} + 2}{\dfrac{9}{x^2} - 4}$

**6.** $\dfrac{\dfrac{9}{x - 4} + \dfrac{6}{x - 5}}{\dfrac{5x - 23}{x^2 - 9x + 20}}$

In Exercises 7–12, use Method II to simplify each complex rational expression.

**7.** $\dfrac{\dfrac{x + 2}{15x}}{\dfrac{2x - 1}{10x}}$

**8.** $\dfrac{\dfrac{8x}{x + y}}{\dfrac{2x^3}{y}}$

**9.** $\dfrac{\dfrac{3}{x^2 - x - 2}}{\dfrac{x + 4}{x - 2}}$

**10.** $\dfrac{2 - \dfrac{a}{b}}{\dfrac{a^2}{b^2} - 4}$

**11.** $\dfrac{3 - \dfrac{7}{x-2}}{\dfrac{8}{x-2} - 5}$

**12.** $\dfrac{\dfrac{x+5}{x} - \dfrac{8}{x-1}}{\dfrac{x+1}{x} + \dfrac{x+1}{x-1}}$

In Exercises 13–26, simplify each complex rational expression using Method I or Method II.

**13.** $\dfrac{7 + \dfrac{1}{x}}{7 - \dfrac{1}{x}}$

**14.** $\dfrac{\dfrac{1}{x} + 3}{\dfrac{1}{x^2} - 9}$

**15.** $\dfrac{\dfrac{x+2}{x-6} - \dfrac{x+12}{x+5}}{x + 82}$

**16.** $\dfrac{\dfrac{6}{x} + \dfrac{5}{y}}{\dfrac{5}{x} - \dfrac{6}{y}}$

**17.** $\dfrac{\dfrac{1}{16} - \dfrac{1}{x^2}}{\dfrac{1}{4} + \dfrac{1}{x}}$

**18.** $\dfrac{\dfrac{8}{x+4} - \dfrac{2}{x+7}}{\dfrac{x+8}{x+4}}$

**19.** $\dfrac{\dfrac{2}{x+13} + \dfrac{1}{x-5}}{2 - \dfrac{x+25}{x+13}}$

**20.** $\dfrac{\dfrac{x-5}{x+5} + \dfrac{x-5}{x-9}}{1 + \dfrac{x+5}{x-9}}$

**21.** $\dfrac{1 + \dfrac{4}{x} - \dfrac{5}{x^2}}{1 - \dfrac{2}{x} - \dfrac{35}{x^2}}$

**22.** $\dfrac{\dfrac{x-4}{x^2-25}}{1 + \dfrac{1}{x-5}}$

**23.** $\dfrac{\dfrac{x+3}{x-3} - \dfrac{x-3}{x+3}}{\dfrac{x-3}{x+3} + \dfrac{x+3}{x-3}}$

**24.** $\dfrac{\dfrac{3}{a^2} - \dfrac{1}{ab} - \dfrac{2}{b^2}}{\dfrac{2}{a^2} - \dfrac{5}{ab} + \dfrac{3}{b^2}}$

**25.** $\dfrac{6x^{-1} + 6y^{-1}}{xy^{-1} - x^{-1}y}$

**26.** $\dfrac{7x^{-1} - 3y^{-1}}{49x^{-2} - 9y^{-2}}$

# 7.6 Solving Rational Equations

## THINGS TO KNOW

Before working through this section, be sure you are familiar with the following concepts:

| VIDEO | ANIMATION | INTERACTIVE |
|---|---|---|

**1.** Solve Linear Equations Containing Fractions (Section 2.2, **Objective 2**)

**2.** Solve a Formula for a Given Variable (Section 2.4, **Objective 3**)

**3.** Factor Polynomials Completely (Section 6.6, **Objective 1**)

**4.** Solve Polynomial Equations by Factoring (Section 6.7, **Objective 2**)

**You Try It**

5. Evaluate Rational Expressions (Section 7.1, **Objective 1**)

**You Try It**

6. Find Restricted Values for Rational Expressions (Section 7.1, **Objective 2**)

**You Try It**

7. Find the Least Common Denominator of Rational Expressions (Section 7.3, **Objective 1**)

## OBJECTIVES

**1** Identify Rational Equations

**2** Solve Rational Equations

**3** Identify and Solve Proportions

**4** Solve a Formula Containing Rational Expressions for a Given Variable

OBJECTIVE 1   IDENTIFY RATIONAL EQUATIONS

As we have seen before, we make a clear distinction between an **equation** and an **expression**. Recall that we *simplify* expressions and *solve* equations.

We have previously defined **linear equations** and **polynomial equations**. Now, let's define a *rational equation*.

---

**Definition   Rational Equation**

A **rational equation** is a statement in which two **rational expressions** are set equal to each other.

---

Some examples of rational equations are

$$\underbrace{\frac{2x}{x-1}}_{\substack{\text{Rational}\\\text{expression}}} \underbrace{=}_{\substack{\text{Equal}\\\text{sign}}} \underbrace{\frac{2x+3}{x}}_{\substack{\text{Rational}\\\text{expression}}} \quad \text{and} \quad \underbrace{\frac{1}{x-2}+\frac{1}{x+2}}_{\substack{\text{Rational}\\\text{expression}}} \underbrace{=}_{\substack{\text{Equal}\\\text{sign}}} \underbrace{\frac{4}{x^2-4}}_{\substack{\text{Rational}\\\text{expression}}}.$$

### Example 1  Identifying Rational Equations

Determine if each statement is a rational equation. If not, state why.

a. $\dfrac{x-4}{x}+\dfrac{4}{x+5}=\dfrac{6}{x}$

b. $\dfrac{5}{y}+\dfrac{7}{y+2}$

c. $\dfrac{\sqrt{k+1}}{k+3}=\dfrac{k-5}{k+4}$

d. $5n^{-1}=3n^{-2}$

### Solutions

a. Both $\dfrac{x-4}{x}+\dfrac{4}{x+5}$ and $\dfrac{6}{x}$ are **rational expressions**. They are set equal to each other, so the statement is a **rational equation**.

b. This statement $\dfrac{5}{y}+\dfrac{7}{y+2}$ does not contain an equal sign, so it is not a rational equation. Instead, it is a rational *expression*.

c. This statement is not a rational equation because $\dfrac{\sqrt{k+1}}{k+3}$ is not a rational expression.
Note that the **numerator** is not a **polynomial** because there is a **variable** under a **radical**.

d. Note that $5n^{-1} = \dfrac{5}{n}$ and $3n^{-2} = \dfrac{3}{n^2}$, so they are both rational expressions. They are set
equal to each other, so this statement is a rational equation.

**You Try It**    Work through this You Try It problem.

Work Exercises 1–4 in this textbook or in the **MyMathLab** Study Plan.

## OBJECTIVE 2    SOLVE RATIONAL EQUATIONS

In **Section 2.2**, we learned to **clear fractions** from **equations** by multiplying both sides of the equation by the **LCD**. We then solved the resulting equation. This idea is the key to solving **rational equations**.

### Example 2  Solving Rational Equations

Solve.

a. $\dfrac{1}{2}x + \dfrac{2}{3} = \dfrac{3}{4}$

b. $\dfrac{1}{x} + \dfrac{1}{2} = \dfrac{1}{3}$

**Solutions**

a. The LCD of the fractions $\dfrac{1}{2}, \dfrac{2}{3},$ and $\dfrac{3}{4}$ is 12.

Write the original equation:    $\dfrac{1}{2}x + \dfrac{2}{3} = \dfrac{3}{4}$

Multiply both sides by the LCD:    $12\left(\dfrac{1}{2}x + \dfrac{2}{3}\right) = 12\left(\dfrac{3}{4}\right)$

Distribute:    $12\left(\dfrac{1}{2}x\right) + 12\left(\dfrac{2}{3}\right) = 12\left(\dfrac{3}{4}\right)$

Simplify:    $6x + 8 = 9$

Subtract 8 from both sides:    $6x = 1$

Divide both sides by 6:    $x = \dfrac{1}{6}$

View this **popup box** to see that this answer checks. The **solution set** is $\left\{\dfrac{1}{6}\right\}$.

*My video summary*    b. The **LCD** of $\dfrac{1}{x}, \dfrac{1}{2},$ and $\dfrac{1}{3}$ is $6x$.

Write the original equation:    $\dfrac{1}{x} + \dfrac{1}{2} = \dfrac{1}{3}$

Multiply both sides by the LCD:    $6x\left(\dfrac{1}{x} + \dfrac{1}{2}\right) = 6x\left(\dfrac{1}{3}\right)$

Distribute:    $6x \cdot \dfrac{1}{x} + 6x \cdot \dfrac{1}{2} = 6x \cdot \dfrac{1}{3}$

Divide out common factors:   $6x \cdot \dfrac{1}{\overset{}{\cancel{x}}} + \overset{3}{\cancel{6}}x \cdot \dfrac{1}{\underset{1}{\cancel{2}}} = \overset{2}{\cancel{6}}x \cdot \dfrac{1}{\underset{1}{\cancel{3}}}$

Simplify:            $6 + 3x = 2x$

Finish solving this equation on your own. Remember to check your result. View the **answer**, or watch this **video** for a complete solution to part b.

**You Try It**    **Work through this You Try It problem.**

**Work Exercises 5–8 in this textbook or in the** *MyMathLab*® **Study Plan.**

Recall that values that cause the **denominator** of a **rational expression** to equal zero are **restricted values**. Restricted values cannot be **solutions** of **rational equations**. However, sometimes when we multiply both sides of a rational equation by the **LCD**, a solution of the resulting **linear** or **polynomial equation** will be a restricted value of the original rational equation. Such "solutions" are called **extraneous solutions**, and they must be excluded from the **solution set**.

For example, let's solve the equation $\dfrac{x+1}{x-2} = \dfrac{3}{x-2}$. Notice that $x = 2$ is a restricted value for these rational expressions. Also, the LCD of the rational expressions is $x - 2$.

Multiply both sides by the LCD:   $(x-2)\left(\dfrac{x+1}{x-2}\right) = (x-2)\left(\dfrac{3}{x-2}\right)$

Divide out common factors:   $\cancel{(x-2)} \cdot \dfrac{x+1}{\cancel{(x-2)}} = \cancel{(x-2)} \cdot \dfrac{3}{\cancel{(x-2)}}$

Simplify:            $x + 1 = 3$

Subtract 1 from both sides:      $x = 2$

The resulting solution is the restricted value, which means it does not make the original equation true. So, we must discard this result as an extraneous solution. This means that the equation $\dfrac{x+1}{x-2} = \dfrac{3}{x-2}$ has no solution.

The following steps can be used to solve rational equations.

---

**Solving Rational Equations**

**Step 1.**  List all **restricted values**.

**Step 2.**  Determine the **LCD** of all **denominators** in the equation.

**Step 3.**  Multiply both sides of the equation by the LCD.

**Step 4.**  Solve the resulting **polynomial equation**.

**Step 5.**  Discard any restricted values and check the remaining **solutions** in the original equation.

---

### Example 3  Solving Rational Equations

Solve $\dfrac{2}{x} - \dfrac{x-3}{2x} = 3$.

**Solution** We follow the **five-step process.**

**Step 1.** To find the **restricted values**, we look for values that make any **denominator** equal to zero.

$$\frac{2}{x} - \frac{x-3}{2x} = \frac{3}{1}$$

Examining the variable factors in the denominators, we see that the only restricted value is $0$.

**Step 2.** There are unique **factors** of $2$ and $x$ in the denominators, each occurring at most once. The **LCD** is $2x$.

**Step 3.** Multiply both sides of the equation by the LCD.

Multiply both sides by the LCD:  $2x\left(\dfrac{2}{x} - \dfrac{x-3}{2x}\right) = 2x(3)$

Distribute:  $2x \cdot \dfrac{2}{x} - 2x \cdot \dfrac{x-3}{2x} = 2x \cdot 3$

Divide out common factors:  $2\!\!\!/x \cdot \dfrac{2}{\!\!\!/x} - 2\!\!\!/x \cdot \dfrac{x-3}{2\!\!\!/x} = 2x \cdot 3$

Multiply:  $4 - (x-3) = 6x$

**Step 4.** Solve the resulting equation.

Distribute:  $4 - x + 3 = 6x$

Simplify:  $7 - x = 6x$

Add $x$ to both sides:  $7 = 7x$

Divide both sides by 7:  $1 = x$

**Step 5.** The potential solution $x = 1$ is not a **restricted value**, so we do not discard it. View this **popup box** to see the check of $x = 1$ in the original equation. The **solution set** is $\{1\}$.

**You Try It**    Work through this **You Try It problem.**

Work Exercises 9 and 10 in this textbook or in the *MyMathLab*® Study Plan.

**Example 4  Solving Rational Equations**

 Solve $\dfrac{4}{5} - \dfrac{3}{x-3} = \dfrac{1}{x}$.

**Solution** We follow the **five-step process.**

**Step 1.** To find the **restricted values**, we look for values that make any **denominator** equal to zero. Set each denominator with a **variable** equal to zero and solve.

$$x - 3 = 0 \quad \text{or} \quad x = 0$$
$$x = 3$$

The restricted values are $0$ and $3$.

**Step 2.** There are unique **factors** of $5$, $x$, and $x - 3$ in the denominators, each occurring at most once. The **LCD** is $5x(x-3)$.

**Step 3.** Multiply both sides of the equation by the LCD.

Multiply both sides by the LCD:
$$5x(x-3)\left(\frac{4}{5}-\frac{3}{x-3}\right)=5x(x-3)\left(\frac{1}{x}\right)$$

Distribute:
$$5x(x-3)\cdot\frac{4}{5}-5x(x-3)\cdot\frac{3}{x-3}=5x(x-3)\cdot\frac{1}{x}$$

Divide out common factors:
$$\cancel{5}x(x-3)\cdot\frac{4}{\cancel{5}}-5x\cancel{(x-3)}\cdot\frac{3}{\cancel{x-3}}=5\cancel{x}(x-3)\cdot\frac{1}{\cancel{x}}$$

Multiply:
$$4x(x-3)-15x=5(x-3)$$

Try to finish solving the equation on your own. View the **answer**, or watch this **video** for a complete solution.

**You Try It**  Work through this **You Try It** problem.

**Work Exercises 11 and 12 in this textbook or in the *MyMathLab*®️ Study Plan.**

## Example 5  Solving Rational Equations

 Solve $\dfrac{m}{m+2}+\dfrac{5}{m-2}=\dfrac{20}{m^2-4}$.

**Solution**  We follow the **five-step process**.

**Step 1.** To find the **restricted values**, look for values that make any denominator equal to zero. First, factor the denominators.

$$\frac{m}{m+2}+\frac{5}{m-2}=\frac{20}{(m-2)(m+2)}$$

Now set each **variable factor** equal to zero and solve.

$$m+2=0 \quad \text{or} \quad m-2=0$$
$$m=-2 \qquad\qquad m=2$$

The restricted values are $-2$ and $2$.

**Step 2.** We have unique **factors** of $m-2$ and $m+2$ in the denominators, each occurring at most once. The **LCD** is $(m-2)(m+2)$.

**Step 3.** Multiply both sides of the equation by the **LCD** and **divide out** common factors.

$$(m-2)(m+2)\left(\frac{m}{m+2}+\frac{5}{m-2}\right)=(m-2)(m+2)\cdot\frac{20}{(m-2)(m+2)}$$

$$(m-2)(m+2)\cdot\frac{m}{m+2}+(m-2)(m+2)\cdot\frac{5}{m-2}=(m-2)(m+2)\cdot\frac{20}{(m-2)(m+2)}$$

$$(m-2)\cancel{(m+2)}\cdot\frac{m}{\cancel{m+2}}+\cancel{(m-2)}(m+2)\cdot\frac{5}{\cancel{m-2}}=\cancel{(m-2)}\cancel{(m+2)}\cdot\frac{20}{\cancel{(m-2)}\cancel{(m+2)}}$$

$$m(m-2)+5(m+2)=20$$

Finish solving the equation on your own, making sure to check for **restricted values**. View the **answer**, or watch this **video** for a complete solution.

**You Try It**  Work through this **You Try It** problem.

**Work Exercises 13 and 14 in this textbook or in the *MyMathLab*®️ Study Plan.**

### Example 6  Solving Rational Equations

*My video summary*   Solve $\dfrac{2}{x-3} - \dfrac{4}{x^2 - 2x - 3} = \dfrac{1}{x+1}$.

**Solution**  Following the **five-step process**, we start by finding the **restricted values**. Factoring the denominators gives

$$\frac{2}{x-3} - \frac{4}{(x-3)(x+1)} = \frac{1}{x+1}.$$

Set each **variable factor** in the denominators equal to zero and solve.

$$x - 3 = 0 \quad \text{or} \quad x + 1 = 0$$
$$x = 3 \qquad\qquad x = -1$$

The restricted values are $-1$ and $3$.

The **LCD** is $(x-3)(x+1)$. Multiplying both sides of the equation by the LCD gives the **polynomial equation**

$$2(x+1) - 4 = x - 3.$$

Solving the polynomial equation leads to $x = -1$. However, $x = -1$ is a restricted value, so it must be discarded. Since no other possible **solution** exists, the equation has no solution. The **solution set** is { } or $\varnothing$. Watch this **video** for a fully worked solution.

**You Try It**  Work through this **You Try It** problem.

**Work Exercises 15 and 16 in this textbook or in the *MyMathLab*® Study Plan.**

---

OBJECTIVE 3   IDENTIFY AND SOLVE PROPORTIONS

A **ratio** is the **quotient** of two numbers or **algebraic expressions**, such as with **slope** $\left(\dfrac{\text{rise}}{\text{run}}\right)$.

A **rational number** is the quotient of two **integers**, so rational numbers are ratios. Likewise, a **rational expression** is the quotient of two **polynomials**, so rational expressions are ratios.

A **proportion** is a statement that two ratios are equal, such as $\dfrac{P}{Q} = \dfrac{R}{S}$. When a **rational equation** is a proportion, we can solve it by first **cross multiplying** to get $P \cdot S = Q \cdot R$ and then solving the resulting **polynomial equation**.

Cross Multiplying

$$\frac{P}{Q} \diagdown\!\!\!\!\diagup \frac{R}{S}$$

$$P \cdot S = Q \cdot R$$

Note that any "solution" that is a **restricted value** in the original equation must be discarded. Proportions can be used to solve many types of applications, as we will see later in Section 7.7.

### Example 7  Solving Proportions

Solve.

**a.** $\dfrac{8}{x+3} = \dfrac{5}{x}$   **b.** $\dfrac{x}{6} = \dfrac{2}{x-1}$

**Solutions**

a. This **rational equation** has the form $\dfrac{P}{Q} = \dfrac{R}{S}$, so it is a **proportion**.

First, we identify the **restricted values** by setting each denominator equal to zero and solving.

$$x + 3 = 0 \quad \text{or} \quad x = 0$$
$$x = -3$$

We have two restricted values, $-3$ and $0$.

Next, we **cross multiply** and solve the resulting **polynomial equation**.

| | |
|---|---|
| Begin with the original equation: | $\dfrac{8}{x + 3} = \dfrac{5}{x}$ |
| Cross multiply: | $8x = 5(x + 3)$ |
| Distribute: | $8x = 5x + 15$ |
| Subtract $5x$ from both sides: | $3x = 15$ |
| Divide both sides by 3: | $x = 5$ |

Because 5 is not a restricted value, the solution set is $\{5\}$. View this **popup box** to see that this result checks.

 b. This **rational equation** has the form $\dfrac{P}{Q} = \dfrac{R}{S}$, so it is a **proportion**.

The denominator $x - 1$ is the only one that contains a variable, so the only **restricted value** is 1.

| | |
|---|---|
| Begin with the original equation: | $\dfrac{x}{6} = \dfrac{2}{x - 1}$ |
| Cross multiply: | $x(x - 1) = 6(2)$ |

Try to solve this resulting equation on your own. Be sure to check for **extraneous solutions**. View the **answer**, or watch this **video** for a complete solution.

**You Try It** **Work through this You Try It problem.**

**Work Exercises 17–20 in this textbook or in the *MyMathLab*®️ Study Plan.**

 Only proportions can be solved by cross multiplying. If a rational equation is not a proportion, do not try to cross multiply.

OBJECTIVE 4    SOLVE A FORMULA CONTAINING RATIONAL EXPRESSIONS
FOR A GIVEN VARIABLE

Recall that a **formula** is an **equation** that describes the relationship between two or more **variables**. In **Section 2.4**, we learned to **solve a formula** for a given variable. If a formula contains rational expressions, we can solve it for a given variable by using the **steps for solving a rational equation**.

## Example 8 Solving a Formula for a Given Variable

Solve each formula for the given variable.

**a.** $I = \dfrac{E}{r + R}$ for $R$   **b.** $\dfrac{1}{f} = \dfrac{1}{c} + \dfrac{1}{d}$ for $d$

**Solutions**

**a.**

| | |
|---|---|
| Write the original formula: | $I = \dfrac{E}{r + R}$ |
| Multiply both sides by $r + R$, the LCD: | $I(r + R) = \left(\dfrac{E}{r + R}\right)(r + R)$ |
| Distribute; divide out common factor: | $Ir + IR = E$ ← $\boxed{\left(\dfrac{E}{r + R}\right)(r + R) = E}$ |
| Subtract $Ir$ from both sides: | $IR = E - Ir$ |
| Divide both sides by $I$: | $R = \dfrac{E - Ir}{I}$ |

Solving the formula $I = \dfrac{E}{r + R}$ for $R$ results in $R = \dfrac{E - Ir}{I}$.

 *My video summary*    **b.** Multiplying both sides of the **formula** by its **LCD** $cdf$ results in

$$cd = df + cf.$$

To **solve** for $d$, we must get both **terms** that contain $d$ on the same side of the equal sign. We subtract $df$ from both sides to result in

$$cd - df = cf.$$

To be solved for $d$, the formula can only contain one $d$, so we **factor** it out.

$$d(c - f) = cf$$

Dividing both sides by $c - f$ gives

$$d = \dfrac{cf}{c - f},$$

which is the desired result. Watch this **video** for a fully worked solution.

**You Try It**    Work through this You Try It problem.

**Work Exercises 21–28 in this textbook or in the *MyMathLab*® Study Plan.**

 When solving a formula for a variable that occurs more than once, you will typically need to move all terms containing the desired variable to the same side of the equal sign and then factor out the variable.

# 7.6 Exercises

In Exercises 1–4, determine if the statement is a rational equation. If not, state why.

**You Try It**

**1.** $\dfrac{x - 3}{x^2 + 1} = \dfrac{\sqrt{x^2 - 4x + 1}}{x - 5}$

**2.** $4y^{-2} = 15 - 17y^{-1}$

3. $\dfrac{x^2 + 3x - 4}{2x + 5} = \dfrac{x - 2}{3x + 1}$

4. $\dfrac{x^3 - 4x^2 + 5x}{2x - 9}$

In Exercises 5–20, solve the rational equation. If there is no solution, state this as your answer.

**You Try It**

5. $\dfrac{1}{5}x - \dfrac{5}{6} = \dfrac{2}{3}$

6. $\dfrac{n - 7}{8} = \dfrac{5n}{3} - \dfrac{7}{12}$

7. $\dfrac{1}{y} + \dfrac{1}{6} = \dfrac{1}{4}$

8. $\dfrac{5}{m} - \dfrac{3}{10} = \dfrac{3}{2m} - \dfrac{9}{4}$

**You Try It**

9. $\dfrac{4}{3x} - \dfrac{x + 1}{x} = \dfrac{2}{5}$

10. $\dfrac{x - 1}{x} = \dfrac{3}{4} - \dfrac{3}{2x}$

**You Try It**

11. $\dfrac{3}{z} - \dfrac{9}{z - 5} = \dfrac{1}{4}$

12. $\dfrac{5}{3p} + \dfrac{2p}{p - 5} = -\dfrac{4}{3}$

**You Try It**

13. $\dfrac{b}{b - 5} + \dfrac{2}{b + 5} = \dfrac{50}{b^2 - 25}$

14. $\dfrac{3}{3m + 2} - \dfrac{2}{m - 6} = \dfrac{1}{3m^2 - 16m - 12}$

15. $\dfrac{1}{n + 4} + \dfrac{3}{n + 5} = \dfrac{1}{n^2 + 9n + 20}$

16. $\dfrac{6}{y^2 - 25} = \dfrac{3}{y^2 - 5y}$

17. $\dfrac{3}{t + 2} = \dfrac{6}{t + 16}$

18. $\dfrac{k}{7} = \dfrac{5}{k - 2}$

19. $\dfrac{x + 2}{x - 5} = \dfrac{x - 6}{x + 1}$

20. $\dfrac{x}{3x + 2} = \dfrac{x + 5}{3x - 1}$

In Exercises 21–28, solve the formula for the given variable.

**You Try It**

21. $E = \dfrac{RC}{T}$ for $C$

22. $L = \dfrac{h}{mv}$ for $v$

23. $z = \dfrac{x - m}{s}$ for $s$

24. $P = \dfrac{A}{1 + rt}$ for $t$

25. $m = \dfrac{y_2 - y_1}{x_2 - x_1}$ for $x_2$

26. $\dfrac{F}{G} = \dfrac{m_1 m_2}{r^2}$ for $G$

27. $R = \dfrac{(E - B) + D}{B}$ for $B$

28. $\dfrac{1}{A} - \dfrac{1}{B} = \dfrac{1}{C}$ for $A$

# 7.7 Applications of Rational Equations

## THINGS TO KNOW

Before working through this section, be sure you are familiar with the following concepts:

|  | VIDEO | ANIMATION | INTERACTIVE |

**You Try It**
1. Use the Problem-Solving Strategy to Solve Direct Translation Problems (Section 2.3, **Objective 2**)

**You Try It**
2. Find the Value of a Non-isolated Variable in a Formula (Section 2.4, **Objective 2**)

**You Try It**
3. Solve Problems Involving Uniform Motion (Section 2.5, **Objective 3**)

**You Try It**
4. Solve Rational Equations (Section 7.6, **Objective 2**)

**You Try It**
5. Identify and Solve Proportions (Section 7.6, **Objective 3**)

## OBJECTIVES

1 Use Proportions to Solve Problems

2 Use Formulas Containing Rational Expressions to Solve Problems

3 Solve Uniform Motion Problems Involving Rational Equations

4 Solve Problems Involving Rate of Work

---

OBJECTIVE 1   USE PROPORTIONS TO SOLVE PROBLEMS

Recall from **Section 7.6** that a **proportion** is a statement that two **ratios** are equal, such as $\frac{P}{Q} = \frac{R}{S}$. Writing proportions is a powerful tool in solving application problems. Given a ratio (or **rate**) of two quantities, a proportion can be used to determine an unknown quantity.

### Example 1  Defective Lightbulbs

A quality-control inspector examined a sample of 200 lightbulbs and found 18 of them to be defective. At this ratio, how many defective bulbs can the inspector expect in a shipment of 22,000 light bulbs?

**Solution**  We follow our usual **problem-solving strategy**.

**Step 1.** We want to find the number of defective lightbulbs expected in a shipment of 22,000 lightbulbs. We know that there were 18 defective bulbs in a sample of 200 bulbs.

**Step 2.** Let $x =$ the number of defective bulbs expected in the shipment.

**Step 3.** The ratio of defective bulbs to total bulbs in the shipment is assumed to be the same as the ratio of defective bulbs to total bulbs in the sample. We can set these two ratios equal to each other to form a proportion.

$$\begin{array}{cc} \text{Shipment} & \text{Sample} \\ \downarrow & \downarrow \end{array}$$

$$\frac{\text{Defective Bulbs}}{\text{Total Bulbs}} = \frac{\text{Defective Bulbs}}{\text{Total Bulbs}}$$

$$\frac{x}{22{,}000} = \frac{18}{200}$$

**Step 4.** Because we have a **proportion**, we can **cross multiply** to solve.

Cross multiply: $\frac{x}{22{,}000} \times \frac{18}{200}$

$$200x = 22{,}000(18)$$

Simplify: $200x = 396{,}000$

Divide both sides by 200: $x = 1980$

**Step 5.** There are 110 times as many lightbulbs in the shipment as in the sample $(110 \cdot 200 = 22{,}000)$. Therefore, there also should be 110 times as many defects in the shipment as in the sample. Because $110 \cdot 18 = 1980$, our result is reasonable and checks.

**Step 6.** The inspector can expect 1980 defective bulbs in a shipment of 22,000 lightbulbs.

**You Try It** **Work through this You Try It problem.**

**Work Exercises 1–4 in this textbook or in the *MyMathLab*® Study Plan.**

## Example 2 Planting Grass Seed

*My video summary*

A landscaper plants grass seed at a general **rate** of 7 pounds for every 1000 square feet. If the landscaper has 25 pounds of grass seed on hand, how many additional pounds of grass seed will he need to purchase for a job to plant grass on a 45,000 square-foot yard?

**Solution** We follow our usual **problem-solving strategy**.

**Step 1.** We need to find how many pounds of grass seed the landscaper must purchase to plant a 45,000 $\text{ft}^2$ yard. We know 7 pounds of seed will plant 1000 $\text{ft}^2$, and we know the landscaper already has 25 pounds of seed on hand.

**Step 2.** Let $x =$ the pounds of additional seed purchased. Then $x + 25 =$ the total pounds of seed needed for this job.

**Step 3.** The rate of *pounds of seed* to *square feet of yard* for this job is the same as the general rate, so we can form a proportion.

$$\begin{array}{cc} \text{General rate} & \text{Job rate} \\ \downarrow & \downarrow \end{array}$$

$$\frac{\text{Pounds}}{\text{Square feet}} = \frac{\text{Pounds}}{\text{Square feet}}$$

$$\frac{7}{1000} = \frac{x + 25}{45{,}000}$$

**Step 4.** **Cross multiply** to solve the **proportion**.

Try to finish solving this problem on your own. View the **answer**, or watch this **video** for the complete solution.

**You Try It** **Work through this You Try It problem.**

**Work Exercises 5 and 6 in this textbook or in the *MyMathLab*® Study Plan.**

**Similar triangles** have the same shape but not necessarily the same size. For similar triangles, **corresponding angles** are equal, and **corresponding sides** have lengths that are **proportional**. So, we can use proportions to find unknown lengths in similar triangles.

### Example 3  Similar Triangles

Find the unknown length $n$ for the following similar triangles.

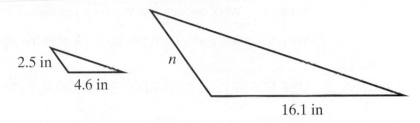

**Solution**  Follow the **problem-solving strategy**. Since the triangles are **similar**, their **corresponding sides** are proportional. Side $n$ on the larger triangle corresponds to the 2.5-inch side on the smaller triangle. The 16.1-inch side on the larger triangle corresponds to the 4.6-inch side on the smaller triangle. So, we have the following proportion:

$$\frac{n}{2.5} = \frac{16.1}{4.6}$$

Cross multiply:  $4.6n = 2.5(16.1)$

Simplify:  $4.6n = 40.25$

Divide both sides by 4.6:  $\dfrac{4.6n}{4.6} = \dfrac{40.25}{4.6}$

Simplify:  $n = 8.75$

The unknown side length $n$ is 8.75 inches long.

**You Try It**    Work through this You Try It problem.

**Work Exercises 7 and 8 in this textbook or in the *MyMathLab*®  Study Plan.**

### Example 4  Height of a Tree

*My video summary*    A forest ranger wants to determine the height of a tree. She measures the tree's shadow as 84 feet long. Her own shadow at the same time is 7.5 feet long. If she is 5.5 feet tall, how tall is the tree?

**Solution** Follow the **problem-solving strategy**. Since the shadows were measured at the same time, the **ratio** of the height-to-shadow length of the tree is **proportional** to the ratio of height-to-shadow length of the forest ranger. Write the proportion, and then use it to finish solving this problem. View the **answer**, or watch this **video** for a complete solution.

**You Try It**   Work through this You Try It problem.

Work Exercises 9 and 10 in this textbook or in the *MyMathLab* ® Study Plan.

OBJECTIVE 2   USE FORMULAS CONTAINING RATIONAL EXPRESSIONS TO SOLVE PROBLEMS

In **Section 7.6**, we saw that **formulas** often involve **rational expressions**. These formulas can be used to solve application problems.

### Example 5  Resistance in Parallel Circuits

In electronics, the total resistance $R$ of a circuit containing two resistors in parallel is given by the formula $\dfrac{1}{R} = \dfrac{1}{R_1} + \dfrac{1}{R_2}$, where $R_1$ and $R_2$ are the two individual resistances. If the total resistance is 10 ohms and one resistor has twice the resistance of the other, find the resistance of each circuit.

### Solution

**Step 1.** We want to find the individual resistances for a circuit containing two resistors in parallel. We know that the total resistance is 10 ohms and one resistor has twice the resistance of the other.

**Step 2.** We are given the resistance of one resistor in terms of the other. If we let $x =$ the resistance of the first resistor, then $2x =$ the resistance of the second.

**Step 3.** Using the given formula, we substitute in the total resistance and expressions for the individual resistances. This gives the equation

$$\frac{1}{10} = \frac{1}{x} + \frac{1}{2x}.$$

**Step 4.** The LCD is $10x$. Note that the only restricted value is $x = 0$.

Begin with the original equation:
$$\frac{1}{10} = \frac{1}{x} + \frac{1}{2x}$$

Multiply both sides by the LCD:
$$10x\left(\frac{1}{10}\right) = 10x\left(\frac{1}{x} + \frac{1}{2x}\right)$$

Distribute:
$$10x \cdot \frac{1}{10} = 10x \cdot \frac{1}{x} + 10x \cdot \frac{1}{2x}$$

Simplify:
$$\cancel{10}x \cdot \frac{1}{\cancel{10}} = 10\cancel{x} \cdot \frac{1}{\cancel{x}} + \overset{5}{\cancel{10}}\cancel{x} \cdot \frac{1}{\cancel{2x}}$$

$$x \cdot 1 = 10 \cdot 1 + 5 \cdot 1$$

$$x = 10 + 5$$

$$x = 15$$

**Step 5.** We can check our answer using our equation $\dfrac{1}{10} = \dfrac{1}{x} + \dfrac{1}{2x}$.

$$\frac{1}{15} + \frac{1}{2(15)} = \frac{1}{15} + \frac{1}{30} = \frac{2}{30} + \frac{1}{30} = \frac{3}{30} = \frac{1}{10}$$

The total resistance is 10 ohms, so this result is reasonable and it checks.

**Step 6.** The resistances are 15 ohms and $2(15) = 30$ ohms.

**You Try It**  Work through this You Try It problem.

Work Exercises **11** and **12** in this textbook or in the **MyMathLab** Study Plan.

OBJECTIVE 3  SOLVE UNIFORM MOTION PROBLEMS INVOLVING RATIONAL EQUATIONS

In **Section 2.5**, we studied **uniform motion problems** and used the **formula** $d = rt$. Solving this formula for $t$ gives $t = \dfrac{d}{r}$, or time $= \dfrac{\text{distance}}{\text{rate}}$. We now use this **rational equation** to solve problems.

### Example 6  Boat Speed

Emalie can travel 16 miles upriver in the same amount of time it takes her to travel 24 miles downriver. If the speed of the current is 4 mph, how fast can her boat travel in still water?

**Solution**  Follow the **problem-solving strategy**.

**Step 1.** We want to find the speed of Emalie's boat in still water. We know she can travel 16 miles upriver in the same amount of time she can travel 24 miles downriver, and we know the speed of the current is 4 mph.

**Step 2.** Let $r$ = the speed of Emalie's boat in still water. Then her speed upriver is $r - 4$ (because she goes against the current) and her speed downriver is $r + 4$ (because she goes with the current).

**Step 3.** We know the amount of time traveled is the same in either direction.

$$\text{time}_{\text{upriver}} = \text{time}_{\text{downriver}}$$

Solving the **distance formula** for $t$ gives us $t = \dfrac{d}{r}$, or time $= \dfrac{\text{distance}}{\text{rate}}$. So, we rewrite our equation as follows:

$$\left(\frac{\text{distance}}{\text{rate}}\right)_{\text{upriver}} = \left(\frac{\text{distance}}{\text{rate}}\right)_{\text{downriver}}$$

Substituting the given distances and our expressions for rates from Step 2, we get the equation

$$\frac{16}{r - 4} = \frac{24}{r + 4}.$$

Try to solve this equation to finish this problem on your own. View the **answer**, or watch this **video** for a complete solution.

**You Try It**     **Work through this You Try It problem.**

Work Exercises 13 and 14 in this textbook or in the *MyMathLab*® Study Plan.

### Example 7   Train Speed

*My video summary*    Fatima rode an express train 223.6 miles from Boston to New York City and then rode a passenger train 218.4 miles from New York City to Washington, DC. If the express train travels 30 miles per hour faster than the passenger train and her total trip took 6.5 hours, what was the average speed of the express train?

**Solution**   Follow the **problem-solving strategy.**

**Step 1.** We want to find the average speed of the express train. We know the express train traveled 30 miles per hour faster than the passenger train. We also know that the distances traveled on the express and passenger trains were 223.6 and 218.4 miles, respectively. Finally, we know the time of the trip was 6.5 hours.

**Step 2.** Let $r$ = the average speed of the express train. Then $r - 30$ = the average speed of the passenger train.

**Step 3.** The total time of Fatima's trip must be the **sum** of her time on the express train and her time on the passenger train.

$$\text{time}_{\text{express}} + \text{time}_{\text{passenger}} = \text{time}_{\text{total}}$$

Solving the **distance formula** for $t$ gives us $t = \dfrac{d}{r}$, or $\text{time} = \dfrac{\text{distance}}{\text{rate}}$.

So, we rewrite our equation as

$$\left(\frac{\text{distance}}{\text{rate}}\right)_{\text{express}} + \left(\frac{\text{distance}}{\text{rate}}\right)_{\text{passenger}} = \text{time}_{\text{total}}.$$

Substituting the known distances, the expressions for train speeds from Step 2, and the total time, we get the equation

$$\frac{223.6}{r} + \frac{218.4}{r - 30} = 6.5.$$

**Step 4.** The **LCD** is $r(r - 30)$. Multiply both sides of the equation by the LCD to obtain a **polynomial equation.**

Write the original equation:
$$\frac{223.6}{r} + \frac{218.4}{r - 30} = 6.5$$

Multiply by the LCD:
$$r(r - 30)\left(\frac{223.6}{r} + \frac{218.4}{r - 30}\right) = r(r - 30)(6.5)$$

Distribute:
$$r(r - 30) \cdot \frac{223.6}{r} + r(r - 30) \cdot \frac{218.4}{r - 30} = r(r - 30)(6.5)$$

Divide out common factors:
$$\cancel{r}(r - 30) \cdot \frac{223.6}{\cancel{r}} + r\cancel{(r - 30)} \cdot \frac{218.4}{\cancel{r - 30}} = r(r - 30)(6.5)$$

Multiply:
$$223.6(r - 30) + 218.4r = r(r - 30)(6.5)$$

Finish solving the equation on your own to answer the question. View the **answer,** or watch this **video** for a complete solution.

You Try It    Work through this You Try It problem.

Work Exercises 15 and 16 in this textbook or in the *MyMathLab* Study Plan.

OBJECTIVE 4    SOLVE PROBLEMS INVOLVING RATE OF WORK

For work problems involving multiple workers, such as people, copiers, pumps, etc., it is often helpful to consider **rate of work**.

---

**Rate of Work**

The **rate of work** is the number of jobs that can be completed in a given unit of time.

If one job can be completed in $t$ units of time, then the rate of work is given by $\dfrac{1}{t}$.

---

We can add **rates** but not times. For example, if it takes Avril 4 hours to paint a room and Anisa 2 hours to paint the same room, it would not take $4 + 2 = 6$ hours for them to paint the room together. It cannot take longer for two people to do a job than it would take either one alone. When dealing with two workers, we use the following formula:

$$\underbrace{\frac{1}{t_1}}_{\substack{\text{Work rate of} \\ \text{1st worker}}} + \underbrace{\frac{1}{t_2}}_{\substack{\text{Work rate of} \\ \text{2nd worker}}} = \underbrace{\frac{1}{t}}_{\substack{\text{Work rate} \\ \text{together}}}$$

Here, $t_1$ and $t_2$ are the individual times to complete one job, and $t$ is the time to complete the job when working together.

## Example 8  Painting a Room

*My video summary*

Avril can paint a room in 4 hours if she works alone. Anisa can paint the same room in 2 hours if she works alone. How long will it take the two women to paint the room if they work together?

**Solution**    Follow the **problem-solving strategy**.

**Step 1.** We want to find how long it will take Avril and Anisa to paint the room if they work together. Working alone, Avril can paint the room in 4 hours, while Anisa can paint the room in 2 hours.

**Step 2.** Let $x$ = the time to paint the room when working together.

**Step 3.** Avril can paint the room in 4 hours, so her **rate of work** is $\dfrac{1}{4}$ room per hour. Anisa can paint the room in 2 hours, so her rate of work is $\dfrac{1}{2}$ room per hour. Together, they can paint the room in $x$ hours, so their combined rate of work is $\dfrac{1}{x}$ room per hour. The **sum** of the two individual work rates results in the combined work rate:

$$\underbrace{\frac{1}{4}}_{\text{Avril's rate}} + \underbrace{\frac{1}{2}}_{\text{Anisha's rate}} = \underbrace{\frac{1}{x}}_{\text{Combined rate}}.$$

**Step 4.** The **LCD** is $4x$.

Write the original equation: $\dfrac{1}{4} + \dfrac{1}{2} = \dfrac{1}{x}$

Multiply both sides by the LCD: $4x\left(\dfrac{1}{4} + \dfrac{1}{2}\right) = 4x\left(\dfrac{1}{x}\right)$

Distribute: $4x \cdot \dfrac{1}{4} + 4x \cdot \dfrac{1}{2} = 4x \cdot \dfrac{1}{x}$

Divide out common factors: $\cancel{4}x \cdot \dfrac{1}{\cancel{4}} + \overset{2}{\cancel{4}}x \cdot \dfrac{1}{\cancel{2}} = 4\cancel{x} \cdot \dfrac{1}{\cancel{x}}$

Simplify: $x + 2x = 4$

Combine like terms: $3x = 4$

Divide both sides by 3: $x = \dfrac{4}{3}$

**Step 5.** The result $\dfrac{4}{3}$ hours is less than the 2 hours needed for the faster worker alone, so the result is reasonable.

**Step 6.** Together, Avril and Anisa can paint the room in $\dfrac{4}{3}$ hours, or 1 hour and 20 minutes.

**You Try It**    Work through this You Try It problem.

**Work Exercises 17 and 18 in this textbook or in the _MyMathLab_® Study Plan.**

### Example 9  Emptying a Pool

*My video summary*    A small pump takes 8 more hours than a larger pump to empty a pool. Together the pumps can empty the pool in 3 hours. How long will it take the larger pump to empty the pool if it works alone?

**Solution**  Follow the **problem-solving strategy**.

**Step 1.** We want to find the time it will take the larger pump to empty the pool by itself. We know that the two pumps take 3 hours to empty the pool and the smaller pump takes 8 more hours than the larger pump to empty the pool by itself.

**Step 2.** Let $x =$ the time for the larger pump to empty the pool. Then $x + 8 =$ the time for the smaller pump to empty the pool.

**Step 3.** Since we are combining **rates of work**, we get

$$\underbrace{\dfrac{1}{t_1}}_{\text{Larger pump rate}} + \underbrace{\dfrac{1}{t_2}}_{\text{Smaller pump rate}} = \underbrace{\dfrac{1}{t}}_{\text{Combined rate}}.$$

Substituting in the given combined time, and expressions for individual times, we get

$$\underbrace{\dfrac{1}{x}}_{\text{Larger pump rate}} + \underbrace{\dfrac{1}{x + 8}}_{\text{Smaller pump rate}} = \underbrace{\dfrac{1}{3}}_{\text{Combined rate}}.$$

**Step 4.** The **LCD** is $3x(x + 8)$. Multiply both sides of the equation by the LCD and solve the resulting **polynomial equation** to finish this problem on your own. View the **answer**, or watch this **video** for a complete solution.

**You Try It**   **Work through this You Try It problem.**

**Work Exercises 19 and 20 in this textbook or in the *MyMathLab*® Study Plan.**

### Example 10  Filling a Pond

*My animation summary*

A garden hose can fill a pond in 2 hours whereas an outlet pipe can drain the pond in 10 hours. If the outlet pipe is accidentally left open, how long would it take to fill the pond?

**Solution**  Watch this **animation** or continue reading for a complete solution.

We want to find the time required to fill the pond if the hose is running and the outlet pipe is open. Let $t$ = the time required to fill the pond.

We are combining **rates of work**, but the rates are not working together. Rates that work towards completion of a task are positive actions and therefore are positive values. Rates that work against the completion of the task are negative actions and therefore are negative values.

$$\underbrace{\frac{1}{t_1}}_{\text{Garden hose rate}} + \underbrace{\frac{-1}{t_2}}_{\text{Outlet pipe rate}} = \underbrace{\frac{1}{t}}_{\text{Combined rate}}$$

Substituting in the given individual times, we get

$$\underbrace{\frac{1}{2}}_{\text{Garden hose rate}} + \underbrace{\frac{-1}{10}}_{\text{Outlet pipe rate}} = \underbrace{\frac{1}{t}}_{\text{Combined rate}}$$

or

$$\underbrace{\frac{1}{2}}_{\text{Garden hose rate}} - \underbrace{\frac{1}{10}}_{\text{Outlet pipe rate}} = \underbrace{\frac{1}{t}}_{\text{Combined rate}}.$$

Because the outlet pipe is working against the garden hose, we end up subtracting the rates. Note that this is similar to the situation when a boat is traveling upstream or downstream. When going downstream (with the current) we added the rates. However, when going upstream (against the current) we subtracted the rates because the current was working against the boat.

The **LCD** is $10t$. Multiply both sides of the equation by the LCD and solve the resulting equation for $t$.

$$\text{Original equation:}\qquad \frac{1}{2} - \frac{1}{10} = \frac{1}{t}$$

$$\text{Multiply by the LCD:}\qquad 10t\left(\frac{1}{2} - \frac{1}{10}\right) = 10t\left(\frac{1}{t}\right)$$

$$\text{Distribute:}\qquad 5t - t = 10$$

$$\text{Simplify:}\qquad 4t = 10$$

$$\text{Divide both sides by 4:}\qquad t = \frac{10}{4} = 2.5$$

It will take 2.5 hours to fill the pond with the outlet pipe open.

**You Try It**    Work through this You Try It problem.

Work Exercises 21 and 22 in this textbook or in the **MyMathLab**® Study Plan.

# 7.7 Exercises

In Exercises 1–6, use a proportion to solve the problem.

**You Try It**

1. **Defective Cell Phones**    A quality control inspector finds 8 defective units in a sample of 50 cellular phones. At this ratio, how many defective units can the inspector expect to find in a batch of 1500 such phones?

2. **Students and Teachers**    If a high school employs 5 teachers for every 76 students, how many students are in the school if it employs 115 teachers?

3. **Ice Cream**    If $\frac{1}{2}$ cup of chocolate chip cookie dough ice cream contains 15 grams of fat, how many grams of fat are in a quart of the ice cream? **Hint:** 1 quart $=$ 4 cups.

4. **Identity Theft**    In 2008, there were 75 identity theft complaints per 100,000 residents in the state of Missouri. How many identity theft complaints were made for the entire state if the population of Missouri was 5,900,000?

**You Try It**

5. **Planting Corn**    To plant a 250-acre field, a farmer used 3500 pounds of corn seed. If the farmer still has 500 pounds of corn seed on hand, how many more pounds of corn seed will he need to buy in order to plant an additional 180-acre field?

6. **Time to Complete an Order**    Olivia works at a craft store and can make 2 stone crafts in 45 minutes. If she has been working for 30 minutes on an order for 7 stone crafts, how much longer does she need to work to complete the order?

In Exercises 7 and 8, find the unknown length $n$ for the given similar triangles.

**You Try It**

7.

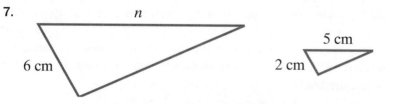

8.

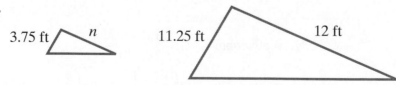

In Exercises 9–22, solve the problem.

**You Try It**

9. **Height of a Lighthouse**    Elena wants to determine the height of a lighthouse. The lighthouse casts an 84 meter shadow. If Elena is 1.8 meters tall and casts a 2.4-meter shadow when standing next to the lighthouse, how tall is the lighthouse?

**10. Tennis Anyone?** At what height must a tennis ball be served in order to just clear the net and land 20 feet past the net? Assume the ball travels a straight path as shown in the figure.

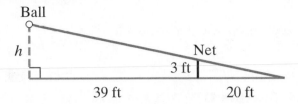

**11. Resistors in Parallel** If the total resistance of two circuits in parallel is 9 ohms and one circuit has three times the resistance of the other, find the resistance of each circuit. Use $\frac{1}{R} = \frac{1}{R_1} + \frac{1}{R_2}$, where $R_1$ and $R_2$ are the two individual resistances.

**12. Focal Length** The focal length $f$ of a lens is given by the equation $\frac{1}{f} = \frac{1}{x_o} + \frac{1}{x_i}$, where $x_o$ is the distance from the object to the lens, and $x_i$ is the distance from the lens to the image of the object. If the focal length of a lens is 12 cm and an object is three times as far from the lens as its image, then how far is the object from the lens?

**13. Current Speed** On a float trip, Kim traveled 30 miles downstream in the same amount of time that it would take her to row 12 miles upstream. If her speed in still water is 7 mph, find the speed of the current.

**14. Plane Speed** A plane can fly 500 miles against the wind in the same amount of time that it can fly 725 miles with the wind. If the wind speed is 45 mph, find the speed of the plane in still air.

**15. Walking Speed** For exercise, Mae jogs 10 miles then walks an additional 2 miles to cool down. If her jogging speed is 4.5 miles per hour faster than her walking speed and her total exercise time is 3 hours, what is her walking speed?

**16. Moving Sidewalk** Choi can walk 108 m in 72 seconds. Standing on a moving sidewalk at Thailand's Suvarnabhumi International Airport, she can travel 108 m in 40 seconds. How long will it take her to travel 108 m if she walks on the moving sidewalk? Round your answer to the nearest tenth.

**17. Making Copies** Beverly can copy final exams in 30 minutes using a new copy machine. Using an old copy machine, the same job takes 45 minutes. If both copy machines are used, how long will it take to copy the final exams?

**18. Mowing the Lawn** By himself, Chris can mow his lawn in 60 minutes. If his daughter Claudia helps, they can mow the lawn together in 40 minutes. How long would it take Claudia to mow the lawn by herself?

**19. Staining a Deck** It takes Shawn 2 more hours to stain a deck than Michelle. Together, it takes them 2.4 hours to complete the work. How long would it take Shawn to stain the deck by himself?

**20. Lining Fields** Together, Sari and Rilee can mark the lines on the soccer fields at a recreation center 18 minutes faster than if Rilee lines the fields on her own. If it takes Sari 56 minutes to line the fields by herself, how long would it take Rilee to do the job alone?

**21. Filling a Hot Tub** A garden hose can fill a hot tub in 180 minutes, whereas the drain on a hot tub can empty it in 300 minutes. If the drain is accidentally left open, how long will it take to fill the tub?

**22. Emptying a Boat's Bilge** A leak in the hull will fill a boat's bilge in 8 hours. The boat's bilge pump can empty a full bilge in 3 hours. How long will it take the pump to empty a full bilge if the boat is leaking?

# 7.8 Variation

## THINGS TO KNOW

Before working through this section, be sure that you are familiar with the following concepts:

VIDEO      ANIMATION      INTERACTIVE

**You Try It**

1. Evaluate a Formula
   (Section 2.4, **Objective 1**)

**You Try It**

2. Find the Value of a Non-isolated Variable in a
   Formula (Section 2.4, **Objective 2**)

## OBJECTIVES

1 Solve Problems Involving Direct Variation

2 Solve Problems Involving Inverse Variation

---

OBJECTIVE 1    SOLVE PROBLEMS INVOLVING DIRECT VARIATION

Often in application problems, we need to know how one quantity varies with respect to other quantities. A **variation equation** allows us to describe how one quantity changes with respect to one or more additional quantities. We will examine two types of variation, *direct* variation and *inverse* variation.

*Direct variation* means that one **variable** is a constant multiple of another variable.

> **Definition**    **Direct Variation**
>
> For an equation of the form
> $$y = kx,$$
> we say that $y$ **varies directly** with $x$, or $y$ is **directly proportional** to $x$. The nonzero constant $k$ is called the **constant of variation** or the **proportionality constant**.

A variation equation such as $y = kx$ is also called a **model** that describes the relationship between the variables $x$ and $y$.

### Example 1  Direct Variation

Suppose $y$ varies directly with $x$, and $y = 20$ when $x = 8$.

**a.** Find the equation that relates $x$ and $y$.

**b.** Find $y$ when $x = 12$.

### Solutions

**a.** We know that $y$ **varies directly** with $x$, so the equation has the form $y = kx$. We use the fact that $y = 20$ when $x = 8$ to find $k$.

$$\text{Write the direct variation equation:} \quad y = kx$$
$$\text{Substitute } y = 20 \text{ and } x = 8: \quad 20 = k(8)$$
$$\text{Divide both sides by 8:} \quad \frac{20}{8} = \frac{k(8)}{8}$$
$$\text{Simplify:} \quad 2.5 = k$$

The **constant of variation** is 2.5, so the equation is $y = 2.5x$.

**b.** We now know that $y$ and $x$ are related by the equation $y = 2.5x$. To find $y$ when $x = 12$, substitute 12 for $x$ in the equation and simplify:

$$y = 2.5(12) = 30$$

Therefore, $y = 30$ when $x = 12$.

**You Try It**    Work through this You Try It problem.

**Work Exercises 1 and 2 in this textbook or in the *MyMathLab*® Study Plan.**

In Example 1, the **model** $y = kx$ is a **linear equation**. However, direct variation equations will not always be linear. For example, if we say that $y$ varies directly with the **cube** of $x$, then the equation will be $y = kx^3$, where $k$ is the constant of variation.

## Example 2  Direct Variation

Suppose $y$ **varies directly** with the cube of $x$, and $y = 375$ when $x = 5$.

**a.** Find the equation that relates $x$ and $y$.

**b.** Find $y$ when $x = 2$.

**Solutions**  Because $y$ varies directly with the cube of $x$, or $x^3$, the variation equation has the form $y = kx^3$. We can use the fact that $y = 375$ when $x = 5$ to find $k$, and then the equation. Try to complete this problem on your own. View the **answer**, or watch this **video** for a complete solution.

**You Try It**    Work through this You Try It problem.

**Work Exercises 3–6 in this textbook or in the *MyMathLab*® Study Plan.**

For direct variation, the **ratio** of the two quantities is constant (the **constant of variation**). For example, consider $y = kx$ and $y = kx^3$.

$$\underbrace{y = kx}_{\substack{y \text{ varies} \\ \text{directly} \\ \text{with } x}} \rightarrow \underbrace{\frac{y}{x}}_{\substack{\text{Ratio of the} \\ \text{quantities} \\ y \text{ and } x}} = \underbrace{k}_{\substack{\text{Constant of} \\ \text{variation}}} \qquad \underbrace{y = kx^3}_{\substack{y \text{ varies directly} \\ \text{with the cube} \\ \text{of } x}} \rightarrow \underbrace{\frac{y}{x^3}}_{\substack{\text{Ratio of the} \\ \text{quantities} \\ y \text{ and } x^3}} = \underbrace{k}_{\substack{\text{Constant} \\ \text{of variation}}}$$

Problems involving variation can generally be solved using the following guidelines.

---

**Solving Variation Problems**

**Step 1.**  Translate the problem into an equation that models the situation.

**Step 2.**  Substitute given values for the variables into the equation and solve for the **constant of variation,** $k$.

**Step 3.**  Substitute the value for $k$ into the equation to form the general **model**.

**Step 4.**  Use the general model to answer the question posed in the problem.

---

## Example 3  Kinetic Energy

The kinetic energy of an object in motion varies directly with the square of its speed. If a van traveling at a speed of 30 meters per second has 945,000 joules of kinetic energy, how much kinetic energy does it have if it is traveling at a speed of 20 meters per second?

**Solution** Follow the guidelines for solving variation problems.

Step 1.  We are told that the kinetic energy of an object in motion varies **directly** with the square of its speed. If we let $K$ = kinetic energy and $s$ = speed, we can translate the problem statement into the model

$$K = ks^2,$$

where $k$ is the **constant of variation**.

Step 2.  To determine the value of $k$, we use the fact that the kinetic energy is 945,000 joules when the velocity is 30 meters per second.

$$\text{Substitute 945,000 for } K \text{ and 30 for } s: \quad 945{,}000 = k(30)^2$$
$$\text{Simplify } 30^2: \quad 945{,}000 = 900k$$
$$\text{Divide both sides by 900:} \quad 1050 = k$$

Step 3.  The constant of variation is 1050, so the general model is $K = 1050s^2$.

Step 4.  We want to determine the kinetic energy of the van if its speed is 20 meters per second. Substituting 20 for $s$, we find

$$K = 1050(20)^2 = 1050(400) = 420{,}000.$$

The van will have 420,000 joules of kinetic energy if it is traveling at a speed of 20 meters per second.

**You Try It**    Work through this You Try It problem.

Work Exercises 7 and 8 in this textbook or in the ***MyMathLab***® Study Plan.

## Example 4  Measuring Leanness

The Ponderal Index measure of leanness states that weight varies directly with the cube of height. If a "normal" person who is 1.2 m tall weighs 21.6 kg, how much will a "normal" person weigh if they are 1.8 m tall?

**Solution** Follow the guidelines for solving variation problems.

Step 1.  We are told that weight varies **directly** with the cube of height. If we let $w$ = weight and $h$ = height, we can translate the problem statement into the **model**

$$w = kh^3,$$

where $k$ is the **constant of variation**.

Step 2.  To determine the value of $k$, we use the fact that a normal person who is 1.2 meters tall has a weight of 21.6 kg.

$$\text{Substitute 1.2 for } h \text{ and 21.6 for } w: \quad 21.6 = k(1.2)^3$$
$$\text{Simplify } (1.2)^3: \quad 21.6 = 1.728k$$
$$\text{Divide both sides by 1.728:} \quad 12.5 = k$$

Step 3.  The constant of variation is 12.5, so the general **model** is $w = 12.5h^3$.

Use the general model to determine the weight of a normal person who is 1.8 m tall. Check your **answer**, or watch this **video** for a detailed solution.

**You Try It**    **Work through this You Try It problem.**

**Work Exercises 9 and 10 in this textbook or in the *MyMathLab*® Study Plan.**

---

OBJECTIVE 2    SOLVE PROBLEMS INVOLVING INVERSE VARIATION

*Inverse variation* means that one variable is a constant multiple of the **reciprocal** of another variable.

---

**Definition    Inverse Variation**

For equations of the form

$$y = \frac{k}{x} \quad \text{or} \quad y = k \cdot \frac{1}{x},$$

we say that $y$ **varies inversely** with $x$, or $y$ is **inversely proportional** to $x$. The constant $k$ is called the **constant of variation**.

---

For inverse variation, the **product** of the two quantities is constant (the constant of variation). For example, consider $y = \dfrac{k}{x}$ and $y = \dfrac{k}{x^2}$.

$$\underbrace{y = \frac{k}{x}}_{\substack{y \text{ varies} \\ \text{inversely} \\ \text{with } x}} \rightarrow \underbrace{xy}_{\substack{\text{Product of the} \\ \text{quantities} \\ y \text{ and } x}} = \underbrace{k}_{\substack{\text{Constant} \\ \text{of} \\ \text{variation}}} \qquad \underbrace{y = \frac{k}{x^2}}_{\substack{y \text{ varies} \\ \text{inversely with} \\ \text{the square of } x}} \rightarrow \underbrace{x^2 y}_{\substack{\text{Product of} \\ \text{the quantities} \\ y \text{ and } x^2}} = \underbrace{k}_{\substack{\text{Constant} \\ \text{of} \\ \text{variation}}}$$

**Example 5  Inverse Variation**

*My video summary*  Suppose $y$ **varies inversely** with $x$, and $y = 72$ when $x = 50$.

a.  Find the equation that relates $x$ and $y$.

b.  Find $y$ when $x = 45$.

**Solutions**

a.  Because $y$ varies inversely with $x$, the equation has the form $y = \dfrac{k}{x}$. We use the fact that $y = 72$ when $x = 50$ to find $k$.

$$\text{Write the inverse variation equation:} \qquad y = \frac{k}{x}$$

$$\text{Substitute } y = 72 \text{ and } x = 50: \qquad 72 = \frac{k}{50}$$

$$\text{Multiply both sides by 50:} \qquad 3600 = k$$

The **constant of variation** is 3600, so the equation is $y = \dfrac{3600}{x}$.

**b.** Try to work this part on your own. View the **answer**, or watch this **video** for a complete solution to both parts.

**You Try It**   Work through this You Try It problem.

**Work Exercises 11–14 in this text book or in the *MyMathLab*® Study Plan.**

Problems involving **inverse variation** can be solved using the same **guidelines for solving variation problems**.

## Example 6  Density of an Object

For a given mass, the density of an object is **inversely proportional** to its volume. If 50 cubic centimeters $(cm^3)$ of an object with a density of $28 \, g/cm^3$ is compressed to $40 \, cm^3$, what would be its new density?

### Solution

**Step 1.** We are told that the density of an object with a given mass varies inversely with its volume. If we let $D$ = density and $V$ = volume, we can translate the problem statement into the **model**

$$D = \frac{k}{V},$$

where $k$ is the **constant of variation**.

**Step 2.** To determine the value of $k$, we use the fact that the density is $28 \, g/cm^3$ when the volume is $50 \, cm^3$.

Substitute 28 for $D$ and 50 for $V$:   $28 = \dfrac{k}{50}$

Multiply both sides by 50:   $1400 = k$

**Step 3.** The constant of variation is 1400, so the general model is $D = \dfrac{1400}{V}$.

**Step 4.** We want to determine the density of the object if the volume is compressed to $40 \, cm^3$. Substituting 40 for $V$, we find

$$D = \frac{1400}{40} = 35 \, g/cm^3.$$

The density of the compressed object would be $35 \, g/cm^3$.

**You Try It**   Work through this You Try It problem.

**Work Exercises 15 and 16 in this textbook or in the *MyMathLab*® Study Plan.**

## Example 7  Shutter Speed

*My video summary*   The shutter speed, $S$, of a camera **varies inversely** as the square of the aperture setting, $f$. If the shutter speed is 125 for an aperture of 5.6, what is the shutter speed if the aperture is 1.4?

**Solution** Follow the guidelines for solving variation problems.

**Step 1.** We are told that shutter speed varies inversely with the square of the aperture setting. Letting $S$ = shutter speed and $f$ = aperture setting, we can translate the problem statement into the **model**

$$S = \frac{k}{f^2},$$

where $k$ is the **constant of variation**.

**Step 2.** To determine the value of $k$, we use the fact that an aperture setting of 5.6 corresponds to a shutter speed of 125.

Substitute 5.6 for $f$ and 125 for $S$: $\quad 125 = \dfrac{k}{(5.6)^2}$

Simplify $(5.6)^2$: $\quad 125 = \dfrac{k}{31.36}$

Multiply both sides by 31.36: $\quad 3920 = k$

**Step 3.** The constant of variation is 3920, so the general model is $S = \dfrac{3920}{f^2}$.

Use the general model to determine the shutter speed for an aperture setting of 1.4. Check your **answer**, or watch this **video** for a detailed solution.

**You Try It** Work through this **You Try It** problem.

Work Exercises 17 and 18 in this textbook or in the **_MyMathLab_**® Study Plan.

# 7.8 Exercises

In Exercises 1–18, solve each variation problem.

**You Try It**

**1.** Suppose $y$ varies directly with $x$, and $y = 14$ when $x = 4$.
 a. Find the equation that relates $x$ and $y$.
 b. Find $y$ when $x = 7$.

**2.** Suppose $S$ in directly proportional to $t$, and $S = 900$ when $t = 25$.
 a. Find the equation that relates $t$ and $S$.
 b. Find $S$ when $t = 4$.

**You Try It**

**3.** Suppose $y$ varies directly with the square of $x$, and $y = 216$ when $x = 3$.
 a. Find the equation that relates $x$ and $y$.
 b. Find $y$ when $x = 5$.

**4.** Suppose $M$ is directly proportional to the square of $r$, and $M = 14$ when $r = \dfrac{1}{6}$.
 a. Find the equation that relates $r$ and $M$.
 b. Find $M$ when $r = \dfrac{2}{3}$.

**5.** Suppose $p$ is directly proportional to the cube of $n$, and $p = 135$ when $n = 6$.
 a. Find the equation that relates $n$ and $p$.
 b. Find $p$ when $n = 10$.

6. Suppose $A$ varies directly with the cube of $m$, and $A = 48$ when $m = 4$.
   a. Find the equation that relates $m$ and $A$.
   b. Find $A$ when $m = 8$.

**You Try It**

7. **Scuba Diving**  The water pressure on a scuba diver is directly proportional to the depth of the diver. If the pressure on a diver is 13.5 psi when she is 30 feet below the surface, how far below the surface will she be when the pressure is 18 psi?

8. **Pendulum Length**  The length of a simple pendulum varies directly with the square of its period. If a pendulum of length 2.25 meters has a period of 3 seconds, how long is a pendulum with a period of 8 seconds?

**You Try It**

9. **Water Flow Rate**  For a fixed water flow rate, the amount of water that can be pumped through a pipe varies directly as the square of the diameter of the pipe. In one hour, a pipe with an 8-inch diameter can pump 400 gallons of water. Assuming the same water flow rate, how much water could be pumped through a pipe that is 12 inches in diameter?

10. **Falling Distance**  The distance an object falls varies directly with the square of the time it spends falling. If a ball falls 19.6 meters after falling for 2 seconds, how far will it fall after 9 seconds?

**You Try It**

11. Suppose $y$ varies inversely with $x$, and $y = 24$ when $x = 25$.
    a. Find the equation that relates $x$ and $y$.
    b. Find $y$ when $x = 30$.

12. Suppose $p$ is inversely proportional to $q$, and $p = 3.5$ when $q = 4.6$.
    a. Find the equation that relates $q$ and $p$.
    b. Find $p$ when $q = 1.4$.

13. Suppose $n$ varies inversely with the square of $m$, and $n = 6.4$ when $m = 9$.
    a. Find the equation that relates $m$ and $n$.
    b. Find $n$ when $m = 4$.

14. Suppose $d$ is inversely proportional to the cube of $t$, and $d = 76.8$ when $t = 2.5$.
    a. Find the equation that relates $t$ and $d$.
    b. Find $d$ when $t = 5$.

**You Try It**

15. **Car Depreciation**  The value of a car is inversely proportional to its age. If a car is worth $8100 when it is 4 years old, how old will it be when it is worth $3600?

16. **Electric Resistance**  For a given voltage, the resistance of a circuit is inversely related to its current. If a circuit has a resistance of 5 ohms and a current of 12 amps, what is the resistance if the current is 20 amps?

**You Try It**

17. **Weight of an Object**  The weight of an object within Earth's atmosphere varies inversely with the square of the distance of the object from Earth's center. If a low Earth orbit satellite weighs 100 kg on Earth's surface (6400 km), how much will it weigh in its orbit 800 km above Earth?

18. **Light Intensity**  The intensity of a light varies inversely as the square of the distance from the light source. If the intensity from a light source 3 feet away is 8 lumens, what is the intensity at a distance of 2 feet?

# Radicals and Rational Exponents

# 8.1 Finding Roots

## THINGS TO KNOW

Before working through this section, be sure you are familiar with the following concepts:

VIDEO     ANIMATION     INTERACTIVE

**You Try It**

1. Find the Absolute Value of a Real Number (Section 1.1, **Objective 4**)  

**You Try It**

2. Evaluate Exponential Expressions (Section 1.4, **Objective 1**)

## OBJECTIVES

1 Find Square Roots

2 Approximate Square Roots

3 Simplify Square Roots Containing Variables

4 Find Cube Roots

5 Find and Approximate $n$th Roots

OBJECTIVE 1   FIND SQUARE ROOTS

We have seen earlier that *squaring* a number means to multiply the number by itself. For example:

$$\text{The square of 3 is } 3^2 = 3 \cdot 3 = 9.$$

$$\text{The square of } -3 \text{ is } (-3)^2 = (-3)(-3) = 9.$$

$$\text{The square of } \frac{1}{5} \text{ is } \left(\frac{1}{5}\right)^2 = \frac{1}{5} \cdot \frac{1}{5} = \frac{1}{25}.$$

The opposite operation of squaring is to find the **square root**. The square root of a **non-negative** real number $a$ is a **real number** $b$ that, when **squared**, results in $a$. So, $b$ is a square root of $a$ if $b^2 = a$.

Every positive real number has two square roots: one positive and one negative. For example, 3 and $-3$ are both square roots of 9 because

$$3^2 = 9 \text{ and } (-3)^2 = 9.$$

3 is the *positive*, or *principal*, *square root* of 9, and $-3$ is the *negative square root* of 9.

We use the **radical sign** $\sqrt{\phantom{x}}$ to denote **positive** or **principal square roots**. We place a negative sign in front of a radical sign, $-\sqrt{\phantom{x}}$, to denote **negative square roots**. For example, $\sqrt{9}$ represents the principal square root of 9, whereas $-\sqrt{9}$ represents the negative square root of 9. So, we write $\sqrt{9} = 3$ and $-\sqrt{9} = -3$.

---

**Definition   Square Roots**

For $a > 0$, $\sqrt{a}$ is the **positive** or **principal square root** of $a$, and $-\sqrt{a}$ is the **negative square root** of $a$.

$$\sqrt{a} = b \text{ only if } b^2 = a \text{ and } b > 0$$

Also, $\sqrt{0} = 0$.

---

The expression $\sqrt{9}$ is called a *radical expression*. A **radical expression** is an expression that contains a **radical sign**. The expression beneath the radical sign is called the **radicand**.

$$\text{Radical expression: } \overset{\text{radical sign}}{\sqrt{a}} \leftarrow \text{radicand}$$

When the **radicand** is a **perfect square**, the **square root** will simplify to a **rational number**.

### Example 1  Finding Square Roots

*My video summary*    Evaluate.

a. $\sqrt{64}$          b. $-\sqrt{169}$          c. $\sqrt{-100}$

d. $\sqrt{\dfrac{9}{25}}$          e. $\sqrt{0.81}$          f. $\sqrt{0}$

### Solution

a. We need to find the **principal square root** of 64. Because $8^2 = 64$, $\sqrt{64} = 8$.

b. We need to find the **negative square root** of 169. Because $(-13)^2 = 169$, $-\sqrt{169} = -13$.

c. There is no **real number** that can be **squared** to result in $-100$ because the square of a real number will always be **non-negative**. So, $\sqrt{-100}$ is not a real number.

**d.–f.** Try to evaluate these square roots on your own. View the **answers to parts d–f**, or watch this **video** to see solutions to all six parts.

 The square root of a **negative number** is not a real number.

**You Try It**    Work through this You Try It problem.

Work Exercises 1–5 in this textbook or in the *MyMathLab*® Study Plan.

OBJECTIVE 2    APPROXIMATE SQUARE ROOTS

In **Example 1**, we saw that a **square root** simplifies to a **rational number** if the **radicand** is a **perfect square**. But what happens when the radicand is not a perfect square? In this case, the square root is an **irrational number**.

Consider $\sqrt{7}$. The radicand 7 is not a perfect square, so $\sqrt{7}$ is an irrational number. For such **radical expressions**, we can find decimal approximations using a calculator. Before doing this, consider what might be a **reasonable** result. Notice that 7 is between the perfect squares 4 and 9.

$$4 < 7 < 9$$

So, the **principal square root** of 7 should be between the principal square roots of 4 and 9.

$$\sqrt{4} < \sqrt{7} < \sqrt{9}, \text{ or}$$
$$2 < \sqrt{7} < 3$$

Figure 1 shows the TI-84 Plus calculator display for $\sqrt{7}$. Rounding to three decimal places, we have $\sqrt{7} \approx 2.646$. This approximation is between 2 and 3 as expected.

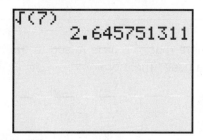

**Figure 1**
TI-84 Plus Display Approximating $\sqrt{7}$

**Example 2  Approximating Square Roots**

Use a calculator to approximate each **square root**. Round to three decimal places. Check that the answer is **reasonable**.

a. $\sqrt{10}$              b. $\sqrt{40}$              c. $\sqrt{72.5}$

**Solution**

a. The **radicand** 10 is between the **perfect squares** $3^2 = 9$ and $4^2 = 16$, so $\sqrt{9} < \sqrt{10} < \sqrt{16}$ or $3 < \sqrt{10} < 4$. The square root of 10 is between 3 and 4. From **Figure 2**, we approximate $\sqrt{10} \approx 3.162$, which is between 3 and 4.

b. 40 is between $6^2 = 36$ and $7^2 = 49$, so $\sqrt{36} < \sqrt{40} < \sqrt{49}$ or $6 < \sqrt{40} < 7$. **Figure 2** shows us that $\sqrt{40} \approx 6.325$, which is between 6 and 7.

c. 72.5 is between $8^2 = 64$ and $9^2 = 81$, so $\sqrt{64} < \sqrt{72.5} < \sqrt{81}$ or $8 < \sqrt{72.5} < 9$. In **Figure 2**, we see that $\sqrt{72.5} \approx 8.515$, which is reasonable.

```
√(10)
          3.16227766
√(40)
          6.32455532
√(72.5)
          8.514693183
```

**Figure 2**
TI-84 Plus Display for Estimating
$\sqrt{10}$, $\sqrt{40}$, and $\sqrt{72.5}$

**You Try It**    Work through this You Try It problem.

Work Exercises 6 and 7 in this textbook or in the *MyMathLab*® Study Plan.

OBJECTIVE 3    SIMPLIFY SQUARE ROOTS CONTAINING VARIABLES

Let's now consider **square roots** with **variables** in the **radicand**. A common misconception is to think that $\sqrt{a^2} = a$, but this is not necessarily true. To see why, substitute $a = -2$ in the expression $\sqrt{a^2}$ and simplify:

$$\text{Substitute } -2 \text{ for } a: \quad \sqrt{a^2} = \sqrt{(-2)^2}$$
$$\text{Simplify } (-2)^2: \quad = \sqrt{4}$$
$$\text{Find the principal square root of 4:} \quad = 2$$

The final result is not $-2$ but $|-2| = 2$. This illustrates the following square root property for simplifying **radical expressions** of the form $\sqrt{a^2}$.

---

**Simplifying Radical Expressions of the Form $\sqrt{a^2}$**

For any **real number** $a$,

$$\sqrt{a^2} = |a|.$$

---

When taking the **square root** of a **base** raised to the second **power**, the result will be the **absolute value** of the base. Therefore,

$$\sqrt{(-2)^2} = |-2| = 2,$$
$$\sqrt{x^2} = |x|,$$
$$\sqrt{(z + 2)^2} = |z + 2|, \quad \text{and so on.}$$

For simplicity, we **will assume that all radicands are non-negative and all variables in the radicand represent non-negative numbers only**. This will allow us to avoid the need for absolute value when simplifying radicals. Therefore,

$$\sqrt{x^2} = |x| = x \leftarrow \text{Assume } x \text{ is non-negative}$$
$$\sqrt{(z + 1)^2} = |z + 1| = z + 1 \leftarrow \text{Assume } z + 1 \text{ is non-negative}$$

To simplify square roots involving variables, we want to write the **radicand** as a **perfect square**. For example,

$$\underbrace{\sqrt{x^6} = \sqrt{(x^3)^2} = x^3}_{x^6 = x^3 \cdot x^3 = (x^3)^2} \quad \text{and} \quad \underbrace{\sqrt{m^2 + 2m + 1} = \sqrt{(m + 1)^2} = m + 1.}_{\substack{\text{Radicand is a} \\ \text{perfect square} \\ \text{trinomial}}}$$

### Example 3  Simplifying Square Roots Containing Variables

 Simplify. Assume all variables and radicands are non-negative.

a. $\sqrt{(3x - 7)^2}$

b. $\sqrt{121y^2}$

c. $\sqrt{9x^{12}}$

d. $\sqrt{25x^2 + 60x + 36}$

### Solution

a.  The **radicand** is written as a perfect square so we get $\sqrt{(3x - 7)^2} = 3x - 7$.

b.  Because $121 = 11^2$, we can write $121y^2$ as $(11y)^2$. Thus, $\sqrt{121y^2} = \sqrt{(11y)^2} = 11y$.

c.–d.  Try to simplify each square root on your own by first writing the radicand as a **base** raised to the second **power**. View the **answers** to parts c–d, or watch this **video** to see solutions to all four parts.

**You Try It**    Work through this You Try It problem.

Work Exercises 8–15 in this textbook or in the *MyMathLab*® Study Plan.

### OBJECTIVE 4  FIND CUBE ROOTS

We can apply the process of finding **square roots** to other types of roots such as the cube root. The **cube root** of a **real number** $a$ is a real number $b$ that, when **cubed**, results in $a$. So, $b$ is the cube root of $a$ if $b^3 = a$. For example, 2 is the cube root of 8 because $2^3 = 8$. To write the cube root of $a$, we use the notation $\sqrt[3]{a}$. The 3 in this **radical expression** is called the **index** and indicates a cube root instead of a square root. Using this notation, we write

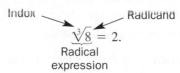

Index → $\sqrt[3]{8} = 2.$ ← Radicand

Radical
expression

---
**Definition    Cube Roots**

A real number $b$ is the **cube root** of a real number $a$, denoted as $b = \sqrt[3]{a}$, if $b^3 = a$.

---

Recall that the cube of a negative number is a negative number. Unlike square roots, cube roots with negative numbers in the **radicand** will have real number solutions. For example, $\sqrt[3]{-64} = -4$ because $(-4)^3 = -64$. Every real number has one real **cube root**. If the radicand is positive, the cube root will be positive. If the radicand is negative, the cube root will be negative.

---
**Simplifying Radical Expressions of the Form $\sqrt[3]{a^3}$**

For any real number $a$,

$$\sqrt[3]{a^3} = a.$$

---

### Example 4  Finding Cube Roots

 Simplify.

a. $\sqrt[3]{125}$

b. $\sqrt[3]{-1000}$

c. $\sqrt[3]{x^{15}}$

d. $\sqrt[3]{0.064}$

e. $\sqrt[3]{\dfrac{8}{27}}$

f. $\sqrt[3]{-64y^9}$

**Solution**

a. $\sqrt[3]{125} = \sqrt[3]{5^3} = 5$.

b. $\sqrt[3]{-1000} = \sqrt[3]{(-10)^3} = -10$.

c. $\sqrt[3]{x^{15}} = \sqrt[3]{(x^5)^3} = x^5$.

**d.–f.** Try to simplify each **cube root** on your own by first writing each **radicand** as a **base** raised to the third **power**. View the **answers** to parts d–f, or watch this **video** to see solutions to all six parts.

**You Try It**    Work through this You Try It problem.

**Work Exercises 16–21 in this textbook or in the *MyMathLab*** ® **Study Plan.**

OBJECTIVE 5    FIND AND APPROXIMATE $n$th ROOTS

We know that 2 is a **square root** of 4 because $2^2 = 4$ and 2 is a **cube root** of 8 because $2^3 = 8$. In fact, 2 is a **4th root** of 16 because $2^4 = 16$, and 2 is a **5th root** of 32 because $2^5 = 32$. We denote each of these roots using **radical signs** as follows: $\sqrt{4} = 2$, $\sqrt[3]{8} = 2$, $\sqrt[4]{16} = 2$, and $\sqrt[5]{32} = 2$. These are called *n*th roots and are written as $\sqrt[n]{a}$.

In the notation $\sqrt[n]{a}$, $n$ is the **index** of the **radical expression**, and it indicates the type of root. For example, a cube root has an index of $n = 3$. If no index is shown, then it is understood to be $n = 2$ for a square root.

If $n$ is an odd integer, then the **radicand** $a$ can be any real number. However, if $n$ is an even integer (as with square roots), then the radicand $a$ must be **non-negative** for the result to be a real number. We will continue to assume that all radicands containing variables represent non-negative numbers.

We simplify these expressions in the same way that we simplified square roots and cube roots. We try to write the radicand as a **base** raised to an **exponent** that is equal to the index.

**Example 5  Finding $n$th Roots**

*My video summary*     Simplify.

a. $\sqrt[4]{16}$          b. $\sqrt[5]{-243}$          c. $\sqrt[5]{y^{30}}$

d. $\sqrt[6]{\dfrac{1}{x^{12}}}$          e. $\sqrt[6]{(z+3)^6}$          f. $\sqrt[4]{-81}$

**Solution**

a. $\sqrt[4]{16} = \sqrt[4]{2^4} = 2$

b. $\sqrt[5]{-243} = \sqrt[5]{(-3)^5} = -3$

c. $\sqrt[5]{y^{30}} = \sqrt[5]{(y^6)^5} = y^6$

**d.–f.** Try to simplify each *n*th **root** on your own by first writing each **radicand** as a **base** raised to the *n*th **power**. View the **answers** to parts d–f, or watch this **video** to see the solutions to all six parts.

**You Try It**    Work through this You Try It problem.

**Work Exercises 22–29 in this textbook or in the *MyMathLab*** ® **Study Plan.**

In **Example 5** we saw that $\sqrt[4]{16} = 2$ because $16 = 2^4$. Because 16 is a **perfect 4th power**, its **4th root** is a **rational number**. But what if the **radicand of an $n$th root** *is not a perfect $n$th power*? Such expressions represent **irrational numbers**. As with square roots, we can use a calculator to approximate these types of $n$th roots.

### Example 6 Approximating $n$th Roots

Use a calculator to approximate each root. Round to three decimal places. Check that the answer is **reasonable**.

a. $\sqrt[3]{11}$    b. $\sqrt[4]{152}$    c. $\sqrt[5]{203}$

**Solution**

a. The radicand 11 is between the perfect cubes $2^3 = 8$ and $3^3 = 27$, so $\sqrt[3]{8} < \sqrt[3]{11} < \sqrt[3]{27}$ or $2 < \sqrt[3]{11} < 3$. From **Figure 3**, we get $\sqrt[3]{11} \approx 2.224$, which is between 2 and 3.

b. 152 is between $3^4 = 81$ and $4^4 = 256$, so $3 < \sqrt[4]{152} < 4$. In **Figure 3**, we see that $\sqrt[4]{152} \approx 3.511$, which is between 3 and 4.

c. 203 is between $2^5 = 32$ and $3^5 = 243$, so $2 < \sqrt[5]{203} < 3$. **Figure 3** shows us that $\sqrt[5]{203} \approx 2.894$, which is reasonable.

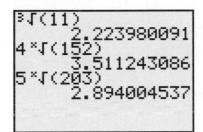

**Figure 3**
TI 84 Plus Display for Estimating $\sqrt[3]{11}$, $\sqrt[4]{152}$, and $\sqrt[5]{203}$

**You Try It**    Work through this You Try It problem.

Work Exercises 30–33 in this textbook or in the **_MyMathLab_**® Study Plan.

# 8.1  Exercises

In Exercises 1–5, evaluate each square root. If the answer is not a real number, state so.

**You Try It**

**1.** $\sqrt{36}$    **2.** $-\sqrt{25}$    **3.** $\sqrt{-25}$    **4.** $\sqrt{\dfrac{16}{81}}$    **5.** $\sqrt{0.16}$

In Exercises 6–7, use a calculator to approximate each square root. Round to three decimal places. Check that the answer is reasonable.

**You Try It**

**6.** $\sqrt{88}$    **7.** $\sqrt{20.2}$

In Exercises 8–15, simplify. Assume all variables and radicands represent non-negative real numbers.

**You Try It**

**8.** $\sqrt{(12y)^2}$    **9.** $\sqrt{(4x + 1)^2}$    **10.** $\sqrt{z^{16}}$    **11.** $\sqrt{49x^2}$

**12.** $\sqrt{16m^8}$    **13.** $-\sqrt{169x^4}$    **14.** $\sqrt{x^2 + 4x + 4}$    **15.** $\sqrt{4y^2 - 20y + 25}$

In Exercises 16–29, simplify. If the answer is not a real number, state so. Assume all variables represent non-negative real numbers.

**You Try It**

**16.** $\sqrt[3]{64}$

**17.** $\sqrt[3]{-343}$

**18.** $\sqrt[3]{-\dfrac{64}{125}}$

**19.** $\sqrt[3]{0.125}$

**20.** $\sqrt[3]{y^{30}}$

**21.** $\sqrt[3]{-27x^{12}}$

**You Try It**

**22.** $\sqrt[5]{3125}$

**23.** $-\sqrt[4]{625}$

**24.** $\sqrt[4]{\dfrac{1}{256}}$

**25.** $\sqrt[5]{-1024}$

**26.** $\sqrt[4]{x^{36}}$

**27.** $\sqrt[6]{(x+5)^6}$

**28.** $\sqrt[4]{-625}$

**29.** $\sqrt[5]{32x^{75}}$

In Exercises 30–33, use a calculator to approximate each root. Round to three decimal places. Check that the answer is reasonable.

**You Try It**

**30.** $\sqrt[3]{36}$

**31.** $\sqrt[4]{212}$

**32.** $\sqrt[5]{45}$

**33.** $\sqrt[6]{423}$

# 8.2  Simplifying Radicals

## THINGS TO KNOW

Before working through this section, be sure you are familiar with the following concepts:

|  |  | VIDEO | ANIMATION | INTERACTIVE |

**You Try It**

**1.** Simplify Exponential Expressions Using the Product Rule (Section 5.1, **Objective 1**)

**You Try It**

**2.** Simplify Exponential Expressions Using the Quotient Rule (Section 5.1, **Objective 2**)

**You Try It**

**3.** Find Square Roots (Section 8.1, **Objective 1**)

**You Try It**

**4.** Simplify Square Roots Containing Variables (Section 8.1, **Objective 3**)

**You Try It**

**5.** Find and Approximate *n*th Roots (Section 8.1, **Objective 5**)

## OBJECTIVES

**1** Simplify Square Roots Using the Product Rule

**2** Simplify Square Roots Using the Quotient Rule

**3** Simplify Square Roots Involving Variables

**4** Simplify Higher Roots

## OBJECTIVE 1  SIMPLIFY SQUARE ROOTS USING THE PRODUCT RULE

A **square root** is simplified if the **radicand** contains no **perfect square** factors other than 1. For example, $\sqrt{6}$ is simplified because the radicand, 6, contains no perfect square factors. However, $\sqrt{32}$ is not simplified because 32 can be factored as $32 = 16 \cdot 2$ and 16 is a perfect square. So how do we simplify such expressions? Consider the following:

$$\sqrt{4 \cdot 25} = \sqrt{100} = 10$$

and

$$\sqrt{4} \cdot \sqrt{25} = 2 \cdot 5 = 10$$

We can conclude that $\sqrt{4 \cdot 25} = \sqrt{4} \cdot \sqrt{25}$. This is an example of the **product rule for square roots**.

---

**Product Rule for Square Roots**

If $\sqrt{a}$ and $\sqrt{b}$ are real numbers, then $\sqrt{a \cdot b} = \sqrt{a} \cdot \sqrt{b}$.

---

This means that the square root of a **product** is equal to the product of the square roots. We can use this rule to simplify **radical expressions** such as $\sqrt{32}$ by splitting up the radicand into the product of two factors, one of which is a perfect square.

$$
\begin{aligned}
\text{Begin with the original expression:} \quad & \sqrt{32} \\
\text{Factor the radicand so one factor is a perfect square:} \quad & = \sqrt{16 \cdot 2} \\
\text{Apply the product rule for square roots:} \quad & = \sqrt{16} \cdot \sqrt{2} \\
\text{Evaluate } \sqrt{16}: \quad & = 4\sqrt{2}
\end{aligned}
$$

Because 2 has no perfect square factors, we cannot simplify any further. Therefore, $\sqrt{32}$ simplifies to $4\sqrt{2}$.

### Example 1  Simplifying Square Roots Using the Product Rule

*My video summary* 🎬 Simplify.

**a.** $\sqrt{8}$       **b.** $\sqrt{10}$       **c.** $\sqrt{48}$       **d.** $\sqrt{125}$

**Solution**

**a.** 8 is not a perfect square, but $8 = 4 \cdot 2$ so we can factor it into two factors such that one of the factors is a perfect square. We then apply the **product rule for square roots** and **simplify**.

$$
\begin{aligned}
\text{Begin with the original expression:} \quad & \sqrt{8} \\
\text{Factor 8 into the product } 4 \cdot 2: \quad & = \sqrt{4 \cdot 2} \\
\text{Use the product rule for square roots:} \quad & = \sqrt{4} \cdot \sqrt{2} \\
\text{Evaluate } \sqrt{4}: \quad & = 2\sqrt{2}
\end{aligned}
$$

**b.** 10 cannot be factored such that one of the factors is a **perfect square** (other than 1). Therefore, $\sqrt{10}$ is already simplified.

**c.** Begin with the original expression:   $\sqrt{48}$

Factor 48 into the product $16 \cdot 3$:   $= \sqrt{16 \cdot 3}$

Now apply the **product rule for square roots** and simplify, then view the **answer**.

**d.** Try **simplifying** this expression on your own. View the **answer**, or watch this **video** for complete solutions to all four parts.

**You Try It**    Work through this You Try It problem.

Work Exercises 1–4 in this textbook or in the *MyMathLab* Study Plan.

 Remember to check the remaining **radical** to see if it can be simplified further. For example, $\sqrt{32} = \sqrt{4 \cdot 8} = \sqrt{4} \cdot \sqrt{8} = 2\sqrt{8}$ is not simplified because we can further simplify $\sqrt{8}$.

### Example 2  Simplifying Square Roots Using the Product Rule

 Simplify.

**a.** $2\sqrt{75}$              **b.** $-3\sqrt{128}$

### Solution

**a.** 

| | |
|---|---|
| Begin with the original expression: | $2\sqrt{75}$ |
| Factor 75 into the product $25 \cdot 3$: | $= 2 \cdot \sqrt{25 \cdot 3}$ |
| Use the **product rule for square roots**: | $= 2 \cdot \sqrt{25} \cdot \sqrt{3}$ |
| Evaluate $\sqrt{25}$: | $= 2 \cdot 5 \cdot \sqrt{3}$ |
| Evaluate $2 \cdot 5$: | $= 10\sqrt{3}$ |

**b.** Try **simplifying** this expression on your own. View the **answer**, or watch this **video** for complete solutions to both parts.

**You Try It**    Work through this You Try It problem.

Work Exercises 5–8 in this textbook or in the *MyMathLab* Study Plan.

OBJECTIVE 2    SIMPLIFY SQUARE ROOTS USING THE QUOTIENT RULE

In addition to the **product rule for square roots**, we also have a rule for quotients. Consider the following:

$$\sqrt{\frac{100}{4}} = \sqrt{25} = 5$$

and

$$\frac{\sqrt{100}}{\sqrt{4}} = \frac{10}{2} = 5.$$

We can conclude that $\sqrt{\dfrac{100}{4}} = \dfrac{\sqrt{100}}{\sqrt{4}}$. This is an example of the **quotient rule for square roots**.

---

**Quotient Rule for Square Roots**

If $\sqrt{a}$ and $\sqrt{b}$ are real numbers and $b \neq 0$, then $\sqrt{\dfrac{a}{b}} = \dfrac{\sqrt{a}}{\sqrt{b}}$.

---

This means that the **square root** of a **quotient** is equal to the quotient of the square roots.

*My video summary*

### Example 3 Simplifying Square Roots Using the Quotient Rule

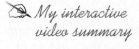 *My video summary*   **Simplify.**

**a.** $\sqrt{\dfrac{25}{16}}$  **b.** $\sqrt{\dfrac{75}{49}}$  **c.** $-\sqrt{\dfrac{64}{121}}$  **d.** $-\sqrt{\dfrac{52}{36}}$

**Solution**

**a.**  Begin with the original expression:  $\sqrt{\dfrac{25}{16}}$

Use the **quotient rule for square roots:**  $= \dfrac{\sqrt{25}}{\sqrt{16}}$

Evaluate $\sqrt{25}$ and $\sqrt{16}$:  $= \dfrac{5}{4} \begin{matrix} \leftarrow \sqrt{25} = 5 \\ \leftarrow \sqrt{16} = 4 \end{matrix}$

**b.**  Begin with the original expression:  $\sqrt{\dfrac{75}{49}}$

Use the quotient rule for square roots:  $= \dfrac{\sqrt{75}}{\sqrt{49}}$

Write 75 as the product $25 \cdot 3$:  $= \dfrac{\sqrt{25 \cdot 3}}{\sqrt{49}}$

Use the product rule for square roots:  $= \dfrac{\sqrt{25} \cdot \sqrt{3}}{\sqrt{49}}$

Evaluate $\sqrt{25}$ and $\sqrt{49}$:  $= \dfrac{5\sqrt{3}}{7} \begin{matrix} \leftarrow \sqrt{25} = 5 \\ \leftarrow \sqrt{49} = 7 \end{matrix}$

**c.–d.** Try simplifying these expressions on your own by applying the **quotient rule for square roots**, then view the **answers to parts c and d.**

**You Try It**  **Work through this You Try It problem.**

**Work Exercises 9–12 in this textbook or in the *MyMathLab*® Study Plan.**

OBJECTIVE 3  **SIMPLIFY SQUARE ROOTS INVOLVING VARIABLES**

To simplify **square roots** involving **variables**, we can use the **product rule** and **quotient rule** in basically the same way as with numbers. We split the **radicand** into two **factors** where one of the factors is the largest **perfect square** factor of the radicand.

If a variable has an even **exponent**, it can be written as a perfect square because its exponent will be divisible by 2. For example, $x^{10} = x^{5 \cdot 2} = (x^5)^2$. If a variable is raised to an odd exponent, we use the greatest even power as part of the perfect square factor and the remaining variable as part of the other factor. For example, we can write $x^{11}$ as $x^{10} \cdot x^1 = x^{10} \cdot x$. We would include $x^{10}$ with the perfect square factor and $x$ with the other factor. Then we use the product rule to **simplify**.

### Example 4 Simplifying Square Roots Containing Variables

*My interactive video summary*  **Simplify.**

**a.** $\sqrt{y^{15}}$  **b.** $\sqrt{20x^5}$  **c.** $\sqrt{\dfrac{24}{x^4}}$  **d.** $\sqrt{\dfrac{98x}{y^8}}$

**Solution**

a. Begin with the original expression: $\sqrt{y^{15}}$

Factor the radicand: $= \sqrt{y^{14} \cdot y}$ ← $y^{14}$ is the largest perfect square factor of the radicand, $y^{14} = (y^7)^2$

Product rule for square roots: $= \sqrt{y^{14}} \cdot \sqrt{y}$

Simplify: $= \sqrt{(y^7)^2} \cdot \sqrt{y}$

$= y^7 \sqrt{y}$

b. Begin with the original expression: $\sqrt{20x^5}$

Factor the radicand: $= \sqrt{4x^4 \cdot 5x}$ ← $4x^4$ is the largest perfect square factor of the radicand, $4x^4 = (2x^2)^2$

Product rule for square roots: $= \sqrt{4x^4} \cdot \sqrt{5x}$

Finish **simplifying** the expression on your own, then view the **answer to part b**. See the complete solutions to all four parts in this **interactive video**.

c. Begin with the original expression: $\sqrt{\dfrac{24}{x^4}}$

Quotient rule for square roots: $= \dfrac{\sqrt{24}}{\sqrt{x^4}}$

Factor the radicands: $= \dfrac{\sqrt{4 \cdot 6}}{\sqrt{x^4}}$ ← 4 is the largest perfect square factor of 24
← $x^4$ is a perfect square, $x^4 = (x^2)^2$

Product rule for square roots: $= \dfrac{\sqrt{4} \cdot \sqrt{6}}{\sqrt{x^4}}$

Finish **simplifying** the expression on your own, then view the **answer to part c**. See the complete solutions to all four parts in this **interactive video**.

d. Try simplifying this expression on your own. View the **answer to part d**, or view the complete solutions to all four parts in this **interactive video**.

**You Try It**   Work through this You Try It problem.

Work Exercises 13–20 in this textbook or in the *MyMathLab*® Study Plan.

OBJECTIVE 4   SIMPLIFY HIGHER ROOTS

The **product rule** and **quotient rule for square roots** can be generalized so we can simplify other roots. For example, to simplify **cube roots**, we factor the **radicand** into two factors where one of the factors is a **perfect cube**. To simplify a **fourth root**, we factor the radicand into two factors where one of the factors is a **perfect fourth power**. And so on. We illustrate this by considering some different roots of 32.

$$\sqrt{32} = \sqrt{16 \cdot 2}$$
16 is the largest
perfect square
factor of 32
$16 = 4^2$

$$= \sqrt{16} \cdot \sqrt{2}$$

$$= 4\sqrt{2}$$

$$\sqrt[3]{32} = \sqrt[3]{8 \cdot 4}$$
8 is the largest
perfect cube
factor of 32
$8 = 2^3$

$$= \sqrt[3]{8} \cdot \sqrt[3]{4}$$

$$= 2 \cdot \sqrt[3]{4}$$

$$\sqrt[4]{32} = \sqrt[4]{16 \cdot 2}$$
16 is the largest
perfect 4th power
factor of 32
$16 = 2^4$

$$= \sqrt[4]{16} \cdot \sqrt[4]{2}$$

$$= 2 \cdot \sqrt[4]{2}$$

 Be careful not to confuse an **exponent** with the **index** of a **radical**, or vice versa. For example, consider the product of $x^3$ and $\sqrt{y}$, written as $x^3\sqrt{y}$. Now, consider the product of $x$ and $\sqrt[3]{y}$, written as $x\sqrt[3]{y}$. Can you see how one expression might be confused with the other? When using such expressions, it is very important to write neatly. Consider using a multiplication symbol to make the expressions as clear as possible: $x^3 \cdot \sqrt{y}$ and $x \cdot \sqrt[3]{y}$.

A radical of the form $\sqrt[n]{a}$ is **simplified** if the **radicand** $a$ has no **factors** that are **perfect nth powers** (other than 1 or $-1$). The index of the radical determines how we should factor the radicand.

General forms of the product and quotient rules are given next.

---

**Product Rule for Radicals**

If $\sqrt[n]{a}$ and $\sqrt[n]{b}$ are **real numbers**, then $\sqrt[n]{ab} = \sqrt[n]{a} \cdot \sqrt[n]{b}$.

---

**Quotient Rule for Radicals**

If $\sqrt[n]{a}$ and $\sqrt[n]{b}$ are **real numbers** and $b \neq 0$, then $\sqrt[n]{\dfrac{a}{b}} = \dfrac{\sqrt[n]{a}}{\sqrt[n]{b}}$.

---

### Example 5  Simplifying Higher Roots

*My interactive video summary*

Simplify.

**a.** $\sqrt[3]{40}$    **b.** $\sqrt[3]{\dfrac{81x^7}{y^{12}}}$    **c.** $\sqrt[4]{80}$    **d.** $\sqrt[4]{\dfrac{512m^9}{81n^4}}$

**Solution**  Work through each of the following, or see this **interactive video** for fully worked solutions.

| Original expression | Write as a product using perfect cube factors | Product rule | Simplify the cube root |

**a.**  $\sqrt[3]{40} = \underset{8\,-\,2^3}{\sqrt[3]{8 \cdot 5}} = \sqrt[3]{8} \cdot \sqrt[3]{5} = 2 \cdot \sqrt[3]{5}$

| Original expression | Quotient rule | Write as a product using perfect cube factors | Product rule | Simplify the cube roots |

**b.**  $\sqrt[3]{\dfrac{81x^7}{y^{12}}} = \dfrac{\sqrt[3]{81x^7}}{\sqrt[3]{y^{12}}} = \dfrac{\sqrt[3]{27x^6 \cdot 3x}}{\sqrt[3]{y^{12}}} = \dfrac{\sqrt[3]{27x^6} \cdot \sqrt[3]{3x}}{\sqrt[3]{y^{12}}} = \dfrac{3x^2 \cdot \sqrt[3]{3x}}{y^4}$

$$27x^6 = (3x^2)^3;\ y^{12} = (y^4)^3$$

| Original expression | Write as a product using perfect fourth factors | Product rule | Simplify the fourth root |

**c.**  $\sqrt[4]{80} = \underset{16\,=\,2^4}{\sqrt[4]{16 \cdot 5}} = \sqrt[4]{16} \cdot \sqrt[4]{5} = 2 \cdot \sqrt[4]{5}$

| Original expression | Quotient rule | Write as a product using perfect fourth factors | Product rule | Simplify the fourth roots |

**d.**  $\sqrt[4]{\dfrac{512m^9}{81n^4}} = \dfrac{\sqrt[4]{512m^9}}{\sqrt[4]{81n^4}} = \dfrac{\sqrt[4]{256m^8 \cdot 2m}}{\sqrt[4]{81n^4}} = \dfrac{\sqrt[4]{256m^8} \cdot \sqrt[4]{2m}}{\sqrt[4]{81n^4}} = \dfrac{4m^2 \cdot \sqrt[4]{2m}}{3n}$

$$256m^8 = (4m^2)^4;\ 81n^4 = (3n)^4$$

**You Try It**    Work through this You Try It problem.

Work Exercises 21–26 in this textbook or in the *MyMathLab*® Study Plan.

# 8.2  Exercises

In Exercises 1–8, simplify using the product rule.

**You Try It**

1. $\sqrt{44}$    2. $-\sqrt{63}$    3. $\sqrt{51}$    4. $\sqrt{169}$

**You Try It**

5. $5\sqrt{147}$    6. $-3\sqrt{250}$    7. $7\sqrt{14}$    8. $-6\sqrt{144}$

In Exercises 9-12, simplify using the quotient rule.

9. $\sqrt{\dfrac{24}{121}}$    10. $-\sqrt{\dfrac{30}{49}}$    11. $\sqrt{\dfrac{64}{25}}$    12. $-\sqrt{\dfrac{45}{144}}$

**You Try It**

In Exercises 13–26, simplify.

13. $\sqrt{x^9}$    14. $\sqrt{25y^3}$    15. $\sqrt{48m^6}$    16. $\sqrt{54x^7}$

**You Try It**

17. $\sqrt{\dfrac{45}{b^8}}$    18. $\sqrt{\dfrac{40m}{n^6}}$    19. $\sqrt{\dfrac{27x^7}{y^{24}}}$    20. $\sqrt{\dfrac{50x^4}{y^{14}}}$

21. $\sqrt[3]{54}$    22. $\sqrt[3]{\dfrac{81}{8}}$    23. $\sqrt[4]{162}$    24. $\sqrt[4]{\dfrac{243}{256}}$

**You Try It**

25. $\sqrt[5]{128}$    26. $\sqrt[5]{\dfrac{64}{243}}$

# 8.3  Adding and Subtracting Radical Expressions

## THINGS TO KNOW

Before working through this section, be sure you are familiar with the following concepts:    VIDEO    ANIMATION    INTERACTIVE

**You Try It**
1. Simplify Algebraic Expressions (Section 1.6, **Objective 2**)

**You Try It**
2. Find Square Roots (Section 8.1, **Objective 1**)

**You Try It**
3. Simplify Square Roots Containing Variables (Section 8.1, **Objective 3**)

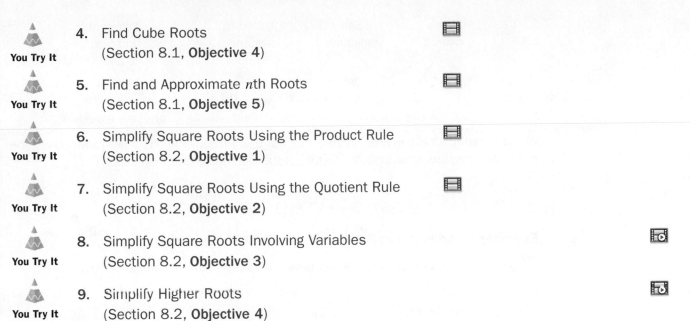

**4.** Find Cube Roots
(Section 8.1, **Objective 4**)

**5.** Find and Approximate $n$th Roots
(Section 8.1, **Objective 5**)

**6.** Simplify Square Roots Using the Product Rule
(Section 8.2, **Objective 1**)

**7.** Simplify Square Roots Using the Quotient Rule
(Section 8.2, **Objective 2**)

**8.** Simplify Square Roots Involving Variables
(Section 8.2, **Objective 3**)

**9.** Simplify Higher Roots
(Section 8.2, **Objective 4**)

## OBJECTIVES

**1** Add and Subtract Like Radicals

**2** Add and Subtract Radical Expressions That Must First Be Simplified

OBJECTIVE 1  ADD AND SUBTRACT LIKE RADICALS

As we learned in **Section 1.6**, to simplify an **algebraic expression**, we **combine like terms**. Recall that **like terms** have the exact same **variable factors**. For example, $9x^2y$ and $4x^2y$ are like terms because they have the same variable factor $x^2y$. We add or subtract like terms to result in a single term by using the **distributive property** in reverse.

$$9x^2y + 4x^2y = (9 + 4)x^2y = 13x^2y$$
$$9x^2y - 4x^2y = (9 - 4)x^2y = 5x^2y$$

The terms $7x^2y$ and $8xy^2$ are not like terms because the variable factors are different, so we cannot add or subtract them to result in a single term. This idea is true as well when **radicals** are involved. We can only add or subtract *like radicals*.

---

**Definition   Like Radicals**

Two or more **radical expressions** are **like radicals** if they have both the same **index** and the same **radicand**.

---

The radical expressions $9\sqrt{7}$ and $4\sqrt{7}$ are **like radicals** because they have the same **index** 2 (which is understood when no index is written) and the same **radicand** 7. The radical expressions $\sqrt[3]{5}$ and $\sqrt[3]{6}$ are **not** like radicals because the radicands are different. Likewise, $\sqrt{2}$ and $\sqrt[3]{2}$ are **not** like radicals because the indices are different.

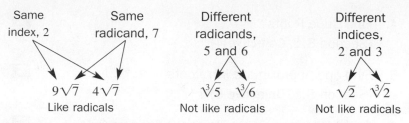

| Same index, 2 | Same radicand, 7 | Different radicands, 5 and 6 | Different indices, 2 and 3 |
|---|---|---|---|

$9\sqrt{7} \quad 4\sqrt{7}$

Like radicals

$\sqrt[3]{5} \quad \sqrt[3]{6}$

Not like radicals

$\sqrt{2} \quad \sqrt[3]{2}$

Not like radicals

We add or subtract like radicals in the same way that we add or subtract **like terms**. Using the **distributive property** in reverse, we factor out the like radical and then simplify.

$$9\sqrt{7} + 4\sqrt{7} = (9 + 4)\sqrt{7} = 13\sqrt{7}$$
$$9\sqrt{7} - 4\sqrt{7} = (9 - 4)\sqrt{7} = 5\sqrt{7}$$

## Example 1  Adding and Subtracting Radical Expressions

 Add or subtract if possible. Assume variables represent **non-negative** values.

**a.** $4\sqrt{13} + 6\sqrt{13}$          **b.** $\sqrt[3]{6x} - 9\sqrt[3]{6x}$

**c.** $5\sqrt{2y} + 6\sqrt{2} + 8\sqrt{2y}$          **d.** $7\sqrt{10} + 2\sqrt[3]{10}$

### Solutions

**a.** The **terms** $4\sqrt{13}$ and $6\sqrt{13}$ have the same **index** and the same **radicand**, so they are **like radicals**. They can be added to result in a single radical expression.

Begin with the original expression: $\quad 4\sqrt{13} + 6\sqrt{13}$

Reverse the **distributive property**: $\quad = (4 + 6)\sqrt{13}$

Simplify: $\quad = 10\sqrt{13}$

**b.** The terms $\sqrt[3]{6x}$ and $-9\sqrt[3]{6x}$ are like radicals. Note that $\sqrt[3]{6x} = 1\sqrt[3]{6x}$.

Begin with the original expression: $\quad \sqrt[3]{6x} - 9\sqrt[3]{6x} = 1\sqrt[3]{6x} - 9\sqrt[3]{6x}$

Reverse the **distributive property**: $\quad\quad\quad\quad\quad\quad\quad = (1 - 9)\sqrt[3]{6x}$

Simplify: $\quad\quad\quad\quad\quad\quad\quad\quad\quad\quad\quad = -8\sqrt[3]{6x}$

**c.–d.** Try to work these problems on your own. View the **answers**, or watch this **video** for complete solutions to all four parts.

**You Try It**  Work through this You Try It problem.

Work Exercises 1–8 in this textbook or in the **MyMathLab**® Study Plan.

Consider the expression $5\sqrt{2} + x\sqrt{2}$. Even though the two terms share the **like radical** $\sqrt{2}$, they cannot be combined fully into a single term because non-radical **factors**, $5$ and $x$, are not **like terms**.

$$5\sqrt{2} + x\sqrt{2} = \underbrace{(5 + x)}_{\substack{\text{Cannot} \\ \text{combine}}}\sqrt{2}$$

Therefore, to add or subtract like radicals into a single term, their non-radical factors must also be like terms.

## Example 2  Adding and Subtracting Radical Expressions

Add or subtract, if possible.

**a.** $5x\sqrt{15} + 7x\sqrt{15}$          **b.** $9x\sqrt{5} + 12x\sqrt{5} - 7\sqrt{5}$

**Solutions**

**a.** The **terms** $5x\sqrt{15}$ and $7x\sqrt{15}$ share the **like radical** $\sqrt{15}$. Also, the non-radical factors $5x$ and $7x$ are **like terms**, so they can be added.

$$\text{Begin with the original expression: } 5x\sqrt{15} + 7x\sqrt{15}$$
$$\text{Reverse the \textbf{distributive property}: } = (5+7)x\sqrt{15}$$
$$\text{Simplify: } = 12x\sqrt{15}$$

**b.** The **terms** $9x\sqrt{5}$, $12x\sqrt{5}$, and $-7\sqrt{5}$ all share the **like radical** $\sqrt{5}$. The non-radical factors $9x$ and $12x$ are **like terms**, but the factor $-7$ is not. This means we can combine $9x\sqrt{5}$ and $12x\sqrt{5}$ but not $-7\sqrt{5}$.

$$\text{Begin with the original expression: } 9x\sqrt{5} + 12x\sqrt{5} - 7\sqrt{5}$$
$$\text{Reverse the \textbf{distributive property}: } = (9+12)x\sqrt{5} - 7\sqrt{5}$$
$$\text{Simplify: } = 21x\sqrt{5} - 7\sqrt{5}$$

Notice that we can also factor out the $\sqrt{5}$ from our final answer $21x\sqrt{5} - 7\sqrt{5}$ to write it as $(21x - 7)\sqrt{5}$. Either answer is acceptable.

**You Try It**   Work through this You Try It problem.

Work Exercises 9 and 10 in this textbook or in the *MyMathLab*® Study Plan.

OBJECTIVE 2   ADD AND SUBTRACT RADICAL EXPRESSIONS THAT MUST FIRST BE SIMPLIFIED

Consider $\sqrt{12}$ and $\sqrt{27}$. The two **square roots** are not **like radicals** because they have different **radicands**. However, we can simplify them using the **product rule for square roots**:

$$\sqrt{12} = \sqrt{4\cdot 3} = \sqrt{4}\cdot\sqrt{3} = 2\sqrt{3} \quad\text{and}\quad \sqrt{27} = \sqrt{9\cdot 3} = \sqrt{9}\cdot\sqrt{3} = 3\sqrt{3}.$$

So, $\sqrt{12}$ and $\sqrt{27}$ simplify to the like radicals $2\sqrt{3}$ and $3\sqrt{3}$. In their simplified forms, they can be added or subtracted.

$$\sqrt{12} + \sqrt{27} = 2\sqrt{3} + 3\sqrt{3} = (2+3)\sqrt{3} = 5\sqrt{3}$$
$$\sqrt{12} - \sqrt{27} = 2\sqrt{3} - 3\sqrt{3} = (2-3)\sqrt{3} = -\sqrt{3}$$

 Sometimes **radical expressions** must be **simplified** first before we can determine if they can be added or subtracted.

**Example 3  Adding and Subtracting Radical Expressions**

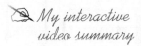
*My interactive video summary*

Add or subtract, if possible, by first simplifying each radical.

**a.** $\sqrt{20} + 9\sqrt{5}$

**b.** $7\sqrt{12} - 4\sqrt{75}$

**c.** $\sqrt{98} + \sqrt{72} - \sqrt{48}$

**d.** $\sqrt{64} + 2\sqrt{6} - \sqrt{54} + \sqrt{8}$

**Solutions**

**a.**

$$\text{Begin with the original expression: } \sqrt{20} + 9\sqrt{5}$$
$$\text{Factor 20 into the product } 4\cdot 5: = \sqrt{4\cdot 5} + 9\sqrt{5}$$
$$\text{Use the \textbf{product rule for square roots}: } = \sqrt{4}\cdot\sqrt{5} + 9\sqrt{5}$$
$$\text{Evaluate } \sqrt{4}: = 2\sqrt{5} + 9\sqrt{5}$$
$$\text{Combine the \textbf{like radicals}: } = 11\sqrt{5} \leftarrow \boxed{2\sqrt{5} + 9\sqrt{5} = (2+9)\sqrt{5} = 11\sqrt{5}}$$

**b.**  Begin with the original expression:  $7\sqrt{12} - 4\sqrt{75}$

Factor $12 = 4 \cdot 3$ and $75 = 25 \cdot 3$:  $= 7\sqrt{4 \cdot 3} - 4\sqrt{25 \cdot 3}$

Use the product rule for square roots:  $= 7 \cdot \sqrt{4} \cdot \sqrt{3} - 4 \cdot \sqrt{25} \cdot \sqrt{3}$

Evaluate $\sqrt{4}$ and $\sqrt{25}$:  $= 7 \cdot 2 \cdot \sqrt{3} - 4 \cdot 5 \cdot \sqrt{3}$

Multiply:  $= 14\sqrt{3} - 20\sqrt{3}$

Combine the like radicals:  $= -6\sqrt{3} \leftarrow \boxed{14\sqrt{3} - 20\sqrt{3} = (14 - 20)\sqrt{3} = -6\sqrt{3}}$

**c.–d.** Try to work these problems on your own. View the **answers**, or watch this **interactive video** for complete solutions to all four parts.

**You Try It**   **Work through this You Try It problem.**

**Work Exercises 11–20 in this textbook or in the *MyMathLab*® Study Plan.**

### Example 4  Adding and Subtracting Radical Expressions

Add or subtract, if possible.

**a.** $\sqrt[3]{54} - \sqrt[3]{16}$    **b.** $3\sqrt[3]{7} - \sqrt[3]{\dfrac{7}{8}}$    **c.** $\sqrt[3]{24} - \sqrt[3]{192} + 4\sqrt[3]{250}$

### Solutions

**a.** Begin with the original expression:  $\sqrt[3]{54} - \sqrt[3]{16}$

Factor $54 = 27 \cdot 2$ and $16 = 8 \cdot 2$:  $= \sqrt[3]{27 \cdot 2} - \sqrt[3]{8 \cdot 2}$

Use the **product rule for radicals**:  $= \sqrt[3]{27} \cdot \sqrt[3]{2} - \sqrt[3]{8} \cdot \sqrt[3]{2}$

Evaluate:  $= 3\sqrt[3]{2} - 2\sqrt[3]{2}$

Combine the **like radicals**:  $= \sqrt[3]{2} \leftarrow \boxed{3\sqrt[3]{2} - 2\sqrt[3]{2} = (3 - 2)\sqrt[3]{2} = \sqrt[3]{2}}$

**b.** Begin with the original expression:  $3\sqrt[3]{7} - \sqrt[3]{\dfrac{7}{8}}$

Use the **quotient rule for radicals**:  $= 3\sqrt[3]{7} - \dfrac{\sqrt[3]{7}}{\sqrt[3]{8}}$

Evaluate $\sqrt[3]{8}$:  $= 3\sqrt[3]{7} - \dfrac{\sqrt[3]{7}}{2}$

Reverse the **distributive property**:  $= \left(3 - \dfrac{1}{2}\right)\sqrt[3]{7} \leftarrow \boxed{\dfrac{\sqrt[3]{7}}{2} = \dfrac{1}{2}\sqrt[3]{7}}$

Simplify:  $= \dfrac{5}{2}\sqrt[3]{7}$ or $\dfrac{5\sqrt[3]{7}}{2} \leftarrow \boxed{3 - \dfrac{1}{2} = \dfrac{6}{2} - \dfrac{1}{2} = \dfrac{5}{2}}$

*My video summary*   📼 **c.** Try to work this problem on your own. View the **answer**, or watch this **video** for the complete solution to part c.

**You Try It**   **Work through this You Try It problem.**

**Work Exercises 21–26 in this textbook or in the *MyMathLab*® Study Plan.**

Recall that when variables are involved in **radical expressions**, we will assume that all **radicands** are **non-negative** and all variables in the radicand represent non-negative numbers.

**Example 5** Adding and Subtracting Radical Expressions Involving Variables

Add or subtract, if possible.

a. $\sqrt{49x^3} - \sqrt{16x^3} + 5\sqrt{x^2}$    b. $7\sqrt[3]{6y^3} - y\sqrt[3]{48} + \sqrt[3]{y^4}$

**Solutions**

a.  Begin with the original expression: $\sqrt{49x^3} - \sqrt{16x^3} + 5\sqrt{x^2}$

Factor the radicands: $= \sqrt{49x^2 \cdot x} - \sqrt{16x^2 \cdot x} + 5\sqrt{x^2}$

Use the **product rule for radicals**: $= \sqrt{49x^2} \cdot \sqrt{x} - \sqrt{16x^2} \cdot \sqrt{x} + 5\sqrt{x^2}$

Simplify: $= \sqrt{(7x)^2} \cdot \sqrt{x} - \sqrt{(4x)^2} \cdot \sqrt{x} + 5\sqrt{x^2}$

$= 7x\sqrt{x} - 4x\sqrt{x} + 5x$

Combine the like radicals: $= 3x\sqrt{x} + 5x$

*My video summary*   b.  Begin with the original expression: $7\sqrt[3]{6y^3} - y\sqrt[3]{48} + \sqrt[3]{y^4}$

Factor the radicands: $= 7\sqrt[3]{y^3 \cdot 6} - y\sqrt[3]{8 \cdot 6} + \sqrt[3]{y^3 \cdot y}$

Next, apply the **product rule for radicals**, and then finish this problem on your own. View the **answer**, or watch this **video** for the complete solution to part b.

**You Try It**    Work through this You Try It problem.

**Work Exercises 27–32 in this textbook or in the *MyMathLab*® Study Plan.**

# 8.3  Exercises

In Exercises 1–32, add or subtract, if possible. Assume that all variables represent non-negative values and be sure to simplify your answer.

**You Try It**

**1.** $5\sqrt{10} + 8\sqrt{10}$

**2.** $2\sqrt[3]{4y} - 7\sqrt[3]{4y}$

**3.** $5\sqrt{3} + 4\sqrt{6}$

**4.** $\sqrt{3x} + 3\sqrt{x} - 9\sqrt{3x}$

**5.** $\sqrt{11} - 3\sqrt[3]{5} + 6\sqrt{11} - 8\sqrt[3]{5}$

**6.** $11\sqrt{5} - 5\sqrt{x} - 8\sqrt{5} + 4\sqrt{x}$

**7.** $6\sqrt{a} + 2\sqrt{b} - a\sqrt{3}$

**8.** $\dfrac{\sqrt{x}}{4} + \dfrac{3}{8}\sqrt{x}$

**You Try It** **9.** $6x\sqrt{11} - 13x\sqrt{11}$

**10.** $4xy\sqrt[3]{2} + 3xy\sqrt[3]{2} - 5x\sqrt[3]{2}$

**11.** $\sqrt{18} + \sqrt{98}$

**12.** $\sqrt{63} - \sqrt{28}$

**You Try It**
 **13.** $\sqrt{50} + 3\sqrt{2}$

**14.** $3\sqrt{48} - \sqrt{24}$

**15.** $7\sqrt{45} - 4\sqrt{125}$

**16.** $\sqrt{\dfrac{3}{16}} + \sqrt{\dfrac{27}{4}}$

**17.** $5\sqrt{12} - \sqrt{75} + 3\sqrt{20}$

**18.** $\sqrt{180} - \sqrt{150} - \sqrt{45}$

**19.** $9\sqrt{7} - 6\sqrt{72} + 4\sqrt{18} - 5\sqrt{112}$

**20.** $3\sqrt{10} + \sqrt{81} - 4\sqrt{2} + \sqrt{160}$

**You Try It** **21.** $\sqrt[3]{72} - \sqrt[3]{9}$

**22.** $3\sqrt[3]{32} + \sqrt[3]{500}$

23. $4\sqrt[3]{40} - \sqrt[3]{320}$

24. $2\sqrt[3]{125} + 4\sqrt[3]{625}$

25. $7\sqrt[3]{54} + 3\sqrt[3]{8} - 5\sqrt[3]{16}$

26. $\sqrt[3]{\dfrac{3}{125}} + \sqrt[3]{81}$

**You Try It** 27. $6\sqrt{20x^3} + 4x\sqrt{45x}$

28. $4\sqrt{81x^5} - 3\sqrt{4x^5}$

29. $\sqrt{100t^3} + 6\sqrt{t^2} - \sqrt{25t^3}$

30. $3a^2\sqrt{a} + \sqrt{16a^5} - 5a\sqrt{a^3}$

31. $6\sqrt[3]{5x^3} + 2x\sqrt[3]{45} - \sqrt[3]{x^6}$

32. $9\sqrt[3]{y^4} - 3y\sqrt[3]{125y} + \sqrt[3]{64y^4}$

# 8.4 Multiplying and Dividing Radicals; Rationalizing Denominators

## THINGS TO KNOW

Before working through this section, be sure you are familiar with the following concepts:

| | | VIDEO | ANIMATION | INTERACTIVE |
|---|---|---|---|---|

 **You Try It**

1. Multiply the Sum and Difference of Two Terms (Section 5.5, **Objective 3**)

 **You Try It**

2. Simplify Square Roots Using the Product Rule (Section 8.2, **Objective 1**)

**You Try It**

3. Simplify Square Roots Using the Quotient Rule (Section 8.2, **Objective 2**)

**You Try It**

4. Simplify Square Roots Involving Variables (Section 8.2, **Objective 3**)

**You Try It**

5. Simplify Higher Roots (Section 8.2, **Objective 4**)

## OBJECTIVES

**1** Multiply Radical Expressions

**2** Divide Radical Expressions

**3** Rationalize Denominators

**4** Use Conjugates to Rationalize Denominators

### OBJECTIVE 1  MULTIPLY RADICAL EXPRESSIONS

In Section 8.2, we used the **product rule for radicals** to simplify radical expressions. This rule works in both directions: $\sqrt[n]{ab} = \sqrt[n]{a}\sqrt[n]{b}$ and $\sqrt[n]{a}\sqrt[n]{b} = \sqrt[n]{ab}$. So, we can also use the rule to multiply radical expressions. Let's rewrite the rule in the multiplication form.

**Product Rule for Radicals**

If $\sqrt[n]{a}$ and $\sqrt[n]{b}$ are **real numbers**, then $\sqrt[n]{a}\sqrt[n]{b} = \sqrt[n]{ab}$.

The **index** on each radical expression must be the same in order to use the product rule for radicals.

### Example 1  Multiplying Radicals

Multiply. Then simplify if possible. Assume all **variables** represent **non-negative** values.

a. $\sqrt{2}\cdot\sqrt{5}$    b. $\sqrt{2}\cdot\sqrt{18}$    c. $\sqrt[3]{2}\cdot\sqrt[3]{4}$    d. $\sqrt[5]{4x^2}\cdot\sqrt[5]{7y^3}$

**Solutions**  In each case, we use the **product rule for radicals** and then **simplify**.

$$
\begin{array}{ccccc}
\overbrace{\text{Original}}^{} & \overbrace{\text{Product rule}}^{} & \overbrace{\text{Simplify the}}^{} & & \\
\text{expression} & \text{for radicals} & \text{radicand} & & \\
\text{a. } \sqrt{2}\cdot\sqrt{5} & = & \sqrt{2\cdot5} & = & \sqrt{10}
\end{array}
$$

$$
\begin{array}{ccccccc}
\text{Original} & & \text{Product rule} & & \text{Simplify the} & & \\
\text{expression} & & \text{for radicals} & & \text{radicand} & & \text{Evaluate} \\
\text{b. } \sqrt{2}\cdot\sqrt{18} & = & \sqrt{2\cdot18} & = & \sqrt{36} & = & 6
\end{array}
$$

$$
\begin{array}{ccccccc}
\text{Original} & & \text{Product rule} & \text{Simplify the} & & \\
\text{expression} & & \text{for radicals} & \text{radicand} & & \text{Evaluate} \\
\text{c. } \sqrt[3]{2}\cdot\sqrt[3]{4} & = & \sqrt[3]{2\cdot4} & = & \sqrt[3]{8} & = & 2
\end{array}
$$

$$
\begin{array}{ccccc}
\text{Original} & & \text{Product rule} & & \text{Simplify the} \\
\text{expression} & & \text{for radicals} & & \text{radicand} \\
\text{d. } \sqrt[5]{4x^2}\cdot\sqrt[5]{7y^3} & = & \sqrt[5]{4x^2\cdot7y^3} & = & \sqrt[5]{28x^2y^3}
\end{array}
$$

**You Try It**    Work through this You Try It problem.

Work Exercises 1–4 in this textbook or in the ***MyMathLab***® Study Plan.

### Example 2  Multiplying Radicals

*My interactive video summary*

Multiply and **simplify**. Assume all **variables** represent **non-negative** values.

a. $3\sqrt{10}\cdot7\sqrt{2}$    b. $2\sqrt[3]{4}\cdot5\sqrt[3]{6}$    c. $\sqrt[4]{18x^3}\cdot\sqrt[4]{45x^2}$

**Solutions**  First, use the **product rule** to multiply **radicals**. Then, simplify.

a.

| | |
|---|---|
| Begin with the original expression: | $3\sqrt{10}\cdot7\sqrt{2}$ |
| Rearrange factors to group radicals together: | $= 3\cdot7\cdot\sqrt{10}\cdot\sqrt{2}$ |
| Use the product rule for radicals: | $= 3\cdot7\cdot\sqrt{10\cdot2}$ |
| Multiply $3\cdot7$; multiply $10\cdot2$: | $= 21\sqrt{20}$ |
| Factor the radicand: | $= 21\sqrt{4\cdot5}$ |
| Use the product rule for radicals: | $= 21\sqrt{4}\cdot\sqrt{5}$ |
| Simplify $\sqrt{4}$: | $= 21\cdot2\sqrt{5}$ |
| Multiply $21\cdot2$: | $= 42\sqrt{5}$ |

4 is the largest factor of 20 that is a perfect square.

**b.–c.** Try to work these problems on your own. View the **answers**, or watch this **interactive video** for complete solutions to all three parts.

**You Try It**    Work through this You Try It problem.

**Work Exercises 5–10 in this textbook or in the _MyMathLab_® Study Plan.**

Next, we multiply **radical expressions** with more than one term. To do this, we follow the same approach as when multiplying **polynomial expressions**. We use the **distributive property** to multiply each term in the first expression by each term in the second. Then we simplify the resulting products and combine **like terms** and **like radicals**.

### Example 3  Multiplying Radical Expressions With More than One Term

Multiply. Assume **variables** represent non-negative values.

**a.** $5\sqrt{2}\left(3\sqrt{2} - \sqrt{3}\right)$    **b.** $\sqrt[3]{2n^2}\left(\sqrt[3]{4n} + \sqrt[3]{5n}\right)$

**Solutions**

**a.** Begin with the original expression:  $5\sqrt{2}\left(3\sqrt{2} - \sqrt{3}\right)$

Use the distributive property:  $= 5\sqrt{2}\cdot 3\sqrt{2} - 5\sqrt{2}\cdot\sqrt{3}$

Rearrange factors:  $= 5\cdot 3\cdot\sqrt{2}\cdot\sqrt{2} - 5\cdot\sqrt{2}\cdot\sqrt{3}$

Multiply:  $= 15\sqrt{4} - 5\sqrt{6}$

Simplify radical:  $= 15\cdot 2 - 5\sqrt{6}$

Simplify:  $= 30 - 5\sqrt{6}$

✎ *My video summary*    ▤ **b.** Try this problem on your own. View the **answer**, or watch this **video** for a detailed solution to part b.

**You Try It**    Work through this You Try It problem.

**Work Exercises 11–14 in this textbook or in the _MyMathLab_® Study Plan.**

### Example 4  Multiplying Radical Expressions With More than One Term

Multiply. Assume **variables** represent non-negative values.

**a.** $\left(7\sqrt{2} - 2\sqrt{3}\right)\left(\sqrt{2} - 5\right)$    **b.** $\left(\sqrt{m} - 4\right)\left(3\sqrt{m} + 7\right)$

**Solutions**

**a.** Use the **FOIL** method to multiply the **radical expressions**. The two first **terms** are $7\sqrt{2}$ and $\sqrt{2}$. The two outside terms are $7\sqrt{2}$ and $-5$. The two inside terms are $-2\sqrt{3}$ and $\sqrt{2}$. The two last terms are $-2\sqrt{3}$ and $-5$.

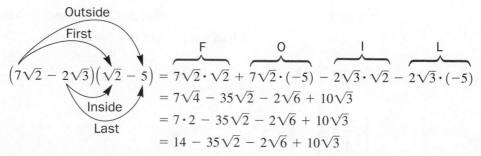

*My video summary*   b. Try this problem on your own. View the answer, or watch this **video** for a detailed solution to part b.

**You Try It**   Work through this You Try It problem.

**Work Exercises 15–20 in this textbook or in the *MyMathLab* Study Plan.**

## Example 5  Using Special Products to Multiply Radical Expressions

Multiply. Assume variables represent non-negative values.

**a.** $\left(\sqrt{y} + 3\right)\left(\sqrt{y} - 3\right)$   **b.** $\left(3\sqrt{x} - 2\right)^2$

**Solutions**

**a.** $(\sqrt{y} + 3)(\sqrt{y} - 3)$ is a **product of conjugates**. We use the rule for $(A + B)(A - B)$ with $A = \sqrt{y}$ and $B = 3$.

$$\text{Write the product of conjugates rule:} \quad (A + B)(A - B) = A^2 - B^2$$
$$\text{Substitute } \sqrt{y} \text{ for } A \text{ and 3 for } B: \quad \left(\sqrt{y} + 3\right)\left(\sqrt{y} - 3\right) = \left(\sqrt{y}\right)^2 - (3)^2$$
$$\text{Simplify:} \quad = y - 9$$

 *My video summary*   b. This **expression** has the form of the **square of a binomial difference** $(A - B)^2$. Use the **special product rule** $(A - B)^2 = A^2 - 2AB + B^2$ with $A = 3\sqrt{x}$ and $B = 2$. Finish working this problem on your own. View the **answer**, or watch this **video** for a detailed solution to part b.

**You Try It**   Work through this You Try It problem.

**Work Exercises 21–26 in this textbook or in the *MyMathLab* Study Plan.**

In Example 5a, the product of **conjugates** involving **square roots** resulted in an expression without any **radicals**. This result is true in general and occurs because the product of conjugates equals the **difference of two squares**. We will make use of this result later in Objective 4.

**OBJECTIVE 2   DIVIDE RADICAL EXPRESSIONS**

In **Section 8.2**, we used the **quotient rule for radicals** to simplify radical expressions involving fractions. Like the **product rule**, the quotient rule works in both directions:

$$\sqrt[n]{\frac{a}{b}} = \frac{\sqrt[n]{a}}{\sqrt[n]{b}} \text{ and } \frac{\sqrt[n]{a}}{\sqrt[n]{b}} = \sqrt[n]{\frac{a}{b}}. \text{ Let's rewrite the rule in the division form.}$$

---

**Quotient Rule for Radicals**

If $\sqrt[n]{a}$ and $\sqrt[n]{b}$ are **real numbers**, and $b \neq 0$, then $\dfrac{\sqrt[n]{a}}{\sqrt[n]{b}} = \sqrt[n]{\dfrac{a}{b}}$.

---

 The **Index** on each radical must be the same in order to use the quotient rule for radicals.

## Example 6  Dividing Radical Expressions

Divide. Then simplify if possible. Assume variables represent positive values.

a. $\dfrac{\sqrt{90}}{\sqrt{10}}$    b. $\dfrac{\sqrt{140}}{\sqrt{5}}$    c. $\dfrac{\sqrt{75x^3}}{\sqrt{3x}}$    d. $\dfrac{\sqrt[3]{750x}}{\sqrt[3]{6x}}$

**Solutions** In each case, we begin by using the **quotient rule for radicals**.

a.

| Original expression | Quotient rule for radicals | Simplify the radicand | Evaluate |
|---|---|---|---|

$$\underbrace{\frac{\sqrt{90}}{\sqrt{10}}}_{} = \underbrace{\sqrt{\frac{90}{10}}}_{} = \underbrace{\sqrt{9}}_{} = \underbrace{3}_{}$$

b.

| Original expression | Quotient rule for radicals | Simplify the radicand | Factor the radicand | Product rule for radicals | Simplify |
|---|---|---|---|---|---|

$$\underbrace{\frac{\sqrt{140}}{\sqrt{5}}}_{} = \underbrace{\sqrt{\frac{140}{5}}}_{} = \underbrace{\sqrt{28}}_{} = \underbrace{\sqrt{4\cdot 7}}_{} = \underbrace{\sqrt{4}\cdot\sqrt{7}}_{} = \underbrace{2\sqrt{7}}_{}$$

c.

| Original expression | Quotient rule for radicals | Simplify the radicand | Simplify |
|---|---|---|---|

$$\underbrace{\frac{\sqrt{75x^3}}{\sqrt{3x}}}_{} = \underbrace{\sqrt{\frac{75x^3}{3x}}}_{} = \underbrace{\sqrt{25x^2}}_{} = \underbrace{5x}_{}$$

d.

| Original expression | Quotient rule for radicals | Simplify the radicand | Evaluate |
|---|---|---|---|

$$\underbrace{\frac{\sqrt[3]{750x}}{\sqrt[3]{6x}}}_{} = \underbrace{\sqrt[3]{\frac{750x}{6x}}}_{} = \underbrace{\sqrt[3]{125}}_{} = \underbrace{5}_{}$$

**You Try It**    Work through this You Try It problem.

Work Exercises 27–32 in this textbook or in the *MyMathLab* ® Study Plan.

OBJECTIVE 3    RATIONALIZE DENOMINATORS

We now add more requirements for a radical expression to be **simplified**.

> **Simplified Radical Expression**
>
> For a **radical expression** to be **simplified**, it must meet the following three conditions:
>
> **Condition 1.**    The **radicand** has no **factor** that is a **perfect power** of the **index** of the radical.
>
> **Condition 2.**    The **radicand** contains no fractions or **negative exponents**.
>
> **Condition 3.**    No **denominator** contains a radical.

We can use the **product rule for radicals** to help resolve issues with Condition 1. We use the **quotient rule for radicals** to resolve issues with Conditions 2 and 3.

In **Example 6** we used the **quotient rule** to divide radicals. However, the quotient rule alone is not always enough to remove all radicals from the denominator. For example, simplifying $\sqrt{\frac{3}{32}}$ using the quotient rule and **product rule** yields the following:

$$\sqrt{\frac{3}{32}} = \overbrace{\frac{\sqrt{3}}{\sqrt{32}}}^{\text{Quotient rule}} = \overbrace{\frac{\sqrt{3}}{\sqrt{16}\cdot\sqrt{2}}}^{\text{Product rule}} = \frac{\sqrt{3}}{4\sqrt{2}}$$

We still have a radical in the denominator, so the expression is not simplified. We must remove all radicals from the denominator so that the denominator is a **rational number**. This process is called **rationalizing the denominator**.

---

**Rationalizing a Denominator with One Term**

To **rationalize a denominator** with a single radical of **index** $n$, multiply the **numerator** and denominator by a radical of index $n$ so that the radicand in the denominator is a **perfect $n$th power**.

---

Multiplying the numerator and denominator by the same radical is equivalent to multiplying by 1 since $\dfrac{\sqrt{2}}{\sqrt{2}} = 1$, $\dfrac{\sqrt{11}}{\sqrt{11}} = 1$, $\dfrac{\sqrt{x}}{\sqrt{x}} = 1$, and so on. We are simply writing an **equivalent expression** without radicals in the denominator. This idea is like writing **equivalent fractions** with a **least common denominator**.

We can continue simplifying $\sqrt{\dfrac{3}{32}}$ by multiplying $\dfrac{\sqrt{3}}{4\sqrt{2}}$ by $\dfrac{\sqrt{2}}{\sqrt{2}}$ as follows:

$$
\frac{\sqrt{3}}{4\sqrt{2}} \;=\; \overset{\substack{\text{Multiply}\\\text{by }1}}{\frac{\sqrt{3}}{4\sqrt{2}}\cdot\frac{\sqrt{2}}{\sqrt{2}}} \;=\; \overset{\substack{\text{Multiply}\\\text{fractions}}}{\frac{\sqrt{3}\cdot\sqrt{2}}{4\sqrt{2}\cdot\sqrt{2}}} \;=\; \overset{\substack{\text{Product}\\\text{rule}}}{\frac{\sqrt{6}}{4\sqrt{4}}} \;=\; \overset{\substack{\text{Simplify the}\\\text{denominator}}}{\frac{\sqrt{6}}{8}} \quad \boxed{4\sqrt{4} = 4\cdot 2 = 8}
$$

All three conditions for a **simplified radical expression** are now met, so $\sqrt{\dfrac{3}{32}} = \dfrac{\sqrt{6}}{8}$ is simplified.

## Example 7  Rationalizing Denominators

**Rationalize the denominator.**

**a.** $\dfrac{\sqrt{5}}{\sqrt{3}}$       **b.** $\sqrt{\dfrac{2}{5x}}$

## Solutions

**a.** Because the **denominator** contains a **square root**, we multiply the **numerator** and denominator by a square root so that the **radicand** in the denominator is a **perfect square**. $3\cdot 3 = 3^2 = 9$ is a perfect square, so we multiply the numerator and denominator by $\sqrt{3}$.

$$
\overset{\substack{\text{Original}\\\text{expression}}}{\frac{\sqrt{5}}{\sqrt{3}}} \;=\; \overset{\substack{\text{Multiply}\\\text{by }1}}{\frac{\sqrt{5}}{\sqrt{3}}\cdot\frac{\sqrt{3}}{\sqrt{3}}} \;=\; \overset{\substack{\text{Product}\\\text{rule}}}{\frac{\sqrt{15}}{\sqrt{9}}} \;=\; \overset{\substack{\text{Simplify}}}{\frac{\sqrt{15}}{3}}
$$

**b.** Using the **quotient rule**, we write $\sqrt{\dfrac{2}{5x}} = \dfrac{\sqrt{2}}{\sqrt{5x}}$. Because the denominator contains a square root, we multiply the numerator and denominator by a **square root** so that the **radicand** in the denominator is a **perfect square**.

$5x \cdot 5x = (5x)^2 = 25x^2$ is a perfect square, so we multiply the numerator and denominator by $\sqrt{5x}$.

$$\overbrace{\sqrt{\frac{2}{5x}}}^{\substack{\text{Original} \\ \text{expression}}} = \overbrace{\frac{\sqrt{2}}{\sqrt{5x}}}^{\substack{\text{Quotient} \\ \text{rule}}} = \overbrace{\frac{\sqrt{2}}{\sqrt{5x}} \cdot \frac{\sqrt{5x}}{\sqrt{5x}}}^{\substack{\text{Multiply} \\ \text{by } 1}} = \overbrace{\frac{\sqrt{10x}}{\sqrt{25x^2}}}^{\substack{\text{Product} \\ \text{rule}}} = \overbrace{\frac{\sqrt{10x}}{5x}}^{\text{Simplify}}$$

**You Try It**     Work through this You Try It problem.

**Work Exercises 33–36 in this textbook or in the *MyMathLab*®** Study Plan.

Why would we want to **rationalize a denominator**? View this popup box to find out.

## Example 8   Rationalizing Denominators with Cube Roots or Fourth Roots

**Rationalize the denominator.**

a. $\sqrt[3]{\dfrac{11}{25x}}$      b. $\dfrac{\sqrt[4]{7x}}{\sqrt[4]{27y^2}}$

## Solutions

a. Using the **quotient rule**, we write $\sqrt[3]{\dfrac{11}{25x}} = \dfrac{\sqrt[3]{11}}{\sqrt[3]{25x}}$. For **roots** with indices greater than 2,

it is helpful to write the **radicand** of the denominator in **exponential form**.

$$\sqrt[3]{25x} = \sqrt[3]{5^2 x^1}$$

Because the denominator contains a **cube root**, we multiply the numerator and denominator by a cube root so that the radicand in the denominator is a **perfect cube**. To do this, we need the **factors** in the radicand to have **exponents** of 3 or **multiples** of 3. We have two 5's and one $x$, so we need one more 5 and two more $x$'s to get $5^2 x^1 \cdot 5^1 x^2 = 5^3 x^3 = (5x)^3$, which is a perfect cube. Therefore, we multiply the numerator and denominator by $\sqrt[3]{5^1 x^2} = \sqrt[3]{5x^2}$.

$$\overbrace{\sqrt[3]{\frac{11}{25x}}}^{\substack{\text{Original} \\ \text{expression}}} = \overbrace{\frac{\sqrt[3]{11}}{\sqrt[3]{25x}}}^{\substack{\text{Quotient} \\ \text{rule}}} = \overbrace{\frac{\sqrt[3]{11}}{\sqrt[3]{25x}} \cdot \frac{\sqrt[3]{5x^2}}{\sqrt[3]{5x^2}}}^{\substack{\text{Multiply} \\ \text{by } 1}} = \overbrace{\frac{\sqrt[3]{55x^2}}{\sqrt[3]{125x^3}}}^{\substack{\text{Product} \\ \text{rule}}} = \overbrace{\frac{\sqrt[3]{55x^2}}{5x}}^{\text{Simplify}}$$

 *My video summary*       b. Try this problem on your own. View the **answer**, or watch this **video** for a detailed solution to part b.

**You Try It**     Work through this You Try It problem.

**Work Exercises 37–38 in this textbook or in the *MyMathLab*®** Study Plan.

Though not required, it is best to **rationalize the denominator** last when **simplifying radical expressions** since the **radicand** in the denominator may then involve simpler expressions.

### Example 9 Rationalizing Denominators

 *My video summary*

 Simplify each expression first and then rationalize the denominator.

a. $\sqrt{\dfrac{3x}{50}}$    b. $\dfrac{\sqrt{18x}}{\sqrt{27xy}}$    c. $\sqrt[3]{\dfrac{-4x^5}{16y^5}}$

**Solutions** Try these problems on your own. See the **answers**, or watch this **video** for detailed solutions to all three parts.

**You Try It**    Work through this You Try It problem.

Work Exercises 39–40 in this textbook or in the *MyMathLab* Study Plan.

## OBJECTIVE 4    USE CONJUGATES TO RATIONALIZE DENOMINATORS

The goal of **rationalizing a denominator** is to write an equivalent expression without **radicals** in the denominator. Earlier in this section, we noted that multiplying **conjugates** involving **square roots** results in an expression without any radicals. So, we can use the **product of conjugates** to rationalize the denominator of a radical expression whose **denominator** contains two **terms** involving one or more square roots.

> **Rationalizing a Denominator with Two Terms**
>
> To rationalize a denominator with two terms involving one or more square roots, multiply the numerator and denominator by the conjugate of the denominator.

### Example 10    Rationalizing Denominators with Two Terms

Rationalize the denominator.

a. $\dfrac{2}{\sqrt{3}+5}$    b. $\dfrac{7}{3\sqrt{x}-4}$    c. $\dfrac{\sqrt{y}-3}{\sqrt{y}+2}$

**Solutions**

a. Because the **denominator** has two **terms** and involves a **square root**, we multiply the **numerator** and denominator by the **conjugate** of the denominator $\sqrt{3}-5$.

Multiply numerator and denominator by $\sqrt{3}-5$: $\dfrac{2}{\sqrt{3}+5} = \dfrac{2}{\sqrt{3}+5}\cdot\dfrac{\sqrt{3}-5}{\sqrt{3}-5}$

Multiply numerators and multiply denominators: $= \dfrac{2(\sqrt{3}-5)}{(\sqrt{3})^2-(5)^2}$

$= \dfrac{2(\sqrt{3}-5)}{3-25}$

Simplify the denominator: $= \dfrac{2(\sqrt{3}-5)}{-22}$

Divide out the common factor of 2: $= \dfrac{\overset{1}{2}(\sqrt{3}-5)}{\underset{11}{-\cancel{22}}}$

Simplify: $= -\dfrac{\sqrt{3}-5}{11}$ or $\dfrac{5-\sqrt{3}}{11}$

**b.** Since the **denominator** has two **terms** and involves a **square root**, we multiply the **numerator** and denominator by the **conjugate** of the denominator, $3\sqrt{x} + 4$.

$$\frac{7}{3\sqrt{x} - 4} = \overbrace{\frac{7}{3\sqrt{x} - 4} \cdot \frac{3\sqrt{x} + 4}{3\sqrt{x} + 4}}^{\substack{\text{Multiply numerator}\\\text{and denominator by}\\3\sqrt{x} + 4}} = \overbrace{\frac{7(3\sqrt{x} + 4)}{(3\sqrt{x})^2 - (4)^2}}^{\substack{\text{Multiply numerators}\\\text{and multiply}\\\text{denominators}}} = \overbrace{\frac{7(3\sqrt{x} + 4)}{9x - 16}}^{\text{Simplify}} \text{ or } \overbrace{\frac{21\sqrt{x} + 28}{9x - 16}}^{\text{Distribute}}$$

We can leave the result in factored form, or we can distribute in the final step. Leaving the numerator factored until the end helps with dividing out **common factors**, if they are present.

*My video summary*  **c.** Work this problem on your own. View the answer, or watch this **video** for a detailed solution to part c.

**You Try It**   **Work through this You Try It problem.**

**Work Exercises 41–46 in this textbook or in the *MyMathLab*® Study Plan.**

# 8.4 Exercises

For Exercises 1–46, assume that all variables represent positive values.

In Exercises 1–10, multiply. Then simplify if possible.

 **You Try It**   **1.** $\sqrt{5} \cdot \sqrt{7}$         **2.** $\sqrt{3} \cdot \sqrt{12}$         **3.** $\sqrt[3]{5x^2} \cdot \sqrt[3]{25x}$         **4.** $\sqrt[4]{5x} \cdot \sqrt[4]{15y^3}$

**5.** $2\sqrt{6} \cdot 4\sqrt{30}$         **6.** $4\sqrt{3x} \cdot 7\sqrt{6x}$         **7.** $\sqrt{20x^3y} \cdot \sqrt{18xy^4}$

 **You Try It**   **8.** $\sqrt[3]{9x^2} \cdot \sqrt[3]{6x^2}$         **9.** $2x\sqrt[3]{5y^2} \cdot 3x\sqrt[3]{25y}$         **10.** $\sqrt[4]{8x^2} \cdot \sqrt[4]{6x^3}$

In Exercises 11–26, multiply.

 **11.** $\sqrt{5}(2x + \sqrt{3})$         **12.** $\sqrt{2x}(\sqrt{6x} - 5)$

 **You Try It** **13.** $2\sqrt{3m}(m - 3\sqrt{5m})$         **14.** $\sqrt[3]{x}(\sqrt[3]{54x^2} - \sqrt[3]{x})$

 **15.** $(8\sqrt{3} - 5)(\sqrt{3} - 1)$         **16.** $(3 + \sqrt{5})(2 + \sqrt{7})$

 **You Try It** **17.** $(\sqrt{x} + 4)(\sqrt{x} + 6)$         **18.** $(4 - 2\sqrt{3x})(5 + \sqrt{3x})$

 **19.** $(3 - \sqrt{x})(4 - \sqrt{y})$         **20.** $(\sqrt[3]{y} + 1)(\sqrt[3]{y} - 2)$

 **You Try It** **21.** $(\sqrt{7z} + 3)(\sqrt{7z} - 3)$         **22.** $(\sqrt{5} + 3y)^2$

**23.** $(\sqrt{x} + 3)(3 - \sqrt{x})$         **24.** $(\sqrt{3a} - \sqrt{2b})(\sqrt{3a} + \sqrt{2b})$

**25.** $(2\sqrt{x} + 5)^2$         **26.** $(2\sqrt{m} - 3\sqrt{n})^2$

In Exercises 27–32, divide. Then simplify if possible.

**27.** $\dfrac{\sqrt{45}}{\sqrt{5}}$
**You Try It**

**28.** $\dfrac{\sqrt{96x^6y^8}}{\sqrt{12x^2y^5}}$

**29.** $\dfrac{\sqrt{120x^9}}{\sqrt{3x}}$

**30.** $\dfrac{\sqrt[3]{5000}}{\sqrt[3]{5}}$

**31.** $\dfrac{\sqrt[3]{-72m^5}}{\sqrt[3]{3m^2}}$

**32.** $\dfrac{\sqrt{3x^5y^3}}{\sqrt{48xy^5}}$

In Exercises 33–46, rationalize the denominator.

**33.** $\dfrac{4}{\sqrt{6}}$

**34.** $\dfrac{\sqrt{3}}{\sqrt{5}}$

**You Try It** **35.** $\sqrt{\dfrac{9}{2x}}$

**36.** $\dfrac{5}{\sqrt{11x}}$

**You Try It** **37.** $\sqrt[3]{\dfrac{9}{4x^2}}$

**38.** $\dfrac{\sqrt[4]{3b^5}}{\sqrt[4]{25a}}$

**39.** $\dfrac{8}{\sqrt{12x^3y^4}}$
**You Try It**

**40.** $\dfrac{\sqrt[3]{2x^2}}{\sqrt[3]{36y^5}}$

**41.** $\dfrac{5}{2-\sqrt{3}}$

**42.** $\dfrac{-4x}{5+\sqrt{6}}$

**43.** $\dfrac{-3}{\sqrt{x}+1}$
**You Try It**

**44.** $\dfrac{\sqrt{a}}{3\sqrt{a}-\sqrt{b}}$

**45.** $\dfrac{\sqrt{m}-4}{\sqrt{m}-7}$

**46.** $\dfrac{5\sqrt{3}+2\sqrt{6}}{4\sqrt{6}-\sqrt{3}}$

# 8.5 Solving Radical Equations

## THINGS TO KNOW

Before working through this section, be sure you are familiar with the following concepts:

|  |  | VIDEO | ANIMATION | INTERACTIVE |
|---|---|---|---|---|

**You Try It**
**1.** Solve Linear Equations Using Both Properties of Equality (Section 2.1, **Objective 5**)  

**You Try It**
**2.** Evaluate a Formula (Section 2.4, **Objective 1**)  

**You Try It**
**3.** Solve a Formula for a Given Variable (Section 2.4, **Objective 3**)  

**You Try It**
**4.** Square a Binomial Sum (Section 5.5, **Objective 1**)

**You Try It**
**5.** Square a Binomial Difference (Section 5.5, **Objective 2**)

**You Try It**

6. Solve Quadratic Equations by Factoring
   (Section 6.7, **Objective 1**)

**You Try It**

7. Multiply Radical Expressions
   (Section 8.4, **Objective 1**)

## OBJECTIVES

**1** Solve Equations Involving One Radical Expression

**2** Solve Equations Involving Two Radical Expressions

**3** Evaluate Formulas Involving Square Roots

**4** Use Radical Equations to Solve Application Problems

OBJECTIVE 1   SOLVE EQUATIONS INVOLVING ONE RADICAL EXPRESSION

Recall that, when **solving** an **equation in one variable**, we find all values of the **variable** that make the equation true. All of these values together form the **solution set** of the equation. Previously, we have solved **linear equations**, **polynomial equations**, and **rational equations**. Now we learn how to solve *radical equations*.

> **Definition   Radical Equation**
>
> A **radical equation** is an equation that contains at least one **radical expression** with a variable in the **radicand**.

In this section, we will focus on radical equations involving **square roots** such as in the following two examples.

$$\sqrt{x+4} = 15 \quad \text{and} \quad \sqrt{x-6} - \sqrt{x} = 3$$

The first equation contains one radical expression, while the second equation contains two radical expressions.

 Not all equations that contain radical expressions are radical equations. If no radical expression contains a variable in the radicand, then the equation is not a radical equation. For example, $2x + 1 = \sqrt{3}$ is not a radical equation.

Consider the equation $\sqrt{x} = 9$, which has one **radical expression** $\sqrt{x}$. The radical expression $\sqrt{x}$ is an **isolated radical expression** because it stands alone on one side of the equal sign. An isolated radical can be eliminated from an equation by raising both sides of the equation to the power of the **index**. In this case, we **square both sides** of the equation because the radical expression is a **square root**.

$$\begin{aligned} \text{Begin with the original equation:} \quad & \sqrt{x} = 9 \\ \text{Square both sides:} \quad & \left(\sqrt{x}\right)^2 = (9)^2 \\ \text{Simplify:} \quad & x = 81 \end{aligned}$$

Checking this answer, we substitute 81 for $x$ in the original equation:

$$\begin{aligned} \text{Substitute 81 for } x: \quad & \sqrt{81} \overset{?}{=} 9 \\ \text{Simplify:} \quad & 9 = 9 \quad \text{True} \end{aligned}$$

Because $x = 81$ **satisfies** the original equation, the **solution set** is $\{81\}$.

---

**Squaring Property of Equality**

If $A$ and $B$ represent **algebraic expressions**, then any **solution** to the equation $A = B$ is also a solution to the equation $A^2 = B^2$.

---

Sometimes when we square both sides of an **equation**, the new equation will have **solutions** that are not solutions to the original equation. Such "solutions" are called **extraneous solutions**. They must be identified and excluded from the **solution set**. View this **popup** for a more in-depth explanation of extraneous solutions.

We can use the following general steps to **solve** equations involving one **radical expression**.

---

**Solving Equations Involving One Radical Expression**

**Step 1.** Isolate the radical expression. Use the **properties of equality** to get the radical expression by itself on one side of the equal sign.

**Step 2.** Eliminate the radical. Identify the **index** of the radical expression and raise both sides of the equation to the index power.

**Step 3.** Solve the resulting equation.

**Step 4.** Check each solution from Step 3 in the original equation. Disregard any **extraneous solutions**.

---

 When solving **radical equations**, answers can fail to check for two reasons: An answer may be an extraneous solution, or an error may have been made while solving. When an answer is disregarded as an extraneous solution, it is important to make sure that no errors were made while solving.

## Example 1  Solving an Equation Involving One Radical Expression

Solve $\sqrt{x + 5} = 3$.

**Solutions**  We follow the **four-step process**.

**Step 1.** The **radical expression** is already **Isolated** on the left.

**Step 2.** Using the **squaring property of equality**, we can eliminate the radical by **squaring both sides**.

$$\text{Square both sides:} \quad \left(\sqrt{x + 5}\right)^2 = (3)^2$$
$$\text{Simplify:} \quad x + 5 = 9$$

**Step 3.** Solve the resulting **linear equation**.

$$\text{Subtract 5 from both sides:} \quad x + 5 - 5 = 9 - 5$$
$$\text{Simplify:} \quad x = 4$$

**Step 4.** Check $x = 4$ in the original equation.

$$\text{Substitute 4 for } x \text{ in the original equation:} \quad \sqrt{4 + 5} \overset{?}{=} 3$$
$$\text{Simplify beneath the radical:} \quad \sqrt{9} \overset{?}{=} 3$$
$$\text{Evaluate the square root:} \quad 3 = 3 \quad \text{True}$$

The answer checks in the original equation, so the solution set is $\{4\}$.

**You Try It**   **Work through this You Try It problem.**

**Work Exercises 1–4 in this textbook or in the *MyMathLab*® Study Plan.**

### Example 2 Solving an Equation Involving One Radical Expression

 *My video summary*  Solve $\sqrt{2-x} - 3 = 4$.

**Solution** We follow the **four-step process**.

**Step 1. Isolate** the **radical expression** by adding 3 to both sides.

$$\text{Begin with the original equation:} \qquad \sqrt{2-x} - 3 = 4$$
$$\text{Add 3 to both sides:} \quad \sqrt{2-x} - 3 + 3 = 4 + 3$$
$$\text{Simplify:} \qquad \sqrt{2-x} = 7$$

**Step 2.** Eliminate the radical by **squaring both sides**.

$$\text{Square both sides:} \quad \left(\sqrt{2-x}\right)^2 = (7)^2$$
$$\text{Simplify:} \qquad 2 - x = 49$$

**Step 3.** Solve the resulting **linear equation**.

$$\text{Subtract 2 from both sides:} \quad 2 - x - 2 = 49 - 2$$
$$\text{Simplify:} \qquad -x = 47$$
$$\text{Multiply both sides by } -1: \qquad x = -47$$

Check this answer on your own to confirm that the **solution set** is $\{-47\}$. Watch this **video** to see the full solution including the check.

**You Try It**   **Work through this You Try It problem.**

**Work Exercises 5–10 in this textbook or in the *MyMathLab*®** Study Plan.

**Note:** In Example 2 we see that it is possible to have negative solutions even though the equation contains a square root. Remember that it is the *radicand* that cannot be negative.

⬦ CAUTION  When solving a **radical equation**, it is important to **isolate the radical** before raising both sides to the **index** power. If this is not done, then the **radical expression** will not be eliminated. View this **popup** to see an example.

### Example 3 Solving an Equation Involving One Radical Expression

Solve $\sqrt{3x-8} + 5 = 3$.

**Solution**

**Step 1.**    Begin with the original equation: $\qquad \sqrt{3x-8} + 5 = 3$
$$\text{Subtract 5 from both sides:} \quad \sqrt{3x-8} + 5 - 5 = 3 - 5$$
$$\text{Simplify:} \qquad \sqrt{3x-8} = -2$$
**Step 2.**    Square both sides: $\quad \left(\sqrt{3x-8}\right)^2 = (-2)^2$
$$\text{Simplify:} \qquad 3x - 8 = 4$$
**Step 3.**    Add 8 to both sides: $\quad 3x = 12$
$$\text{Divide both sides by 3:} \quad x = 4$$
**Step 4.** Check $x = 4$ in the original equation: $\qquad \sqrt{3(4)-8} + 5 \overset{?}{=} 2$
$$\text{Simplify beneath the radical:} \qquad \sqrt{4} + 5 \overset{?}{=} 2$$
$$\text{Evaluate the square root:} \qquad 2 + 5 \overset{?}{=} 2$$
$$\text{Add:} \qquad 7 = 2 \quad \text{False}$$

The answer $x = 4$ does not check in the original equation. It is an **extraneous solution**. Because this is the only possible **solution**, the equation has no real solution. The **solution set** is { } or $\varnothing$.

**Note:** We might have noticed sooner that this equation would have no real solution. In Step 1, when we **isolated the radical**, the resulting equation was $\sqrt{3x - 8} = -2$. This equation states that a **principal square root** equals a negative value, but this is impossible because principal square roots must be non-negative. So, this equation has no real solution.

**You Try It**    Work through this You Try It problem.

**Work Exercises 11 and 12 in this textbook or in the *MyMathLab*® Study Plan.**

### Example 4 Solving an Equation Involving One Radical Expression

 *My video summary*    Solve $\sqrt{x^2 + 3x - 6} = x$.

**Solution** We follow the **four-step process**.

**Step 1.** The **radical expression** is already **Isolated** on the left.

**Step 2.** Eliminate the radical by **squaring both sides**.

$$\text{Square both sides:} \quad \left(\sqrt{x^2 + 3x - 6}\right)^2 = (x)^2$$
$$\text{Simplify:} \quad x^2 + 3x - 6 = x^2$$

Finish solving this problem on your own. View the **answer**, or watch this **video** to see the full solution including the check.

**You Try It**    Work through this You Try It problem.

**Work Exercises 13 and 14 in this textbook or in the *MyMathLab*® Study Plan.**

### Example 5 Solving an Equation Involving One Radical Expression

 *My video summary*    Solve $\sqrt{3x + 7} - x = 1$.

**Solution** We follow the **four-step process**.

**Step 1.** Isolate the radical expression by adding $x$ to both sides.

$$\text{Begin with the original equation:} \quad \sqrt{3x + 7} - x = 1$$
$$\text{Add } x \text{ to both sides:} \quad \sqrt{3x + 7} - x + x = 1 + x$$
$$\text{Simplify each side:} \quad \sqrt{3x + 7} = x + 1$$

**Step 2.** We eliminate the radical by **squaring both sides** of the equation.

$$\text{Square both sides:} \quad \left(\sqrt{3x + 7}\right)^2 = (x + 1)^2 \quad \leftarrow \quad \text{Right side is the square}$$
$$\text{Simplify each side:} \quad 3x + 7 = x^2 + 2x + 1 \qquad \text{of a binomial sum}$$

Try to solve this **quadratic equation** on your own by using the **four-step process** from **Section 6.7**. Then finish solving the original equation by checking for **extraneous solutions**. View the **answer**, or watch this **video** for the complete solution.

**You Try It**    Work through this You Try It problem.

**Work Exercises 15–18 in this textbook or in the *MyMathLab*® Study Plan.**

Be careful when squaring a side containing more than one term. It is not correct to just square each term. For example, $(x + 1)^2 \neq x^2 + 1^2$. Sometimes it is helpful to write the square as a product. For example, we could write $(x + 1)^2$ as $(x + 1)(x + 1)$.

OBJECTIVE 2   SOLVE EQUATIONS INVOLVING TWO RADICAL EXPRESSIONS

**Solving** an equation involving two **radical expressions** is similar to solving an equation involving one radical expression. However, it may be necessary to repeat the process for eliminating a radical in order to eliminate both radicals from the equation.

---

**Solving Equations Involving Two Radical Expressions**

**Step 1.**   **Isolate** one of the radical expressions. Use the **properties of equality** to get a radical expression by itself on one side of the equal sign.

**Step 2.**   Eliminate the radical from the isolated radical expression. Identify the **index** of the isolated radical expression and raise both sides of the equation to this index **power**.

**Step 3.**   If all the radicals have been eliminated, then solve the resulting equation. Otherwise, repeat Steps 1 and 2.

**Step 4.**   Check each solution from Step 3 in the original equation. Disregard any **extraneous solutions**.

---

**Example 6  Solving an Equation Involving Two Radical Expressions**

Solve $\sqrt{3x - 7} = \sqrt{x + 3}$.

**Solution**   We follow the **four-step process**.

**Step 1.** We note that $\sqrt{3x - 7}$ is already isolated on the left.

**Step 2.** We **square both sides** to eliminate the isolated radical.

$$\text{Square both sides:} \quad \left(\sqrt{3x - 7}\right)^2 = \left(\sqrt{x + 3}\right)^2$$
$$3x - 7 = x + 3$$

**Step 3.** There are no radicals remaining so we solve the resulting equation.

$$\text{Resulting equation:} \quad 3x - 7 = x + 3$$
$$\text{Subtract } x \text{ from both sides:} \quad 2x - 7 = 3$$
$$\text{Add 7 to both sides:} \quad 2x = 10$$
$$\text{Divide both sides by 2:} \quad x = 5$$

Check this answer on your own to confirm that the **solution set** is $\{5\}$.

**You Try It**   **Work through this You Try It problem.**

**Work Exercises 19 and 20 in this textbook or in the *MyMathLab*®** **Study Plan.**

### Example 7 Solving an Equation Involving Two Radical Expressions

*My video summary*  Solve $\sqrt{x + 9} - \sqrt{x} = 1$.

**Solution** We follow the four-step process.

**Step 1.** We choose to isolate $\sqrt{x + 9}$.

Begin with the original equation: $\sqrt{x + 9} - \sqrt{x} = 1$

Add $\sqrt{x}$ to both sides: $\sqrt{x + 9} - \sqrt{x} + \sqrt{x} = 1 + \sqrt{x}$

Simplify: $\sqrt{x + 9} = 1 + \sqrt{x}$

**Step 2.** Eliminate the isolated radical by **squaring both sides**. The right side will have the form of the **square of a binomial sum** $(A + B)^2$, so we use the **special product rule** with $A = 1$ and $B = \sqrt{x}$.

Square both sides: $\left(\sqrt{x + 9}\right)^2 = \left(1 + \sqrt{x}\right)^2$

$$x + 9 = \underbrace{(1)^2 + 2(1)\left(\sqrt{x}\right) + \left(\sqrt{x}\right)^2}_{(A + B)^2 = A^2 + 2AB + B^2}$$

Simplify: $x + 9 = 1 + 2\sqrt{x} + x$

**Step 3.** The equation still contains a **radical expression**, so we repeat Steps 1 and 2. Try to finish solving this equation on your own using Examples 1–5 to guide you. View the **answer**, or watch this **video** for the complete solution.

**You Try It** Work through this You Try It problem.

Work Exercises 21–26 in this textbook or in the *MyMathLab*® Study Plan

---

## OBJECTIVE 3  EVALUATE FORMULAS INVOLVING SQUARE ROOTS

We evaluate formulas involving **square roots** the same way we evaluated formulas in **Section 2.4**. We substitute a value for each variable and simplify. In the case of square roots, we need to **simplify** using the **product rule** and **quotient rule for square roots**.

From the **Pythagorean Theorem**, the **hypotenuse** of a **right triangle** is given by the formula

$$c = \sqrt{a^2 + b^2}$$

where $a$ and $b$ are the lengths of the **legs** of the right triangle.

### Example 8  Using the Pythagorean Theorem

A right triangle has legs of length 10 inches and 5 inches. Determine the exact length of the hypotenuse.

**Solution**

Hypotenuse formula: $c = \sqrt{a^2 + b^2}$

Substitute 10 for $a$ and 5 for $b$: $c = \sqrt{(10)^2 + (5)^2}$

Evaluate $10^2$ and $5^2$: $c = \sqrt{100 + 25}$

Add: $c = \sqrt{125}$

Factor the **radicand**: $c = \sqrt{25 \cdot 5}$

Product rule for square roots: $c = \sqrt{25} \cdot \sqrt{5}$

Evaluate $\sqrt{25}$: $c = 5\sqrt{5}$

The length of the hypotenuse is $5\sqrt{5}$ inches.

**You Try It**  Work through this You Try It problem.

Work Exercises 27 and 28 in this textbook or in the *MyMathLab*® Study Plan.

We can also use the **Pythagorean Theorem** to develop a **formula** for finding the distance between two **points**. To find the distance $d$ between the points $(x_1, y_1)$ and $(x_2, y_2)$, we use the **distance formula**.

---

**Distance Formula**

The distance $d$ between two points $(x_1, y_1)$ and $(x_2, y_2)$ is given by

$$d = \sqrt{(x_2 - x_1)^2 + (y_2 - y_1)^2}.$$

---

 In the distance formula, it does not matter which point is $(x_1, y_1)$ and which is $(x_2, y_2)$.

### Example 9  Using the Distance Formula

Find the distance $d$ between points $(0, 5)$ and $(3, 14)$.

**Solution**  Let $(x_1, y_1) = (0, 5)$ and $(x_2, y_2) = (3, 14)$ and use the distance formula.

$$\begin{aligned}
\text{Distance formula:} \quad d &= \sqrt{(x_2 - x_1)^2 + (y_2 - y_1)^2} \\
\text{Substitute the coordinates:} \quad &= \sqrt{(3 - 0)^2 + (14 - 5)^2} \\
\text{Simplify:} \quad &= \sqrt{(3)^2 + (9)^2} \\
&= \sqrt{90} \quad \leftarrow 3^2 + 9^2 = 9 + 81 = 90 \\
\text{Simplify the square root:} \quad &= 3\sqrt{10} \leftarrow \sqrt{90} = \sqrt{9} \cdot \sqrt{10} = 3\sqrt{10}
\end{aligned}$$

The distance between the points is $3\sqrt{10}$ units.

**You Try It**  Work through this You Try It problem.

Work Exercises 29 and 30 in this textbook or in the *MyMathLab*® Study Plan.

OBJECTIVE 4    USE RADICAL EQUATIONS TO SOLVE APPLICATION PROBLEMS

In addition to applications of the **Pythagorean Theorem**, we can use **radical equations** to solve a variety of application problems in many different disciplines. In Example 10, a radical equation is used to model the readability of written text.

### Example 10  Assessing the Readability of Written Text

*My video summary*  A **SMOG** grade for written text is a minimum reading grade level $G$ that a reader must possess in order to fully understand the written text being graded. If $w$ is the number of words that have three or more syllables in a sample of 30 sentences from a given text, then the SMOG grade for that text is given by the **formula** $G = \sqrt{w} + 3$. Use the SMOG grade formula to answer the following questions. (Source: readabilityformulas.com).

**a.** If a sample of 30 sentences contains 18 words with three or more syllables, then what is the SMOG grade for the text? If necessary, round to a **whole number** for the grade level.

**b.** If a text must have a tenth-grade reading level, then how many words with three or more syllables would be needed in the sample of 30 sentences?

## Solution

**a.** There are 18 words with three or more syllables, so we substitute 18 for $w$ in the given formula.

$$\begin{aligned}
\text{Begin with the original formula:} \quad & G = \sqrt{w} + 3 \\
\text{Substitute 18 for } w: \quad & G = \sqrt{18} + 3 \\
\text{Approximate the square root:} \quad & G \approx 4.24 + 3 \\
\text{Simplify:} \quad & G \approx 7.24
\end{aligned}$$

Rounding to the nearest whole number, we get $G = 7$. So, the SMOG grade for this text is a seventh-grade reading level.

**b.** The text must have a tenth-grade reading level, so we substitute 10 for $G$ in the **formula** and solve for $w$.

$$\begin{aligned}
\text{Begin with the original formula:} \quad & G = \sqrt{w} + 3 \\
\text{Substitute 10 for } G: \quad & 10 = \sqrt{w} + 3
\end{aligned}$$

Try to finish solving this problem on your own. View the **answer to part b**, or watch this **video** for a complete solution to both parts.

**You Try It**    **Work through this You Try It problem.**

**Work Exercises 31–34 in this textbook or in the *MyMathLab*® Study Plan.**

## Example 11  Punting a Football

An important component of a good punt in football is **hang time**, which is the length of time that the punted ball remains in the air. If wind resistance is ignored, the relationship between the hang time $t$, in seconds, and the vertical height $h$, in feet, that the ball reaches can be modeled by the formula $t = \dfrac{\sqrt{h}}{2}$. Use this formula to answer the following questions.

**a.** If the average hang time for an NFL punt is 4.6 seconds, then what is the vertical height for an average NFL punt? Round to the nearest foot.

**b.** Cowboys Stadium in Arlington, Texas has a huge high-definition screen centered over most of the football field. The bottom of the screen is 90 feet above the field. What hang time would result in the ball hitting the screen? Round to the nearest hundredth of a second.

## Solution

**a.** We substitute the hang time 4.6 for $t$ in the formula and solve for $h$.

$$\begin{aligned}
\text{Begin with the original formula:} \quad & t = \frac{\sqrt{h}}{2} \\
\text{Substitute 4.6 for } t: \quad & 4.6 = \frac{\sqrt{h}}{2} \\
\text{Multiply both sides by 2:} \quad & 9.2 = \sqrt{h} \\
\text{Square both sides:} \quad & (9.2)^2 = \left(\sqrt{h}\right)^2 \\
\text{Simplify:} \quad & 84.64 = h
\end{aligned}$$

Rounding, the average NFL punt reaches a vertical height of about 85 feet.

**b.** We substitute the vertical height, 90 feet, for $h$ and simplify.

Begin with the original formula: $\quad t = \dfrac{\sqrt{h}}{2}$

Substitute 90 for $h$: $\quad t = \dfrac{\sqrt{90}}{2}$

Approximate the square root: $\quad t \approx \dfrac{9.4868}{2}$

Divide: $\quad t \approx 4.7434$

Rounding, a hang time of about 4.74 seconds will result in a punt that hits the screen.

**You Try It**    Work through this You Try It problem.

Work Exercises **35** and **36** in this textbook or in the *MyMathLab* ® Study Plan.

# 8.5 Exercises

In Exercises 1–26, solve each radical equation.

**You Try It**

**1.** $\sqrt{x} = 12$

**2.** $\sqrt{x + 15} = 3$

**3.** $\sqrt{2x - 1} = 3$

**4.** $\sqrt{3 - x} = 4$

**You Try It**

**5.** $\sqrt{r - 3} = -1$

**6.** $3\sqrt{m} + 5 = 7$

**7.** $\sqrt{z + 5} + 2 = 9$

**8.** $\sqrt{1 - x} + 1 = 4$

**9.** $\sqrt{4p + 1} - 3 = 2$

**10.** $5\sqrt{w - 9} - 2 = 8$

**You Try It** **11.** $\sqrt{q + 10} = -2$

**12.** $\sqrt{2x - 1} + 6 = 1$

**13.** $\sqrt{x^2 + 4x - 5} = x$

**14.** $\sqrt{y^2 - 2y + 1} = 3$

**You Try It**

**15.** $\sqrt{2m + 1} - m = -1$

**16.** $p - \sqrt{4p + 9} + 3 = 0$

**You Try It** **17.** $\sqrt{32 - 4x} - x = 0$

**18.** $\sqrt{7z + 2} + 3z = 2$

**You Try It**

**19.** $\sqrt{7x - 4} = \sqrt{4x + 11}$

**20.** $\sqrt{x + 6} = \sqrt{2 - 3x}$

**You Try It**

**21.** $\sqrt{x + 15} - \sqrt{x} = 3$

**22.** $\sqrt{x} - 1 = \sqrt{x - 5}$

**23.** $\sqrt{x - 64} + 8 = \sqrt{x}$

**24.** $\sqrt{6 - x} + \sqrt{5x + 6} = 6$

**25.** $\sqrt{3x + 1} - \sqrt{x + 4} = 1$

**26.** $\sqrt{6 + 5x} + \sqrt{3x + 4} = 2$

In Exercises 27 and 28, find the length of the hypotenuse of a right triangle whose legs have the given lengths.

**You Try It** **27.** $a = 9$ m; $b = 3$ m        **28.** $a = 2$ ft; $b = 8$ ft

In Exercises 29–30, find the distance between the given points.

**You Try It** **29.** $(0, 9)$; $(4, 13)$        **30.** $(-1, 6)$; $(5, 9)$

In Exercises 31–36, use the given model to solve each application problem.

**You Try It** **31. SMOG Grade** Recall from Example 10 that the formula $G = \sqrt{w} + 3$ gives the minimum reading grade level $G$ needed to fully understand written text containing $w$ words with three or more syllables in a sample of 30 sentences.

     **a.** If a sample of 30 sentences contains 51 words with three or more syllables, then what is the SMOG grade for the text? If necessary, round your answer to a whole number to find the grade level.

     **b.** If a text must have a ninth-grade reading level, then how many words with three or more syllables are needed in a sample of 30 sentences?

**32. Measure of Leanness** The *ponderal index* is a measure of the "leanness" of a person. A person who is $h$ inches tall and weighs $w$ pounds has a ponderal index $I$ given by $I = \dfrac{h}{\sqrt[3]{w}}$.

     **a.** Compute the ponderal index for a person who is 75 inches tall and weighs 190 pounds. Round to the nearest hundredth.

     **b.** What is a man's weight if he is 70 inches tall and has a ponderal index of 12.35? Round to the nearest whole number.

**33. Body Surface Area** The Mosteller Formula is used in the medical field to estimate a person's body surface area. The formula is $A = \sqrt{\dfrac{hw}{3600}}$, where $A$ is body surface area in square meters, $h$ is height in centimeters, and $w$ is weight in kilograms.

     **a.** Compute the body surface area of a person who is 178 cm tall and weighs 90.8 kg. Round to the nearest hundredth.

     **b.** If a woman is 165 cm tall and has a body surface area of 1.72 m², how much does she weigh? Round to the nearest tenth.

**34. Skid Marks** Under certain road conditions, the length of a skid mark $S$, in feet, is related to the velocity $v$, in miles per hour, by the formula $v = \sqrt{10S}$. Assuming these same road conditions, answer the following:

     **a.** Compute a car's velocity if it leaves a skid mark of 360 feet.

     **b.** If a car is traveling at 50 miles per hour when it skids, what will be the length of the skid mark?

**You Try It** **35. Hang Time** Recall from Example 11 that hang time $t$, in seconds, and the vertical height $h$, in feet, can be modeled by $t = \dfrac{\sqrt{h}}{2}$. This same formula can be used to model an athlete's hang time when jumping.

     **a.** If LeBron James has a vertical leap of 3.7 feet, what is his hang time? Round to the nearest hundredth.

     **b.** When Mark "Wild Thing" Wilson of the Harlem Globetrotters slam-dunked a regulation basketball on a 12-foot rim in front of an Indianapolis crowd, his hang time for the shot was approximately 1.16 seconds. What was the vertical distance of his jump? Round to the nearest tenth.

**36. Distance to the Horizon** From a boat, the distance $d$, in miles, that a person can see to the horizon is modeled by the formula $d = \dfrac{3\sqrt{h}}{2}$, where $h$ is the height, in feet, of eye level above the sea.

     **a.** From his ship, how far can a sailor see to the horizon if his eye level is 36 feet above the sea?

     **b.** How high is the eye level of a sailor who can see 12 miles to the horizon?

# 8.6 Rational Exponents

## THINGS TO KNOW

Before working through this section, be sure you are familiar with the following concepts:

|  |  | VIDEO | ANIMATION | INTERACTIVE |

 **You Try It**
1. Simplify Exponential Expressions Using the Product Rule (Section 5.1, **Objective 1**)

 **You Try It**
2. Simplify Exponential Expressions Using the Quotient Rule (Section 5.1, **Objective 2**)

 **You Try It**
3. Use the Power-to-Power Rule (Section 5.1, **Objective 4**)

 **You Try It**
4. Use the Product-to-Power Rule (Section 5.1, **Objective 5**)

 **You Try It**
5. Use the Quotient-to-Power Rule (Section 5.1, **Objective 6**)

 **You Try It**
6. Use the Negative-Power Rule (Section 5.6, **Objective 1**)

 **You Try It**
7. Find Square Roots (Section 8.1, **Objective 1**)

**You Try It**
8. Simplify Square Roots Containing Variables (Section 8.1, **Objective 3**)

**You Try It**
9. Find and Approximate $n$th Roots (Section 8.1, **Objective 5**)

## OBJECTIVES

**1** Use the Definition for Rational Exponents of the Form $a^{\frac{1}{n}}$

**2** Use the Definition for Rational Exponents of the Form $a^{\frac{m}{n}}$

**3** Simplify Exponential Expressions Involving Rational Exponents

**4** Use Rational Exponents to Simplify Radical Expressions

---

OBJECTIVE 1   USE THE DEFINITION FOR RATIONAL EXPONENTS OF THE FORM $a^{\frac{1}{n}}$

In addition to the radical notation we have used throughout this chapter, **radical expressions** can also be expressed using exponents that are **rational numbers**. This alternate notation, developed by using rules for exponents, is often more efficient and easier to work with than radical notation. Consider the following:

$$\underbrace{\left(\sqrt{5}\right)^2 = 5}_{\substack{\text{Definition of} \\ \text{square roots}}} = \underbrace{5^1}_{\substack{\text{Implied} \\ \text{exponent of 1}}} = \underbrace{5^{\frac{1}{2}\cdot 2}}_{\substack{\text{Rewrite 1} \\ \text{as } \frac{1}{2}\, 2}} = \underbrace{\left(5^{\frac{1}{2}}\right)^2}_{\substack{\text{Power-to-} \\ \text{power rule}}}$$

From this we can conclude that $\sqrt{5} = 5^{\frac{1}{2}}$. This result can be generalized as

$$a^{\frac{1}{2}} = \sqrt{a},$$

and can be used to define a rational exponent of the form $a^{\frac{1}{n}}$.

---

**Definition   Rational Exponent of the Form $a^{\frac{1}{n}}$**

If $n$ is an **integer** such that $n \geq 2$ and if $\sqrt[n]{a}$ is a **real number**, then $a^{\frac{1}{n}} = \sqrt[n]{a}$.

---

The **denominator** $n$ of the **rational exponent** is the **index** of the **root**. The **base** $a$ of the **exponential expression** is the **radicand** of the root. If $n$ is odd, then $a$ can be any real number. If $n$ is even, then $a$ must be **non-negative**.

### Example 1  Converting Exponential Expressions to Radical Expressions

 *My video summary*     Write each exponential expression as a **radical expression**. **Simplify** if possible.

**a.** $25^{\frac{1}{2}}$    **b.** $(-64x^3)^{\frac{1}{3}}$    **c.** $-100^{\frac{1}{2}}$    **d.** $(-81)^{\frac{1}{4}}$    **e.** $(7x^3y)^{\frac{1}{5}}$

**Solution**

**a.** The index of the root is 2, the denominator of the rational exponent. So, the radical expression is a **square root**. The radicand is 25.

$$25^{\frac{1}{2}} = \sqrt{25} = 5$$

**b.** From the denominator of the rational exponent, the index is 3. So, we have a **cube root**. The radicand is $-64x^3$.

$$(-64x^3)^{\frac{1}{3}} = \sqrt[3]{-64x^3} = \sqrt[3]{(-4x)^3} = -4x$$

**c.** As in part (a), the **radical expression** is a **square root**. The negative sign is not part of the **radicand** because it is not part of the **base**. We can tell this because there are no grouping symbols around $-100$.

$$-100^{\frac{1}{2}} = -\sqrt{100} = -10$$

**d.–e.** Try to convert each expression on your own. View the **answers** to parts d and e, or watch this **video** for the complete solutions to all five parts.

**You Try It**    Work through this **You Try It** problem.

Work Exercises 1–6 in this textbook or in the ***MyMathLab***® Study Plan.

### Example 2  Converting Radical Expressions to Exponential Expressions

 *My video summary*     Write each radical expression as an **exponential expression**.

**a.** $\sqrt{5y}$    **b.** $\sqrt[3]{7x^2y}$    **c.** $\sqrt[4]{\dfrac{2m}{3n}}$

**Solution**

**a.** The radical expression is a square root, so the **index** is 2 and the **denominator** of the **rational exponent** is 2. The base of the exponential expression is the **radicand** $5y$.

$$\sqrt{5y} = (5y)^{\frac{1}{2}}$$

b. The **index** is 3 and the **radicand** is $7x^2y$, so the **denominator** of the **rational exponent** is 3 and the **base** of the **exponential expression** is $7x^2y$. Finish writing the exponential expression. View the **answer**, or watch this **video** for the complete solutions to all three parts.

c. Try to convert this expression on your own. View the **answer**, or watch this **video** for the complete solutions to all three parts.

**You Try It**    Work through this You Try It problem.

**Work Exercises 13–15 in this textbook or in the *MyMathLab*®** **Study Plan.**

OBJECTIVE 2    USE THE DEFINITION FOR RATIONAL EXPONENTS OF THE FORM $a^{\frac{m}{n}}$

What if the **numerator** of the **rational exponent** is not 1? For example, consider the **exponential expression** $a^{\frac{2}{3}}$. From the **power-to-power rule** for exponents, we write

$$a^{\frac{2}{3}} = a^{\frac{1}{3}\cdot 2} = \left(a^{\frac{1}{3}}\right)^2 = \left(\sqrt[3]{a}\right)^2 \quad \text{or} \quad a^{\frac{2}{3}} = a^{2\cdot\frac{1}{3}} = \left(a^2\right)^{\frac{1}{3}} = \sqrt[3]{a^2}.$$

This idea can be used to define a rational exponent of the form $a^{\frac{m}{n}}$.

---

**Definition**    **Rational Exponent of the Form $a^{\frac{m}{n}}$**

If $\dfrac{m}{n}$ is a **rational number** in **lowest terms**, $m$ and $n$ are integers such that $n \geq 2$, and $\sqrt[n]{a}$ is a **real number**, then

$$a^{\frac{m}{n}} = \left(\sqrt[n]{a}\right)^m = \sqrt[n]{a^m}.$$

---

To **simplify** $a^{\frac{m}{n}}$ using the form $\left(\sqrt[n]{a}\right)^m$, we find the **root** first and the **power** second. Using the form $\sqrt[n]{a^m}$, we find the power first and the root second. Both forms may be used, but the form $\left(\sqrt[n]{a}\right)^m$ is usually easier because it involves smaller numbers.

**Note:** As we have done earlier in this chapter, we assume that all variable **factors** of the **radicand** of a **radical expression** with an even **index** will be **non-negative** real numbers. Likewise, all variable factors of the **base** of an **exponential expression** containing a rational exponent with an even **denominator** will be non-negative **real numbers**. This assumption allows us to avoid using **absolute value** symbols when simplifying radical expressions.

**Example 3**   **Converting Exponential Expressions to Radical Expressions**

 Write each exponential expression as a radical expression. **Simplify** if possible.

a. $16^{\frac{3}{2}}$    b. $\left(\dfrac{y^3}{1000}\right)^{\frac{2}{3}}$    c. $-81^{\frac{3}{4}}$    d. $(-36)^{\frac{5}{2}}$    e. $(x^2y)^{\frac{2}{5}}$

**Solution**

a. We use the form $a^{\frac{m}{n}} = \left(\sqrt[n]{a}\right)^m$. Since the denominator of the **rational exponent** is 2, the radical expression is a **square root**.

$$
\begin{aligned}
\text{Begin with the original exponential expression:} \quad & 16^{\frac{3}{2}} \\
\text{Use } a^{\frac{m}{n}} = \left(\sqrt[n]{a}\right)^m \text{ to rewrite as a radical expression:} \quad & = \left(\sqrt{16}\right)^3 \\
\text{Simplify } \sqrt{16}: \quad & = (4)^3 \\
\text{Simplify:} \quad & = 64
\end{aligned}
$$

**b.** $\left(\dfrac{y^3}{1000}\right)^{\frac{2}{3}} = \left(\sqrt[3]{\dfrac{y^3}{1000}}\right)^2 = \left(\sqrt[3]{\left(\dfrac{y}{10}\right)^3}\right)^2 = \left(\dfrac{y}{10}\right)^2 = \dfrac{y^2}{100}$

**c.** Note that the negative sign is not part of the **base**, so it goes in front of the **radical expression**.

$$-81^{\frac{3}{4}} = -\left(\sqrt[4]{81}\right)^3 = -(3)^3 = -27$$

**d.–e.** Try to convert each expression on your own. View the **answers** to parts d and e, or watch this **video** for the complete solutions to all five parts.

**You Try It**   **Work through this You Try It problem.**

**Work Exercises 7–12 in this textbook or in the *MyMathLab*® Study Plan.**

## Example 4  Converting Radical Expressions to Exponential Expressions

 Write each radical expression as an **exponential expression**.

**a.** $\sqrt[8]{x^5}$        **b.** $\left(\sqrt[5]{2ab^2}\right)^3$        **c.** $\sqrt[4]{(10x)^3}$

### Solution

**a.** The **index** is 8, so the **denominator** of the **rational exponent** is 8. The **radicand** is a **power** of 5, so the **numerator** is 5.

$$\sqrt[8]{x^5} = x^{\frac{5}{8}}$$

**b.** The index is 5, so the denominator of the rational exponent is 5. The expression is raised to a power of 3, so the numerator is 3. Finish writing the exponential expression. View the **answer**, or watch this **video** for the complete solutions to all three parts.

**c.** Try to convert this expression on your own. View the **answer**, or watch this **video** for the complete solutions to all three parts.

**You Try It**   **Work through this You Try It problem.**

**Work Exercises 16–18 in this textbook or in the *MyMathLab*® Study Plan.**

If a rational exponent is negative, we can first use the **negative-power rule** from **Section 5.6** to rewrite the expression with a positive exponent.

## Example 5  Using the Negative-Power Rule with Negative Rational Exponents

 Write each **exponential expression** with positive exponents. **Simplify** if possible.

**a.** $1000^{-\frac{1}{3}}$        **b.** $\dfrac{1}{81^{-\frac{1}{4}}}$        **c.** $125^{-\frac{2}{3}}$        **d.** $\dfrac{1}{8^{-\frac{4}{3}}}$        **e.** $(-25)^{-\frac{3}{2}}$

### Solution

**a.**        Begin with the original exponential expression:   $1000^{-\frac{1}{3}}$

Use the negative-power rule $a^{-n} = \dfrac{1}{a^n}$:   $= \dfrac{1}{1000^{\frac{1}{3}}}$

Rewrite the rational exponent as a radical expression:   $= \dfrac{1}{\sqrt[3]{1000}}$

Evaluate the root:   $= \dfrac{1}{10}$

**b.**    Begin with the original exponential expression:    $\dfrac{1}{81^{-\frac{1}{4}}}$

Use the negative-power rule $\dfrac{1}{a^{-n}} = a^n$:    $= 81^{\frac{1}{4}}$

Rewrite the rational exponent as a radical expression:    $= \sqrt[4]{81}$

Evaluate the root:    $= 3$

**c.–e.** Try to rewrite and simplify each expression on your own. View the **answers** to parts c–e, or watch this **video** for the complete solutions to all five parts.

**You Try It**    Work through this You Try It problem.

Work Exercises 19–21 in this textbook or in the *MyMathLab*® Study Plan.

OBJECTIVE 3    SIMPLIFY EXPONENTIAL EXPRESSIONS INVOLVING RATIONAL EXPONENTS

In **Example 5** we used the **negative-power rule** for exponents to rewrite and simplify **exponential expressions** containing negative **rational** exponents. In **Section 5.1** we used several other **rules** to **simplify exponential expressions** involving **integer** exponents. We can use these rules to simplify expressions involving rational exponents. For convenience, we repeat them here.

---

**Rules for Exponents**

Product Rule            $a^m \cdot a^n = a^{m+n}$

Quotient Rule           $\dfrac{a^m}{a^n} = a^{m-n} \quad (a \neq 0)$

Zero-Power Rule         $a^0 = 1 \quad (a \neq 0)$

Power-to-Power Rule     $(a^m)^n = a^{m \cdot n}$

Product-to-Power Rule   $(ab)^n = a^n b^n$

Quotient-to-Power Rule  $\left(\dfrac{a}{b}\right)^n = \dfrac{a^n}{b^n} \quad (b \neq 0)$

Negative-Power Rule     $a^{-n} = \dfrac{1}{a^n} \quad \text{or} \quad \dfrac{1}{a^{-n}} = a^n \quad (a \neq 0)$

---

Recall that an exponential expression is **simplified** when

- No parentheses or **grouping symbols** are present.
- No zero or **negative exponents** are present.
- No **powers** are raised to powers.
- Each **base** occurs only once.

### Example 6  Simplifying Expressions Involving Rational Exponents

*My interactive video summary*

▶ Use the **rules for exponents** to simplify each expression. Assume all **variables** represent positive values.

**a.** $x^{\frac{3}{8}} \cdot x^{\frac{1}{6}}$    **b.** $\dfrac{49^{\frac{7}{10}}}{49^{\frac{1}{5}}}$    **c.** $\left(64^{\frac{4}{9}}\right)^{\frac{3}{2}}$    **d.** $\left(32x^{\frac{5}{6}}y^{\frac{10}{9}}\right)^{\frac{3}{5}}$    **e.** $\left(\dfrac{125x^{\frac{5}{4}}}{y^8 z^{\frac{9}{4}}}\right)^{\frac{4}{3}}$    **f.** $\left(4x^{\frac{1}{6}}y^{\frac{3}{4}}\right)^2\left(3x^{\frac{5}{9}}y^{-\frac{3}{2}}\right)$

## Solution

**a.**  Begin with the original expression:  $x^{\frac{3}{8}} \cdot x^{\frac{1}{6}}$

Use the **product rule** for exponents:  $= x^{\frac{3}{8}+\frac{1}{6}}$

Add:  $= x^{\frac{13}{24}} \leftarrow \frac{3}{8} + \frac{1}{6} = \frac{9}{24} + \frac{4}{24} = \frac{13}{24}$

**b.**  Begin with the original expression:  $\dfrac{49^{\frac{7}{10}}}{49^{\frac{1}{5}}}$

Use the **quotient rule** for exponents:  $= 49^{\frac{7}{10}-\frac{1}{5}}$

Subtract:  $= 49^{\frac{1}{2}} \leftarrow \frac{7}{10} - \frac{1}{5} = \frac{7}{10} - \frac{2}{10} = \frac{5}{10} = \frac{1}{2}$

Use $a^{\frac{1}{n}} = \sqrt[n]{a}$ to rewrite as a radical expression:  $= \sqrt{49}$

Simplify:  $= 7$

**c.**  Begin with the original expression:  $\left(64^{\frac{4}{9}}\right)^{\frac{3}{2}}$

Use the **power-to-power rule** for exponents:  $= 64^{\frac{4}{9}\cdot\frac{3}{2}}$

Multiply:  $= 64^{\frac{2}{3}} \leftarrow \frac{4}{9} \cdot \frac{3}{2} = \frac{\overset{2}{\cancel{4}}}{\underset{3}{\cancel{9}}} \cdot \frac{\overset{1}{\cancel{3}}}{\underset{1}{\cancel{2}}} = \frac{2}{3}$

Use $a^{\frac{m}{n}} = \left(\sqrt[n]{a}\right)^m$ to rewrite as a radical expression:  $= \left(\sqrt[3]{64}\right)^2$

Simplify $\sqrt[3]{64}$:  $= (4)^2$

Simplify:  $= 16$

**d.–f.**  Try to simplify each expression on your own. View the **answers** to parts d–f, or work through this **interactive video** for complete solutions to all six parts.

**You Try It**   Work through this You Try It problem.

**Work Exercises 22–28** in this textbook or in the *MyMathLab* Study Plan.

---

OBJECTIVE 4   USE RATIONAL EXPONENTS TO SIMPLIFY RADICAL EXPRESSIONS

Some **radical expressions** can be simplified by first writing them with **rational exponents**. We can use the following process.

> **Using Rational Exponents to Simplify Radical Expressions**
>
> **Step 1.** Convert each radical expression to an **exponential expression** with rational exponents.
>
> **Step 2.** Simplify by writing fractions in **lowest terms** or using the **rules of exponents**, as necessary.
>
> **Step 3.** Convert any remaining rational exponents back to a radical expression.

### Example 7  Simplifying Radical Expressions

*My interactive video summary*

Use rational exponents to simplify each radical expression. Assume all **variables** represent positive values.

**a.**  $\sqrt[6]{y} \cdot \sqrt[3]{y}$      **b.**  $\sqrt{\sqrt[3]{x}}$      **c.**  $\sqrt[6]{x^4}$

**d.**  $\sqrt[8]{25x^2y^6}$      **e.**  $\sqrt[4]{49}$      **f.**  $\dfrac{\sqrt[3]{x}}{\sqrt[4]{x}}$

**Solution**

a. Rewrite each radical using **rational exponents**: $\sqrt[6]{y} \cdot \sqrt[3]{y} = y^{\frac{1}{6}} \cdot y^{\frac{1}{3}}$

Use the **product rule** for exponents: $= y^{\frac{1}{6}+\frac{1}{3}}$

Add: $= y^{\frac{1}{2}}$

Convert back to a **radical expression**: $= \sqrt{y}$

b. Rewrite the **radicand** $\sqrt[3]{x}$ using a rational exponent: $\sqrt{\sqrt[3]{x}} = \sqrt{x^{\frac{1}{3}}}$

Rewrite the square root using a rational exponent: $= \left(x^{\frac{1}{3}}\right)^{\frac{1}{2}}$

Use the **power-to-power** rule for exponents: $= x^{\frac{1}{3}\cdot\frac{1}{2}}$

Multiply: $= x^{\frac{1}{6}}$

Convert back to a radical expression: $= \sqrt[6]{x}$

c. Convert to a rational exponent: $\sqrt[6]{x^4} = x^{\frac{4}{6}}$

Write the rational exponent in **lowest terms**: $= x^{\frac{2}{3}}$

Convert back to a radical expression: $= \sqrt[3]{x^2}$

**d.–f.** Try to simplify each radical expression on your own. View the **answers** to parts d–f, or work through this **interactive video** for complete solutions to all six parts.

**You Try It** Work through this You Try It problem.

Work Exercises 29–34 in this textbook or in the *MyMathLab*® Study Plan.

# 8.6 Exercises

In Exercises 1–12, write each exponential expression as a radical expression. Simplify if possible.

**You Try It**

**1.** $36^{\frac{1}{2}}$

**2.** $(-125)^{\frac{1}{3}}$

**3.** $-81^{\frac{1}{4}}$

**4.** $(27x^3)^{\frac{1}{3}}$

**5.** $(7xy)^{\frac{1}{5}}$

**6.** $(13xy^2)^{\frac{1}{3}}$

**You Try It**

**7.** $100^{\frac{3}{2}}$

**8.** $-8^{\frac{4}{3}}$

**9.** $(-16)^{\frac{3}{4}}$

**10.** $\left(\dfrac{x^3}{64}\right)^{\frac{2}{3}}$

**11.** $\left(\dfrac{1}{8}\right)^{\frac{5}{3}}$

**12.** $(xy^2)^{\frac{3}{7}}$

In Exercises 13–18, write each radical expression as an exponential expression.

**You Try It**

**13.** $\sqrt{10}$

**14.** $\sqrt[3]{7m}$

**15.** $\sqrt[5]{\dfrac{2x}{y}}$

**You Try It**

**16.** $\sqrt[7]{x^4}$

**17.** $\left(\sqrt[5]{3xy^3}\right)^4$

**18.** $\sqrt[9]{(2xy)^4}$

In Exercises 19–21, write each exponential expression with positive exponents. Simplify if possible.

**You Try It**

**19.** $25^{-\frac{1}{2}}$

**20.** $\dfrac{1}{625^{-\frac{3}{4}}}$

**21.** $\left(\dfrac{27}{8}\right)^{-\frac{2}{3}}$

For Exercises 22–34, assume that all variables represent positive values.

In Exercises 22–28, use the rules for exponents to simplify. Write final answers in exponential form when necessary.

**You Try It**

**22.** $x^{\frac{1}{3}} \cdot x^{\frac{1}{2}}$

**23.** $9^{\frac{3}{10}} \cdot 9^{\frac{1}{5}}$

**24.** $\dfrac{36^{\frac{3}{4}}}{36^{\frac{1}{4}}}$

**25.** $\left(x^{\frac{5}{3}}\right)^{\frac{9}{10}}$

**26.** $\left(9m^4 n^{-\frac{3}{2}}\right)^{\frac{1}{2}}$

**27.** $(32x^5 y^{-10})^{\frac{1}{5}}\left(xy^{-\frac{1}{2}}\right)$

**28.** $\left(\dfrac{16x^{\frac{2}{3}}}{81x^{\frac{5}{4}}y^{\frac{2}{3}}}\right)^{\frac{3}{4}}$

In Exercises 29–34, use rational exponents to simplify each radical expression. Write final answers in radical form when necessary.

**You Try It**

**29.** $\sqrt[5]{x^2} \cdot \sqrt[4]{x}$

**30.** $\sqrt[4]{\sqrt[3]{x^2}}$

**31.** $\sqrt[6]{x^3}$

**32.** $\dfrac{\sqrt{5}}{\sqrt[6]{5}}$

**33.** $\sqrt[8]{81}$

**34.** $\sqrt[10]{49x^6 y^8}$

# 8.7  Complex Numbers

## THINGS TO KNOW

Before working through this section, be sure you are familiar with the following concepts:

|  |  | VIDEO | ANIMATION | INTERACTIVE |

**You Try It**

1. Multiply Two Binomials
   (Section 5.4, **Objective 3**)

**You Try It**

2. Find Square Roots
   (Section 8.1, **Objective 1**)

**You Try It**

3. Simplifying Square Roots Using the
   Product Rule (Section 8.2, **Objective 1**)

## OBJECTIVES

1  Simplify Powers of $i$

2  Add and Subtract Complex Numbers

3  Multiply Complex Numbers

4  Divide Complex Numbers

5  Simplify Radicals with Negative Radicands

## OBJECTIVE 1 SIMPLIFY POWERS OF $i$

So far we have learned about **real number** solutions to **equations**, but not every equation has real number **solutions**. For example, consider the equation $x^2 + 1 = 0$. We can check if a value is a solution by substituting the value for $x$ and seeing if a true statement results. Is $x = -1$ a solution to this equation?

$$\text{Begin with the original equation:} \qquad x^2 + 1 = 0$$
$$\text{Substitute } -1 \text{ for } x: \qquad (-1)^2 + 1 \overset{?}{=} 0$$
$$\text{Simplify:} \qquad 1 + 1 \overset{?}{=} 0$$
$$2 = 0 \quad \text{False}$$

Because $2 = 0$ is not a true statement, $x = -1$ is not a solution to the equation $x^2 + 1 = 0$. In fact, this equation has no real solution. See **why.**

To find the solution to equations such as $x^2 + 1 = 0$, we introduce a new number called the **imaginary unit** $i$.

---

**Definition    Imaginary Unit $i$**

The **imaginary unit** $i$ is defined as

$$i = \sqrt{-1}, \text{ where } i^2 = -1.$$

---

When working with the **imaginary unit**, we will encounter various powers of $i$. Let's consider some powers of $i$ and look for patterns that will help us simplify them.

$$\underbrace{i^1 = i = \sqrt{-1}}_{\text{Defined}} \qquad\qquad \underbrace{i^2 = -1}_{\text{Defined}}$$

$$i^3 = \underbrace{i^2 \cdot i}_{\substack{\text{Product rule} \\ \text{for exponents}}} = \underbrace{(-1)}_{i^2 = -1} \cdot\, i = -i \qquad i^4 = \underbrace{i^2 \cdot i^2}_{\substack{\text{Product rule} \\ \text{for exponents}}} = \underbrace{(-1)}_{i^2 = -1} \cdot \underbrace{(-1)}_{i^2 = -1} = 1$$

$$i^5 = \underbrace{i^4 \cdot i}_{\substack{\text{Product rule} \\ \text{for exponents}}} = \underbrace{(1)}_{i^4 = 1} \cdot\, i = i \qquad i^6 = \underbrace{i^4 \cdot i^2}_{\substack{\text{Product rule} \\ \text{for exponents}}} = \underbrace{(1)}_{i^4 = 1} \cdot \underbrace{(-1)}_{i^2 = -1} = -1$$

$$i^7 = \underbrace{i^4 \cdot i^3}_{\substack{\text{Product rule} \\ \text{for exponents}}} = \underbrace{(1)}_{i^4 = 1} \cdot \underbrace{(-i)}_{i^3 = -i} = -i \qquad i^8 = \underbrace{i^4 \cdot i^4}_{\substack{\text{Product rule} \\ \text{for exponents}}} = \underbrace{(1)}_{i^4 = 1} \cdot \underbrace{(1)}_{i^4 = 1} = 1$$

$$\vdots$$

Notice that the powers of $i$ follow the pattern $i, -1, -i, 1$. Based on this pattern, what is the value of $i^0$? **Find out.** We can use the following procedure to simplify powers of $i$.

---

**Simplifying $i^n$ for $n > 4$**

**Step 1.** Divide $n$ by 4 and find the **remainder** $r$.

**Step 2.** Replace the **exponent** (power) on $i$ by the remainder, $i^n = i^r$.

**Step 3.** Use the results $i^0 = 1$, $i^1 = i$, $i^2 = -1$, and $i^3 = -i$ to simplify if necessary.

---

### Example 1  Simplifying Powers of *i*

*My video summary*   Simplify.

**a.** $i^{17}$    **b.** $i^{60}$    **c.** $i^{39}$    **d.** $-i^{90}$    **e.** $i^{14} + i^{29}$

**Solution**

**a. Step 1.** Divide the exponent by 4 and find the remainder:

$$4)\overline{17} \quad \longleftarrow \text{ exponent, } n$$
$$\underline{16}$$
$$1 \quad \longleftarrow \text{ remainder, } r$$

**Step 2.** Replace *n* by *r*: $i^{\overset{n}{17}} = i^{\overset{r}{1}}$

**Step 3.** Simplify if necessary: $i^{17} = i^1 = i$

**b.** Divide the **exponent** 60 by 4 and find the **remainder**.

$$4)\overline{60} \quad \longleftarrow \text{ exponent, } n$$
$$\underline{4}$$
$$20$$
$$\underline{20}$$
$$0 \quad \longleftarrow \text{ remainder, } r$$

Replace *n* by *r* and simplify if necessary.

$$i^{\overset{n}{60}} = i^{\overset{r}{0}} = 1$$

**c.–e.** Try to simplify these **powers of** *i* on your own. View the **answers** to parts c–e, or watch this **video** to see detailed solutions to all five parts.

**You Try It**   **Work through this You Try It problem.**

**Work Exercises 1–5 in this textbook or in the *MyMathLab*® Study Plan.**

With the **imaginary unit**, we can now expand our number system from the set of **real numbers** to the set of **complex numbers**.

---

**Complex Numbers**

The set of all numbers of the form

$$a + bi,$$

where *a* and *b* are real **numbers** and *i* is the **imaginary unit**, is called the set of **complex numbers**. The number *a* is called the **real part**, and the number *b* is called the **imaginary part**.

---

If $b = 0$, then the complex number is a purely real number. If $a = 0$, then the complex number is a purely **imaginary number**. **Figure 4** illustrates the relationships between complex numbers.

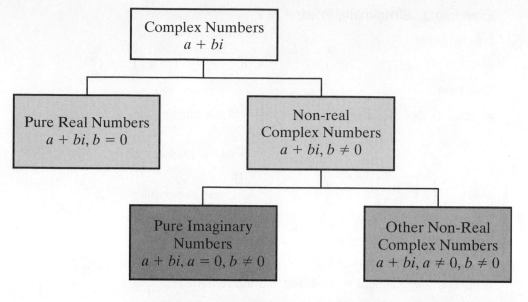

**Figure 4**
The Complex Number System

Figure 4 shows us that all **real numbers** are **complex numbers**, but not all complex numbers are real numbers. This distinction is important, particularly when solving equations. For example, $x^2 + 1 = 0$ has no *real* solutions, but it does have two *complex* solutions.

A complex number of the form $a + bi$ is written in **standard form**, and is the typical way to write complex numbers. Below are examples of complex numbers written in **standard form**.

| | Standard form | | Real part, $a = 6$ | | Imaginary part, $b = 0$ | |
|---|---|---|---|---|---|---|
| Real Number: | 6 | = | 6 | + | 0 | $i$ |

| | Standard form | | Real part, $a = 0$ | | Imaginary part, $b = 9$ | |
|---|---|---|---|---|---|---|
| Imaginary Number: | $9i$ | = | 0 | + | 9 | $i$ |

| | Standard form | | Real part, $a = 7$ | | Imaginary part, $b = 2$ | |
|---|---|---|---|---|---|---|
| Non-real Complex Number: | $7 + 2i$ | = | 7 | + | 2 | $i$ |

| | Standard form | | Real part, $a = \dfrac{3}{5}$ | | Imaginary part, $b = \dfrac{8}{5}$ | |
|---|---|---|---|---|---|---|
| | $\dfrac{3}{5} + \dfrac{8}{5}i$ | = | $\dfrac{3}{5}$ | + | $\dfrac{8}{5}$ | $i$ |

## OBJECTIVE 2   ADD AND SUBTRACT COMPLEX NUMBERS

To add or subtract **complex numbers**, we combine the **real parts** and combine the **imaginary parts**.

---

**Adding and Subtracting Complex Numbers**

To add complex numbers, add the real parts and add the imaginary parts.

$$(a + bi) + (c + di) = (a + c) + (b + d)i$$

To subtract complex numbers, subtract the real parts and subtract the imaginary parts.

$$(a + bi) - (c + di) = (a - c) + (b - d)i$$

---

## Example 2  Adding and Subtracting Complex Numbers

 *My video summary*

 Perform the indicated operations.

**a.** $(3 + 5i) + (2 - 7i)$    **b.** $(3 + 5i) - (2 - 7i)$    **c.** $(-3 - 4i) + (2 - i) - (3 + 7i)$

### Solution

**a.**

| | |
|---|---|
| Original expression: | $(3 + 5i) + (2 - 7i)$ |
| Remove parentheses: | $= 3 + 5i + 2 - 7i$ |
| Collect like terms: | $= 3 + 2 + \overbrace{5i}^{\text{Real}} \overbrace{- 7i}^{\text{Imaginary}}$ |
| Combine real parts: | $= \underbrace{5}_{3+2} + 5i - 7i$ |
| Combine imaginary parts: | $= 5 - \underbrace{2i}_{5i-7i}$ |

**b.**

| | |
|---|---|
| Original expression: | $(3 + 5i) - (2 - 7i)$ |
| Change to add the opposite: | $= (3 + 5i) + (-2 + 7i)$ |
| Remove parentheses: | $= 3 + 5i - 2 + 7i$ |

Finish working the problem on your own. View the **answer**, or watch this **video** to see the full solutions to all three parts.

**c.** Try to work the problem on your own. View the **answer**, or watch this **video** for detailed solutions to all three parts.

**You Try It**   Work through this You Try It problem.

**Work Exercises 6–13 in this textbook or in the *MyMathLab*® Study Plan.**

OBJECTIVE 3  MULTIPLY COMPLEX NUMBERS

When multiplying **complex numbers**, we use the **distributive property** and the **FOIL method** just as when multiplying **binomials**. Remember that $i^2 = -1$ when simplifying.

### Example 3  Multiplying Complex Numbers

Multiply.

**a.** $-4i(3 - 8i)$    **b.** $(3 - 4i)(6 + 11i)$

### Solution

**a.** Multiply using the distributive property.

| | |
|---|---|
| Original expression: | $-4i(3 - 8i)$ |
| Distribute: | $= -4i \cdot 3 + (-4i)(-8i)$ |
| Multiply: | $= -12i + 32i^2$ |
| Replace $i^2$ with $-1$: | $= -12i + 32\underbrace{(-1)}_{i^2 = -1}$ |
| Simplify: | $= -12i - 32$ |
| Write in **standard form**: | $= -32 - 12i$ |

**b.** Multiply using the **FOIL method**.

$$(3 - 4i)(6 + 11i) = \overbrace{3 \cdot 6}^{F} + \overbrace{3 \cdot 11i}^{O} - \overbrace{4i \cdot 6}^{I} - \overbrace{4i \cdot 11i}^{L}$$

$$= 18 \underbrace{+ 33i - 24i}_{\text{Collect like terms}} - 44i^2$$

$$= 18 + 9i \underbrace{- 44i^2}_{i^2 = -1}$$

$$= 18 + 9i - 44(-1)$$

$$= \underbrace{18 + 9i + 44}_{\text{Collect like terms}}$$

$$= 62 + 9i$$

## Example 4  Multiplying Complex Numbers

*My video summary*   Multiply $(5 + 4i)^2$.

**Solution**  We can multiply $(5 + 4i)^2 = (5 + 4i)(5 + 4i)$ by FOILing, or by using the special product rule for the **square of a binomial sum**.

Square of a binomial sum rule:  $(A + B)^2 = A^2 + 2AB + B^2$

Substitute 5 for $A$ and $4i$ for $B$:  $(5 + 4i)^2 = (5)^2 + 2(5)(4i) + (4i)^2$

Simplify:  $= 25 + 40i + 16i^2$

Finish simplifying on your own. View the **answer**, or watch this **video** for a complete solution.

**You Try It**  Work through this You Try It problem.

**Work Exercises 14–21 in this textbook or in the *MyMathLab*® Study Plan.**

In **Section 5.5** we learned that a **binomial sum** and a **binomial difference** made from the same two **terms** are **conjugates** of each other. We also saw that the **product of conjugates** resulted in the **difference of two squares**. These results extend to our discussion of **complex numbers** as follows.

---

**Complex Conjugates**

The complex numbers $(a + bi)$ and $(a - bi)$ are called **complex conjugates** of each other. A complex conjugate is obtained by changing the sign of the imaginary part in a complex number. Also, $(a + bi)(a - bi) = a^2 + b^2$.

---

Notice that the product of complex conjugates is a *sum* of two **squares** rather than a difference and is always a **real number**. See **why**.

## Example 5  Multiplying Complex Conjugates

Multiply $(5 + 9i)(5 - 9i)$.

**Solution**  Since the two **complex numbers** are **conjugates**, we find the product using the result for the **product of complex conjugates**.

Identifying $a = 5$ and $b = 9$ we get

$$\underbrace{\frac{(a + bi)(a - bi)}{(5 + 9i)(5 - 9i)}} = \overbrace{(5)^2}^{a^2} + \overbrace{(9)^2}^{b^2} = 25 + 81 = 106.$$

We can also find the same result using the **FOIL method**. See **how**.

**You Try It**   Work through this **You Try It** problem.

**Work Exercises 22–26 in this textbook or in the *MyMathLab*® Study Plan.**

OBJECTIVE 4   DIVIDE COMPLEX NUMBERS

When dividing **complex numbers**, the goal is to eliminate the imaginary part from the denominator and to express the **quotient** in **standard form**, $a + bi$. To do this, we multiply the numerator and denominator by the **complex conjugate** of the denominator.

**Example 6  Dividing Complex Numbers**

Divide. Write the quotient in standard form.

$$\frac{5 + 3i}{6 + 7i}$$

**Solution**  The **denominator** is $6 + 7i$, so its **complex conjugate** is $6 - 7i$. We multiply both the **numerator** and denominator by the **complex conjugate** and simplify to **standard form**. Work through the following, or watch this **video** for a detailed solution.

Multiply numerator and denominator by $6 - 7i$:   $\dfrac{5 + 3i}{6 + 7i} = \dfrac{5 + 3i}{6 + 7i} \cdot \dfrac{6 - 7i}{6 - 7i}$

Multiply numerators and multiply denominators:
(remember that $(a + bi)(a - bi) = a^2 + b^2$)   $= \dfrac{30 - 35i + 18i - 21i^2}{(6)^2 + (7)^2}$

Simplify exponents:
(remember that $i^2 = -1$)   $= \dfrac{30 - 35i + 18i - 21(-1)}{36 + 49}$

Simplify the numerator and denominator:   $= \dfrac{51 - 17i}{85}$

Write in standard form, $a + bi$:   $= \dfrac{3}{5} - \dfrac{1}{5}i$

**You Try It**   Work through this **You Try It** problem.

**Work Exercises 27–30 in this textbook or in the *MyMathLab*® Study Plan.**

**Note:** Remember that multiplying the numerator and denominator of an expression by the same quantity is the same as multiplying the expression by 1.

**Example 7  Dividing Complex Numbers**

*My video summary*   Divide. Write the quotient in standard form.

$$\frac{5 + 7i}{2i}$$

**Solution** First, we multiply the **numerator** and **denominator** by the **complex conjugate** of the denominator. The denominator is $2i = 0 + 2i$, so its complex conjugate is $0 - 2i = -2i$.

$$\frac{5 + 7i}{2i} = \frac{5 + 7i}{2i} \cdot \frac{-2i}{-2i} = \frac{-10i - 14i^2}{-4i^2}$$

- Multiply numerator and denominator by $-2i$
- Multiply numerators and multiply denominators

Finish simplifying on your own, and write the answer in standard form. View the **answer**, or watch this **video** for the complete solution.

**You Try It**   Work through this You Try It problem.

Work Exercise 31 in this textbook or in the *MyMathLab*® Study Plan.

In Example 7, we can get the same result if we multiply the numerator and denominator by $2i$. Find out **why**.

OBJECTIVE 5   SIMPLIFY RADICALS WITH NEGATIVE RADICANDS

In the next chapter, we will solve equations involving solutions with radicals having a negative **radicand**. Thus, we must first learn how to simplify a **radical** with a negative radicand such as $\sqrt{-49}$. By remembering that $\sqrt{-1} = i$, we can use the following rule to simplify this expression.

---

**Square Root of a Negative Number**

For any positive real number $a$,

$$\sqrt{-a} = \sqrt{-1} \cdot \sqrt{a} = i\sqrt{a}$$

---

So, $\sqrt{-49} = \sqrt{-1} \cdot \sqrt{49} = i \cdot 7 = 7i$.

At this point, you might want to review how to simplify radicals using the **product rule** in Section 8.2.

**Example 8  Simplifying a Square Root with a Negative Radicand**

   Simplify.

a. $\sqrt{-81}$        b. $\sqrt{-48}$        c. $\sqrt{-108}$

**Solution**

a. $\sqrt{-81} = \underbrace{\sqrt{-1}}_{i} \cdot \sqrt{81} = i \cdot 9 = 9i$

b. $\sqrt{-48} = \underbrace{\sqrt{-1}}_{i} \cdot \sqrt{48} = i \cdot \sqrt{16} \cdot \sqrt{3}$
$= i \cdot 4\sqrt{3} = 4i\sqrt{3}$

> Note: Since $4\sqrt{3}i$ can be confused with $4\sqrt{3i}$, we write $4\sqrt{3}i$ as $4i\sqrt{3}$.

c. Try this problem on your own. View the **answer**, or watch this **video** for detailed solutions to all three parts.

**You Try It**   **Work through this You Try It problem.**

**Work Exercise 32 in this textbook or in the *MyMathLab*® Study Plan.**

When simplifying or performing operations involving **radicals** with a negative **radicand** and an even **index**, it is important to first write the numbers in terms of the **imaginary unit** $i$ if possible.

The property $\sqrt{a} \cdot \sqrt{b} = \sqrt{ab}$ is only true when $a \geq 0$ and $b \geq 0$ so that $\sqrt{a}$ and $\sqrt{b}$ are **real numbers**. This property does not apply to non-real numbers. To find the correct answer if $a$ or $b$ are negative, we must first write each number in terms of the imaginary unit $i$.

$$\sqrt{-3} \cdot \sqrt{-12} = \sqrt{(-3)(-12)} = \sqrt{36} = 6 \quad \text{False}$$
$$\sqrt{-3} \cdot \sqrt{-12} = \underbrace{\sqrt{-1}}_{i} \cdot \sqrt{3} \cdot \underbrace{\sqrt{-1}}_{i} \cdot \sqrt{12} = i\sqrt{3} \cdot i\sqrt{12} = \underbrace{i^2}_{i^2 = -1}\sqrt{36} = -6 \quad \text{True}$$

Notice that in order to get the correct answer, we had to first write each number in terms of the **imaginary unit** $i$.

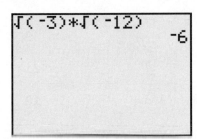

Figure 6

We can use a graphing calculator to check the result. See Figure 5.

### Example 9 Simplifying Expressions with Negative Radicands

   Simplify.

a. $\sqrt{-8} + \sqrt{-18}$    b. $\sqrt{-8} \cdot \sqrt{-18}$    c. $\dfrac{6 + \sqrt{(6)^2 - 4(2)(5)}}{2}$    d. $\dfrac{4 - \sqrt{-12}}{4}$

**Solution**

a. $\sqrt{-8} + \sqrt{-18} = \underbrace{\sqrt{-1}}_{i} \cdot \sqrt{8} + \underbrace{\sqrt{-1}}_{i} \cdot \sqrt{18}$

$\qquad = i \cdot 2\sqrt{2} + i \cdot 3\sqrt{2}$

$\qquad = 2i\sqrt{2} + 3i\sqrt{2}$

$\qquad = 5i\sqrt{2}$

b. $\sqrt{-8} \cdot \sqrt{-18} = \left(\underbrace{\sqrt{-1}}_{i} \cdot \sqrt{8}\right) \cdot \left(\underbrace{\sqrt{-1}}_{i} \cdot \sqrt{18}\right)$

$\qquad = (i \cdot 2\sqrt{2}) \cdot (i \cdot 3\sqrt{2})$

$\qquad = \underbrace{i^2}_{i^2 = -1} \cdot 6\sqrt{4}$

$\qquad = -1 \cdot 6 \cdot 2$

$\qquad = -12$

**c.–d.** Try these problems on your own. View the **answers** to parts c and d, or watch this **video** for detailed solutions to all four parts.

**You Try It**   Work through this You Try It problem.

Work Exercises 33–40 in this textbook or in the *MyMathLab*®  Study Plan.

# 8.7  Exercises

In Exercises 1–5, write each power of $i$ as $i$, $-1$, $-i$, or 1.

**You Try It**

**1.** $i^{41}$          **2.** $i^{28}$          **3.** $-i^{19}$          **4.** $(-i)^7$          **5.** $i^{22} + i^{13}$

In Exercises 6–13, find the sum or difference. Write each answer in standard form, $a + bi$.

**You Try It**

**6.** $(3 - 2i) + (-7 + 9i)$                    **7.** $(3 - 2i) - (-7 + 9i)$

**8.** $i - (1 + i)$                              **9.** $5 + (2 - 3i)$

**10.** $(2 + 5i) - (2 - 5i)$                    **11.** $(2 + 5i) + (2 - 5i)$

**12.** $[(-1 + 8i) - (3 - 4i)] + (9 - 4i)$      **13.** $(6 + 3i) - [(2 + 4i) + (5 - 2i)]$

In Exercises 14–21, perform the indicated operations. Write each answer in standard form.

**You Try It**

**14.** $3i(7i)$                    **15.** $2i(4 - 3i)$                    **16.** $-i(1 - i)$

**17.** $(3 - 2i)(6 + i)$          **18.** $(-2 - i)(3 - 4i)$              **19.** $(5 + i)(2 + 3i)$

**You Try It**

**20.** $(2 + 7i)^2$                **21.** $(6 - 2i)^2$

In Exercises 22–26, find the product of the complex number and its conjugate.

**You Try It**

**22.** $5 - 2i$      **23.** $1 - i$      **24.** $\dfrac{1}{2} - 3i$      **25.** $\sqrt{5} + i$      **26.** $4i$

In Exercises 27–31, write each quotient in standard form.

**You Try It**

**27.** $\dfrac{2 - i}{3 + 4i}$                    **28.** $\dfrac{1}{2 - i}$                    **29.** $\dfrac{3i}{2 + 2i}$

**You Try It**

**30.** $\dfrac{5 + i}{5 - i}$                    **31.** $\dfrac{2 - 3i}{5i}$

In Exercises 32–40, simplify.

**You Try It**

**32.** $\sqrt{-320}$

**33.** $\sqrt{-36} - \sqrt{49}$

**34.** $\sqrt{-1} + 3 - \sqrt{-64}$

**You Try It**

**35.** $\sqrt{-2} \cdot \sqrt{-18}$

**36.** $\left(\sqrt{-8}\right)^2$

**37.** $\left(i\sqrt{-4}\right)^2$

**38.** $\dfrac{-4 - \sqrt{-20}}{2}$

**39.** $\dfrac{-3 - \sqrt{-81}}{6}$

**40.** $\dfrac{4 + \sqrt{-8}}{4}$

CHAPTER NINE
# Quadratic Equations

## CHAPTER NINE CONTENTS

# 9.1 Solving Quadratic Equations Using the Square Root Property

## THINGS TO KNOW

Before working through this section, be sure you are familiar with the following concepts:

VIDEO    ANIMATION    INTERACTIVE

You Try It
1. Solve Quadratic Equations by Factoring (Section 6.7, **Objective 1**)

You Try It
2. Find Square Roots (Section 8.1, **Objective 1**)

You Try It
3. Approximate Square Roots (Section 8.1, **Objective 2**)

You Try It
4. Simplify Square Roots Containing Variables (Section 8.1, **Objective 3**)

You Try It
5. Simplify Square Roots Using the Product Rule (Section 8.2, **Objective 1**)

You Try It
6. Simplify Square Roots Using the Quotient Rule (Section 8.2, **Objective 2**)

You Try It
7. Rationalize Denominators (Section 8.4, **Objective 3**)

You Try It
8. Simplify Radicals with Negative Radicands (Section 8.7, **Objective 5**)

## OBJECTIVES

**1** Use the Square Root Property to Solve Quadratic Equations with Real-Number Solutions

**2** Use the Square Root Property to Solve Quadratic Equations with Non-Real Solutions

**3** Use Formulas Involving Squared Expressions to Solve Problems

OBJECTIVE 1   USE THE SQUARE ROOT PROPERTY TO SOLVE QUADRATIC EQUATIONS WITH REAL-NUMBER SOLUTIONS

Recall that a **quadratic equation** is an equation that can be written in the standard form $ax^2 + bx + c = 0$ where $a$, $b$, and $c$ are **real numbers** and $a \neq 0$. In **Section 6.7**, we solved quadratic equations by **factoring** and then using the **zero product property**. For example, we can solve the equation $x^2 = 100$ as follows:

| | |
|---|---|
| Write the original equation: | $x^2 = 100$ |
| Subtract 100 from both sides: | $x^2 - 100 = 0$ |
| Factor the **difference of squares**: | $(x - 10)(x + 10) = 0$ |
| Set each **factor** equal to 0: | $x - 10 = 0$ or $x + 10 = 0$ |
| Solve each resulting equation: | $x = 10$ or $x = -10$ |

Both values **check**, so the **solution set** is $\{-10, 10\}$.

Some quadratic equations are not easy to solve by factoring, and other quadratic equations cannot be solved by factoring at all. Therefore, in this chapter, we discuss other methods for solving quadratic equations, beginning with the *square root property*.

To understand this property, look again at the equation $x^2 = 100$. Notice that the two solutions, $-10$ and $10$, are the negative and positive **square roots** of the number 100. Looking at a similar equation, $x^2 = 36$, do you see that the solutions will be the negative and positive square roots of 36? Therefore, $x = -\sqrt{36} = -6$ or $x = \sqrt{36} = 6$. This illustrates the **square root property**.

---

**Square Root Property**

If $u$ is an **algebraic expression** and $k$ is a **real number**, then $u^2 = k$ is equivalent to $u = -\sqrt{k}$ or $u = \sqrt{k}$. Equivalently, if $u^2 = k$, then $u = \pm\sqrt{k}$.

---

See this **popup box** to learn more about the **plus or minus symbol** $(\pm)$.

**Note:** If $k \geq 0$, then the solutions to the equation $u^2 = k$ will be **real numbers**. However, if $k < 0$, then the solutions to $u^2 = k$ will be non-real **complex numbers**. At first, we will consider only equations with real solutions ($k \geq 0$). When we reach Objective 2, we will consider non-real solutions ($k < 0$).

### Example 1  Using the Square Root Property

*My video summary*   Use the square root property to solve each **quadratic equation**. Write each answer in **simplest form**.

**a.** $x^2 = 144$          **b.** $x^2 = 48$

**Solutions** For each equation, we apply the square root property and simplify. Work through each of the following, or watch this **video** for fully worked solutions.

a.   Write the original equation:   $x^2 = 144$

Apply the square root property:   $x = \pm\sqrt{144}$

Evaluate:   $x = \pm 12$

These results **check**, so the **solution set** is $\{-12, 12\}$.

b.   Write the original equation:   $x^2 = 48$

Apply the **square root property**:   $x = \pm\sqrt{48}$

Factor the **radicand**:   $x = \pm\sqrt{16 \cdot 3}$

Simplify:   $x = \pm 4\sqrt{3}$

View the **check**. The solution set is $\{-4\sqrt{3}, 4\sqrt{3}\}$.

**You Try It**   **Work through this You Try It problem.**

**Work Exercises 1–4 in this textbook or in the MyMathLab® Study Plan.**

 When solving equations of the form $u^2 = k$, we often simply say that we are taking the square root of both sides. However, because the **square root** of a number yields only one value, the **principal root**, many students forget to include the $\pm$. Remember that applying the square root property for $k \neq 0$ will result in *two* values: the positive and negative square roots of $k$.

To solve **quadratic equations** using the **square root property**, we can use the following guidelines.

---

**Solving Quadratic Equations Using the Square Root Property**

**Step 1.**   Write the equation in the form $u^2 = k$ to isolate the quantity being squared.

**Step 2.**   Apply the square root property.

**Step 3.**   Solve the resulting equations.

**Step 4.**   Check the solutions in the original equation.

---

## Example 2  Solving Quadratic Equations Using the Square Root Property

Solve each **quadratic equation**. Write each answer in **simplest form**.

a.  $x^2 - 289 = 0$          b.  $(x - 1)^2 = 9$

## Solutions

a.  In this case, we must first **isolate** $x^2$ to get the form $u^2 = k$. Then we can apply the **square root property**.

Begin with the original equation:   $x^2 - 289 = 0$

Add 289 to both sides:   $x^2 = 289$ ← Form $u^2 = k$

Apply the square root property:   $x = \pm\sqrt{289}$

Simplify:   $x = \pm 17$

View the **check**. The solution set is $\{-17, 17\}$.

*My video summary*    **b.** In this case, we have the square of an **algebraic expression**. Because the squared expression is already isolated, we first apply the square root property and then solve the resulting equation.

Begin with the original equation: $(x - 1)^2 = 9 \leftarrow$ Form $u^2 = k$

Apply the square root property: $x - 1 = \pm\sqrt{9}$

Try to finish solving this equation on your own. See the **answer**, or watch this **video** for a complete solution to part b.

**You Try It**   **Work through this You Try It problem.**

**Work Exercises 5–12 in this textbook or in the *MyMathLab*® Study Plan.**

### Example 3  Solving Quadratic Equations Using the Square Root Property

*My video summary*   Solve $3x^2 + 7 = 9$. Write the answer in **simplest form**.

**Solution**   We must first **isolate** $x^2$ to get the form $u^2 = k$. Then we can apply the **square root property** and simplify. Work through the following, or watch this **video** for a fully worked solution.

Write the original equation: $3x^2 + 7 = 9$

Subtract 7 from both sides: $3x^2 = 2$

Divide both sides by 3: $x^2 = \dfrac{2}{3} \leftarrow$ Form $u^2 = k$

Apply the square root property: $x = \pm\sqrt{\dfrac{2}{3}}$

Rationalize the denominator: $x = \pm\dfrac{\sqrt{2}}{\sqrt{3}} \cdot \dfrac{\sqrt{3}}{\sqrt{3}} \leftarrow$ Multiplying by 1

Simplify: $x = \pm\dfrac{\sqrt{6}}{3}$

These answers **check**, so the solution set is $\left\{ -\dfrac{\sqrt{6}}{3}, \dfrac{\sqrt{6}}{3} \right\}$.

**You Try It**   **Work through this You Try It problem.**

**Work Exercises 13–16 in this textbook or in the *MyMathLab*® Study Plan.**

### Example 4  Solving Quadratic Equations Using the Square Root Property

*My video summary*   Solve $2(x + 1)^2 - 17 = 23$. Write the answer in **simplest form**.

**Solution**   Try to work this problem on your own. Remember to first **isolate** the squared expression to get the form $u^2 = k$. Then apply the **square root property** and simplify. See the **answer**, or watch this **video** for the complete solution.

**You Try It**   **Work through this You Try It problem.**

**Work Exercises 17–20 in this textbook or in the *MyMathLab*® Study Plan.**

OBJECTIVE 2   USE THE SQUARE ROOT PROPERTY TO SOLVE QUADRATIC EQUATIONS WITH NON-REAL SOLUTIONS

In **Section 8.7**, we defined the **imaginary unit $i$** as $i = \sqrt{-1}$, where $i^2 = -1$, and then used it to simplify **radical expressions** with negative **radicands**. For example,

$$\sqrt{-25} = \sqrt{25} \cdot \sqrt{-1} = 5i.$$

When a **quadratic equation** has the form $u^2 = k$ with $k < 0$, the solutions will be non-real **complex numbers**. Consider the equation $x^2 = -9$. We can use the **square root property** to solve this equation as follows:

Write the original equation:   $x^2 = -9$

Apply the square root property:   $x = \pm\sqrt{-9}$

Simplify:   $x = \pm\sqrt{9} \cdot \sqrt{-1}$

$x = \pm 3i$

We can check these answers in the same way that we check real-number answers, by substituting them into the original equation to see if a true statement results:

|  | Checking $x = -3i$ | Checking $x = 3i$ |
|---|---|---|
| Write the original equation: | $x^2 = -9$ | $x^2 = -9$ |
| Substitute $-3i$ or $3i$ for $x$: | $(-3i)^2 \overset{?}{=} -9$ | $(3i)^2 \overset{?}{=} -9$ |
| Evaluate the exponent: | $9i^2 \overset{?}{=} -9$ | $9i^2 \overset{?}{=} -9$ |
| Use $i^2 = -1$: | $9(-1) \overset{?}{=} -9$ | $9(-1) \overset{?}{=} -9$ |
| Multiply: | $-9 = -9$  True | $-9 = -9$  True |

Both answers check, so the **solution set** is $\{-3i, 3i\}$.

## Example 5  Solving Quadratic Equations with Non-Real Solutions

Solve each **quadratic equation**. Write each answer in **simplest form**.

a. $(x - 3)^2 = -16$    b. $-\dfrac{1}{2}x^2 = 6$    c. $(2x - 10)^2 = -72$

## Solutions

a. Begin with the original equation:   $(x - 3)^2 = -16$ ← Form $u^2 = k$

Apply the **square root property**:   $x - 3 = \pm\sqrt{-16}$

Simplify:   $x - 3 = \pm 4i$

Add 3 to both sides:   $x = 3 \pm 4i$

View the **check**. The solution set is $\{3 - 4i, 3 + 4i\}$.

b. In this case, we must first isolate $x^2$ to get the form $u^2 = k$. Then we can apply the **square root property**.

Begin with the original equation:   $-\dfrac{1}{2}x^2 = 6$

Multiply both sides by $-2$:   $x^2 = -12$ ← Form $u^2 = k$

Apply the square root property:   $x = \pm\sqrt{-12}$

Simplify:   $x = \pm\sqrt{4} \cdot \sqrt{-1} \cdot \sqrt{3}$

$x = \pm 2i\sqrt{3}$

View the **check**. The solution set is $\{-2i\sqrt{3}, 2i\sqrt{3}\}$.

 *My video summary*  **c.** This equation is already in the form $u^2 = k$. Try to solve this equation on your own. Apply the **square root property** and simplify the radical expression. Then solve the resulting equations. View the **answer**, or watch this **video** for a complete solution to part c.

**You Try It**      **Work through this You Try It problem.**

**Work Exercises 21–30 in this textbook or in the *MyMathLab*®️ Study Plan.**

OBJECTIVE 3    USE FORMULAS INVOLVING SQUARED EXPRESSIONS TO SOLVE PROBLEMS

Many **formulas** involve squared expressions. We can use such formulas along with the square root property to solve application problems.

Recall from **Section 6.8**, if $a$ and $b$ are the **legs** and $c$ is the **hypotenuse** of a **right triangle**, then by the **Pythagorean Theorem**, $a^2 + b^2 = c^2$. If we know two of the lengths, then we can find the third.

### Example 6   Finding the Length of a Support Wire

To keep an electric pole vertical, a lineworker must install a support wire. The top of the support wire will be attached to the pole 20 feet above ground and the bottom of the wire will be attached to the ground 10 feet away from the pole. How long will the support wire be? Give both the exact length and an approximate length, rounded to the nearest tenth of a foot.

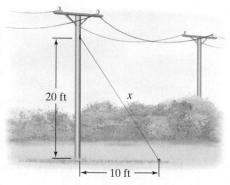

20 ft      $x$

|← 10 ft →|

**Solution**    We follow the **problem-solving strategy**.

**Step 1.** We want to find the length of the support wire. The pole, ground, and support wire form a right triangle with legs 20 feet and 10 feet long. We must find the length of the hypotenuse.

**Step 2.** Let $x$ = the length of the support wire.

**Step 3.** Use the **Pythagorean Theorem** to form an equation:

$$(\text{leg})^2 + (\text{leg})^2 = (\text{hypotenuse})^2$$
$$10^2 + 20^2 \quad = x^2$$

**Step 4.** Begin with the original equation:    $10^2 + 20^2 = x^2$

Evaluate the exponents:    $100 + 400 = x^2$

Add:    $500 = x^2$

Apply the **square root property**:    $x = \pm\sqrt{500}$

Factor the **radicand**:    $x = \pm\sqrt{100 \cdot 5}$

Simplify:    $x = \pm 10\sqrt{5}$

Approximate:    $x \approx \pm 22.4$

**Step 5.** The length must be positive, so we discard the negative result. An approximate length of 22.4 feet is slightly longer than the distance of 20 feet where the wire is connected to the top of the pole. This is a reasonable answer.

**Step 6.** The exact length of the support wire is $10\sqrt{5}$ feet. The approximate length is 22.4 feet.

**You Try It**    Work through this You Try It problem.

Work Exercises 31–34 in this textbook or in the *MyMathLab* Study Plan.

### Example 7  Finding the Girth of a Trout

 The weight $w$ of a trout, in pounds, can be approximated by the formula $w = \dfrac{lg^2}{800}$, where $l$ is the length and $g$ is the girth (distance around) of the trout, both in inches.

Girth, $g$

Length, $l$

In 1959, Alan Dow caught Wyoming's record-winning cutthroat trout. The trout weighed 15 pounds and was 32 inches long, but the girth was not recorded. Use the formula to estimate the trout's girth, rounded to the nearest tenth of an inch. (Source: Wyoming Department of Natural Resources)

**Solution**    We follow the **problem-solving strategy**.

**Step 1.** We want to find the girth of the record-winning trout. We know the weight is 15 pounds and the length is 32 inches. The formula

$$w = \frac{lg^2}{800}.$$

**Step 2.** Let $g = $ the record-winning trout's girth.

**Step 3.** Substitute $w = 15$ and $l = 32$ into the formula to form an equation:

$$15 = \frac{32g^2}{800}$$

**Step 4.** Try to solve this equation and finish this problem on your own. Note that girth must be positive, so any negative values found for $g$ must be discarded. View the **answer**, or watch this **video** for the complete solution.

**You Try It**    Work through this You Try It problem.

Work Exercises 35–40 in this textbook or in the *MyMathLab* Study Plan.

# 9.1 Exercises

In Exercises 1–30, solve each quadratic equation. Write each answer in simplest form.

**You Try It**

**1.** $x^2 = 256$

**2.** $y^2 = 80$

**3.** $m^2 = \dfrac{25}{64}$

**4.** $p^2 = \dfrac{20}{49}$

**5.** $x^2 - 121 = 0$

**6.** $z^2 - 45 = -13$

**You Try It**

**7.** $3x^2 = 84$

**8.** $5p^2 = 7$

**9.** $(x - 7)^2 = 625$

**10.** $(n + 5)^2 = 13$

**11.** $(z + 3)^2 = 54$

**12.** $\left(y - \dfrac{1}{2}\right)^2 = \dfrac{9}{4}$

**You Try It**

**13.** $7x^2 - 35 = 0$

**14.** $16t^2 + 3 = 4$

**15.** $5x^2 - 7 = -4$

**16.** $\dfrac{1}{3}x^2 + 8 = 14$

**17.** $(x + 3)^2 - 16 = 0$

**18.** $(n - 13)^2 + 11 = 79$

**You Try It**

**19.** $3(t - 11)^2 + 19 = 544$

**20.** $(2m + 3)^2 - 5 = 76$

**21.** $x^2 = -4$

**22.** $y^2 + 45 = 0$

**23.** $2w^2 + 23 = 7$

**24.** $(z + 9)^2 = -196$

**You Try It**

**25.** $(n - 1)^2 = -5$

**26.** $-\dfrac{3}{4}t^2 = \dfrac{1}{75}$

**27.** $(p - 2)^2 + 32 = 0$

**28.** $(2x - 14)^2 = -100$

**29.** $25 - (t - 15)^2 = 169$

**30.** $(3x - 15)^2 = -162$

In Exercises 31–32, use the Pythagorean Theorem to find the unknown length. Give the exact answer in simplified form. Then give an approximate answer rounded to two decimal places.

**31.**

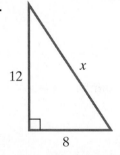

**32.**

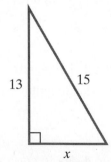

In Exercises 33–40, solve each problem.

**You Try It**

**33. Washing Windows**  A homeowner leans a 15-foot ladder against his house to wash windows. If the base of the ladder is 6 feet from the house, how high does the ladder reach? Assume the ground is level. Find the height, rounded to the nearest tenth of a foot.

**34. Television**  The rectangular flat screen of a television is 48 inches long and 27 inches wide. What is the screen's diagonal measurement? Find the measurement, rounded to the nearest whole inch.

**You Try It**

35. **Finding the Girth of a Salmon** The weight $w$ of a salmon, in pounds, is approximated by the formula $w = \dfrac{lg^2}{800}$, where $l$ is the length and $g$ is the girth (distance around) of the salmon, both in inches. If a particular salmon weighs 7.5 pounds and is 24 inches long, use the formula to find its girth, rounded to the nearest tenth of an inch.

36. **Finding the Length of a Bass** The weight $w$ of a bass, in pounds, is approximated by the formula $w = \dfrac{l^2 g}{1200}$, where $l$ is the length and $g$ is the girth of the bass, both in inches. A particular bass weighs 4.5 pounds and has a girth of 13.5 inches. Use the formula to find its length, rounded to the nearest tenth of an inch.

37. **Freefall Time** The distance $d$, in feet, that a free-falling object travels in $t$ seconds is given by the formula $d = 16t^2$ (ignoring wind resistance). If a stone is dropped from a bridge 100 feet above the water's surface, how long will it take to hit the water?

38. **Cliff Divers** The distance $d$, in meters, that a free-falling object travels in $t$ seconds is given by the formula $d = 4.9t^2$ (ignoring wind resistance). If a diver jumps from a cliff 35 meters above the water, how long will it take him to hit the water? Round to the nearest tenth of a second.

39. **Area of a Circle** The area $A$ of a circle with radius $r$ is given by the formula $A = \pi r^2$. Find the radius of a circle with an area of 706.5 square feet. Use $\pi \approx 3.14$.

40. **Volume of a Cone** The volume $V$ of a cone with radius $r$ and height $h$ is given by the formula $V = \dfrac{1}{3}\pi r^2 h$. Find the radius of a cone with a volume of 2826 cubic meters and a height of 12 meters. Use $\pi \approx 3.14$.

# 9.2 Solving Quadratic Equations by Completing the Square

## THINGS TO KNOW

Before working through this section, be sure you are familiar with the following concepts:

| | | VIDEO | ANIMATION | INTERACTIVE |

**You Try It**

1. Find Square Roots (Section 8.1, **Objective 1**)

**You Try It**

2. Simplify Square Roots Using the Product Rule (Section 8.2, **Objective 1**)

**You Try It**

3. Simplify Square Roots Using the Quotient Rule (Section 8.2, **Objective 2**)

**You Try It**

4. Simplify Radicals with Negative Radicands (Section 8.7, **Objective 5**)

**You Try It**

5. Use the Square Root Property to Solve Quadratic Equations with Real-Number Solutions (Section 9.1, **Objective 1**)

**You Try It**

6. Use the Square Root Property to Solve Quadratic Equations with Non-Real Solutions (Section 9.1, **Objective 2**)

## OBJECTIVES

**1** Write Perfect Square Trinomials by Completing the Square

**2** Solve Equations of the Form $x^2 + bx + c = 0$ by Completing the Square

**3** Solve Equations of the Form $ax^2 + bx + c = 0$ by Completing the Square

**4** Use Completing the Square with Non-Real Solutions

.......................................................................................................................................................................

### OBJECTIVE 1   WRITE PERFECT SQUARE TRINOMIALS BY COMPLETING THE SQUARE

Consider the following **perfect square trinomials:**

$$x^2 + 6x + 9 = (x + 3)^2 \qquad x^2 - 8x + 16 = (x - 4)^2 \qquad x^2 + 3x + \frac{9}{4} = \left( x + \frac{3}{2} \right)^2$$

$$\left( \frac{1}{2} \cdot 6 \right)^2 = 9 \qquad\qquad \left( \frac{1}{2} \cdot (-8) \right)^2 = 16 \qquad\qquad \left( \frac{1}{2} \cdot (3) \right)^2 = \frac{9}{4}$$

In each case, notice the relationship between the **coefficient** of the **linear term** ($x$-term) and the **constant** term. The constant term of a perfect square trinomial is equal to the square of $\frac{1}{2}$ the **linear coefficient**.

---

**Completing the Square**

To *complete the square* means to add an appropriate constant so that a binomial of the form $x^2 + bx$ becomes a perfect square trinomial. The appropriate constant is the square of half the linear coefficient, $\left( \frac{1}{2} \cdot b \right)^2$. For example, to complete the square given $x^2 + 12x$, we add $\left( \frac{1}{2} \cdot 12 \right)^2 = 6^2 = 36$ so we can write $x^2 + 12x + 36 = (x + 6)^2$.

---

When writing the **perfect square trinomial** as a **binomial** squared, note that the first term of the binomial is the variable (such as $x$) and the second term is $\frac{1}{2}$ times the **linear coefficient** from the trinomial. Consider the perfect square trinomials we saw earlier:

$$x^2 + 6x + 9 = (x + 3)^2 \qquad x^2 - 8x + 16 = (x - 4)^2 \qquad x^2 + 3x + \frac{9}{4} = \left( x + \frac{3}{2} \right)^2$$

$$\left( \frac{1}{2} \cdot 6 \right) = 3 \qquad\qquad \left( \frac{1}{2} \cdot (-8) \right) = -4 \qquad\qquad \left( \frac{1}{2} \cdot 3 \right) = \frac{3}{2}$$

**Example 1  Completing the Square**

Add the appropriate constant to complete the square, then factor the resulting perfect square trinomial.

**a.** $x^2 + 14x$ 
**b.** $x^2 - 10x$ 
**c.** $x^2 - \frac{3}{2}x$

**Solutions**

**a.** The linear coefficient is 14, so we must add $\left( \frac{1}{2} \cdot 14 \right)^2 = (7)^2 = 49$ to **complete the square.** Thus, the expression $x^2 + 14x + 49$ is a perfect square trinomial. Factoring gives $x^2 + 14x + 49 = (x + 7)^2$.

**b.** The **linear coefficient** is $-10$, so we must add $\left(\frac{1}{2}\cdot(-10)\right)^2 = (-5)^2 = 25$ to complete the square. Factoring gives $x^2 - 10x + 25 = (x-5)^2$.

*My video summary*    **c.** Try to work this problem on your own. View the **answer**, or watch this **video** for the complete solution.

**You Try It**    Work through this You Try It problem.

Work Exercises 1–4 in this textbook or in the **MyMathLab**® Study Plan.

OBJECTIVE 2    SOLVE EQUATIONS OF THE FORM $x^2 + bx + c = 0$ BY COMPLETING THE SQUARE

We can solve a **quadratic equation** of the form $x^2 + bx + c = 0$ by **completing the square**, and then applying the **square root property** as shown in the next example.

### Example 2 Solving a Quadratic Equation by Completing the Square

Solve $x^2 - 6x = -5$ by completing the square.

**Solution**   We start by completing the square on the left-hand side. Multiply $\frac{1}{2}$ times $b$ (the coefficient of the $x$-term), square the result, and add this to both sides of the equation.

$$\left(\frac{1}{2}\cdot(-6)\right)^2 = (-3)^2 = 9 \rightarrow x^2 - 6x + 9 = -5 + 9$$

$$x^2 - 6x + 9 = 4$$

Now write the left-hand side as a **perfect square**:   $(x-3)^2 = 4$

Apply the square root property and solve for $x$.

$$\begin{aligned}
\text{Use the square root property:} \quad & x - 3 = \pm\sqrt{4} \\
\text{Simplify } \sqrt{4}: \quad & x - 3 = \pm 2 \\
\text{Add 3 to both sides:} \quad & x = 3 \pm 2 \\
\text{Simplify:} \quad & x = 3 - 2 = 1 \quad \text{or} \quad x = 3 + 2 = 5
\end{aligned}$$

The **solution set** is $\{1, 5\}$.

### Example 3 Solving a Quadratic Equation by Completing the Square

*My video summary*    Solve $x^2 - 5x + 3 = 0$ by **completing the square**.

**Solution**   Subtract 3 from both sides to isolate the variable terms on the left-hand side.

$$x^2 - 5x + 3 = 0$$
$$x^2 - 5x + 3 - 3 = 0 - 3$$
$$x^2 - 5x = -3$$

Complete the square on the left-hand side. Multiply $\frac{1}{2}$ times $b$ (the **coefficient** of the $x$-term), square the result, and add this to both sides of the equation.

$$\left(\frac{1}{2}\cdot(-5)\right)^2 = \left(-\frac{5}{2}\right)^2 = \frac{25}{4} \quad \rightarrow \quad x^2 - 5x + \frac{25}{4} = -3 + \frac{25}{4}$$

$$x^2 - 5x + \frac{25}{4} = \frac{13}{4} \quad \leftarrow -\frac{3}{1} + \frac{25}{4} = -\frac{12}{4} + \frac{25}{4} = \frac{13}{4}$$

Finish solving this problem on your own. View the **answer**, or watch this **video** for the complete solution.

**You Try It**    Work through this You Try It problem.

Work Exercises 5–12 in this textbook or in the **MyMathLab**® Study Plan.

### Example 4  Solving a Quadratic Equation by Completing the Square

Solve $(x + 4)(x - 2) = 7$ by **completing the square**.

**Solution**

Original equation:  $(x + 4)(x - 2) = 7$

FOIL:  $x^2 - 2x + 4x - 8 = 7$

Simplify:  $x^2 + 2x - 8 = 7$

Add 8 to both sides:  $x^2 + 2x = 15$

Complete the square:  $x^2 + 2x + 1 = 15 + 1 \leftarrow \left(\frac{1}{2} \cdot 2\right)^2 = (1)^2 = 1$

Rewrite as a binomial squared:  $(x + 1)^2 = 16$

Use the **square root property**:  $x + 1 = \pm 4$

Subtract 1 from both sides:  $x = -1 \pm 4$

Simplify:  $x = -1 - 4 = -5$  or  $x = -1 + 4 = 3$

The **solution set** is $\{-5, 3\}$.

**You Try It**    Work through this You Try It problem.

Work Exercises 13 and 14 in this textbook or in the **MyMathLab**® Study Plan.

OBJECTIVE 3    SOLVE EQUATIONS OF THE FORM $ax^2 + bx + c = 0$ BY COMPLETING THE SQUARE

To solve a **quadratic equation** of the form $ax^2 + bx + c = 0$ with $a \neq 1$ by **completing the square**, we must first divide both sides of the equation by $a$ so that the **coefficient** on $x^2$ is 1.

---

**Solving $ax^2 + bx + c = 0$ with $a \neq 0$ by Completing the Square**

**Step 1.**  If $a \neq 1$, divide both sides of the equation by $a$.

**Step 2.**  Move all constants to the right-hand side.

**Step 3.**  Find $\frac{1}{2}$ times the coefficient of the $x$-term, **square** it, and add it to both sides of the equation.

**Step 4.**  The left-hand side is now a **perfect square**. Rewrite it as a binomial squared.

**Step 5.**  Use the **square root property** and solve for $x$.

---

### Example 5  Solving a Quadratic Equation by Completing the Square

Solve $4x^2 + 40x + 28 = 0$ by completing the square.

**Solution**

**Step 1.**  Original equation:  $4x^2 + 40x + 28 = 0$

Divide both sides by 4:  $\dfrac{4x^2}{4} + \dfrac{40x}{4} + \dfrac{28}{4} = \dfrac{0}{4}$

Simplify:  $x^2 + 10x + 7 = 0$

**Step 2.** Subtract 7 from both sides:  $x^2 + 10x = -7$

**Step 3.**  $\left(\dfrac{1}{2} \cdot 10\right)^2 = (5)^2 = 25 \rightarrow x^2 + 10x + 25 = -7 + 25$

$$x^2 + 10x + 25 = 18$$

**Step 4.** Write the left-hand side as a **binomial** squared:  $(x + 5)^2 = 18$

**Step 5.**  Use the **square root property**:  $x + 5 = \pm\sqrt{18}$

Use the **product rule for square roots**:  $x + 5 = \pm\sqrt{9} \cdot \sqrt{2}$

Simplify:  $x + 5 = \pm 3 \cdot \sqrt{2}$

Subtract 5 from both sides:  $x = -5 \pm 3\sqrt{2}$

The **solution set** is $\{-5 - 3\sqrt{2}, -5 + 3\sqrt{2}\}$.

### Example 6  Solving a Quadratic Equation by Completing the Square

 *My video summary*  Solve $2x^2 - 7x + 1 = 0$ by **completing the square.**

**Solution**

Original equation:  $2x^2 - 7x + 1 = 0$

Divide both sides by 2:  $x^2 - \dfrac{7}{2}x + \dfrac{1}{2} = 0$

Subtract $\dfrac{1}{2}$ from both sides:  $x^2 - \dfrac{7}{2}x = -\dfrac{1}{2}$

Complete the square:  $x^2 - \dfrac{7}{2}x + \dfrac{49}{16} = -\dfrac{1}{2} + \dfrac{49}{16} \leftarrow \left(\dfrac{1}{2} \cdot \left(-\dfrac{7}{2}\right)\right)^2 = \left(-\dfrac{7}{4}\right)^2 = \dfrac{49}{16}$

Try finishing this example on your own. View the **answer**, or watch this **video** for the complete solution.

**You Try It**  Work through this You Try It problem.

**Work Exercises 15–22 in this textbook or in the *MyMathLab*®  Study Plan.**

OBJECTIVE 4  USE COMPLETING THE SQUARE WITH NON-REAL SOLUTIONS

As we saw in **Section 9.1**, not all quadratic equations have **real-number** solutions, but this does not mean there are *no* solutions.

### Example 7  Solving a Quadratic Equation by Completing the Square

Solve $2x^2 + 12x + 42 = 0$ by **completing the square.**

## Solution

| | |
|---:|:---|
| Original equation: | $2x^2 + 12x + 42 = 0$ |
| Divide both sides by 2: | $x^2 + 6x + 21 = 0$ |
| Subtract 21 from both sides: | $x^2 + 6x = -21$ |
| Complete the square: | $x^2 + 6x + 9 = -21 + 9 \leftarrow \left(\frac{1}{2} \cdot 6\right)^2 = (3)^2 = 9$ |
| Rewrite as a binomial squared: | $(x + 3)^2 = -12$ |
| Use the **square root property**: | $x + 3 = \pm\sqrt{-12}$ |
| Use the **product rule for square roots**: | $x + 3 = \pm\sqrt{-4} \cdot \sqrt{3}$ |
| Simplify: | $x + 3 = \pm 2i \cdot \sqrt{3} \leftarrow \sqrt{-4} = \sqrt{4} \cdot \sqrt{-1} = 2 \cdot i$ |
| Subtract 3 from both sides: | $x = -3 \pm 2i\sqrt{3}$ |

The **solution set** is $\{-3 - 2i\sqrt{3},\, -3 + 2i\sqrt{3}\}$.

### Example 8  Solving a Quadratic Equation by Completing the Square

*My video summary*   Solve $4x^2 + 8x = -5$ by **completing the square**.

## Solution

| | |
|---:|:---|
| Original equation: | $4x^2 - 8x = -5$ |
| Divide both sides by 4: | $x^2 - 2x = -\dfrac{5}{4}$ |
| Complete the square: | $x^2 - 2x + 1 = -\dfrac{5}{4} + 1 \leftarrow \left(\dfrac{1}{2} \cdot (-2)\right)^2 = (-1)^2 = 1$ |
| Rewrite as a binomial squared: | $(x - 1)^2 = -\dfrac{1}{4} \quad \leftarrow -\dfrac{5}{4} + 1 = -\dfrac{5}{4} + \dfrac{4}{4} = -\dfrac{1}{4}$ |
| Use the **square root property**: | $x - 1 = \pm\sqrt{-\dfrac{1}{4}}$ |

Try finishing this example on your own. View the **answer**, or watch this **video** for the complete solution.

**You Try It**    Work through this You Try It problem.

Work Exercises 23–28 in this textbook or in the *MyMathLab*® Study Plan.

# 9.2  Exercises

In Exercises 1–4, add the appropriate constant to complete the square, then factor the resulting perfect square trinomial.

**You Try It**

**1.** $x^2 + 8x$        **2.** $y^2 - 20y$        **3.** $z^2 + 3z$        **4.** $x^2 - \dfrac{5}{3}x$

In Exercises 5–22, solve each quadratic equation by completing the square.

**5.** $x^2 + 2x = 24$        **6.** $x^2 - 14x = -33$        **7.** $z^2 - 14x + 9 = 0$

**You Try It**   **8.** $x^2 - 8x - 2 = 0$        **9.** $y^2 + 3 = 6y$        **10.** $x^2 - 3x = -1$

**11.** $x^2 + 7x + 14 = 0$          **12.** $z^2 + 4z - 8 = 0$          **13.** $x(x + 6) = 2$

**You Try It**
**14.** $(x - 4)(x - 6) = 31$          **15.** $3x^2 + 6x - 3 = 0$          **16.** $2x^2 - 2x + 1 = 0$

**17.** $3y^2 + 12y + 3 = 0$          **18.** $4z^2 + 10 = 16z$          **19.** $2x^2 - 3x = 7$

**20.** $3x^2 = 7 - 24x$          **21.** $4y^2 + 1 = 9y$          **22.** $4x^2 - 7x = 2$

In Exercises 23–28, solve each quadratic equation in the complex number system by completing the square.

**You Try It**
**23.** $x^2 - 12x + 40 = 0$          **24.** $x^2 + 14 = -6x$          **25.** $x^2 - 4x + 13 = 0$

**26.** $5y^2 - 10y + 7 = 0$          **27.** $5x^2 + 8x + 4 = 0$          **28.** $16x^2 + 73 = 56x$

# 9.3 Solving Quadratic Equations Using the Quadratic Formula

## THINGS TO KNOW

Before working through this section, be sure you are familiar with the following concepts:

|  |  | VIDEO | ANIMATION | INTERACTIVE |
|--|--|-------|-----------|-------------|

**You Try It**
**1.** Find Square Roots (Section 8.1, **Objective 1**)    

**You Try It**
**2.** Approximate Square Roots (Section 8.1, **Objective 2**)

**You Try It**
**3.** Simplify Square Roots Using the Product Rule (Section 8.2, **Objective 1**)    

**You Try It**
**4.** Simplify Radicals with Negative Radicands (Section 8.7, **Objective 5**)    

**You Try It**
**5.** Solve Equations of the form $x^2 + bx + c = 0$ by Completing the Square (Section 9.2, **Objective 2**)    

**You Try It**
**6.** Solve Equations of the form $ax^2 + bx + c = 0$ by Completing the Square (Section 9.2, **Objective 3**)

**You Try It**
**7.** Use Completing the Square with Non-Real Solutions (Section 9.2, **Objective 4**)    

## OBJECTIVES

**1** Use the Quadratic Formula to Solve Quadratic Equations with Real-Number Solutions

**2** Use the Quadratic Formula to Solve Quadratic Equations with Non-Real Solutions

**3** Use the Discriminant to Determine the Number and Type of Solutions to a Quadratic Equation

**4** Solve Applications Involving Quadratic Equations

OBJECTIVE 1  USE THE QUADRATIC FORMULA TO SOLVE QUADRATIC EQUATIONS
WITH REAL-NUMBER SOLUTIONS

*My animation summary*

 We can solve any **quadratic equation** by **completing the square**. However, this process can be very tedious. If we solve the general quadratic equation $ax^2 + bx + c = 0$, where $a$, $b$, and $c$ are real numbers, and $a \neq 0$, by completing the square, we obtain a useful result known as the **quadratic formula**. The quadratic formula can be used to solve any quadratic equation and typically is more efficient than directly solving by completing the square. Work through this **animation** to see how the quadratic formula is derived.

---

**Quadratic Formula**

The **solutions** to the quadratic equation $ax^2 + bx + c = 0$, $a \neq 0$, are given by the following formula:

$$x = \frac{-b \pm \sqrt{b^2 - 4ac}}{2a}$$

---

The quadratic equation must be written in **standard form** before identifying the coefficients $a$, $b$, and $c$.

### Example 1  Solving a Quadratic Equation Using the Quadratic Formula

Use the **quadratic formula** to solve $3x^2 - 11x - 4 = 0$.

**Solution**  The equation is written in **standard form**, so we can identify the **coefficients**.

$$3x^2 - 11x - 4 = 0$$

$$a = 3, \quad b = -11, \quad c = -4$$

Substitute 3 for $a$, $-11$ for $b$, and $-4$ for $c$ in the quadratic formula.

Write down the quadratic formula:   $x = \dfrac{-b \pm \sqrt{b^2 - 4ac}}{2a}$

Substitute values for $a, b,$ and $c$:   $= \dfrac{-(-11) \pm \sqrt{(-11)^2 - 4(3)(-4)}}{2(3)}$

Simplify:   $= \dfrac{11 \pm \sqrt{121 + 48}}{6}$

$= \dfrac{11 \pm \sqrt{169}}{6}$

Evaluate the square root:   $= \dfrac{11 \pm 13}{6}$

There are two **solutions** to the equation, one from the $+$ sign and one from the $-$ sign.

$$x = \frac{11 + 13}{6} = \frac{24}{6} = 4 \quad \text{or} \quad x = \frac{11 - 13}{6} = \frac{-2}{6} = -\frac{1}{3}$$

View the **check**. The **solution set** is $\left\{ -\dfrac{1}{3}, 4 \right\}$.

**You Try It**   Work through this You Try It problem.

Work Exercises 1–3 in this textbook or in the *MyMathLab*® Study Plan.

(eText Screens 9.3-1–9.3-22)

### Example 2  Solving a Quadratic Equation Using the Quadratic Formula

*My video summary*  Use the **quadratic formula** to solve $x^2 + 4x = -1$.

**Solution**  Work through the following, or watch this **video** for the complete solution.

First, we write the equation in **standard form** by adding 1 to both sides. The standard form is $x^2 + 4x + 1 = 0$. Next, we identify the **coefficients**. No coefficient is written on the $x^2$-term, so it is understood to be 1.

$$1x^2 + 4x + 1 = 0$$

$$a = 1, \; b = 4, \; c = 1$$

Substitute 1 for $a$, 4 for $b$, and 1 for $c$ in the quadratic formula.

Write down the quadratic formula:  $x = \dfrac{-b \pm \sqrt{b^2 - 4ac}}{2a}$

Substitute values for $a, b$, and $c$:  $= \dfrac{-(4) \pm \sqrt{(4)^2 - 4(1)(1)}}{2(1)}$

Simplify:  $= \dfrac{-4 \pm \sqrt{12}}{2}$ ← $\boxed{\sqrt{(4)^2 - 4(1)(1)} = \sqrt{16 - 4} = \sqrt{12}}$

Use the **product rule for radicals**:  $= \dfrac{-4 \pm 2\sqrt{3}}{2}$ ← $\boxed{\sqrt{12} = \sqrt{4} \cdot \sqrt{3} = 2\sqrt{3}}$

Factor out 2 in the numerator:  $= \dfrac{2(-2 \pm \sqrt{3})}{2}$

Divide out the common factor:  $= -2 \pm \sqrt{3}$  ← $\boxed{\dfrac{2(-2 \pm \sqrt{3})}{2}}$

View the **check**. The **solution set** is $\{-2 - \sqrt{3}, -2 + \sqrt{3}\}$.

**You Try It**  Work through this You Try It problem.

**Work Exercises 4–6 in this textbook or in the *MyMathLab*® Study Plan.**

### Example 3  Solving a Quadratic Equation Using the Quadratic Formula

*My video summary*  Solve $3x^2 + 2x - 2 = 0$ using the **quadratic formula**.

**Solution**  The equation is in **standard form**, so we can identify the **coefficients**.

$$3x^2 + 2x - 2 = 0$$

$$a = 3, \; b = 2, \; c = -2$$

Substitute 3 for $a$, 2 for $b$, and $-2$ for $c$ in the quadratic formula.

Write down the quadratic formula:  $x = \dfrac{-b \pm \sqrt{b^2 - 4ac}}{2a}$

Substitute values for $a, b$, and $c$:  $= \dfrac{-(2) \pm \sqrt{(2)^2 - 4(3)(-2)}}{2(3)}$

Try to finish finding the two solutions on your own. Remember to **simplify the radical** and **divide out** any common factors. View the **answer**, or watch this **video** for a complete solution.

**You Try It**   Work through this You Try It problem.

Work Exercises 7–9 in this textbook or in the *MyMathLab*®  Study Plan.

### Example 4  Solving a Quadratic Equation Using the Quadratic Formula

  Solve $14x^2 - 5x = 5x^2 + 7x - 4$ using the **quadratic formula**.

**Solution**  Try to solve this equation on your own. Remember to write the equation in **standard form** first. See this **popup box** to check your standard form. Use the quadratic formula to finish solving the equation. View the **answer**, or watch this **video** for the complete solution.

**You Try It**   Work through this You Try It problem.

Work Exercises 10–12 in this textbook or in the *MyMathLab*®  Study Plan.

When a **quadratic equation** contains fractions, it is good practice to **clear the fractions** before using the quadratic formula. Doing so makes the work more manageable.

### Example 5  Solving a Quadratic Equation Using the Quadratic Formula

  Solve $\dfrac{1}{12}x^2 + \dfrac{1}{6} = \dfrac{1}{2}x$ using the quadratic formula.

**Solution**

Write the original equation:  $\dfrac{1}{12}x^2 + \dfrac{1}{6} = \dfrac{1}{2}x$

Multiply both sides by the **LCD**, 12:  $12\left(\dfrac{1}{12}x^2 + \dfrac{1}{6}\right) = 12\left(\dfrac{1}{2}x\right)$

Simplify:  $x^2 + 2 = 6x$

We now have an equation without fractions. Try to finish solving this equation on your own using the quadratic formula. Remember to write it in **standard form** first. View the **answer**, or watch this **video** for the complete solution.

**You Try It**   Work through this You Try It problem.

Work Exercises 13–15 in this textbook or in the *MyMathLab*®  Study Plan.

OBJECTIVE 2   USE THE QUADRATIC FORMULA TO SOLVE QUADRATIC EQUATIONS WITH NON-REAL SOLUTIONS

We have seen in **Sections 9.1** and **9.2** that **quadratic equations** sometimes have solutions that are non-real **complex numbers**. We can also solve such equations using the **quadratic formula**.

### Example 6  Using the Quadratic Formula with Non-Real Solutions

Solve $x^2 - 2x + 3 = 0$ using the quadratic formula.

**Solution** The equation is in **standard form**, so we can identify the **coefficients**. No coefficient is written on the $x^2$-term, so it is understood to be 1.

$$1x^2 - 2x + 3 = 0$$

$$a = 1, \ b = -2, \ c = 3$$

Substitute 1 for $a$, $-2$ for $b$, and 3 for $c$ in the quadratic formula.

Begin with the quadratic formula: $\quad x = \dfrac{-b \pm \sqrt{b^2 - 4ac}}{2a}$

Substitute values for $a, b,$ and $c$: $\quad = \dfrac{-(-2) \pm \sqrt{(-2)^2 - 4(1)(3)}}{2(1)}$

Simplify: $\quad = \dfrac{2 \pm \sqrt{-8}}{2} \leftarrow \boxed{\sqrt{(-2)^2 - 4(1)(3)} = \sqrt{4 - 12} = \sqrt{-8}}$

Notice that the **radicand** is negative. So, the solutions will be non-real complex numbers, as follows.

Simplify $\sqrt{-8}$: $\quad = \dfrac{2 \pm 2i\sqrt{2}}{2} \leftarrow \boxed{\sqrt{-8} = \sqrt{4} \cdot \sqrt{-1} \cdot \sqrt{2} = 2i\sqrt{2}}$

Factor out 2 in the numerator: $\quad = \dfrac{2(1 \pm i\sqrt{2})}{2}$

Divide out the common factor: $\quad = 1 \pm i\sqrt{2} \leftarrow \boxed{\dfrac{\cancel{2}(1 \pm i\sqrt{2})}{\cancel{2}}}$

View the **check**. The **solution set** is $\{1 - i\sqrt{2}, \ 1 + i\sqrt{2}\}$.

**Example 7  Using the Quadratic Formula with Non-Real Solutions**

*My video summary*  Solve $16x^2 - 8x + 5 = 0$ using the quadratic formula.

**Solution**  Try to solve this equation on your own using the quadratic formula. Remember to write the equation in **standard form** first. View the **answer**, or watch this **video** for the complete solution.

**You Try It**  Work through this You Try It problem.

Work Exercises 16–21 in this textbook or in the **MyMathLab**® Study Plan.

OBJECTIVE 3  **USE THE DISCRIMINANT TO DETERMINE THE NUMBER AND TYPE OF SOLUTIONS TO A QUADRATIC EQUATION**

In **Example 6**, the **quadratic equation** $x^2 - 2x + 3 = 0$ had two non-real solutions. The solutions were non-real **complex numbers** because the expression $b^2 - 4ac$ under the radical of the **quadratic formula** had a negative value. This expression $b^2 - 4ac$ is called the **discriminant**.

Quadratic Formula: $\quad x = \dfrac{-b \pm \sqrt{\boxed{b^2 - 4ac}}}{2a} \leftarrow$ Discriminant

We can use the value of the discriminant to determine the number and type of solutions to a quadratic equation.

---

**Discriminant**

Given a quadratic equation $ax^2 + bx + c = 0$, $a \neq 0$, the expression $D = b^2 - 4ac$ is called the **discriminant**.

If $D > 0$, then the quadratic equation has two real solutions.
If $D < 0$, then the quadratic equation has two non-real solutions.
If $D = 0$, then the quadratic equation has exactly one real solution.

---

### Example 8  Using the Discriminant

Use the **discriminant** to determine the number and type of solutions to the equation $x^2 - 2x + 3 = 0$.

**Solution**  The equation is written in **standard form**, so the **coefficients** are $a = 1$, $b = -2$, and $c = 3$. The discriminant is

$$D = b^2 - 4ac = (-2)^2 - 4(1)(3) = 4 - 12 = -8.$$

Because the discriminant is negative $(D < 0)$, the equation has two non-real solutions.

**Note:** We solved $x^2 - 2x + 3 = 0$ in **Example 6**. Its solutions are $1 - i\sqrt{2}$ and $1 + i\sqrt{2}$, two non-real solutions.

### Example 9  Using the Discriminant

*My video summary*  Use the discriminant to determine the number and type of solutions to each quadratic equation.

**a.** $2x^2 - 5x + 1 = 0$      **b.** $25x^2 + 9 = 30x$      **c.** $x^2 = 4x - 5$

**Solutions**

**a.**  The equation $2x^2 - 5x + 1 = 0$ is written in standard form, so $a = 2$, $b = -5$, and $c = 1$. Then

$$D = b^2 - 4ac = (-5)^2 - 4(2)(1) = 25 - 8 = 17.$$

The discriminant is positive $(D > 0)$, so the equation has two real solutions.

**b.**  The equation $25x^2 + 9 = 30x$ is not written in **standard form**. Subtract $30x$ from both sides to result in the standard form $25x^2 - 30x + 9 = 0$. Use the coefficients $a = 25$, $b = -30$, and $c = 9$ to find the discriminant. Then use the discriminant to determine the number and type of solutions. View the **answers** to parts b and c, or watch this **video** for complete solutions to all three parts.

**c.**  Try working this problem on your own. View the **answers** to parts b and c, or watch this **video** for complete solutions to all three parts.

**You Try It**    Work through this You Try It problem.

Work Exercises 22–27 in this textbook or in the *MyMathLab*® Study Plan.

---

OBJECTIVE 4    SOLVE APPLICATIONS INVOLVING QUADRATIC EQUATIONS

Application problems sometimes involve **quadratic equations**. For such problems, we follow the same **problem-solving strategy** that we have used throughout this text. Depending on the quadratic equation used to model the problem, we may solve it by one of the four methods we have now studied for solving quadratic equations: (1) **factoring** and using the **zero product property**, (2) using the **square root property**, (3) **completing the square**, or (4) using the **quadratic formula**.

### Example 10  Launching an Air Pump Toy Rocket

A toy rocket system uses compressed air from a pump to launch a rocket. If the rocket is launched from 24 feet above ground with an initial velocity of 40 feet per second, then the height $h$ of the rocket above ground, in feet, at any time $t$ seconds after launch is given by $h = -16t^2 + 40t + 24$. How long after launch will it take the rocket to hit the ground?

### Solution

**Step 1.** We want to find the time when the rocket hits the ground ($h = 0$). We know the equation $h = -16t^2 + 40t + 24$ models the height $h$ of the rocket at time $t$ after launch.

**Step 2.** We know $h$ = height in feet and $t$ = time in seconds.

**Step 3.** Using the equation $h = -16t^2 + 40t + 24$, set $h = 0$:

$$0 = -16t^2 + 40t + 24$$

**Step 4.** Begin with the original equation:  $\quad 0 = -16t^2 + 40t + 24$

Divide both sides by $-8$:  $\quad 0 = 2t^2 - 5t - 3$

Factor:  $\quad 0 = (2t + 1)(t - 3)$

Apply the **zero product property**:  $\quad 2t + 1 = 0 \quad \text{or} \quad t - 3 = 0$

Solve each part for $t$:  $\quad t = -\dfrac{1}{2} \quad \text{or} \quad t = 3$

**Step 5.** Because $t$ represents the time after launch, its value cannot be negative. So, we discard $t = -\dfrac{1}{2}$. The only reasonable solution is $t = 3$ seconds.

**Step 6.** The rocket will hit the ground 3 seconds after launch.

**You Try It**    Work through this You Try It problem.

**Work Exercises 28 and 29 in this textbook or in the *MyMathLab*® Study Plan.**

### Example 11  Dimensions of a Television

*My video summary*    The size of a television is the length of the **diagonal** of its rectangular screen. If a 55-inch television is 21 inches longer than it is wide, find the length and width of the screen, rounded to the nearest inch.

### Solution

**Step 1.** We want to find the length and width of the television. We know that the length is 21 inches more than the width, and the diagonal of the rectangular screen is 55 inches.

**Step 2.** Let $w$ = the width of the television. Then $w + 21$ = the length.

**Step 3.** The **diagonal** forms a **right triangle** with **legs** $w$ and $w + 21$, and **hypotenuse** 55. Use the **Pythagorean Theorem** to form a **quadratic equation:**

$$(\text{leg})^2 + (\text{leg})^2 = (\text{hypotenuse})^2$$
$$w^2 + (w + 21)^2 = 55^2$$

**Step 4.** Begin with the original equation: $\qquad w^2 + (w + 21)^2 = 55^2$

**Square the binomial:** $\quad w^2 + w^2 + 42w + 441 = 3025$

**Combine like terms:** $\qquad 2w^2 + 42w - 2584 = 0$

Divide both sides by 2: $\qquad w^2 + 21w - 1292 = 0$

Because this equation does not factor, use the quadratic formula to finish solving it. Then use the result to find the dimensions of the television. View the **answer**, or watch this **video** for the complete solution.

**You Try It**    Work through this You Try It problem.

Work Exercises 30–35 in this textbook or in the *MyMathLab*® Study Plan.

# 9.3 **Exercises**

In Exercises 1–21, solve each quadratic equation using the quadratic formula.

**You Try It**

**1.** $x^2 - 7x - 144 = 0$    **2.** $3x^2 + 8x - 3 = 0$    **3.** $2x^2 - 21x + 40 = 0$

**4.** $x^2 - 8x - 2 = 0$    **5.** $2x^2 + 6x = 5$    **6.** $3x^2 = 1 + 4x$

**You Try It**

**7.** $x^2 + x - 3 = 0$    **8.** $2x^2 + 8x + 5 = 0$    **9.** $x^2 + 4x + 1 = 0$

**You Try It**

**10.** $2x^2 - 7x = x^2 + x - 11$    **11.** $3x^2 + 5x + 1 = 7x + 4$    **12.** $9x^2 - 6x = -1$

**You Try It**

**13.** $\dfrac{1}{2}x^2 - x - \dfrac{1}{3} = 0$    **14.** $\dfrac{3}{16}x^2 - \dfrac{1}{8} = \dfrac{1}{4}x$    **15.** $2x^2 + 5.5x - 1.5 = 0$

**You Try It**

**16.** $x^2 - 4x + 29 = 0$    **17.** $x^2 + 20 = 5x$    **18.** $4x^2 - x + 8 = 0$

**19.** $5x^2 + 3x + 1 = 0$    **20.** $9x^2 = 12x - 5$    **21.** $\dfrac{1}{10}x^2 + \dfrac{1}{2} = \dfrac{2}{5}x$

In Exercises 22–27, find the discriminant. Then use it to determine the number and type of solutions to each quadratic equation. Do not solve the equations.

**You Try It**

**22.** $x^2 + 3x + 1 = 0$    **23.** $4x^2 + 4x + 1 = 0$    **24.** $2x^2 + x = 5$

**25.** $4x^2 + 25 = 12x$    **26.** $9x^2 + 49 = 42x$    **27.** $25x - 6x^2 = 24$

In Exercises 28–34, solve each problem. If necessary, approximate the answers by rounding to two decimal places.

**28. Rocket Launch** A toy rocket is launched from a platform 2.8 meters above the ground in such a way that its height, $h$ (in meters), after $t$ seconds is given by the equation $h = -4.9t^2 + 18.9t + 2.8$. How long will it take for the rocket to hit the ground?

**You Try It**

**29. Projectile Motion** Shawn threw a rock straight up from a cliff that was 24 feet above the water. If the height of the rock $h$, in feet, after $t$ seconds is given by the equation $h = -16t^2 + 20t + 24$, how long will it take for the rock to hit the water?

**30. Computer Screen** The size of a computer screen is the length of its diagonal. If the 12-inch screen of a computer is 3.5 inches longer than it is wide, find the length and width of the screen, rounded to the nearest hundredth of an inch.

**You Try It**

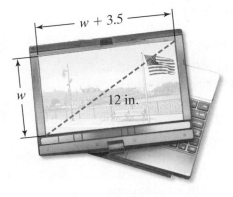

**31. Loading Ramp** A loading ramp in a steel yard has a horizontal run that is 26 feet longer than its vertical rise. If the ramp is 30 feet long, what is the vertical rise, rounded to the nearest hundredth of a foot?

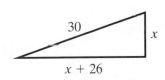

**32. Dimensions of a Rectangle** The length of a rectangle is 1 inch less than twice the width. If the diagonal is 2 inches more than the length, find the dimensions of the rectangle, rounded to the nearest hundredth of an inch.

**33. Dimensions of a Rectangle** The length of a rectangle is 1 cm less than three times the width. If the area of the rectangle is $30\,\text{cm}^2$, find the dimensions of the rectangle.

**34. Building a Walkway** A 35- by 20-ft rectangular swimming pool is surrounded by a walkway of uniform width. If the total area of the walkway is $434\,\text{ft}^2$, how wide is the walkway?

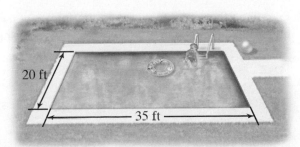

# 9.4 Graphs of Quadratic Equations

## THINGS TO KNOW

Before working through this section, be sure you are familiar
with the following concepts:

| | | VIDEO | ANIMATION | INTERACTIVE |

**You Try It**
1. Graph Linear Equations by Plotting Points
(Section 3.2, **Objective 3**)

**You Try It**
2. Find $x$- and $y$-Intercepts
(Section 3.2, **Objective 4**)

**You Try It**
3. Solve Quadratic Equations by Factoring
(Section 6.7, **Objective 1**)

**You Try It**
4. Use the Square Root Property to Solve
Quadratic Equations with Real-Number
Solutions (Section 9.1, **Objective 1**)

**You Try It**
5. Solve Equations of the Form $ax^2 + bx + c = 0$ by
Completing the Square (Section 9.2, **Objective 3**)

**You Try It**
6. Use the Quadratic Formula to Solve
Quadratic Equations with Real-Number
Solutions (Section 9.3, **Objective 1**)

## OBJECTIVES

**1** Identify the Characteristics of a Parabola

**2** Find the Intercepts of a Parabola

**3** Find the Vertex of a Parabola

**4** Graph Equations of the Form $y = ax^2 + bx + c$

**5** Use the Vertex to Solve Application Problems

---

### OBJECTIVE 1  IDENTIFY THE CHARACTERISTICS OF A PARABOLA

In previous sections, we learned several **methods** that could be used to solve **quadratic
equations**. In this section, we learn about graphing quadratic equations.

Equations that can be written in the form $y = ax^2 + bx + c$, where $a \neq 0$, are said to be
**quadratic equations in two variables**. The following are some examples:

$$y = 3x^2 + 4x - 2, \quad y = x^2 - 7x, \quad y = -\frac{1}{2}x^2 + 6$$

The graph of a quadratic equation in two variables is a "u-shaped" graph called a **parabola**.

The equation $y = x^2$ is a quadratic equation with $a = 1$, $b = 0$, and $c = 0$. Its graph is shown
in **Figure 1a**. The equation $y = -x^2$ is a quadratic equation with $a = -1$, $b = 0$, and $c = 0$.
Its graph is shown in **Figure 1b**. Notice that both graphs have the characteristic "u-shape"
of a parabola.

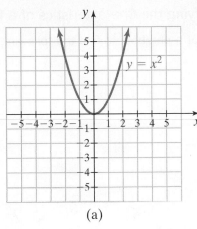

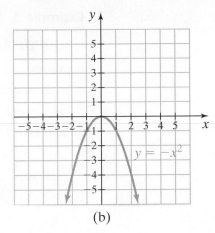

(a)                                    (b)

**Figure 1** Graphs of $y = x^2$ and $y = -x^2$

A **parabola** either opens up or opens down depending on the **leading coefficient**, $a$. If $a > 0$, as in Figure 1a, the parabola will "open up." If $a < 0$, as in Figure 1b, the parabola will "open down."

Before we can sketch graphs of **quadratic equations in two variables**, we must be able to identify four basic characteristics of a **parabola**: *vertex, axis of symmetry, y-intercept,* and *x-intercepts.*

Every parabola has a **vertex**. If the parabola "opens up," the vertex is the lowest point on the graph and the graph has a **minimum** $y$-value. If the parabola "opens down," the vertex is the highest point on the graph and the graph has a **maximum** $y$-value. In either case, the $y$-coordinate of the vertex indicates the maximum or minimum value and the $x$-coordinate indicates *where* this occurs. The vertex is denoted by the ordered pair $(h, k)$.

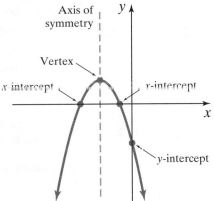

The **axis of symmetry** is an imaginary **vertical line** that passes through the vertex and divides the graph into two mirror images. Points on the graph that are the same distance from the axis of symmetry will have the same $y$-coordinate. The equation of the axis of symmetry is $x = h$, where $h$ is the $x$ coordinate of the vertex.

The graph of every quadratic equation crosses the $y$-axis, so every parabola has a $y$-**intercept**. However, the graph may or may not cross the $x$-axis, so the parabola may or may not have $x$-**intercepts**.

We summarize these four basic characteristics with the figure shown.

*My animation summary*

1. Vertex
2. Axis of symmetry
3. $y$-Intercept
4. $x$-Intercept(s)

Work through this **animation**, selecting each characteristic to get a detailed description.

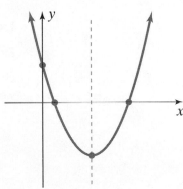

## Example 1  Identifying the Characteristics of a Parabola

Use the given graph of a parabola to find the following:

a. Vertex

b. Axis of symmetry

c. $y$-intercept

d. $x$-intercept(s)

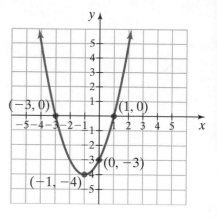

### Solutions

a. Because the graph "opens up," the vertex is the lowest point on the graph, $(h, k) = (-1, -4)$.

b. The axis of symmetry is the vertical line that passes through the vertex. Because the $x$-coordinate of the vertex is $h = -1$, the equation of the axis of symmetry is $x = -1$.

c. The $y$-intercept is the $y$-coordinate of the point where the graph crosses the $y$-axis. The graph crosses the $y$-axis at the point $(0, -3)$, so the $y$-intercept is $-3$.

d. The $x$-intercepts are the $x$-coordinates of points where the graph crosses or touches the $x$-axis. The graph crosses the $x$-axis at the points $(-3, 0)$ and $(1, 0)$, so the $x$-intercepts are $-3$ and $1$.

**You Try It**    **Work through this You Try It problem.**

**Work Exercises 1–4 in this textbook or in the *MyMathLab*® Study Plan.**

OBJECTIVE 2    FIND THE INTERCEPTS OF A PARABOLA

In Chapter 3, we saw that **intercepts** were helpful when graphing **linear equations**. They will also be helpful in graphing **quadratic equations**. Recall from **Section 3.2** that we find $y$-intercepts of the graph of an equation by setting $x = 0$ and solving for $y$. Similarly, we find $x$-intercepts by setting $y = 0$ and solving the resulting equation for $x$.

## Example 2  Finding the $x$- and $y$-Intercepts of a Parabola

Find the $x$- and $y$-intercepts of the graph of $y = x^2 - 3x - 4$.

**Solution**  To find the $x$-intercepts, we let $y = 0$ and solve for $x$. To find the $y$-intercept, we let $x = 0$ and solve for $y$.

| *x*-intercepts: | *y*-intercept: |
|---|---|

**x-intercepts:**

Let $y = 0$: $\quad 0 = x^2 - 3x - 4$

Factor the right-hand side: $\quad 0 = (x - 4)(x + 1)$

Apply the **zero product property.**
Set each factor equal to 0 and solve.

$x - 4 = 0 \quad$ or $\quad x + 1 = 0$

$\qquad x = 4 \qquad\qquad\quad x = -1$

The *x*-intercepts are $-1$ and 4.

**y-intercept:**

Let $x = 0$: $\quad y = (0)^2 - 3(0) - 4$

Simplify: $\quad y = 0 - 0 - 4$

$\qquad\qquad\quad y = -4$

The *y*-intercept is $-4$.

Notice that the *y*-intercept is same as the constant term, *c*, in the quadratic equation. This result is true in general.

### Example 3  Finding the *x*- and *y*-Intercepts of a Parabola

 *My video summary.*   Find the *x*- and *y*-intercepts of the graph of $y = 3x^2 - 6x + 2$.

**Solution**  The constant term is $c = 2$, so the *y*-intercept is 2.

To find the *x*-intercepts, we set $y = 0$ and solve for *x*.

$$0 = 3x^2 - 6x + 2$$

The right-hand side cannot be factored over the integers, so we solve the equation by using the **quadratic formula** with $a = 3$, $b = -6$, and $c = 2$.

Quadratic formula: $\quad x = \dfrac{-b \pm \sqrt{b^2 - 4ac}}{2a}$

Substitute 3 for *a*, $-6$ for *b*, and 2 for *c*. $\quad x = \dfrac{-(-6) + \sqrt{(-6)^2 - 4(3)(2)}}{2(3)}$

Simplify: $\quad x = \dfrac{6 \pm \sqrt{36 - 4(3)(2)}}{2(3)}$

Finish simplifying to find the *x*-intercepts. View the **answer**, or watch this **video** for the complete solution.

**You Try It**  Work through this **You Try It** problem.

**Work Exercises 5–10 in this textbook or in the *MyMathLab*® Study Plan.**

---

OBJECTIVE 3   FIND THE VERTEX OF A PARABOLA

The most important point to find when graphing a **quadratic equation** is the **vertex**. Recall that the vertex is either the lowest point on the parabola if the graph opens up, or the highest point if the graph opens down.

We can use the symmetry of a **parabola** and the quadratic formula to show that the *x*-coordinate of the vertex is equal to $-\dfrac{b}{2a}$. View the **details**.

---

**Finding the Vertex of a Parabola**

Given a quadratic equation of the form $y = ax^2 + bx + c$, $a \neq 0$, the $x$-coordinate of the vertex, $h$, is given by $h = -\dfrac{b}{2a}$. To find the $y$-coordinate of the vertex, $k$, substitute $-\dfrac{b}{2a}$ for $x$ and find $y$.

---

### Example 4  Finding the Vertex of a Parabola

Find the vertex of the graph of the given equation.

**a.** $y = 4x^2 - 8x - 21$    **b.** $y = 2x^2 + 12x + 18$    **c.** $y = -x^2 - 3x + 4$

### Solutions

**a.** We first compare the given equation to the general form $y = ax^2 + bx + c$ to determine the values for $a$ and $b$.

$$y = 4x^2 - 8x - 21$$
$$y = ax^2 + bx + c$$

So we have $a = 4$ and $b = -8$. The $x$-coordinate of the **vertex** is

$$h = -\frac{b}{2a} = -\frac{(-8)}{2(4)} = \frac{8}{8} = 1.$$

To find the $y$-coordinate, we substitute 1 for $x$ in the equation and determine $y$.

$$\begin{aligned}
\text{Original equation:} \quad & y = 4x^2 - 8x - 21 \\
\text{Substitute 1 for } x: \quad & y = 4(1)^2 - 8(1) - 21 \\
\text{Simplify:} \quad & y = 4(1) - 8(1) - 21 \\
& y = 4 - 8 - 21 \\
& y = -25 \;\rightarrow\; k = -25
\end{aligned}$$

The vertex is $(h, k) = (1, -25)$.

**b.** Compare the given equation to the general form $y = ax^2 + bx + c$ to determine the values for $a$ and $b$.

$$y = 2x^2 + 12x + 18$$
$$y = ax^2 + bx + c$$

So we have $a = 2$ and $b = 12$. The $x$-coordinate of the vertex is

$$h = -\frac{b}{2a} = -\frac{12}{2(2)} = -\frac{12}{4} = -3.$$

Substituting $-3$ for $x$ in the equation, we find the $y$-coordinate is $k = 0$. View the details.

The vertex is $(h, k) = (-3, 0)$.

*My video summary*     **c.** Try finding the **vertex** for this problem on your own. View the **answer**, or watch this **video** for the complete solution.

**You Try It**    **Work through this You Try It problem.**

**Work Exercises 11–16 in this textbook or in the *MyMathLab*® Study Plan.**

OBJECTIVE 4    GRAPH EQUATIONS OF THE FORM $y = ax^2 + bx + c$

Building on the information presented in this section, we can use the following guidelines to graph **quadratic equations** of the form $y = ax^2 + bx + c$.

---

**Graphing Equations of the Form $y = ax^2 + bx + c$**

**Step 1.**    Determine if the graph opens up or down.

**Step 2.**    Find the $y$- and $x$-intercepts.

**Step 3.**    Find the vertex and **axis of symmetry**.

**Step 4.**    Plot the vertex and **intercepts** and add the axis of symmetry.

**Step 5.**    Plot additional points and sketch the graph. Use symmetry to help.

---

### Example 5  Graphing an Equation of the Form $y = ax^2 + bx + c$

Graph $y = -x^2 - 4x - 3$.

### Solution

**Step 1.** The coefficient on $x^2$ is $-1$. Because this is less than 0, the graph will open down.

**Step 2.** To find the $y$-intercept, set $x = 0$ in the original equation and find $y$.

$$y = -(0)^2 - 4(0) - 3$$
$$= -3$$

The $y$-intercept is $-3$, so the point $(0, -3)$ is on the graph.

To find the $x$-intercepts, set $y = 0$ in the original equation and solve for $x$.

Substitute 0 for $y$:  $\qquad\qquad 0 = -x^2 - 4x - 3$

Move terms to the left-hand side:  $\qquad x^2 + 4x + 3 = 0$

Factor:  $\quad (x + 1)(x + 3) = 0$

Apply the **zero product property**.

$$x + 1 = 0 \quad \text{or} \quad x + 3 = 0$$
$$x = -1 \qquad\qquad x = -3$$

The $x$-intercepts are $-1$ and $-3$. The points $(-1, 0)$ and $(-3, 0)$ are on the graph.

**Step 3.** In the equation $y = -x^2 - 4x - 3$, we have $a = -1$ and $b = -4$. Therefore, the $x$-coordinate of the vertex is

$$h = -\frac{b}{2a} = -\frac{(-4)}{2(-1)} = -\frac{4}{2} = -2.$$

Substituting $-2$ for $x$ in the original equation gives the following:

$$y = -(-2)^2 - 4(-2) - 3$$
$$= -(4) - 4(-2) - 3$$
$$= -4 + 8 - 3$$
$$= 1$$

So $k = 1$ and the vertex is $(h, k) = (-2, 1)$. The **axis of symmetry** is $x = -2$.

**Step 4.** We plot the points $(-2, 1)$, $(-1, 0)$, $(-3, 0)$, and $(0, -3)$, and graph the vertical line $x = -2$ using a dashed line. See **Figure 2**.

**Step 5.** We can find an additional point by using the *y*-intercept and symmetry. The point $(0, -3)$ lies two units to the right of the **axis of symmetry**. There is a corresponding point on the graph that lies two units to the *left* of the axis of symmetry with the same *y*-coordinate. Thus, from symmetry, the point $(-4, -3)$ must also be on the graph. We plot this additional point and sketch the graph. See Figure 3. (Note: Even more points could be determined by creating a table of values and using symmetry.)

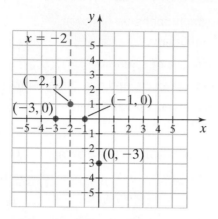

**Figure 2**

**Figure 3** Graph of $y = -x^2 - 4x - 3$

## Example 6 Graphing an Equation of the Form $y = ax^2 + bx + c$

*My video summary*

Graph $y = x^2 - 4x - 1$.

**Solution** Work through the following solution, or watch this **video** for the complete solution.

**Step 1.** The coefficient on $x^2$ is 1. Because this is greater than 0, the graph will open up.

**Step 2.** To find the *y*-intercept, set $x = 0$ in the original equation and find $y = -1$. The *y*-intercept is $-1$, so the point $(0, -1)$ is on the graph. To find the *x*-intercepts, set $y = 0$ in the original equation and solve for $x$.

$$0 = x^2 - 4x - 1$$

The right-hand side cannot be factored over the integers. Use the **quadratic formula** to solve the equation with $a = 1$, $b = -4$, and $c = -1$.

The *x*-intercepts are $2 - \sqrt{5}$ and $2 + \sqrt{5}$. (Approximating with a calculator, the *x*-intercepts are approximately $-0.24$ and $4.24$). The points $(2 - \sqrt{5}, 0)$ and $(2 + \sqrt{5}, 0)$ are on the graph.

**Step 3.** $h = -\dfrac{b}{2a} = -\dfrac{(-4)}{2(1)} = \dfrac{4}{2} = 2.$

Substituting 2 for $x$ in the original equation gives $k = -5$ so the **vertex** is $(h, k) = (2, -5)$. The **axis of symmetry** is $x = 2$.

**Step 4.** We plot the points $(2, -5)$, $(2 - \sqrt{5}, 0)$, $(2 + \sqrt{5}, 0)$, and $(0, -1)$, and graph the vertical line $x = 2$ using a dashed line. See **Figure 4**.

**Step 5.** The *y*-intercept $(0, -1)$ lies two units to the left of the **axis of symmetry**. There is a corresponding point on the graph that lies two units to the *right* with the same *y*-coordinate. Thus, from symmetry, the point $(4, -1)$ must also be on the graph. If we let $x = -1$ in the original equation, we find $y = 4$ so the point $(-1, 4)$ is on the graph. Using symmetry, we also find the point $(5, 4)$ must be on the graph. See **Figure 5**.

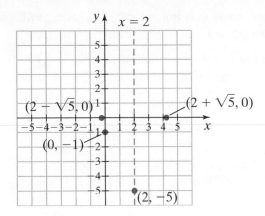

**Figure 4**

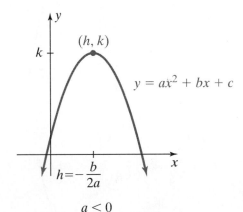

**Figure 5** Graph of $y = x^2 - 4x - 1$

**You Try It**  Work through this You Try It problem.

Work Exercises 17–24 in this textbook or in the *MyMathLab*® Study Plan.

OBJECTIVE 5  USE THE VERTEX TO SOLVE APPLICATION PROBLEMS

With application problems involving **quadratic equations in two variables**, we often need to find a *maximum* or *minimum* value. For example, a builder with a fixed amount of fencing may wish to maximize the area enclosed, or an economist may want to minimize **cost** or maximize **profit**. To solve these types of problems, we can use the **vertex**. Recall the graph of a quadratic equation is a **parabola** and if $a > 0$, the parabola opens *up* and the graph has a **minimum** value at the vertex. If $a < 0$, the parabola opens *down* and the graph has a **maximum** value at the vertex. See Figure 6. Note that the $y$-coordinate of the vertex, $k$, is the maximum or minimum value. The $x$-coordinate, $h$, tells us *where* the value occurs.

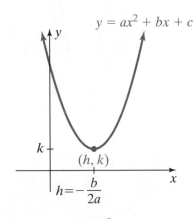

$a > 0$
Minimum value at vertex, $(h, k)$

$a < 0$
Maximum value at vertex, $(h, k)$

**Figure 6**

### Example 7  Vinyl Record Sales

While sales of music CDs have been continually declining, sales of vinyl records are on the rise. For the years 2000–2010, the equation $y = 0.125x^2 - 1.25x + 3.94$ can be used to approximate the number of vinyl albums sold, $y$ (in millions), $x$ years after 2000. In which year were vinyl record sales at a **minimum**? How many vinyl records were sold that year? (*Source:* computerworld.com)

**Solution** Because the equation is **quadratic** with $a = 0.125 > 0$, we know that the graph is a **parabola** that opens up. So, the graph has a minimum value at the **vertex**. The $x$-coordinate of the vertex is

$$-\frac{b}{2a} = -\frac{(-1.25)}{2(0.125)} = \frac{1.25}{0.25} = 5.$$

Therefore, vinyl record sales were a minimum in 2005 ($x = 5$ years after 2000). According to the model, $y = 0.125(5)^2 - 1.25(5) + 3.94 = 0.815$ million (or 815,000) vinyl records were sold in 2005.

**You Try It**    **Work through this You Try It problem.**

**Work Exercises 25 and 26 in this textbook or in the *MyMathLab*® Study Plan.**

# 9.4 Exercises

In Exercises 1–4, use the given graph of a quadratic function to find the following:

**a.** Vertex        **b.** Axis of symmetry        **c.** $y$-intercept        **d.** $x$-intercept(s)

**1.**

**You Try It**

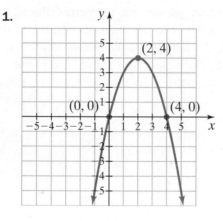

**2.**

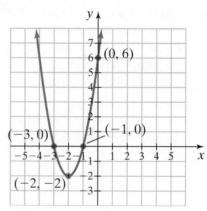

**3.**

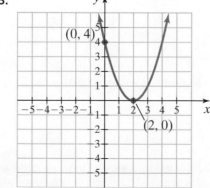

**4.**

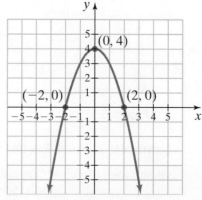

In Exercises 5–10, find the $x$- and $y$-intercepts of the graph of the given equation.

**You Try It**

**5.** $y = x^2 + 4x - 21$

**6.** $y = x^2 - 10x + 24$

**7.** $y = 9x^2 + 12x - 5$

**8.** $y = x^2 - 10x + 22$

**9.** $y = x^2 + 2x - 4$

**10.** $y = 3x^2 - 2x - 4$

In Exercises 11–16, find the vertex of the graph of the given equation.

**You Try It**

**11.** $y = x^2 + 7x + 12$

**12.** $y = 3x^2 - 6x + 2$

**13.** $y = x^2 - 4x$

**14.** $y = -2x^2 - 12x - 10$

**15.** $y = \dfrac{1}{2}x^2 + 6x + 1$

**16.** $y = -3x^2 + 4x - \dfrac{4}{3}$

In Exercises 17–24, graph the given equation. Include the axis of symmetry and label the vertex and intercepts.

**You Try It**

**17.** $y = x^2 + 2x - 3$

**18.** $y = -x^2 + 6x - 5$

**19.** $y = x^2 - 6x + 2$

**20.** $y = x^2 - 2x - 11$

**21.** $y = -4x^2 + 12x$

**22.** $y = x^2 - x - 3$

**23.** $y = \dfrac{1}{2}x^2 + 6x + 1$

**24.** $y = x^2 - 4x + 5$

In Exercises 25 and 26, use the vertex to solve each application problem.

**You Try It**

**25. Minimizing Cost**   A manufacturer of military jackets has daily production costs of $C = 0.3x^2 - 96x + 10,200$, where $C$ is the total cost (in dollars) and $x$ is the number of jackets produced. How many jackets should be produced each day in order to minimize costs? What is the minimum daily cost?

**26. Maximum Height**   A baseball player swings and hits a pop fly straight up in the air to the catcher. The height of the baseball in feet $t$ seconds after it is hit is given by the quadratic equation $h = -16t^2 + 48t + 3$. How long does it take for the baseball to reach its maximum height? What is the baseball's maximum height?

# APPENDIX A
# Relations and Functions

## THINGS TO KNOW

Before working through this section, be sure you are familiar with the following concepts:

VIDEO      ANIMATION      INTERACTIVE

You Try It

1. Evaluate Algebraic Expressions
(Section 1.5, **Objective 1**)

You Try It

2. Solve a Formula for a Given Variable
(Section 2.4, **Objective 3**)

You Try It

3. Write the Solution Set of an
Inequality in Set-Builder Notation
(Section 2.7, **Objective 1**)

You Try It

4. Use Interval Notation to Express
the Solution Set of an Inequality
(Section 2.7, **Objective 3**)

## OBJECTIVES

**1** Identify Independent and Dependent Variables

**2** Find the Domain and Range of a Relation

**3** Determine If Relations Are Functions

**4** Determine If Graphs Are Functions

**5** Express Equations of Functions Using Function Notation

**6** Evaluate Functions

······································································································································

### OBJECTIVE 1    IDENTIFY INDEPENDENT AND DEPENDENT VARIABLES

We have seen that applying math to everyday life often involves situations in which one quantity is related to another. For example, the **cost** to fill a gas tank is related to the number of gallons purchased. The amount of **simple interest** paid on a loan is related to the amount owed. The total cost to manufacture 3D televisions is related to the number of televisions produced, and so on.

In **Section 2.4** we solved **formulas** for a given **variable**. In doing so, we had to express a relationship between the given variable and any remaining variables. When we solve for a variable, that variable is called the **dependent variable** because its value *depends on* the value(s) of the remaining variable(s). Any remaining variables are called **independent variables** because we are free to select their values.

If the average price per gallon of regular unleaded gas is \$3.24 on a given day, then two gallons would cost \$3.24(2) = \$6.48, three gallons would cost \$3.24(3) = \$9.72, and so on. We can model the situation with the **equation**

$$y = 3.24x,$$

where $y$ = cost in dollars and $x$ = gallons of gas purchased.

Since the equation is solved for $y$, we identify $y$ (cost) as the **dependent variable** and $x$ (gallons of gas) as the **independent variable**. When an equation involving $x$ and $y$ is not solved for either variable, like the linear equation $2x + 3y = 12$, we will call $x$ the independent variable and $y$ the dependent variable.

### Example 1 Identifying Independent and Dependent Variables

For each of the following equations, identify the dependent variable and the independent variable(s).

**a.** $y = 3x + 5$    **b.** $w = ab + 3c^2$    **c.** $3x^2 + 9y = 12$

### Solutions

**a.** Since the equation is solved for $y$, we identify $y$ as the **dependent variable**. The remaining variable, $x$, is the **independent variable**.

**b.** Since the equation is solved for $w$, we identify $w$ as the dependent variable. The remaining variables, $a$, $b$, and $c$, are independent variables.

**c.** Since the equation is not solved for either variable, we identify $x$ as the independent variable and $y$ as the dependent variable.

**You Try It**    **Work through this You Try It problem.**

**Work Exercises 1–4 in this textbook or in the *MyMathLab*®** Study Plan.

In Example 1 we see that it is possible to have more than one independent variable in an equation. However, we will limit our discussion mainly to situations involving one dependent and one independent variable. Our gasoline cost **model**, $y = 3.24x$, is an example of such a situation.

OBJECTIVE 2    FIND THE DOMAIN AND RANGE OF A RELATION

A **relation** is a correspondence between two sets of numbers that can be represented by a set of **ordered pairs**.

---
**Definition**

A **relation** is a set of ordered pairs.

---

In **Section 3.2** we learned that **equations in two variables** define a set of **ordered pair solutions**. When writing ordered pair solutions, we write the value of the **independent variable** first, followed by the value of the **dependent variable**. For example, if 1 gallon of regular unleaded gas is purchased, the total cost is $3.24(1) = \$3.24$, which gives the ordered pair $(1, 3.24)$. If 2 gallons are purchased, the total cost is $3.24(2) = \$6.48$, which gives the ordered pair $(2, 6.48)$. We can create more ordered pair solutions by following this process. See a **table of values**.

Since a **graph** is a visual representation of the ordered pair solutions to an equation, we consider equations and graphs to be relations because they define sets of ordered pairs.

A relation shows the connection between the set of values for the independent variable, called the **domain** (or *input values*), and the set of values for the dependent variable, called the **range** (or *output values*).

At this point, you may wish to review **set-builder notation** and **interval notation** in **Section 2.7**.

---

**Definitions**

The **domain** is the set of all values for the **independent variable**. These are the first coordinates in the set of **ordered pairs** and are also known as *input values*.

The **range** is the set of all values for the **dependent variable**. These are the second coordinates in the set of ordered pairs and are also known as *output values*.

---

### Example 2 Finding the Domain and Range of a Relation

*My interactive video summary*

Find the domain and range of each **relation**.

a. $\{(-5, 7), (3, 5), (6, 7), (12, -4)\}$

b.
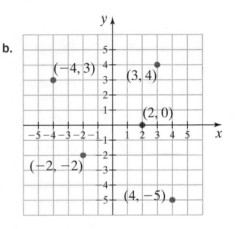

**Solutions** For parts a and b, identify the first **coordinates** of each **ordered pair** to find the **domain** and the second coordinates of each ordered pair to find the **range**. View the **answers**, or watch this **interactive video** for detailed solutions.

**You Try It** Work through this You Try It problem.

Work Exercises 5–6 in this textbook or in the *MyMathLab* Study Plan.

### Example 3 Finding the Domain and Range of a Relation

*My interactive video summary*

Find the domain and range of each **relation**.

a.
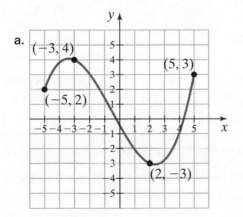

b.
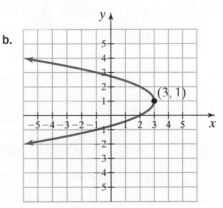

c. $y = |x - 1|$

**Solutions** Try to find the domain and range of each relation on your own. View the answers, or watch this interactive video for detailed solutions.

**You Try It** Work through this You Try It problem.

**Work Exercises 7–8** in this textbook or in the *MyMathLab*® Study Plan.

When working with application problems, we often need to restrict the **domain** to use only those values that make sense within the context of the situation. This restricted domain is called the **feasible domain**. The **feasible** domain is the set of values for the **independent variable** that make sense, or are *feasible*, in the context of the application. For example, in our gasoline cost model, $y = 3.24x$, the domain of the equation is all **real numbers**. However, it doesn't make sense to use negative numbers in the domain since $x$ represents the number of gallons of gas purchased. Therefore, the **feasible domain** would be all real numbers greater than or equal to 0, written as $\{x | x \geq 0\}$ or $[0, \infty)$.

OBJECTIVE 3   DETERMINE IF RELATIONS ARE FUNCTIONS

A **relation** *relates* one set of numbers, the **domain**, to another, the **range**. When each value in the domain corresponds to (is paired with) exactly one value in the range, we have a special type of relation called a **function**.

---

**Definition**

A **function** is a special type of relation in which each value in the domain corresponds to exactly one value in the range.

---

Given a set of ordered pairs, we can determine if the relation is a function by looking at the **x-coordinates**. If no *x*-coordinate is repeated, then the relation is a function because each **input value** corresponds to exactly one **output value**. If the same *x*-coordinate corresponds to two or more different **y-coordinates**, then the relation is not a function.

Given an equation, we can test input values to see if, when substituted into the equation, there is more than one output value. In order for an equation to be a function, each input value must correspond to one and only one output value.

**Example 4  Determining If Relations Are Functions**

*My interactive video summary*

 Determine if each of the following relations is a function.

a.  $\{(-3, 6), (2, 5), (0, 6), (17, -9)\}$          b.  $\{(4, 5), (7, -3), (4, 10), (-6, 1)\}$
c.  $\{(-2, 3), (0, 3), (4, 3), (6, 3), (8, 3)\}$          d.  $|y - 5| = x + 3$
e.  $y = x^2 - 3x + 2$          f.  $4x - 8y = 24$

**Solutions** Try to determine which relations are functions on your own. View the answers, or watch this interactive video for detailed explanations.

**You Try It** Work through this You Try It problem.

**Work Exercises 9–16** in this textbook or in the *MyMathLab*® Study Plan.

OBJECTIVE 4    DETERMINE IF GRAPHS ARE FUNCTIONS

If a **relation** appears as a graph, we can determine if the relation is a **function** by using the **vertical line test**.

> **Vertical Line Test**
>
> If a vertical line intersects (crosses or touches) the graph of a relation at more than one point, then the relation is not a function. If every vertical line intersects the graph of a relation at no more than one point, then the relation is a function.

*My animation summary*

Why does the vertical line test work? Watch this **animation** to find out.

### Example 5   Determining If Graphs Are Functions

*My animation summary*

Use the vertical line test to determine if each graph is a function.

a.
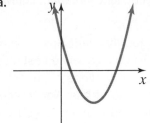

b.
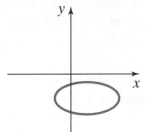

c.
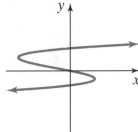

d.
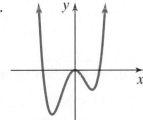

e.
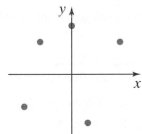

f.
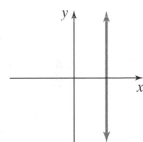

**Solutions**  Apply the vertical line test to each graph on your own. View the **answers**, or work through this **animation** for complete solutions.

**You Try It**    Work through this You Try It problem.

Work Exercises 17–24 in this textbook or in the **_MyMathLab_**® Study Plan.

How many *x*- and *y*-intercepts can the graph of a function have? See the **answer**.

OBJECTIVE 5    EXPRESS EQUATIONS OF FUNCTIONS USING FUNCTION NOTATION

**Functions** expressed as equations are often named using letters such as $f$, $g$, and $h$. The symbol $f(x)$ is read as "$f$ of $x$" and is an example of **function notation**. We can use function notation in place of the **dependent variable** in the equation of a function. For example, the function $y = 2x + 3$ may be written as $f(x) = 2x + 3$.

The symbol $f(x)$ represents the value of the dependent variable (output) for a given value of the **independent variable** (input). For $y = f(x)$, we can interpret $f(x)$ as follows: $f$ is the name of the function that relates the independent variable $x$ to the dependent variable $y$.

 Do not confuse function notation with multiplication. $f(x)$ does not mean $f \cdot x$.

When using function notation, any symbol can be used to name the function, and any other symbol can be used to represent the independent variable. For example, consider the function $y = 2x + 3$, which tells us that the value of the dependent variable is obtained by multiplying the value of the independent variable by 2 and then adding 3. We can express this function in many ways:

$$f(x) = 2x + 3 \qquad \text{Function name: } f \qquad \text{Independent variable: } x$$
$$H(t) = 2t + 3 \qquad \text{Function name: } H \qquad \text{Independent variable: } t$$
$$P(r) = 2r + 3 \qquad \text{Function name: } P \qquad \text{Independent variable: } r$$
$$\Phi(n) = 2n + 3 \qquad \text{Function name: } \Phi \qquad \text{Independent variable: } n$$

These four functions are equivalent even though they have different function names and different letters representing the **independent variable**. **Equivalent functions** represent the same set of **ordered pairs**.

When possible, we choose letters for the **variables** to provide meaning. For example, instead of using the variables $x$ and $y$ in our gasoline cost model from earlier, we might use $C$ to represent "cost" and $g$ to represent "gallons of gas." The **function notation** $C(g)$ represents the cost for purchasing $g$ gallons of gas. For this function, $g$ is the independent variable, $C$ is the dependent variable, and $C(g) = 3.24g$ is the **function** that tells us how to find the cost $C$ from the gallons of gas $g$.

One benefit of using function notation is that it clearly shows the relationship between the independent and dependent variables of an equation. Any equation of a function can be written in function notation using the following procedure:

---

**Expressing Equations of Functions Using Function Notation**

**Step 1.** Choose an appropriate name for the function.

**Step 2.** Solve the equation for the dependent variable.

**Step 3.** Replace the dependent variable with equivalent function notation.

---

 When using letters to name functions, the use of lowercase or uppercase matters. For example, $f$ and $F$ are different symbols, so $f(x)$ and $F(x)$ represent different functions. Lowercase and uppercase letters should not be switched within a problem.

### Example 6 Expressing Equations of Functions Using Function Notation

Write each function using function notation. Let $x$ be the independent variable and $y$ be the dependent variable.

**a.** $y = 2x^2 - 4$        **b.** $y - \sqrt{x} = 0$        **c.** $3x + 2y = 6$

### Solutions

**a.** We name the function $F$. The equation is already solved for $y$, so we replace $y$ with $F(x)$:

$$\text{Begin with the original formula:} \qquad y = 2x^2 - 4$$
$$\text{Replace } y \text{ with } F(x): \quad F(x) = 2x^2 - 4$$

**b.** We name the function $g$. We solve for $y$ and then replace $y$ with $g(x)$:

$$\text{Begin with the original equation:} \quad y - \sqrt{x} = 0$$
$$\text{Solve for } y \text{ by adding } \sqrt{x} \text{ to both sides:} \quad y = \sqrt{x}$$
$$\text{Replace } y \text{ with } g(x): \quad g(x) = \sqrt{x}$$

 *My video summary*   **c.** Try working through this process yourself. Name the function $f$. View the **answer**, or watch this **video** for a complete solution to part c.

**You Try It**  Work through this You Try It problem.

Work Exercises 25–34 in this textbook or in the **MyMathLab** Study Plan.

OBJECTIVE 6  EVALUATE FUNCTIONS

For $y = f(x)$, the symbol $f(x)$ represents the value of the **dependent variable** $y$ for a given value of the **independent variable** $x$. For this reason, we call $f(x)$ the **value of the function**. For example, $f(2)$ represents the value of the function $f$ when $x = 2$. When we determine such a function value, we *evaluate the function*.

To **evaluate a function**, we substitute the given value for the independent variable and simplify.

### Example 7  Evaluating Functions

*My interactive video summary*  If $f(x) = 4x - 5$, $g(t) = 3t^2 - 2t + 1$, and $h(r) = \sqrt{r} - 9$, evaluate each of the following.

**a.** $f(3)$  **b.** $g(-1)$  **c.** $h(16)$  **d.** $f\left(\dfrac{1}{2}\right)$

### Solutions

**a.** The notation $f(3)$ represents the value of the function $f$ when $x$ is 3. We substitute 3 for $x$ in the function $f$ and simplify:

$$\text{Substitute 3 for } x \text{ in the function } f: \quad f(3) = 4(3) - 5$$
$$\text{Simplify:} \quad = 12 - 5$$
$$= 7$$

So, $f(3) = 7$, meaning that the value of $f$ is 7 when $x$ is 3.

**b.** The notation $g(-1)$ represents the value of the function $g$ when $t$ is $-1$. We substitute $-1$ for $t$ in the function $g$ and simplify.

$$\text{Substitute } -1 \text{ for } t \text{ in the function } g: \quad g(-1) = 3(-1)^2 - 2(-1) + 1$$

Finish this problem by simplifying the right side. Check your **answer**, or watch this **interactive video** for complete solutions to all four parts.

**c.** The notation $h(16)$ represents the value of the function $h$ when $r$ is 16. Substitute 16 for $r$ in function $h$ and simplify. Try working through this solution on your own. Check your **answer**, or watch this **interactive video** for complete solutions to all four parts.

**d.** Try to evaluate $f\left(\dfrac{1}{2}\right)$ on your own. View the **answer** or watch this **interactive video** for complete solutions to all four parts.

**You Try It**  Work through this You Try It problem.

Work Exercises 35–42 in this textbook or in the *MyMathLab*® Study Plan.

# Appendix A **Exercises**

In Exercises 1–4, identify the independent and dependent variables.

**You Try It**

**1.** $y = 3x^2 - 7x + 5$    **2.** $k = 12v^2$    **3.** $4x + 5y = 18$    **4.** $e = \dfrac{\sqrt{a^2 - b^2}}{a}$

In Exercises 5–8, find the domain and range of each relation.

**You Try It**

**5.** $\{(4, -2), (1, -1), (0, 0), (1, 1), (4, 2)\}$

**6.**

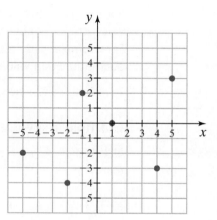

**You Try It**

**7.**

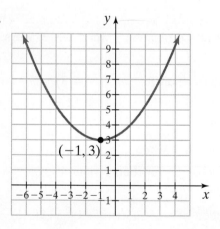

**8.**

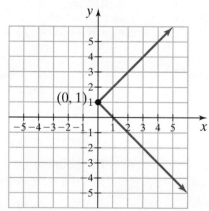

In Exercises 9–16, determine if the relation is a function. Assume $x$ is the independent variable.

**You Try It**

**9.** $\{(2, 9), (3, 2), (7, 4), (8, 1), (10, -8)\}$    **10.** $\{(3, 4), (4, 5), (6, 3), (4, 1), (9, 12)\}$

**11.** $\{(3, 7), (5, 7), (8, 10), (11, 10), (15, 6)\}$    **12.** $\{(4, -5), (4, 0), (4, 3), (4, 7), (4, 11)\}$

**13.** $3x - y = 5$    **14.** $y = |x + 3| - 1$

**15.** $|y - 1| + x = 2$    **16.** $x^2 + y = 1$

In Exercises 17–24, use the vertical line test to determine if each graph is a function.

**17.**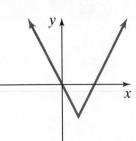

**18.**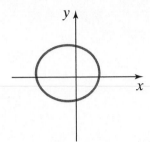

**19.**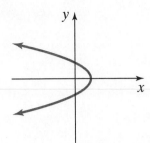

**20.**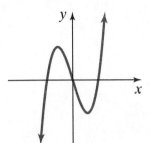

**21.**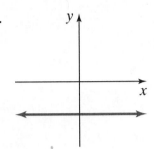

**22.**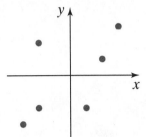

**23.**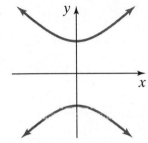

**24.**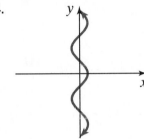

In Exercises 25–34, write each function using function notation. Let $x$ be the independent variable and $y$ be the dependent variable. Use the letter $f$ to name each function.

**25.** $y = |2x - 5|$

**26.** $y = 3x^2 + 2x - 5$

**27.** $y + \sqrt{x} = 1$

**28.** $2x + y = 3$

**29.** $3x + 4y = 12$

**30.** $-6x + 18y = 12$

**31.** $4x^2 - 2y = 10$

**32.** $3y + 6\sqrt{x - 5} = 0$

**33.** $\dfrac{3y - 7}{2} = 3x^2 + 1$

**34.** $\dfrac{5y - 8}{3} = \dfrac{10x^2 + 4}{6}$

In Exercises 35–42, evaluate each function.

**35.** $f(x) = 3x - 5$;  $f(6)$

**36.** $h(x) = 2x^2 + 5x - 17$;  $h(-4)$

**37.** $F(z) = 2|z - 3| - 5$;  $F(0)$

**38.** $T(t) = \dfrac{5}{6}t + \dfrac{1}{3}$;  $T(8)$

**39.** $r(x) = 3 + \sqrt{x - 5}$;  $r(9)$

**40.** $c(x) = \sqrt{25 - x^2}$;  $c(3)$

**41.** $\Phi(p) = (p - 1)p^3$;  $\Phi(3)$

**42.** $R(x) = 8x^2 - 2x + 1$;  $R\left(-\dfrac{1}{2}\right)$

# Mean, Median, and Mode

## THINGS TO KNOW

Before working through this section, be sure you are familiar with the following concepts:

VIDEO      ANIMATION      INTERACTIVE

**1.** Add Two Real Numbers With the Same Sign (Section 1.2, **Objective 1**)

You Try It

**2.** Divide Real Numbers (Section 1.3, **Objective 2**)

You Try It

## OBJECTIVES

**1** Find the Mean

**2** Find the Median

**3** Find the Mode

### OBJECTIVE 1   FIND THE MEAN

When given a set of data values, it is convenient to summarize the data with a single "middle" number called a **measure of central tendency**. For example, your GPA (grade point average) is a single number that summarizes all your grades at a given school.

The three most common measures of central tendency are the (arithmetic) *mean*, the *median*, and the *mode*.

The **mean**, denoted by $\bar{x}$, is often referred to as the *average* and is computed by adding up all the data values and then dividing by the number of values.

### Example 1  Finding the Mean

Valerie keeps track of her gasoline purchases to help her budget monthly expenses. The following data shows the price per gallon (in dollars) for regular unleaded gasoline on her last 10 fill-ups.

*My video summary*

| | | | | |
|---|---|---|---|---|
| 2.97 | 3.00 | 3.08 | 3.10 | 3.04 |
| 2.96 | 3.06 | 3.07 | 2.95 | 3.07 |

a. Find the mean.

b. How many values are below the mean? How many are above the mean?

c. Suppose the value 3.10 was mistakenly recorded as 31.0. Find the mean using this value.

d. How did changing the value in (c) affect the value of the mean?

**Solutions** Work through the following, or watch this **video** for the complete solutions.

**a.** To find the **mean**, $\overline{x}$, add the data values and divide by 10, the number of values.

$$\overline{x} = \frac{2.97 + 3.00 + 3.08 + 3.10 + 3.04 + 2.96 + 3.06 + 3.07 + 2.95 + 3.07}{10}$$

$$= \frac{30.3}{10} = 3.03$$

The average price per gallon is $3.03.

**b.** There are four values below the mean (2.97, 3.00, 2.96, 2.95). There are six values above the mean.

Incorrect
value

**c.** $\overline{x} = \dfrac{2.97 + 3.00 + 3.08 + \overbrace{31.0} + 3.04 + 2.96 + 3.06 + 3.07 + 2.95 + 3.07}{10}$

$$= \frac{58.2}{10} = 5.82$$

Using the incorrect value, the mean is $5.82.

**d.** Using the incorrect value, the mean increased by $2.79.

**You Try It** Work through this You Try It problem.

Work Exercises **1–4** in this textbook or in the **MyMathLab**® Study Plan.

OBJECTIVE 2   FIND THE MEDIAN

As shown in Example 1, the **mean** can be greatly affected by a few extreme values. Another **measure of central tendency**, one that is not as greatly affected by extreme values, is the **median**. Much like a highway median is in the middle of the road, the median of an *ordered* set of data is in the middle of the data. If the number of data values in the ordered list is odd, the median is the middle value. If the number of values is even, the median is the **average** of the two middle values.

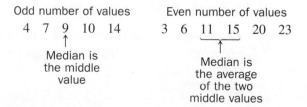

**Example 2  Finding the Median**

The following data show the number of apps downloaded on 8 randomly selected iPhones.

48   39   43   34   27   22   37   46

**a.** Find the median.

**b.** How many values are below the median? How many are above the median?

c. Suppose the value 48 was mistakenly recorded as 84. Find the median using this value.

d. How did changing the value in (c) affect the value of the median?

**Solutions** Work through the following, or watch this **video** for the complete solution.

a. To find the **median**, begin by ordering the data.

$$22 \quad 27 \quad 34 \quad \underbrace{37 \quad 39}_{\substack{\text{Two middle} \\ \text{values}}} \quad 43 \quad 46 \quad 48$$

There is an even number of values so the median is the **average** of the two middle values.

$$\text{median} = \frac{37 + 39}{2} = \frac{76}{2} = 38$$

The median number of apps downloaded is 38.

b. There are four values below the median (22   27   34   37). There are four values above the median (39   43   46   48).

c. Order the data, replacing 48 with 84.

$$22 \quad 27 \quad 34 \quad \underbrace{37 \quad 39}_{\substack{\text{Two middle} \\ \text{values}}} \quad 43 \quad 46 \quad \overset{\substack{\text{Incorrect} \\ \text{value}}}{84}$$

There is an even number of values so the median is the average of the two middle values.

$$\text{median} = \frac{37 + 39}{2} = \frac{76}{2} = 38$$

Using the incorrect value, the median number of apps downloaded is 38.

d. Using the incorrect value, the median remained the same.

**You Try It**   Work through this You Try It problem.

**Work Exercises 5–10 in this textbook or in the *MyMathLab*® Study Plan.**

 Before finding the median, the data values must be ordered.

OBJECTIVE 3   FIND THE MODE

A third **measure of central tendency** is the **mode**. The mode is the value that appears in the data set most often. It is possible for a data set to have no mode, one mode, or more than one mode. Like the **median**, the mode is not greatly affected by a few extreme values.

### Example 3  Finding the Mode

 The following data are the snowfall readings (in inches) at Boston's Logan International Airport for Boston's top twelve winter storms. (Source: cbsboston.com)

$$27.5 \quad 27.1 \quad 26.3 \quad 25.4 \quad 22.5 \quad 21.4$$
$$19.8 \quad 19.4 \quad 18.7 \quad 18.2 \quad 18.2 \quad 18.2$$

a. Find the mode.

b. Suppose the value 27.5 was mistakenly recorded as 72.5. Find the mode using this value.

c. How did changing the value in (c) affect the value of the mode?

**Solutions** Work through the following, or watch this **video** for the complete solution.

a. To find the **mode**, locate the value that occurs most often.

27.5  27.1  26.3  25.4  22.5  21.4

19.8  19.4  18.7  18.2  18.2  18.2

The mode is 18.2 since it occurs most often (3 times).

b. Replace 27.5 with 72.5 and locate the value that occurs most often.

Incorrect
value

72.5  27.1  26.3  25.4  22.5  21.4

19.8  19.4  18.7  18.2  18.2  18.2

Using the incorrect value, the mode is 18.2.

c. Using the incorrect value, the mode remained the same.

**You Try It** Work through this You Try It problem.

Work Exercises 11–15 in this textbook or in the *MyMathLab*® Study Plan.

**Note:** Of the three **measures of central tendency**, the mode is the only one that must always be one of the data values.

# Appendix B **Exercises**

In Exercises 1–4, find the mean of the given data set.

**You Try It**

**1.** 21, 24, 32, 19, 25, 23

**2.** 4.4 cm, 6.5 cm, 2.6 cm, 8.1 cm, 4.1 cm

**3.** 534, 413, 201, 397, 369, 294

**4.** $93, $157, $140, $87, $140, $62, $140

In Exercises 5–10, find the median of the given data set.

**You Try It**

**5.** 17, 40, 16, 47, 83

**6.** 28, 14, 40, 40, 34, 23

**7.** 10.4, 1.6, 3.5, 7.4, 9.1, 3.8

**8.** 6.5 h, 4.3 h, 8.2 h, 17.4 h, 10.3 h, 12.7 h, 8.2 h

**9.** 120 mi, 214 mi, 186 mi, 312 mi, 249 mi

**10.** 350, 198, 412, 376, 343, 581

In Exercises 11–15, find the mode of the given data set.

**You Try It**

**11.** 12, 6, 8, 4, 4, 6, 4

**12.** 9, 12, 3, 7, 9, 4, 10

**13.** 25, 23, 23, 30, 22, 23, 24

**14.** 10, 13, 4, 8, 12, 10, 4, 6

**15.** 58 m, 60 m, 29 m, 75 m, 80 m, 63 m, 20 m

16. **Chicago Snowfall**   The following data are the snowfall readings (in inches) for Chicago's top ten winter storms. (Source: National Weather Service)

    | | | | | |
    |---|---|---|---|---|
    | 23 | 21.6 | 20.2 | 19.2 | 18.8 |
    | 16.2 | 15 | 14.9 | 14.9 | 14.3 |

    a. Find the mean, median, and mode for the data.

    b. Suppose the value 14.3 was actually recorded as 41.3. Compute the mean, median, and mode again using this value. Comment on any differences from the results in part (a).

17. **Basketball Players**   The following data are the heights (in inches) for a sample of 12 basketball players.

    | | | | | | |
    |---|---|---|---|---|---|
    | 71 | 78 | 78 | 76 | 83 | 75 |
    | 78 | 79 | 75 | 80 | 80 | 83 |

    a. Find the mean, median, and mode for the data.

    b. Suppose the value 83 was actually recorded as 38. Compute the mean, median, and mode again using this value. Comment on any differences from the results in part (a).

18. **Exam Scores**   The following data are the final exam scores for an online algebra course.

    | | | | | | | | |
    |---|---|---|---|---|---|---|---|
    | 88 | 80 | 73 | 95 | 98 | 81 | 86 | 93 |
    | 95 | 90 | 90 | 96 | 90 | 92 | 88 | 87 |

    a. Find the mean, median, and mode for the data.

    b. A student taking the exam late scored a 97. Add this value to the data set and find the new mean, median, and mode.

19. **Facebook Friends**   The following data are the number of Facebook friends for 9 random Facebook users.

    | | | | | | | | | |
    |---|---|---|---|---|---|---|---|---|
    | 75 | 192 | 135 | 240 | 196 | 60 | 24 | 271 | 100 |

    a. Find the mean, median, and mode for the data.

    b. A new user opens a Facebook account starting with 10 friends. Add this data value to the data set and find the new mean, median, and mode for the data.

20. **Texting Habits**   The following data represent the average number of texts per day for 12 random 18–24 year-olds.

    | | | | | | |
    |---|---|---|---|---|---|
    | 57 | 56 | 75 | 82 | 64 | 46 |
    | 82 | 44 | 111 | 49 | 75 | 112 |

    Find the mean, median, and mode for the data.

# Geometry Review

## REVIEWS

### REVIEW 1   LINES AND ANGLES

| | |
|---|---|
| A **line** is a straight set of points that extend indefinitely in both directions. | |
| A **line segment** is a piece of a line that has two endpoints. | |
| A **ray** is a part of a line that has one endpoint and goes on forever in one direction. | |
| An **angle** consists of two rays that share the same endpoint, called the **vertex** of the angle. The two rays are the **sides** of the angle. | Vertex   Side   Side |
| Typically, angles are measured in **degrees** (°). One complete revolution equals 360°. | 360° |
| A **right angle** measures 90°. | 90° |
| A **straight angle** measures 180°. | 180° |
| An **acute angle** measures between 0° and 90°. | |

*continued*

| | |
|---|---|
| An **obtuse angle** measures between 90° and 180°. | 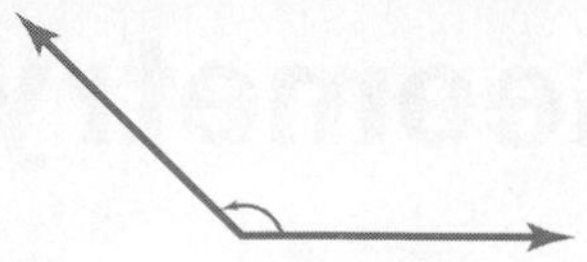 |
| **Complementary angles** have measures that add to 90°. In the figure, angles $A$ and $B$ are **complements**. | 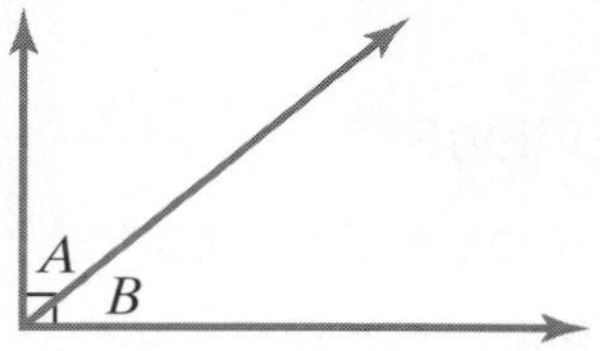 |
| **Supplementary angles** have measures that add to 180°. In the figure, angles $A$ and $B$ are **supplements**. | 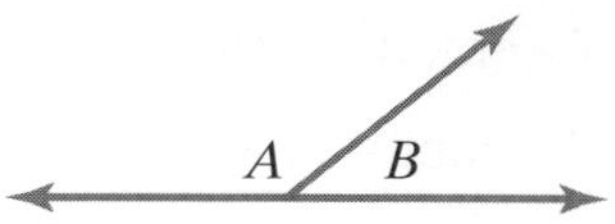 |
| A **plane** is an infinitely large two-dimensional flat surface with zero thickness. To visualize a plane, picture a tabletop that extends forever in the four directions of its surface. | 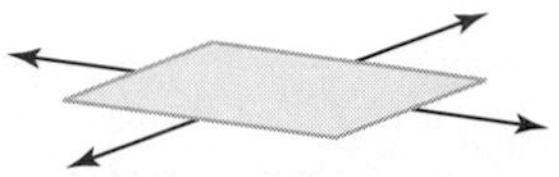 |
| **Parallel lines** lie in the same plane but never cross. | 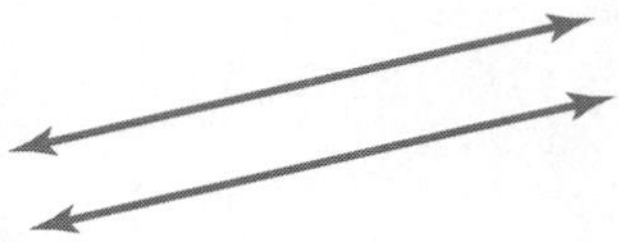 |
| **Intersecting lines** cross at a single point called the **intersection point**. | 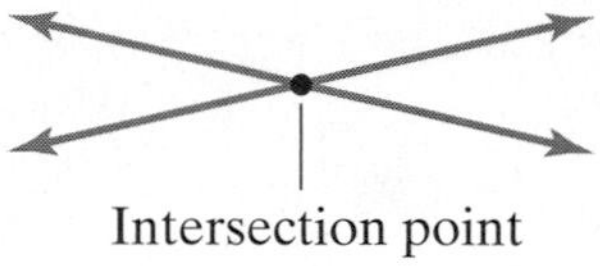 Intersection point |
| **Perpendicular lines** intersect at a right angle. | 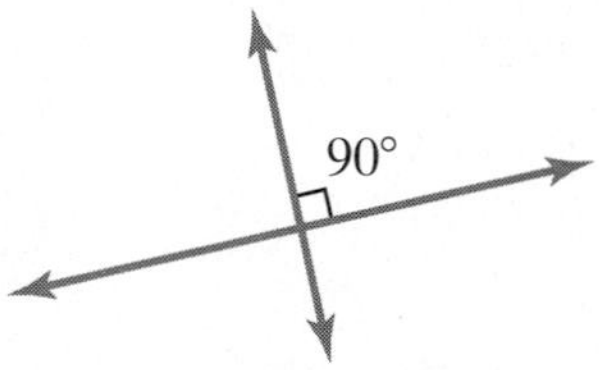 90° |

## REVIEW 2   PLANE FIGURES

**Plane figures** are two-dimensional flat figures with length and width but zero thickness.

| | |
|---|---|
| A **polygon** is a plane figure with three or more sides. The figure shows a five-sided polygon (called a **pentagon**). | 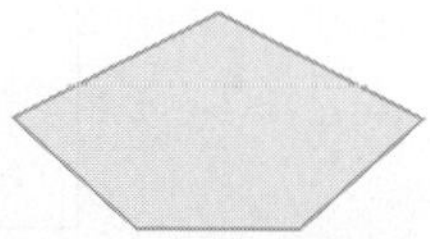 |
| A **triangle** is a three-sided polygon. The measures of the angles add to 180°. | 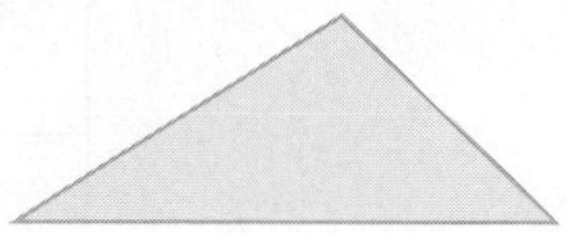 |
| An **acute triangle** has three acute angles. | 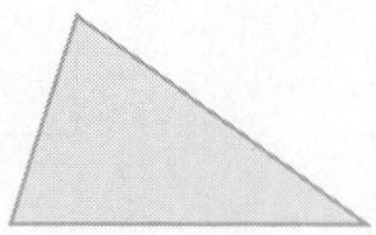 |

| | |
|---|---|
| An **obtuse triangle** contains an obtuse angle. | Obtuse angle |
| A **right triangle** has a right angle. The side opposite to the right angle is the **hypotenuse**. The other two sides are the **legs**. | Leg  Hypotenuse  90°  Leg |
| A **scalene triangle** has three different side lengths and three different angle measures. | |
| An **isosceles triangle** has two sides with equal lengths and two angles with equal measures. | |
| An **equilateral triangle** has three equal sides and three equal angles. | |
| A **quadrilateral** is a four-sided polygon. The measures of the angles add to 360°. | |
| A **trapezoid** is a quadrilateral with exactly one pair of opposite sides that are parallel. The two parallel sides are called **bases**. The two non-parallel sides are the **legs**. | Base  Leg  Leg  Base |
| An **isosceles trapezoid** has legs that are equal in length. Both pairs of adjacent base angles are equal in measure. | |
| A **parallelogram** is a quadrilateral with both pairs of opposite sides parallel. Its opposite sides are equal in length and its opposite angles are equal in measure. | |
| A **rhombus** is a parallelogram with four equal sides. | |
| A **rectangle** is a parallelogram with four right angles. Its opposite sides are equal in length. | |

*continued*

| A **square** is a rectangle with four equal sides (or a rhombus with four right angles). | |
| A **circle** consists of all points on a plane that are located at an equal distance from a fixed point called the **center**. A **radius** is a line segment that extends from the center to the edge of a circle. A **diameter** is a line segment that extends from one side of the circle to the other, through the center. | |

## REVIEW 3   GEOMETRIC SOLIDS

A **solid** is a three-dimensional figure that is bounded by surfaces. Solids have length, width, and thickness (or height).

| A **rectangular solid** (or **rectangular prism**) is a solid that consists of six rectangular surfaces. Each rectangular surface is called a **face** of the solid. A line where two faces touch (or intersect) is an **edge** of the solid. A corner point where three faces touch (or intersect) is a **vertex**. Opposite faces have the same area. | |
| A **cube** is rectangular solid in which all six faces are squares. Each square has the same area. | |
| A **right circular cylinder** consists of two parallel circles and a curved rectangular side. If positioned vertically, the centers of the circles align one directly above the other so that the circles are perpendicular to the height. | |
| A **sphere** consists of all points in three dimensions that are located at an equal distance from a fixed point called the **center**. A **radius** is a line segment that extends from the center to the surface of a sphere. A **diameter** is a line segment that extends from one side of the sphere to the other, through the center. | |

| | |
|---|---|
| A **right cone** contains a circular base and a curved surface with one vertex. If positioned vertically, the vertex aligns directly above the center of the circle so that the circle is perpendicular to the height. | |
| A **pyramid** contains a base that is a polygon and sides that are triangles. In the figure, the base of the pyramid is a square. | |

## REVIEW 4    PERIMETER AND AREA

The **perimeter** of a polygon is the distance around that polygon. Perimeter is measured in units of length or distance such as inches, feet, yards, miles, millimeters, centimeters, meters, and kilometers.

The distance around a circle is called the **circumference.**

The **area** of a plane figure is the amount of surface contained within the sides of the figure. Surface is two-dimensional, so area is measured in units that are also two-dimensional, **square units.** For example, a **square inch** ($in^2$) is a square with 1-inch sides. A **square centimeter** ($cm^2$) is a square with 1-centimeter sides. See figures.

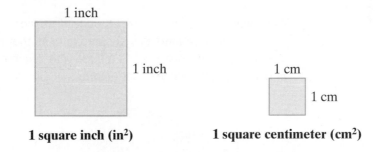

**1 square inch (in²)**          **1 square centimeter (cm²)**

To find the area of a plane figure, we find the number of square units required to cover the figure.

 Area is measured in *square units* regardless of the shape of the plane figure.

The following table gives formulas for computing the perimeter and area of common plane figures.

| Figure | Formulas | Figure | Formulas |
|---|---|---|---|
| **Square** | **Perimeter:** $P = 4s$ <br> **Area:** $A = s^2$ | **Triangle** | **Perimeter:** $P = a + b + c$ <br> **Area:** $A = \dfrac{1}{2}bh$ |
| **Rectangle** | **Perimeter:** $P = 2l + 2w$ <br> **Area:** $A = lw$ | **Trapezoid** | **Perimeter:** $P = a + b + c + d$ <br> **Area:** $A = \dfrac{1}{2}h(a + b)$ |
| **Parallelogram** | **Perimeter:** $P = 2a + 2b$ <br> **Area:** $A = bh$ | **Circle** | **Circumference:** $C = 2\pi r = \pi d$ <br> **Area:** $A = \pi r^2$ |

**Note:** The constant $\pi$ is often approximated by 3.14 or $\dfrac{22}{7}$.

## REVIEW 5   VOLUME AND SURFACE AREA

The **volume** of a solid is the amount of space contained within the solid. Solids are three-dimensional, so volume is measured in units that are also three-dimensional, **cubic units.** For example, a **cubic inch** $(\text{in}^3)$ is a cube with 1-inch edges. A **cubic centimeter** $(\text{cm}^3)$ is a cube with 1-centimeter edges. See figures.

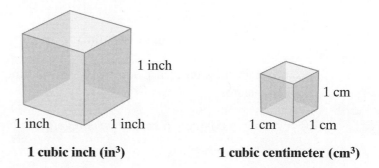

**1 cubic inch (in³)**          **1 cubic centimeter (cm³)**

To find the volume of a solid, we find the number of cubic units required to fill up the figure.

**Surface area** is the total area of the surface of a solid. For example, the surface area of a rectangular solid is the sum of the areas of six rectangles that form its surface. We measure surface area in **square units.** To find the surface area of a solid, we find the number of square units required to cover the entire surface of the figure.

The volume of a solid is measured in *cubic units* and the surface area is measured in *square units* regardless of the shape of the solid.

The following table gives formulas for computing the volume and surface area of common solids.

| Figure | Formulas | Figure | Formulas |
|---|---|---|---|
| **Cube** <br> | Volume: <br> $V = s^3$ <br> Surface Area: <br> $SA = 6s^2$ | **Sphere** <br> | Volume: <br> $V = \dfrac{4}{3}\pi r^3$ <br> Surface Area: <br> $SA = 4\pi r^2$ |
| **Rectangular Solid** <br> | Volume: <br> $V = lwh$ <br> Surface Area: <br> $A = 2lw + 2lh + 2wh$ | **Right Cone** <br> | Volume: <br> $V = \dfrac{1}{3}\pi r^2 h$ <br> Surface Area: <br> $SA = \pi rs + \pi r^2$ |
| **Right Circular Cylinder** <br> | Volume: <br> $V = \pi r^2 h$ <br> Surface Area: <br> $SA = 2\pi r^2 + 2\pi rh$ | **Pyramid** <br> | Volume: <br> $V = \dfrac{1}{3}Bh$, where $B$ is the area of the base. <br> Surface Area: <br> $SA = B + \dfrac{1}{2}Pl$, where $P$ is the perimeter of the base. |

**Note:** The constant $\pi$ is often approximated by 3.14 or $\dfrac{22}{7}$.

# Glossary

**Absolute Value**   The absolute value of a real number $a$, written as $|a|$ is the distance from 0 to $a$ on the number line.

**Absolute Value of Opposites**   The absolute value of a number tells us the distance of the number from 0 on the number line. Because opposites are the same distance away from 0 on the number line, they have the same absolute value. That is, for any real number $a$, $|a| = |-a|$.

**Addition**   Addition is a mathematical operation in which two numbers are combined to form a sum.

**Addition Property of Equality**   If $a, b$, and $c$ be real numbers or algebraic expressions, then, $a = b$ and $a + c = b + c$ are equivalent equations.

**Addition Property of Inequality**   Let $a, b$, and $c$ be real numbers. If $a < b$, then $a + c < b + c$ and $a - c < b - c$. Adding or subtracting the same quantity from both sides of an inequality results in an equivalent inequality. The inequality symbol $<$ can be replaced with $>, \leq, \geq$, or $\neq$.

**Additive Identity**   If 0 is added to any real number, then the sum is the original number.

**Additive Inverse**   The opposite of a real number is also called the *additive inverse* of the number because the sum of a number and its inverse is zero.

**Algebraic Equation**   An algebraic equation is a statement that two quantities are equal where one or both quantities contain at least one variable.

**Algebraic Expression**   A variable or a combination of variables, constants, operations, and grouping symbols.

**Angle**   An angle is a figure formed when two lines extend from a common point.

**Area**   The measure of a surface.

**Area of a Triangle**   The area of a triangle is given by the formula $A = \dfrac{1}{2}bh$, where $b$ is the base of the triangle and $h$ is the height of the triangle.

**Associative Property of Addition**   States that regrouping terms does not affect the sum. Thus, if $a, b$, and $c$ are real numbers, then $(a + b) + c = a + (b + c)$.

**Associative Property of Multiplication**   States that regrouping factors does not affect the product. Thus, if $a, b$, and $c$ are real numbers, then $(a \cdot b) \cdot c = a \cdot (b \cdot c)$.

**Axis of Symmetry**   In a graph of a quadratic equation, an imaginary vertical line that passes through the vertex of the graph and divides the graph into two mirror images.

**Binomial**   A simplified polynomial with two terms.

**Binomial Difference**   A binomial in which one term is subtracted from the other such as $A - B$.

**Binomial Sum**   A binomial in which the two terms are added such as $A + B$.

**Coefficient (or Numerical Coefficient)**   A coefficient (numerical coefficient) is the numeric factor of a term. For example, the coefficient in $4x^2$ is 4, the coefficient in $-3y$ is $-3$, and the coefficient in 13 is 13 (constant coefficient).

**Coefficient of a Monomial**   The coefficient of a monomial is the constant factor.

**Coinciding Lines**   Lines that have the same slope and the same $y$-intercept.

**Collect Like Terms**   Collecting like terms means to rearrange terms in an algebraic expression so that like terms are grouped together.

**Combine Like Terms**   To combine like terms means to add or subtract the like terms. To do this, reverse the distributive property to add or subtract the coefficients of the terms. The variable part remains unchanged.

**Common Factor**   A factor that two or more nonzero whole numbers have in common is called a common factor of the two numbers. For example, $24 = 2 \cdot 12$ and $14 = 2 \cdot 7$, so 24 and 14 have a common factor of 2.

**Common Multiple**   A common multiple of two real numbers is a quantity that is evenly divisible by both numbers. For example, 24 is a common multiple of 6 and 12 because 24 is evenly divisible by both 6 and 12.

**Common Prime Factor**   A prime number that is a factor of each integer in a group of integers is called a common prime factor of the group of integers. For example, $6 = 2 \cdot 3$, $9 = 3 \cdot 3$, and $15 = 3 \cdot 5$, so 3 is a common prime factor of 6, 9, and 15.

**Common Variable Factors for a GCF**   If a variable is a common factor of a group of monomials, then the lowest power on that variable in the group will be the power on the variable in the GCF.

**Commutative Property of Addition**   States that the order of the terms in addition does not affect the sum. Thus, if $a$ and $b$ are real numbers, then $a + b = b + a$.

**Commutative Property of Multiplication**   States that the order of the factors does not affect the product. Thus if $a$ and $b$ are real numbers, then $a \cdot b = b \cdot a$.

**Compare Real Numbers**   To compare real numbers means to determine which number is larger than the other, or if the two numbers are equal.

**Complex Conjugate**   A complex conjugate is obtained by changing the sign of the imaginary part in a complex number. For example, the complex numbers $(a + bi)$ and $(a - bi)$ are complex conjugates of each other.

**Complex Rational Expression**   A rational expression in which the numerator and/or denominator contain(s) rational expressions. Also called a *complex fraction*.

**Compound Inequality**   Two inequalities that are joined together using the words *and* or *or*.

**Concentration**   The concentration of a solution refers to the strength of the solution. The higher the concentration, the stronger the solution. Concentration of a component in a solution is determined by the formula $\text{Concentration} = \dfrac{\text{Amount of component}}{\text{Amount of mixture}}$.

**Conjugates**   Two expressions of the form $u + v$ and $u - v$ are conjugates of each other.

**Consecutive Even Integers**   Consecutive even integers are even integers that appear next to each other in an ordered list of all even integers, for example, 6, 8, and 10 are consecutive even integers.

**Consecutive Integers**   Consecutive integers are integers that appear next to each other in an ordered list of all integers. For example, 3, 4, and 5 are consecutive integers.

**Consistent System**   A system with at least one solution.

**Constant**   A constant is a value that never changes.

**Contradiction**   An equation for which no value of the variable can make the equation true. It has no solution, and its solution set is the empty $\{\}$ or null set $\varnothing$.

**Coordinates**   The $x$- and $y$-coordinates of a point or ordered pair. The coordinates provide the position of the point in the coordinate plane.

**Coordinate Plane**   The plane represented by a rectangular coordinate system, also known as the *Cartesian plane* or *xy-plane*.

**Counting Number**   A counting number, or *natural number*, is an element of the set $N = \{1, 2, 3, 4, 5, \dots \}$.

**Cube Roots**   The cube root of a real number $a$ is a real number $b$ that, when cubed, results in $a$. We denote this as $b = \sqrt[3]{a}$, if $b^3 = a$.

**Degree of a Monomial**   The degree of a monomial is the sum of the exponents on the variables.

**Degree of a Polynomial**   The degree of a polynomial is the largest degree of its terms. For example, the degree of is $10x^5 - 3x^4 + x^3 - 2x^2 + 8x + 9$, is 5, the largest degree of any term in the polynomial.

**Denominator**   The expression below the fraction bar in a fraction or quotient.

**Dependent System**   A system that has an infinite number of solutions.

**Dependent Variable**   A variable whose value depends on the value(s) of other variables. These are also called *output variables* because they represent the value that we get *out* after inputting values for the independent variables.

**Diagonal**   A diagonal of a rectangle is a line that extends between two opposite corners of the rectangle.

**Diameter**   The diameter of a circle is the distance across the circle and through its center. The length of the diameter is two times the length of the radius.

**Difference**   The result of subtracting two real numbers. The number being subtracted is called the *subtrahend*, and the number being subtracted from is called the *minuend*.

**Direct Variation**   For an equation of the form $y = kx$, we say that $y$ varies directly with $x$, or $y$ is directly proportional to $x$. The nonzero constant $k$ is called the *constant of variation* or the *proportionality constant*.

**Discriminant**   Given a quadratic equation of the form $ax^2 + bx + c = 0$, the expression $b^2 - 4ac$ is called the discriminant.

**Distance Formula**   The formula used to compute the distance between two points given by $d = \sqrt{(x_2 - x_1)^2 + (y_2 - y_1)^2}$ where $(x_1, y_1)$ and $(x_2, y_2)$ are the coordinates of the two points.

**Distributive Property**   States that multiplication distributes over addition (or subtraction). Thus, if $a, b,$ and $c$ are real numbers, then $a(b + c) = ab + ac$.

**Dividend**   The quantity that you divide into in a division problem. This is also the numerator of a fraction.

**Divisor**   The quantity that you divide by in a division problem. This is also the denominator of a fraction.

**Division of Two Real Numbers**   If $a$ and $b$ represent two real numbers and $b \neq 0$, then $a \div b = \dfrac{a}{b} = a \cdot \dfrac{1}{b}$.

**Domain**   The set of all values for the independent variable. These are the first coordinates in a set of ordered pairs and are also known as *input values*.

**Double-Negative Rule**   If $a$ is a real number, then $-(-a) = a$.

**Elimination Method**   The elimination method for solving a system of linear equations in two variables involves adding two equations together in a way that will eliminate one of the variables.

**Equivalent Equations**   Two or more equations with the same solution set.

**Equivalent Inequalities**   Two inequalities are equivalent inequalities if they have the same solution set. For example, the equivalent inequalities $4x > 12, 2x > 6,$ and $x > 3$ all have the same solution set: $\{x | x > 3\}$, or $(3, \infty)$ in interval notation.

**Equivalent Numeric Expressions**   Two numeric expressions are equivalent if they simplify to the same value. For example, $\dfrac{2}{4}$ is equivalent to $\dfrac{1}{2}$ and $3(4)$ is equivalent to 12.

**Evaluate a Formula**   To evaluate a formula means to find the value of the isolated variable by substituting values for all other variables in the formula.

**Exponent**   Also called *power*, it is a superscripted number that tells how many times a base expression is multiplied by itself.

**Exponential Expression**   A constant or algebraic expression that is raised to a power. The constant or algebraic expression makes up the base, and the power is the exponent on the base.

**Factor**   To factor a nonzero whole number means to write it as the product of two or more whole numbers.

**Feasible Solution**   When solving an application problem, a feasible solution to an equation is a solution that satisfies any given conditions and makes sense within the context of the problem.

**FOIL Method**   FOIL is the acronym for the method of multiplying binomials, that is, to multiply the two First terms, the two Outside terms, the two Inside terms, and the two Last terms, then add the resulting products together.

**Formula**   An equation that describes the relationship between two or more variables.

**Fraction Bar**   A fraction bar is a horizontal line used to separate the numerator and denominator in a fraction.

**Function**   A special type of relation in which each value in the domain corresponds to exactly one value in the range.

**Greatest Common Factor (GCF) of a Group of Integers**   The largest integer that divides evenly into each integer in the group.

**Greatest Common Factor (GCF) of a Group of Monomials**   The monomial with the largest coefficient and highest degree that divides each monomial evenly.

**Greatest Common Factor (GCF) of a Polynomial**   The expression with the largest coefficient and highest degree that divides each term of the polynomial evenly.

**Grouping Symbols**   Mathematical symbols used to group operations so they are treated as a single quantity. When simplifying, all operations within grouping symbols must be done first. Some examples of grouping symbols are parentheses ( ), brackets [ ], and braces { }. Operators such as absolute value | |; fraction bars, –; and radicals, $\sqrt{\ }$ are also treated as grouping symbols.

**Hypotenuse**   In a right triangle, the side opposite the right angle is called the hypotenuse of the triangle. It is the longest of the three sides.

**Identity**   An equation that is always true for all defined values of its variable.

**Identity Property of Addition**   If $a$ is a real number, then $a + 0 = a$ and $0 + a = a$.

**Identity Property of Multiplication**   If $a$ is a real number, then $a \cdot 1 = a$ and $1 \cdot a = a$.

**Imaginary Unit $i$**   The imaginary unit $i$ is defined as $i = \sqrt{-1}$, where $i^2 = -1$.

**Improper Fraction**   An improper fraction is a fraction in which the absolute value of the numerator is greater than or equal to the absolute value of the denominator. The absolute value of an improper fraction is greater than or equal to 1.

**Inclusive** Means to include the extremes as well as everything between. For example, the set of numbers between 4 and 7, *inclusive*, includes 4 and 7 along with all numbers in between them.

**Inconsistent System of Inequalities** An inconsistent system of inequalities is a system that has no solution. That is, there is no ordered pair that satisfies every inequality in the system.

For example, the system $\begin{cases} x + y < 2 \\ x + y > 3 \end{cases}$ is inconsistent because the two inequalities have no ordered pairs in common (the sum of and cannot be both less than 2 and greater than 3).

**Independent System** A system of two equations in two variables is independent if there is exactly one solution to the system or no solution. In these cases, both equations give new information about the variables. If the two lines in the system intersect or are parallel, then the system is independent.

**Independent Variable** A variable for which we can select values. These are also called *input variables* because they represent the set of values that we can arbitrarily *input* into a relation.

**Index [of a radical expression]** In a radical expression $\sqrt[n]{a}$, $n$ is called the index, and it indicates the type of root.

**Integers** An integer is an element of the set $\mathbb{Z} = \{\ldots, -4, -3, -2, -1, 0, 1, 2, 3, 4, \ldots\}$.

**Intercept** A point where a graph crosses or touches a coordinate axis.

**Inverse Variation** For equations of the form $y = \dfrac{k}{x}$ or $y = k \cdot \dfrac{1}{x}$, we say that $y$ varies inversely with $x$, or $y$ is inversely proportional to $x$. The constant $k$ is called the *constant of variation* or the *proportionality constant*.

**Irrational Number** A number that cannot be written as a fraction $\dfrac{p}{q}$, where $p$ and $q$ are integers. In its decimal form, an irrational number does not repeat or terminate.

**Isosceles Triangle** An isosceles triangle is a triangle that has two equal sides and a third side of a different length.

**Leading Coefficient of a Polynomial in One Variable** When a polynomial in one variable is written in standard form, the coefficient of the first term (the term with the highest degree) is called the *leading coefficient*.

**Least Common Denominator (LCD)** The least common denominator (LCD) of a group of fractions is the smallest number that is divisible by all the denominators in the group.

**Like Fractions** Two fractions are like fractions if they have the same denominator. For example, the fractions $\dfrac{2}{15}$ and $\dfrac{8}{15}$ are like fractions because they have a common denominator of 15.

**Like Radicals** Two or more radical expressions are like radicals if they have both the same index and the same radicand. For example, $3\sqrt{5}$ and $4\sqrt{5}$ are like radicals because they have the same index 2 and the same radicand 5. However, $3\sqrt{5}$ and $4\sqrt[3]{5}$ and are not like radicals because their indices, 2 and 3, are different. Also, $\sqrt[3]{5}$ and $\sqrt[3]{6}$ are not like radicals because their radicands, 5 and 6, are different.

**Like Terms** Two terms are like terms if their variable factors are exactly the same.

**Linear Equation in One Variable** An equation that can be written in the form $ax + b = c$, where $a, b$, and $c$ are real numbers and $a \neq 0$.

**Linear Equation in Two Variables (Standard Form)** An equation that can be written in the standard form $Ax + By = C$, where $A, B$, and $C$ are real numbers, and $A$ and $B$ are not both equal to 0.

**Linear Inequality in One Variable** An inequality that can be written in the form $ax + b < c$, where $a, b$, and $c$ are real numbers and $a \neq 0$. The inequality symbol $<$ can be replaced with $>, \leq, \geq$, or $\neq$.

**Long Division** Long division is a process for dividing two numbers that uses a sequence of steps involving division, multiplication, subtraction, and dropping down the next digit. The steps are repeated until the remainder can no longer be divided.

**Lower Bound** The lower bound of an interval is the smallest value in the interval. No values in the interval can be less than the lower bound. For example, the interval $[6, 9]$ has a lower bound of 6. No value in the interval is less than 6.

**Lowest Terms** A fraction is written in lowest terms, or *simplest form*, if the numerator and denominator have no common factors other than 1.

**Markup** A markup is an amount added by a merchant to the cost of an item to obtain the selling price of the item.

**Mathematical Expression** A mathematical expression is a statement containing symbols (including numbers) or mathematical operations (such as addition or subtraction).

**Mathematical Symbol** A mathematical symbol is a character that represents a mathematical relation or operation.

**Maximum** The maximum value of a graph is the largest $y$-value for any point on the graph.

**Minimum** The minimum value of a graph is the smallest $y$-value for any point on the graph.

**Minor Rational Expressions** The rational expressions within the numerator and denominator of a complex rational expression. Also called *minor fractions*.

**Mixed Number** A mixed number is a number of the form $a\dfrac{b}{c}$, where $a, b$, and $c$ are nonzero whole numbers, and $b < c$. $a\dfrac{b}{c}$ means $a + \dfrac{b}{c}$, but the $+$ symbol is not written. For example, $2\dfrac{3}{4}$ and $5\dfrac{7}{8}$ are mixed numbers.

**Monomial** A monomial is a simplified term in which all variables are raised to non-negative integer powers and no variables appear in any denominator.

**Multiplication Property of Equality** If $a, b$, and $c$ be real numbers or algebraic expressions with $c \neq 0$, then $a = b$ and $a \cdot c = b \cdot c$ are equivalent equations.

**Multiplication Property of Inequality** Let $a, b$, and $c$ be real numbers. If $a < b$ and $c > 0$, then $ac < bc$ and $\dfrac{a}{c} < \dfrac{b}{c}$. Multiplying or dividing both sides of an inequality by a positive number $c$ results in an equivalent inequality. If $a < b$ and $c < 0$, then $ac > bc$ and $\dfrac{a}{c} > \dfrac{b}{c}$. Multiplying or dividing both sides of an inequality by a negative number $c$, and switching the direction of the inequality, results in an equivalent inequality. The inequality symbol $<$ can be replaced with $>, \leq, \geq$, or $\neq$.

**Multiplication Property of Zero** If $a$ is a real number, then $a \cdot 0 = 0$ and $0 \cdot a = 0$.

**Multiplicative Identity** If 1 is multiplied by a real number, then the product is the original number. That is, $a \cdot 1 = a$ and $1 \cdot a = a$.

**Multiplicative Inverse** The reciprocal of a real number is called the *multiplicative inverse* of the number because the product of a number and its reciprocal is 1.

**Natural Numbers** A natural number, or counting number, is an element of the set $\mathbb{N} = \{1, 2, 3, 4, 5 \ldots\}$.

**Negative Exponent** A negative exponent is an exponent, or power, whose value is less than zero.

**Negative Slope** The slope of a line that slants downward, or falls, from left to right has a negative slope.

**Negative-Power Rule** To remove a negative exponent, switch the location of the base (numerator or denominator) and change the exponent to be positive. That is, $a^{-n} = \dfrac{1}{a^n}$ and $\dfrac{1}{a^{-n}} = a^n \, (a \neq 0)$.

**Non-negative**   A real number, $x$, is non-negative if it is 0 or larger. That is, $x \geq 0$.

**Non-strict Inequality**   It contains one or both of the following inequality symbols: $\leq, \geq$. The possibility of equality is included in a non-strict inequality.

**Nonzero**   A real number is nonzero if it is not equal to 0. All positive or negative numbers are nonzero.

**Null Set**   A set that contains no elements. Also called *empty set*.

**Numerator**   The expression above the fraction bar in a fraction or quotient.

**Opposites**   Two numbers are opposites if they are located the same distance away from 0 on the number line but lie on opposite sides of 0.

**Opposite of an Expression**   If a negative sign appears in front of an expression within parentheses, removing the parentheses and changing the sign of each term in the expression results in the opposite of the expression.

**Opposite Reciprocals**   Two numbers are said to be opposite reciprocals if their product equals $-1$. This occurs when the two numbers have opposite signs and their absolute values are reciprocals.

**Ordered Pair**   Each position, or point, on the coordinate plane can be identified using an ordered pair of numbers in the form $(x, y)$. The first number $x$ is called the *x-coordinate* or *abscissa* and indicates the point's horizontal position. The second number $y$ is called the *y-coordinate* or *ordinate* and indicates the point's vertical position.

**Ordered Pair Solution**   An ordered pair solution to an equation in two variables, written as $(x, y)$, gives values for the variables that, when substituted in the equation, result in a true statement. For example, $(2, 5)$ is an ordered pair solution to the equation $2x + y = 9$ because $2(2) + (5) = 4 + 5 = 9$.

**Origin (Rectangular Coordinate System)**   The point at which the $x$- and $y$-axes intersect in the rectangular coordinate system. The origin is represented by the ordered pair $(0, 0)$.

**Origin (Real Number Line)**   The point 0 is called the origin of a real number line. Numbers located to the left of the origin are negative numbers, and numbers located to the right of the origin are positive numbers.

**Paired Data**   Data represented as ordered pairs.

**Parabola**   The U-shaped graph of a quadratic equation in two variables.

**Parallel Lines**   Two lines in the coordinate plane that never touch (have no points in common). The distance between two parallel lines remains the same across the plane. Parallel lines have the same slope but different $y$-intercepts.

**Percent of Change**   The amount of change expressed as a percent of the original amount.

**Percent of Decrease**   The amount of decrease expressed as a percent of the original amount.

**Percent of Increase**   The amount of increase expressed as a percent of the original amount.

**Perfect Square Trinomial**   A perfect square trinomial is a trinomial that can be written as the square of a binomial sum, $(A + B)^2$, or the square of a binomial difference, $(A - B)^2$.

**Perimeter**   The distance around a two-dimensional object.

**Perimeter of a Rectangle**   The perimeter of a rectangle is the distance around the rectangle. If the length of the rectangle is $L$ and the width is $W$, then the perimeter $P$ is given by the formula $P = 2L + 2W$.

**Perpendicular Lines**   Two lines that intersect at a $90°$ (or right) angle.

**Plus or Minus Symbol**   The symbol $\pm$ is read as "plus or minus" and can be used to indicate two values, one positive and one negative. For example, $x = \pm 6$ is read as "$x$ equals plus or minus six" and means $x = 6$ or $x = -6$. The symbol can also be used to

indicate addition and subtraction. For example $x = 5 \pm 3$ is read "$x$ equals five plus or minus three." It means $x = 5 + 3 = 8$ or $x = 5 - 3 = 2$.

**Point-Slope Form**   Given the slope $m$ of a line and a point $(x_1, y_1)$ on the line, the point-slope form of the equation of the line is given by $y - y_1 = m(x - x_1)$.

**Polynomial**   A polynomial in $x$ is a monomial or a finite sum of monomials of the form $ax^n$, where $a$ is any real number and $n$ is any whole number.

**Polynomial in Several Variables**   A polynomial containing two or more variables.

**Positive Slope**   The slope of a line line that slants upward, or rises, from left to right has a positive slope.

**Power**   Also called *exponent*, it is a superscripted number that tells how many times a base number is multiplied by itself.

**Power-to-Power Rule**   When an exponential expression is raised to a power, multiply the exponents, that is, $(a^m)^n = a^{m \cdot n}$.

**Prime Number**   A whole number greater than 1 whose only whole number factors are 1 and itself.

**Prime Polynomial**   A polynomial is a prime polynomial if its only factors over the integers are 1 and itself.

**Product Rule for Exponents**   When multiplying exponential expressions with the same base, add the exponents and keep the common base. That is, $a^m \cdot a^n = a^{m+n}$.

**Product Rule for Radicals**   If $\sqrt[n]{a}$ and $\sqrt[n]{b}$ are real numbers, then $\sqrt[n]{ab} = \sqrt[n]{a} \cdot \sqrt[n]{b}$.

**Product Rule for Square Roots**   If $\sqrt{a}$ and $\sqrt{b}$ are real numbers, then $\sqrt{a \cdot b} = \sqrt{a} \cdot \sqrt{b}$.

**Product-to-Power Rule**   When raising a product to a power, raise each factor of the base to the common exponent. That is, $(ab)^n = a^n b^n$.

**Proper Fraction**   A proper fraction is a fraction in which the absolute value of the denominator is greater than the absolute value of the numerator. The absolute value of an improper fraction is less than 1.

**Property of Equivalent Fractions**   If $a, b$, and $c$ are numbers, then $\dfrac{a}{b} = \dfrac{a \cdot c}{b \cdot c}$ and $\dfrac{a}{b} = \dfrac{a \div c}{b \div c}$ as long as $b$ and $c$ are not equal to 0.

**Proportion**   A proportion is a statement that two ratios are equal, such as $\dfrac{P}{Q} = \dfrac{R}{S}$.

**Pythagorean Theorem**   For right triangles, the sum of the squares of the lengths of the legs of the triangle equals the square of the length of the hypotenuse. That is, $a^2 + b^2 = c^2$.

**Quadrants**   The four regions into which the $x$-axis and the $y$-axis divide the rectangular coordinate system.

**Quadratic Equation in Two Variables**   A quadratic equation in two variables is an equation that can be written in the form $y = ax^2 + bx + c$ where $a \neq 0$.

**Quadratic Equation**   A quadratic equation in standard form is written as $ax^2 + bx + c = 0$, where $a, b$, and $c$ are real numbers and $a \neq 0$.

**Quotient**   The result of division.

**Quotient Rule for Exponents**   When dividing exponential expressions with the same nonzero base, subtract the denominator exponent from the numerator exponent and keep the common base. That is, $\dfrac{a^m}{a^n} = a^{m-n} (a \neq 0)$.

**Quotient Rule for Radicals**   If $\sqrt[n]{a}$ and $\sqrt[n]{b}$ are real numbers and $b \neq 0$, then $\sqrt[n]{\dfrac{a}{b}} = \dfrac{\sqrt[n]{a}}{\sqrt[n]{b}}$.

**Quotient Rule for Square Roots**   If $\sqrt{a}$ and $\sqrt{b}$ are real numbers and $b \neq 0$, then $\sqrt{\dfrac{a}{b}} = \dfrac{\sqrt{a}}{\sqrt{b}}$.

**Quotient-to-Power Rule**   When raising a quotient to a power, raise both the numerator and denominator to the common exponent. That is, $\left(\dfrac{a}{b}\right)^n = \dfrac{a^n}{b^n} (b \neq 0)$.

**Radical Equation**   A radical equation is an equation that contains at least one radical expression with a variable in the radicand.

**Radicand**   The expression beneath the radical sign is called the radicand.

**Radius**   The radius of a circle is the distance from the center to the edge of the circle. The length of the radius is half the length of the diameter.

**Range**   The set of all values for the dependent variable. These are the second coordinates in a set of ordered pairs and are also known as *output values*.

**Rate of Change**   The ratio of the vertical change to the horizontal change when moving from one point on the graph to another. The slope of the line connecting the two points is equal to the rate of change.

**Rate of Work**   The rate of work is the number of jobs that can be completed in a given unit of time. If one job can be completed in $t$ units of time, then the rate of work is given by $\dfrac{1}{t}$.

**Ratio**   A ratio is a comparison of two quantities, usually in the form of a quotient.

**Rational Equation**   A rational equation is a statement in which two rational expressions are set equal to each other.

**Rational Exponent of the Form** $a^{\frac{m}{n}}$   If $\dfrac{m}{n}$ is a rational number in lowest terms, $m$ and $n$ are integers such that $n \geq 2$, and $\sqrt[n]{a}$ is a real number, then $a^{\frac{m}{n}} = (\sqrt[n]{a})^m = \sqrt[n]{a^m}$.

**Rational Exponent of the Form** $a^{\frac{1}{n}}$   If $n$ is an integer such that $n \geq 2$ and if $\sqrt[n]{a}$ is a real number, then $a^{\frac{1}{n}} = \sqrt[n]{a}$.

**Rational Expression**   An expression that can be written as the quotient $\dfrac{P}{Q}$ of two polynomials $P$ and $Q$ as long as $Q \neq 0$.

**Rational Number**   A rational number is a number that can be written as a fraction $\dfrac{p}{q}$, where $p$ and $q$ are integers and $q \neq 0$.

**Rationalizing a Denominator**   The process of removing radicals so that the denominator contains only a rational number is called rationalizing the denominator.

**Real Number**   A real number is any number that is either rational or irrational. Combining the set of rational numbers with the set of irrational numbers forms the set of real numbers, represented by $\mathbb{R}$.

**Real Number Line**   A graph that represents the set of all real numbers, also known simply as the *number line*.

**Reciprocals (or Multiplicative Inverses)**   Two numbers are reciprocals, or multiplicative inverses, if their product is 1.

**Rectangular Coordinate System**   Consists of two perpendicular real number lines called the *coordinate axes*. The horizontal axis is called the *x-axis*, and the vertical axis is called the *y-axis*. The two axes intersect at a point called the *origin*. Also known as the *Cartesian coordinate system* in honor of its inventor, René Descartes.

**Relation**   A correspondence between two sets of numbers that can be represented by a set of ordered pairs.

**Remainder**   The amount that is left over after division, when the divisor can no longer divide into the dividend wholly.

**Restricted Value**   A restricted value for an algebraic expression is a value that makes the expression undefined.

**Restricted Values for Rational Expressions**   Any value that causes the denominator of a rational expression to equal zero is a restricted value for that rational expression.

**Right Angle**   A right angle is an angle that contains 90 degrees.

**Right Triangle**   A triangle with a 90 degree angle (right angle).

**Scatter Plot**   Paired data graphed as points on a coordinate plane.

**Scientific Notation**   A number is written in scientific notation if it has the form $a \times 10^n$, where $a$ is a real number, such that $1 \leq a < 10$, and $n$ is an integer.

**Simplified Algebraic Expression**   An algebraic expression is simplified if grouping symbols are removed and like terms have been combined.

**Simplified Polynomial**   A polynomial is a simplified polynomial if all of its terms are simplified and none of its terms are like terms.

**Slope**   The ratio of the vertical change in $y$, or *rise*, to the horizontal change in $x$, or *run*.

**Slope Formula**   Given two points, $(x_1, y_1)$ and $(x_2, y_2)$, on the graph of a line, the slope $m$ of the line containing the two points is given by the formula
$$m = \frac{Change\ in\ y}{Change\ in\ x} = \frac{Rise}{Run} = \frac{y_2 - y_1}{x_2 - x_1}, \text{ where } x_1 \neq x_2.$$

**Slope-Intercept Form**   A linear equation in two variables of the form $y = mx + b$ is written in slope-intercept form, where $m$ is the slope of the line and $b$ is the $y$-intercept.

**Slope of a Horizontal Line**   All horizontal lines (which have equations of the form $y = b$) have slope 0.

**Slope of a Vertical Line**   All vertical lines (which have equations of the form $x = a$) have undefined slope.

**Solution**   A value that, when substituted for a variable, makes the equation or inequality true.

**Solution Set**   The collection of *all* solutions to an equation or inequality.

**Solution Region of a System of Linear Inequalities**   The common region that all of the inequalities in a system of linear inequalities share.

**Solve**   To solve an equation or inequality means to find its solution set, or the set of all values that make the equation or inequality true.

**Solved for a Variable**   An equation is solved for a variable if the variable is by itself on one side of the equation and is the only time that the variable occurs in the equation.

**Square Both Sides**   To square both sides of an equation means to raise both sides of the equation to the second power. For example, squaring both sides of the equation $x + 2 = \sqrt{x}$, we results in $(x + 2)^2 = (\sqrt{x})^2$.

**Square Root**   The square root of a non-negative real number $a$ is a real number $b$ that, when squared, results in $a$. For $a > 0$, $\sqrt{a}$ is the positive or principal square root of $a$, and $-\sqrt{a}$ is the negative square root of $a$. Also, $\sqrt{0} = 0$.

**Square Root of a Negative Number**   For any positive real number $a$, $\sqrt{-a} = \sqrt{-1} \cdot \sqrt{a} = i\sqrt{a}$.

**Square Root Property**   If $u$ is an algebraic expression and $k$ is a real number, then $u^2 = k$ is equivalent to $u = -\sqrt{k}$ or $u = \sqrt{k}$. Equivalently, if $u^2 = k$, then $u = \pm\sqrt{k}$.

**Squaring Property of Equality**   If $A$ and $B$ represent algebraic expressions, then any solution to the equation $A = B$ is also a solution to the equation $A^2 = B^2$.

**Straight Angle**   A straight angle has a measure of 180 degrees.

**Strict Inequality**   It contains one or more of the following inequality symbols: $<, >, \neq$. There is no possibility of equality in a strict inequality.

**Substitution Method**   Involves solving one of the equations for one variable, substituting the resulting expression into the other equation, and then solving for the remaining variable.

**Sum and Difference of Two Terms Rule (Product of Conjugates Rule)**   The product of the sum and difference of two terms (conjugates) is equal to the difference of the squares of the two terms, for example, $(A + B)(A - B) = A^2 - B^2$.

**Sum**   The result of adding two real numbers. The numbers being added are called *terms*, or *addends*.

**Surface Area**   The total area on the surface of a three-dimensional object.

**System of Linear Equations in Two Variables**   A collection of two or more linear equations in two variables considered together.

**System of Linear Inequalities in Two Variables**   A collection of two or more linear inequalities in two variables considered together.

**Term**   The terms of an algebraic expression are the quantities being added. Terms containing variable factors are called *variable terms*, while terms without variables are called *constant terms*.

**Terms of the Polynomial**   The monomials that make up a polynomial.

**Test Point**   A point used to find the half-plane that contains the ordered pair solutions to a linear inequality in two variables.

**Trapezoid**   A trapezoid is a four-sided figure with two sides parallel.

**Trinomial**   A simplified polynomial with three terms.

**Uniform Motion Problem**   Problems that involve two rates, which means that there will also be two distances and two times.

**Unit Distance**   The distance from 0 to 1 represents the unit distance (the length of one unit) for the number line.

**Upper Bound**   The upper bound of an interval is the largest value in the interval. No values in the interval can be more than the upper bound. For example, the interval $[6, 9]$ has an upper bound of 9. No value in the interval is more than 9.

**Variable**   A symbol (usually a letter) that represents a changing value.

**Vertex**   The vertex is the highest or lowest point on the graph of a quadratic equation. If the parabola "opens up," the vertex is the lowest point on the graph and if the parabola "opens down," the vertex is the highest point on the graph.

**Vertical Line Test**   If a vertical line intersects (crosses or touches) the graph of a relation at more than one point, then the relation is not a function. If every vertical line intersects the graph of a relation at no more than one point, then the relation is a function.

**Volume**   The measure of space occupied by a three-dimensional object.

**Whole Numbers**   A whole number is an element of the set $\mathbb{W} = \{0, 1, 2, 3, 4, 5 \ldots \}$.

**$x$-Axis**   The horizontal axis in the rectangular coordinate system.

**$x$-Coordinate**   The first number in an ordered pair. Also called the *abscissa*.

**$x$-intercept**   is the $x$-coordinate of a point where a graph crosses or touches the $x$-axis.

**$y$-Axis**   The vertical axis in a rectangular coordinate system.

**$y$-Coordinate**   The second number in an ordered pair. Also called the *ordinate*.

**$y$-intercept**   The $y$-coordinate of a point where a graph crosses or touches the $y$-axis.

**Zero Product Property**   If $A$ and $B$ are real numbers or algebraic expressions and $A \cdot B = 0$, then or $A = 0$ or $B = 0$.

**Zero-Power Rule**   A nonzero base raised to the 0 power equals 1. That is, $a^0 = 1(a \neq 0)$.

# CHAPTER R

## R.1 Exercises

**1. a.** $\dfrac{68}{9}$   **b.** $30\dfrac{5}{8}$   **c.** $\dfrac{101}{12}$   **d.** 26   **3.** $\dfrac{2}{12}$   **5.** $\dfrac{7}{28}$   **7.** $\dfrac{3}{4}$   **9.** $\dfrac{16}{45}$   **11.** $\dfrac{11}{15}$   **13.** $\dfrac{15}{56}$

**15.** $\dfrac{21}{4}$   **17.** 20   **19.** $\dfrac{15}{16}$   **21.** $1\dfrac{1}{35}$   **23.** 16   **25.** $\dfrac{9}{16}$   **27.** $\dfrac{29}{30}$   **29.** $\dfrac{1}{2}$   **31.** $5\dfrac{5}{36}$

## R.2 Exercises

**1.** 154.21   **3.** 82.767   **5.** 21.06   **7.** 16.38   **9.** 0.0792   **11.** 8.7   **13.** 2.2   **15.** $\dfrac{57}{100}$   **17.** $\dfrac{9}{125}$

**19.** 0.375   **21.** 2.3125   **23.** 0.075, $\dfrac{3}{40}$   **25.** 36%, $\dfrac{9}{25}$   **27.** $106.\overline{6}\,\%,\ 1.0\overline{6}$

# CHAPTER 1

## 1.1 Exercises

**1.** integer and rational number   **3.** irrational number   **5.** integer and rational number   **7.** rational number

**9.**

**11.** 17   **13.** −12.45   **15.** $-4.\overline{27}$   **17.** 12.5   **19.** $\dfrac{15}{28}$   **21.** >

**23.** =   **25.** =   **27.** >   **29.** false   **31.** true   **33.** false   **35.** $45 \le 50$   **37.** $4 \ne 5$

**39.**

| percentage of teens texting friends in 9/09 | is greater than | percentage of teens texting friends in 2/08 |
|---|---|---|
| ↓ | ↓ | ↓ |
| 54 | > | 38 |

**41.**

| average high temperature in Phoenix for January (°F) | is less than | average high temperature in Phoenix for February (°F) |
|---|---|---|
| ↓ | ↓ | ↓ |
| 66 | < | 70 |

## 1.2 Exercises

**1.** 15   **3.** −71   **5.** −4.03   **7.** $-\dfrac{11}{3}$   **9.** $-\dfrac{46}{15}$   **11.** 4   **13.** −2   **15.** 0   **17.** 0   **19.** 0

**21.** 6   **23.** $\dfrac{40}{21}$   **25.** −4   **27.** $-\dfrac{1}{2}$   **29.** $\dfrac{11}{8}$   **31.** $-7 + 5$   **33.** $(3.7 + 2.1) + 8.1$   **35.** $-15 + 12$

**37.** $12 - (-13)$   **39.** $(2 - 7) + 5$   **41.** $\dfrac{5}{9} - \dfrac{2}{3}$   **43.** a loss of 181.46 points, or −181.46 points

**45.** a difference of 2319 m

## 1.3 Exercises

**1.** −6   **3.** 6   **5.** −70   **7.** 936   **9.** $\dfrac{4}{3}$   **11.** 0   **13.** $\dfrac{32}{3}$   **15.** −13   **17.** 0   **19.** 18   **21.** −7

**23.** 16   **25.** $\dfrac{5}{12}$   **27.** $\dfrac{4}{3}$   **29.** $\dfrac{25}{18}$   **31.** 0.15   **33.** −3   **35.** 0.22(18)   **37.** $\dfrac{2}{3}(-20)$

**39.** $(19 - 12)\dfrac{7}{8}$   **41.** $0.7(20 - 4) + 18$   **43.** $\dfrac{5 \text{ chaperones}}{60 \text{ students}}$   **45.** $\dfrac{3.76}{5}$   **47.** $\dfrac{(30 + 12)}{\frac{7}{8}}$   **49.** \$2.52

**51.** 39,610 people   **53.** 12 liters

## 1.4 Exercises

**1.** 256    **3.** $\dfrac{81}{256}$    **5.** 1    **7.** −81    **9.** $\dfrac{27}{125}$    **11.** −0.0001    **13.** 10    **15.** −1    **17.** $\dfrac{7}{4}$    **19.** 1

**21.** −4    **23.** 11    **25.** 54    **27.** 23    **29.** 4    **31.** $\dfrac{17}{13}$    **33.** $\dfrac{3}{2}$    **35.** $\dfrac{40}{7}$    **37.** 32    **39.** 88

**41.** 1    **43.** 9    **45.** $\dfrac{31}{5}$    **47.** 5

## 1.5 Exercises

**1.** 20    **3.** 144    **5.** $-\dfrac{41}{4}$    **7.** 38    **9.** −1    **11.** −4    **13.** $\dfrac{26}{5}$    **15.** 0    **17.** −23    **19.** −95

**21.** 29    **23.** 33    **25.** $9 + 5$    **27.** $x + 11$    **29.** $-3(8.3)$    **31.** $3 + (6 + 15)$    **33.** $7(y \cdot 4)$
**35.** $1.45 + (a + b)$    **37.** $(-6 + 6) + c$    **39.** $5 + x$    **41.** $24z$    **43.** $3a + 3b$    **45.** $8y - 20$    **47.** $32x + 12$

**49.** $-12x - 8y$    **51.** $-7 + n$    **53.** $\dfrac{3x}{4} - 4$    **55.** $0.3 + 0.68x$    **57.** $12x + 48 - 60y$    **59.** $15x + 3y - 21$

**61.** $4(x + 1)$    **63.** $\dfrac{1}{2}(y - 6)$    **65.** Identity Property of Multiplication

**67.** Inverse Property of Multiplication

## 1.6 Exercises

**1.** 2 terms, coefficients: 3 and −5    **3.** 2 terms, coefficients: $-\dfrac{3}{4}$ and −9    **5.** 3 terms, coefficients: 1, $-\dfrac{1}{2}$, and 8

**7.** $2a$ and $-4a$, $-b$ and $2b$    **9.** $6.3m$ and $-8m$, $-4.5n^2$ and $2.2n^2$    **11.** $7z$    **13.** $5x^2 + 4x - 8$
**15.** $7x + 10x^2$    **17.** $22a - 2b$    **19.** 12.3    **21.** $-1.35c^2$    **23.** $-4x^2 - x - 2$    **25.** $-4x + 15$    **27.** $19x - 8$

**29.** $7w - 17$    **31.** $22x^2 + 18x + 8$    **33.** $x + 8$    **35.** $\dfrac{10}{x} + 2$    **37.** $\dfrac{3}{5}x + 1$    **39.** $3s + 18$ cm

**41.** $3n - 1$ satellites    **43. a.** $12x + 4$ ft    **b.** 64 ft    **45. a.** $0.49x + 175$ gallons    **b.** 180.88 gallons

# CHAPTER 2

## 2.1 Exercises

**1.** No, because the algebraic expression is not equated to a second expression.
**3.** No, because the algebraic equation is nonlinear (the variable is raised to an exponent other than 1).
**5.** No, because there are two different variables in the equation.    **7.** $x = 3$ is a solution    **9.** $z = -4$ is a solution

**11.** $x = \dfrac{9}{4}$ is a solution    **13.** $a = 5$    **15.** $y = -3$    **17.** $x = 7$    **19.** $x = -\dfrac{3}{4}$    **21.** $x = 3$    **23.** $m = 5$

**25.** $w = -1$    **27.** $x = 27$    **29.** $m = 32$    **31.** $z = 14$    **33.** $x = \dfrac{5}{6}$    **35.** $w = \dfrac{17}{5}$    **37.** $x = 5$    **39.** $x = 5$

**41.** $z = 2$    **43.** $x = 8.2$    **45.** $y = 5$    **47.** $b = -2$    **49.** $z = 0$    **51.** $y = -\dfrac{18}{7}$

## 2.2 Exercises

**1.** $\{-1\}$    **3.** $\{-3\}$    **5.** $\{2\}$    **7.** $\{-2\}$    **9.** $\{0\}$    **11.** $\{1\}$    **13.** $\{-5\}$    **15.** $\{-6\}$    **17.** $\{-2\}$

**19.** $\left\{\dfrac{17}{30}\right\}$    **21.** $\{1\}$    **23.** $\left\{\dfrac{3}{2}\right\}$    **25.** $\left\{-\dfrac{11}{5}\right\}$    **27.** $\left\{-\dfrac{8}{15}\right\}$    **29.** $\{4\}$    **31.** $\{7\}$    **33.** $\{2\}$

**35.** $\{-3\}$    **37.** $\left\{\dfrac{19}{20}\right\}$    **39.** $\left\{\dfrac{1073}{230}\right\}$    **41.** $\{\varnothing\}$    **43.** $\{\mathbb{R}\}$    **45.** $\{\varnothing\}$    **47.** $N = 22$ sliders

**49.** $h \approx 57.71$ inches

## 2.3 Exercises

**1.** $x + 13 = 38$    **3.** $\dfrac{x}{4} - 5 = 31$    **5.** $\dfrac{x + 2.7}{6} = \dfrac{x}{9}$    **7.** $\dfrac{x}{2} = x - 19$    **9.** 114    **11.** 3    **13.** $-\dfrac{11}{4}$

**15.** 11    **17.** 14.4 inches, 57.6 inches    **19.** Taipei 101: 508 meters, Burj Khalifa: 828 meters

**21.** Kareem Abdul-Jabbar: 6 times, Michael Jordan: 5 times, Bill Russell: 5 times    **23.** 345, 346    **25.** 57, 59

**27.** 158, 159    **29.** 235 pounds    **31.** security guard: 25 hours, landscaper: 15 hours

## 2.4 Exercises

**1.** 350 miles    **3.** $w = 110 + 5.06(70 - 60) = 160.6$ pounds    **5.** \$15    **7.** 19.0%    **9.** 12 feet

**11.** $P = 381$ cm, $A = 7543.8$ cm$^2$    **13.** \$3675    **15.** \$24.50    **17.** \$3750    **19.** 3.5 hours    **21.** 15 feet

**23.** $r = \dfrac{C}{2\pi}$    **25.** $y = \dfrac{C - Ax}{B}$    **27.** $R = \dfrac{E}{I} - r$    **29.** $C = \dfrac{5}{9}(F - 32)$    **31.** 54.1 gallons    **33.** \$940.50

**35.** 42.6 cubic yards

## 2.5 Exercises

**1.** $w = 4$ feet, $l = 22$ feet    **3.** 2100 square yards    **5.** 15 feet, 20 feet, 25 feet    **7.** 5760 cubic inches

**9.** 20°, 70°    **11.** 30°    **13.** 34°, 73°, 73°    **15.** 50°    **17.** 0.5 hour    **19.** 148.1 miles

## 2.6 Exercises

**1.** 28    **3.** 85%    **5.** 3640    **7.** 216 pounds    **9.** 21.9%    **11.** \$349.93    **13.** 45%    **15.** 150%

**17.** 8%    **19.** \$10,196    **21.** 6.2%    **23.** 27.7%    **25.** 21.6%    **27.** 3.6 liters    **29.** 21 cups

## 2.7 Exercises

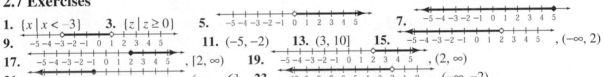

**1.** $\{x \mid x < -3\}$    **3.** $\{z \mid z \ge 0\}$    **5.**    **7.**

**9.**    **11.** $(-5, -2)$    **13.** $(3, 10]$    **15.** , $(-\infty, 2)$

**17.** , $[2, \infty)$    **19.** , $(2, \infty)$

**21.** , $(-\infty, -6]$    **23.** , $(-\infty, -2)$

**25.** , $(-\infty, 5)$    **27.** , $(12, \infty)$

**29.** , $(3, \infty)$    **31.** $(2, \infty)$    **33.** $\left(-\dfrac{9}{8}, \infty\right)$    **35.** $\varnothing$

**37.** , $\left(-\dfrac{3}{2}, 3\right)$    **39.** , $(-1, 2]$

**41.** , $[2, 7)$    **43.** , $[-3, 1)$    **45.** 159 bracelets

**47.** 105 guests

# CHAPTER 3

## 3.1 Exercises

**1. a.** 4    **b.** June    **c.** May, 20    **d.** January, February, December    **3. a.** $A$: $(3, 1)$; Quadrant I

**b.** $B$: $(5, -2)$; Quadrant IV    **c.** $C$: $(-2, 0)$; $x$-axis    **d.** $D$: $(-2, -4)$; Quadrant III

**5-9.**    $(-4, 1)$: Quadrant II; $(-6, -2)$: Quadrant III; $(3.5, 0)$: $x$-axis

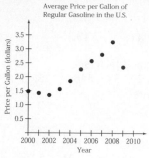

Average Price per Gallon of Regular Gasoline in the U.S.

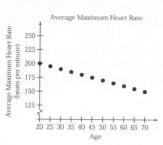

Average Maximum Heart Rate

**11.** **13.**    The scatter plot shows that average maximum heart rate steadily decreases as age increases.

## 3.2 Exercises

**1. a.** no    **b.** yes    **c.** yes    **3. a.** no    **b.** yes    **c.** no    **5. a.** $x = 0$    **b.** $y = \dfrac{3}{4}$

**7.**      **9.**      **11.**      **13.**      **15.**

**17.** the $x$-intercept is 3, the $y$-intercept is 2    **19.** the $x$-intercepts are $-2$, 0, and 1, the $y$-intercept is 0

**21.** the $x$-intercept is $\dfrac{1}{2}$, the $y$-intercept is $-2$    **23.**      **25.**      **27.**

**29.**      **31.**      **33.**      **35.**

$y = -335x + 24{,}120$

**37. a.**      **b.** 15,745    **c.** Natalie still owes \$15,745 toward the loan after 25 months.
**d.** The $y$-intercept is \$24,120, which is the total cost of the automobile at the time of purchase.
**e.** The $x$-intercept represents the number of months that are needed for the entire payment to be made.

## 3.3 Exercises

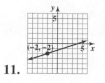 

**1.** $m = \dfrac{3}{4}$    **3.** $m = -\dfrac{6}{5}$    **5.** $m = \dfrac{1}{2}$    **7.** $m = \dfrac{4}{3}$    **9.** $m = 0$    **11.**      **13.**

  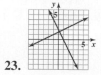

**15.**      **17.** $m_2 = \dfrac{2}{3}$, $m_3 = -\dfrac{3}{2}$    **19.** $m_2 = 6$, $m_3 = -\dfrac{1}{6}$    **21.**      **23.**

**25.** grade $= \dfrac{1}{8} = 0.125$, or 12.5%     **27.** $m = 206$, or an annual increase in tuition and fees of \$206.

**29.** $m \approx 30,518$, or an annual increase in the cases of autism among people ages 3 to 22 of 30,518.

## 3.4 Exercises

**1.** $m = -3,\ b = 9$     **3.** $m = -\dfrac{6}{5},\ b = -\dfrac{4}{5}$     **5.** $m = -\dfrac{5}{4},\ b = 0$     **7.** $m = 0,\ b = 10$

**9.**        **11.**        **13.**        **15.** $y = 4x + 8$    **17.** $y = \dfrac{1}{3}x + 4$    **19.** $y = \dfrac{5}{2}$

**21.** $y = -3x + 11$    **23.** $y = -\dfrac{2}{3}x + 9$    **25.** $y = \dfrac{2}{3}x$    **27.** $y = 1$    **29.** $y = \dfrac{3}{2}x + 3$    **31.** $y = -\dfrac{2}{9}x$

**33.** parallel    **35.** perpendicular    **37.** parallel    **39.** $y = -6x - 11$    **41.** $y = -\dfrac{1}{6}x + \dfrac{7}{3}$

**43. a.** $y = 1.3x - 0.65$    **b.** 5.59%    **45. a.** $y \approx -1.29x + 372.50$    **b.** approximately 179,000 units

## 3.5 Exercises

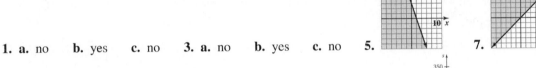

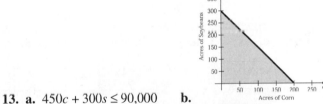

**1. a.** no   **b.** yes   **c.** no    **3. a.** no   **b.** yes   **c.** no    **5.**        **7.**

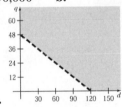

**9.**        **11.**        **13. a.** $450c + 300s \le 90,000$    **b.**

**c.** no ($93,750 > 90,000$)    **15. a.** $0.10d + 0.25q > 12.00$    **b.**        **c.** yes ($14.75 > 12.00$)

# CHAPTER 4

## 4.1 Exercises

**1. a.** yes   **b.** no    **3. a.** no   **b.** yes    **5. a.** yes   **b.** no    **7.** no solution    **9.** no solution

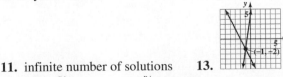       

**11.** infinite number of solutions    **13.**        **15.**        **17.**

**19.**        **21.**

## 4.2 Exercises

**1.** $(3, 7)$ **3.** $(1, 1)$ **5.** $(2, -3)$ **7.** $(3, -2)$ **9.** $(3, 1)$ **11.** $(-2, -2)$ **13.** $(0, 0)$ **15.** $\left(-\dfrac{5}{3}, \dfrac{26}{9}\right)$

**17.** $(-1, 2)$ **19.** $(2.2, -1.1)$ **21.** no solution **23.** $\left\{(x, y) \mid y = \dfrac{1}{3}x + \dfrac{2}{3}\right\}$ **25.** $x, y \mid y \quad \dfrac{1}{3}x \quad \dfrac{1}{3}$

**27.** $x, y \mid y \quad \dfrac{2}{3}x \quad \dfrac{1}{3}$

## 4.3 Exercises

**1.** $(6, 2)$ **3.** $(7, 2)$ **5.** $(5, 0)$ **7.** $(-2, 2)$ **9.** $(-1, 3)$ **11.** $(6, 5)$ **13.** $(-6, 3)$ **15.** $(-4, 7)$

**17.** $(0, 0)$ **19.** $\left(\dfrac{1}{2}, -1\right)$ **21.** $(6, -4)$ **23.** $\left\{(x, y) \mid y = \dfrac{1}{3}x - 2\right\}$ **25.** $\{\}$, or $\varnothing$

## 4.4 Exercises

**1.** Clayton: 12 years, Josh: 14 years **3.** Isaiah: 180 pounds, Geoff: 160 pounds
**5.** Statue of Liberty: 305.5 feet, Gateway Arch: 630 feet **7.** width: 75 cm, length: 124 cm
**9.** equal sides: 30 cm long, short side: 20 cm long **11.** angle 1: 25°, angle 2: 65°
**13.** swimming time: 0.5 hr, running time: 1.0 hr **15.** plane speed: 210 miles/hr, wind speed: 14 miles/hr
**17.** Coke: 161 cans, Dr. Pepper: 110 cans **19.** mango: 1.5 pounds, kiwi: 2.5 pounds
**21.** 18-carat gold: 140 ounces, 12-carat gold: 210 ounces **23.** 1 glass of milk: 11 carbs, 1 snack bar: 13 carbs
**25.** 1 cheeseburger with onion: 480 calories, 1 order of french fries: 400 calories

## 4.5 Exercises

**1. a.** yes **b.** yes **c.** no **3. a.** yes **b.** no **c.** yes **5.** **7.**

**9.** **11.** **13.** **15.** **17.**

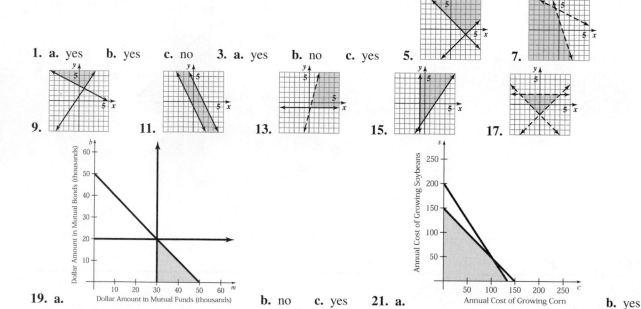

**19. a.** **b.** no **c.** yes **21. a.** **b.** yes
**c.** no

# CHAPTER 5

## 5.1 Exercises

**1.** $7^6$    **3.** $w^6$    **5.** $30a^{10}$    **7.** $18m^{13}n^6$    **9.** $b^4$    **11.** $(-10)^3 = -1000$    **13.** $4q^{12}$    **15.** $x^2y^6z$

**17.** $1$    **19.** $-1$    **21.** $1$    **23.** $x^{24}$    **25.** $(-2)^{12}$    **27.** $p^6q^6$    **29.** $a^{21}b^7$    **31.** $49r^6s^{18}$    **33.** $\dfrac{a^8}{b^8}$

**35.** $\dfrac{t^3}{4^3} = \dfrac{t^3}{64}$    **37.** $\dfrac{u^{35}}{2^7} = \dfrac{u^{35}}{128}$    **39.** $x^{26}$    **41.** $4p^{12}$    **43.** $\dfrac{x^{12}}{y^7}$

## 5.2 Exercises

**1.** trinomial    **3.** none of these    **5.** none of these    **7.** coefficient: 4; degree: 4    **9.** coefficient: 1; degree: 7
**11.** coefficient: –9; degree: 1    **13.** $5x + 4$; degree: 1; leading coefficient: 5    **15.** $-3x + 9$; degree: 1; leading
coefficient: –3    **17.** $4.1x^3 + x^2 - 6.7x + 3.8$; degree: 3; leading coefficient: 4.1    **19.** 7    **21.** –7    **23.** 7.44
**25.** 39    **27.** $-5x + 5$    **29.** $16x + 1$    **31.** $1.3x^2 + 3.2x - 9.1$    **33.** $7x^3 - 8$

## 5.3 Exercises

**1.** $6x + 14$    **3.** $3m^2 + 10$    **5.** $8p^3 - 2p^2 - 5p - 11$    **7.** $1.7w^2 - 3.5w - 0.9$    **9.** $-7x^2 + 24$
**11.** $8z^3 - 3z^2 - 4z + 13$    **13.** $3y + 22$    **15.** $5t^2 + 2t - 4$    **17.** $-5x^3 + 12x^2 - 2x$    **19.** $2.5n^2 - 3.6n + 5.9$

## 5.4 Exercises

**1.** $6x^8$    **3.** $6y^{14}$    **5.** $\dfrac{4}{5}x^5$    **7.** $15x^2 + 10x$    **9.** $3x^3 - 3x^2 + 3x$    **11.** $-3x^6 - 6x^4 + 21x^3$

**13.** $x^2 + 7x + 10$    **15.** $y^2 - \dfrac{11}{15}y - \dfrac{4}{5}$    **17.** $x^2 + 7x - 18$    **19.** $56z^2 + 27z - 18$    **21.** $x^3 - 27$

**23.** $2x^2 - x^2 + x + 1$    **25.** $2x^5 + 7x^3 - 3x^3 + 3x - 9$    **27.** $x^5 + x^4 - 2x^3 + 2x - 1$    **29.** $30x^3 + 23x^3 - 30x$

## 5.5 Exercises

**1.** $y^2 + 4y + 4$    **3.** $x^2 + 7x + 12.25$    **5.** $\dfrac{9}{16}z^2 + \dfrac{3}{2}z + 1$    **7.** $36x^2 + \dfrac{36}{5}x + \dfrac{9}{25}$    **9.** $x^2 - 12x + 36$

**11.** $z^4 - 8.4z^2 + 17.64$    **13.** $64 - 16x + x^2$    **15.** $\dfrac{25}{4}z^2 - 20z + 16$    **17.** $x^2 - 81$    **19.** $9z^2 - 16$

**21.** $4.41 - 9y^2$    **23.** $16x^4 - 49$

## 5.6 Exercises

**1.** $\dfrac{1}{y^3}$    **3.** $\dfrac{6}{w^5}$    **5.** 64    **7.** $\dfrac{1}{2^3} + \dfrac{1}{4} = \dfrac{3}{8}$    **9.** $z^5$    **11.** $(-3)^4 = 81$    **13.** $-12t^3$    **15.** $\dfrac{2^5}{9^2} = \dfrac{32}{81}$

**17.** $\dfrac{32}{x^4}$    **19.** $5t^6$    **21.** $\dfrac{x^6}{216y^{15}}$    **23.** $\dfrac{9b^2}{625a^4c^2}$    **25.** $\dfrac{q^{56}}{p^{16}}$    **27.** $\dfrac{9a^{27}}{b^{23}}$    **29.** $8.9 \times 10^{10}$

**31.** $4.06 \times 10^{-7}$    **33.** $2.54 \times 10^4$    **35.** $5 \times 10^{-9}$    **37.** 0.00000297    **39.** 0.0000000008
**41.** 5,878,000,000,000    **43.** 0.000053    **45.** $6.9 \times 10^{15}$    **47.** $3 \times 10^5$

## 5.7 Exercises

**1.** $6x^8$    **3.** $\dfrac{1}{6}$    **5.** $2x - 5$    **7.** $8y^5 - 4y^3 + 1$    **9.** $2a - 7 - \dfrac{3}{a}$    **11.** $x + 12$    **13.** $x^2 - 5x + 7$

**15.** $x^2 - 3x + 6 - \dfrac{34}{x+5}$    **17.** $3x + 5 + \dfrac{5}{5-2x}$    **19.** $4m^2 - 8m + 21 - \dfrac{49}{m+2}$

## 5.8 Exercises

**1.** 3, 7; –4, 4; 12, 2; –5, 0; 7 **3.** 6, 5; –2.5, 5; 7, 8; –4.3, 0; 8 **5.** –7 **7.** $-18a+18b$
**9.** $-2m^2+16mn+6n$ **11.** $-3x-2xy+2y$ **13.** $3x^3-3xy-4$ **15.** $-6mn^2+m+4n$ **17.** $3x^3y-9x^2y$
**19.** $3x^3y+3x^2y^2$ **21.** $15x^2+16xy-15y^2$ **23.** $0.6a^3-0.08a^2b^2+9ab-1.2b^3$ **25.** $a^2+6ab+9b^2$
**27.** $16m^2-8mn+n^2$ **29.** $25x^2-20xy+4y^2$ **31.** $2x^3+5x^2y-5xy^2+y^3$ **33.** $-6x^4+6x^3y-2x^2y+2xy^2$

# CHAPTER 6

## 6.1 Exercises

**1.** 7 **3.** 9 **5.** $x^2$ **7.** $a^3b$ **9.** $6x^5$ **11.** 5 **13.** $18mn^4$ **15.** $6(5x+2)$ **17.** $4w^3(5w-7)$
**19.** $4y^3(y^2-4y+7)$ **21.** $5ab^3(5b^2-2ab+a^2)$ **23.** $-8x(x-3)$ **25.** $-3(3x^2+9x-4)$
**27.** $(y-4)(5x+8)$ **29.** $(m-2n)(m-3n)$ **31.** $(y+3)(5x+6)$ **33.** $(3b-4)(3a-2)$
**35.** $(6x-11y)(x-1)$ **37.** $(x+6)(x^2+7)$ **39.** $(x-5)(x^2+1)$

## 6.2 Exercises

**1.** $(x+7)(x+3)$ **3.** $(w+9)(w+8)$ **5.** $(n-7)(n-2)$ **7.** $(p+9)(p-6)$ **9.** prime
**11.** $(x-18)(x-6)$ **13.** $(x-33)(x+2)$ **15.** $(w+12)(w-8)$ **17.** $(x+2y)(x+4y)$
**19.** $(a-3b)(a-8b)$ **21.** $5(x+2)(x+8)$ **23.** $2x(x-15)(x+9)$ **25.** $20(m-25)(m-1)$
**27.** $-(x-3)(x+1)$ **29.** $-(x-5y)(x-2y)$

## 6.3 Exercises

**1.** $(3x+1)(x+5)$ **3.** $(5x+2)(x+4)$ **5.** $(3n-7)(2n-1)$ **7.** $(12t-7)(t+1)$ **9.** prime
**11.** $(10p+1)(2p-1)$ **13.** $(2w-3)(w+13)$ **15.** $(5t-3)(3t-8)$ **17.** $(z-6)(4z+3)$
**19.** $(2x+3)(6x-1)$ **21.** $(7y-8)(y+1)$ **23.** $(2w-9)(w-3)$ **25.** $(2m-3)(12m+5)$
**27.** $(3x+y)(3x-8y)$ **29.** $(2a-7b)(a-10b)$

## 6.4 Exercises

**1.** $(5x+2)(x+3)$ **3.** $(4t-5)(t+3)$ **5.** $(3p-2)(p-8)$ **7.** $(2w-11)(w-3)$ **9.** $(10y-3)(y+5)$
**11.** $(2q-1)(9q-2)$ **13.** $(5x+1)(4x+5)$ **15.** $(2t-1)(4t+5)$ **17.** $(z-4)(6z+5)$ **19.** prime
**21.** $(x+2y)(4x+3y)$ **23.** $(2a+3b)(7a-3b)$ **25.** $x^2(x+5)(10x-6)$ **27.** $-(m+4)(3m-2)$
**29.** $-3ab(12a-49+4b)$

## 6.5 Exercises

**1.** $(y-4)(y+4)$ **3.** $(x-y)(x+y)$ **5.** $(x^5-2)(x^5+2)$ **7.** $(3y-1)(3y+1)$ **9.** $(4-5m)(4+5m)$
**11.** $3(2x-3)(2x+3)$ **13.** $5(4x^2+9)$ **15.** $3y(x-2)(x+2)$ **17.** $(x^2+1)(x+1)(x-1)$
**19.** $(x^2+y^2)(x+y)(x-y)$ **21.** $5(x^4+1)(x^2+1)(x+1)(x-1)$ **23.** $(x-1)^2$ **25.** $(2x+7)^2$
**27.** prime **29.** $(m-7n)^2$ **31.** $x(5y+7)^2$ **33.** $(z^3+5)^2$ **35.** $(y-5)(y^2+5y+25)$
**37.** $(2x-1)(4x^2+2x+1)$ **39.** prime **41.** $(xy-3)(x^2y^2+3xy+9)$ **43.** $(x-1)(x^2+x+1)(x^6+x^3+1)$

## 6.6 Exercises

**1.** $3(x-2)(x+2)$    **3.** $(a+2b)(a^2-2ab+4b^2)$    **5.** $(x+3)(x-2)$    **7.** $5(z-3)^2$    **9.** $(2x-5)^2$

**11.** $x(5x-1)(x+3)$    **13.** prime    **15.** $(3x+2)^2$    **17.** $-3(p-2q)(2p+q)$    **19.** $9y(1+2y)$

**21.** $(5m+4n)^2$    **23.** $(b+2a)(b-a)(b+a)$    **25.** $(x^2+5)(x-2)(x+2)$    **27.** $3y(2y+1)(8y-3)$

**29.** $(x-3)(x+3)(x-1)(x+1)$

## 6.7 Exercises

**1.** $\{-2,1\}$    **3.** $\left\{-\dfrac{1}{7},\dfrac{1}{3}\right\}$    **5.** $\{2,3\}$    **7.** $\{-3,-2\}$    **9.** $\{-7,-2\}$    **11.** $\left\{2,\dfrac{1}{3}\right\}$    **13.** $\{-5,5\}$

**15.** $\left\{-\dfrac{4}{3}\right\}$    **17.** $\left\{-3,\dfrac{1}{2}\right\}$    **19.** $\{-3,5\}$    **21.** $\{-3,8\}$    **23.** $\left\{\dfrac{5}{4}\right\}$    **25.** $\{-12,1,3\}$    **27.** $\{-2,7,8\}$

**29.** $\{-2,0,1\}$    **31.** $\left\{-2,-\dfrac{3}{7},2\right\}$    **33.** $\left\{-\dfrac{5}{2},-\dfrac{1}{3},\dfrac{5}{2}\right\}$    **35.** $\{1,2,3\}$    **37.** $\left\{-2,-\dfrac{5}{3},5\right\}$    **39.** $\{-1,1,2\}$

## 6.8 Exercises

**1.** The room numbers are 80 and 82.    **3.** The numbers are 7, 8, and 9.    **5.** 20% of community colleges provide access to grades through mobile devices, and 16% provide access to transcripts.    **7.** The width of the frame is 4 inches.    **9.** The sail has a base of 8 yards and a height of 12 yards.    **11.** The corner post is 7.5 feet tall. **13.** The hypotenuse is 20.    **15.** It will take the object 3 seconds.    **17.** The ball will hit the ground in 3 seconds.    **19.** The city had a population of 7000 in 2006.    **21.** There are 42 occupied units.

# CHAPTER 7

## 7.1 Exercises

**1.** 4    **3.** $\dfrac{1}{2}$    **5.** $-10$    **7.** 1    **9.** $\dfrac{2}{19}$    **11.** $x \neq 8$    **13.** $t \neq \dfrac{4}{7}$    **15.** $p \neq -2, 3$    **17.** $y \neq -\dfrac{5}{3}, 6$

**19.** No restricted values    **21.** $\dfrac{3}{5}$    **23.** $\dfrac{6}{y+6}$    **25.** $\dfrac{z-5}{z+7}$    **27.** $\dfrac{t+8}{2t-5}$    **29.** $\dfrac{1}{x-4}$    **31.** $\dfrac{x-8y}{x-4y}$

**33.** $\dfrac{1}{x+2}$    **35.** $-1$    **37.** $-\dfrac{w+2}{w+8}$    **39.** $\dfrac{n^2+7}{n-5}$

## 7.2 Exercises

**1.** $\dfrac{5b}{a}$    **3.** $\dfrac{z}{y^2}$    **5.** $\dfrac{5x}{21}$    **7.** $\dfrac{3}{5}$    **9.** $\dfrac{1}{(x+2)(x-1)}$    **11.** $\dfrac{3r(r+4)(r-4)}{r+2}$    **13.** $\dfrac{x+1}{x-1}$

**15.** $\dfrac{(x+1)(3x-2)}{(3x+2)(x-1)}$    **17.** 1    **19.** $\dfrac{x-2}{x+1}$    **21.** $\dfrac{x-2}{(x+2)(x+1)}$    **23.** $-\dfrac{5x+1}{2x+1}$    **25.** 1    **27.** $\dfrac{x+3y}{3x-y}$

**29.** $\dfrac{y}{x^2+xy+y^2}$    **31.** $\dfrac{5xy}{2}$    **33.** $\dfrac{(3x+2)(x+3)}{(x+1)(2x+3)}$    **35.** $\dfrac{x^2}{21(x-1)}$    **37.** $\dfrac{1}{3x}$

**39.** $\dfrac{(x+2)(x+3)}{(x^2+3x+9)(x^2+2x+4)}$    **41.** $(x+3)(x-1)$    **43.** $\dfrac{(2x+3)(3x+2)}{(x+2)(x+5)}$    **45.** $\dfrac{3(h+4)(h-2)}{(7h-2)(6h+1)}$

**47.** $\dfrac{(x-y)}{2}$    **49.** $\dfrac{15x^3 y}{16}$    **51.** $\dfrac{x+4}{x-2}$    **53.** $\dfrac{x+4}{x-2}$

## 7.3 Exercises

**1.** $18x^5$    **3.** $14a^3b^5c^3$    **5.** $12x(2x-1)(3x+1)$    **7.** $-(x-3)(x+2)$    **9.** $-(x-3)(x-1)(x+3)$

**11.** $(y-3)(y+3)^2$    **13.** $(2x+1)(x-3)(3x-2)$    **15.** $6x^2(3x-2)(x+5)(2x+1)$

**17.** $(2x+3)(x-1)(3x-2)(x+5)$    **19.** $\dfrac{9x^2y^2}{15x^4y^5}$    **21.** $\dfrac{(x+1)(x-1)}{2(x-1)(x+3)}$    **23.** $\dfrac{(3x-5)(x+2)}{(2x+3)(x+2)}$

**25.** $\dfrac{-(x+1)(x-1)(x-2)}{3x(x-1)(x-2)(x-3)}$    **27.** $\dfrac{5(3x-5)(x-2)}{10(x-3)(x-2)(x+1)}$

## 7.4 Exercises

**1.** $\dfrac{6x}{7}$    **3.** $\dfrac{a^2+b^2+3}{5c^2}$    **5.** $\dfrac{x}{3a^3}$    **7.** $\dfrac{3x+2}{2x+3}$    **9.** $-\dfrac{4}{x}$    **11.** $\dfrac{2}{x+3}$    **13.** $\dfrac{x+3}{x-3}$    **15.** $\dfrac{x+1}{x+2}$

**17.** $\dfrac{13}{6x}$    **19.** $\dfrac{8x}{(x-2)(x+2)}$    **21.** $\dfrac{11x^2-5x-10}{10x(x-2)}$    **23.** $\dfrac{x^2+4x+10}{2x(x+5)}$    **25.** $\dfrac{1}{x(x-1)(x+1)}$

**27.** $\dfrac{2x^2+5x+2}{(2x-5)(x+2)}$    **29.** $\dfrac{6}{x-3}$    **31.** $\dfrac{3(2x+3)}{(x-2)(x+2)}$    **33.** $\dfrac{2x-1}{2x+1}$    **35.** $\dfrac{x^2}{(x-2)(x-1)(x+1)}$

**37.** $\dfrac{5}{x+2}$    **39.** $\dfrac{x-3}{x-1}$    **41.** $\dfrac{x^2-2x-1}{x(x+1)(x-1)}$    **43.** $\dfrac{(2x+3)(x-15)}{(x+6)^2(2x-9)}$    **45.** $\dfrac{2}{x+2}$    **47.** $\dfrac{4}{x}$

## 7.5 Exercises

**1.** $6$    **3.** $\dfrac{5}{(x-3)(x-1)}$    **5.** $-\dfrac{x}{2x-3}$    **7.** $\dfrac{2(x+2)}{3(2x-1)}$    **9.** $\dfrac{3}{(x+1)(x+4)}$    **11.** $-\dfrac{3x-13}{5x-18}$    **13.** $\dfrac{7x+1}{7x-1}$

**15.** $\dfrac{1}{(x-6)(x+5)}$    **17.** $\dfrac{x-4}{4x}$    **19.** $\dfrac{3}{x-5}$    **21.** $\dfrac{x-1}{x-7}$    **23.** $\dfrac{6x}{x^2+9}$    **25.** $\dfrac{6}{x-y}$

## 7.6 Exercises

**1.** No, because $\dfrac{\sqrt{x^2-4x+1}}{x-5}$ is not a rational expression.    **3.** Yes    **5.** $\left\{\dfrac{15}{2}\right\}$    **7.** $\{12\}$    **9.** $\left\{\dfrac{5}{21}\right\}$

**11.** $\{-15,-4\}$    **13.** $\{-12\}$    **15.** $\{\ \}$    **17.** $\{12\}$    **19.** $\{2\}$    **21.** $C=\dfrac{ET}{R}$    **23.** $s=\dfrac{x-m}{z}$

**25.** $x_2=\dfrac{y_2-y_1+mx_1}{m}$    **27.** $B=\dfrac{E+D}{R+1}$

## 7.7 Exercises

**1.** The inspector can expect to find 240 defective units.    **3.** There are 120 grams of fat in a quart of the ice cream.
**5.** He will need 2020 more pounds of corn seed.    **7.** 15 cm    **9.** The lighthouse is 63 meters tall.
**11.** One circuit has a resistance of 12 ohms and the other circuit has a resistance of 36 ohms.
**13.** The current's speed was 3 mph.    **15.** Mae's walking speed is 1.5 mph.
**17.** It will take 18 minutes for both copy machines to copy the final exams.
**19.** It would take Shawn 6 hours to stain the deck by himself.    **21.** It will take 450 minutes to fill the tub.

## 7.8 Exercises

**1. a.** $y = 3.5x$    **b.** $y = 24.5$    **3. a.** $y = 24x^2$    **b.** $y = 600$    **5. a.** $p = \dfrac{5}{8}n^3$    **b.** $p = 625$

**7.** She will be 40 feet below the surface.    **9.** 900 gallons could be pumped through the pipe.    **11. a.** $y = \dfrac{600}{x}$

**b.** $y = 20$    **13. a.** $n = \dfrac{518.4}{m^2}$    **b.** $p = 32.4$    **15.** The car will be 9 years old.

**17.** The satellite will weigh approximately 79.01 kg.

## CHAPTER 8

### 8.1 Exercises

**1.** 6    **3.** Not a real number    **5.** 0.4    **7.** 4.494    **9.** $4x+1$    **11.** $7x$    **13.** $-13x^2$    **15.** $2y-5$
**17.** $-7$    **19.** 0.5    **21.** $-3x^4$    **23.** $-5$    **25.** $-4$    **27.** $x+5$    **29.** $2x^{15}$    **31.** 3.816    **33.** 2.740

### 8.2 Exercises

**1.** $2\sqrt{11}$    **3.** $\sqrt{51}$ is already simplified    **5.** $35\sqrt{3}$    **7.** $7\sqrt{14}$ is already simplified    **9.** $\dfrac{2\sqrt{6}}{11}$    **11.** $\dfrac{8}{5}$

**13.** $x^4\sqrt{x}$    **15.** $4m^3\sqrt{3}$    **17.** $\dfrac{3\sqrt{5}}{b^4}$    **19.** $\dfrac{3x^3\sqrt{3x}}{y^{12}}$    **21.** $3\sqrt[3]{2}$    **23.** $3\sqrt[4]{2}$    **25.** $2\sqrt[5]{4}$

### 8.3 Exercises

**1.** $13\sqrt{10}$    **3.** $5\sqrt{3}+4\sqrt{6}$    **5.** $7\sqrt{11}-11\sqrt[3]{5}$    **7.** $6\sqrt{a}+2\sqrt{b}-a\sqrt{3}$    **9.** $-7x\sqrt{11}$    **11.** $10\sqrt{2}$
**13.** $8\sqrt{2}$    **15.** $\sqrt{5}$    **17.** $5\sqrt{3}+6\sqrt{5}$    **19.** $-11\sqrt{7}-24\sqrt{2}$    **21.** $\sqrt[3]{9}$    **23.** $4\sqrt[3]{5}$    **25.** $11\sqrt[3]{2}+6$
**27.** $24x\sqrt{5x}$    **29.** $5t\sqrt{t}+6t$    **31.** $6x\sqrt[3]{5}+2x\sqrt[3]{45}-x^2$

### 8.4 Exercises

**1.** $\sqrt{35}$    **3.** $5x$    **5.** $48\sqrt{5}$    **7.** $6x^2y^2\sqrt{10y}$    **9.** $30x^2y$    **11.** $2x\sqrt{5}+\sqrt{15}$
**13.** $2m\sqrt{3m}-6m\sqrt{15}$    **15.** $29-13\sqrt{3}$    **17.** $x+10\sqrt{x}+24$    **19.** $12-4\sqrt{x}-3\sqrt{y}+\sqrt{xy}$    **21.** $7z-9$

**23.** $9-x$    **25.** $4x+20\sqrt{x}+25$    **27.** 3    **29.** $2x^4\sqrt{10}$    **31.** $-2m\sqrt[3]{3}$    **33.** $\dfrac{2\sqrt{6}}{3}$    **35.** $\dfrac{3\sqrt{2x}}{2x}$

**37.** $\dfrac{\sqrt[3]{18x}}{2x}$    **39.** $\dfrac{4\sqrt{3x}}{3x^2y^2}$    **41.** $10+5\sqrt{3}$    **43.** $\dfrac{3-3\sqrt{x}}{x-1}$    **45.** $\dfrac{m+3\sqrt{m}-28}{m-49}$

### 8.5 Exercises

**1.** $\{144\}$    **3.** $\{5\}$    **5.** $\{4\}$    **7.** $\{44\}$    **9.** $\{6\}$    **11.** $\{\ \}$ or $\varnothing$    **13.** $\left\{\dfrac{5}{4}\right\}$    **15.** $\{4\}$    **17.** $\{4\}$

**19.** $\{5\}$    **21.** $\{1\}$    **23.** $\{64\}$    **25.** $\{5\}$    **27.** $3\sqrt{10}$ m    **29.** $4\sqrt{2}$    **31. a.** The SMOG grade for
this sample is a tenth-grade reading level.    **b.** 36 words with three or more syllables are needed.
**33. a.** The person's body surface area is about 2.12 square meters.    **b.** The woman weighs about 64.5 kilograms.
**35. a.** His hang time is about 0.96 second.    **b.** The vertical distance of his jump was about 5.4 feet.

## 8.6 Exercises

**1.** $\sqrt{36}=6$ **3.** $-\sqrt[4]{81}=-3$ **5.** $\sqrt[5]{7xy}$ **7.** $\left(\sqrt{100}\right)^3=1000$ or $\sqrt{100^3}=1000$ **9.** $\left(\sqrt[4]{-16}\right)^3$ or $\sqrt[4]{(-16)^3}$

**11.** $\left(\sqrt[3]{\dfrac{1}{8}}\right)^5=\dfrac{1}{32}$ or $\sqrt[3]{\left(\dfrac{1}{8}\right)^5}=\dfrac{1}{32}$ **13.** $10^{\frac{1}{2}}$ **15.** $\left(\dfrac{2x}{y}\right)^{\frac{1}{5}}$ **17.** $\left(3xy^3\right)^{\frac{4}{5}}$ **19.** $\dfrac{1}{25^{\frac{1}{2}}}=\dfrac{1}{5}$

**21.** $\left(\dfrac{8}{27}\right)^{\frac{2}{3}}=\dfrac{4}{9}$ **23.** $9^{\frac{1}{2}}=3$ **25.** $x^{\frac{3}{2}}$ **27.** $\dfrac{2x^2}{y^{\frac{5}{2}}}$ **29.** $\sqrt[20]{x^{13}}$ **31.** $\sqrt{x}$ **33.** $\sqrt{3}$

## 8.7 Exercises

**1.** $i$ **3.** $i$ **5.** $-1+i$ **7.** $10-11i$ **9.** $7-3i$ **11.** $4$ **13.** $-1+i$ **15.** $6+8i$ **17.** $20-9i$

**19.** $7+17i$ **21.** $32-24i$ **23.** $2$ **25.** $6$ **27.** $\dfrac{2}{25}-\dfrac{11}{25}i$ **29.** $\dfrac{3}{4}+\dfrac{3}{4}i$ **31.** $-\dfrac{3}{5}-\dfrac{2}{5}i$

**33.** $-7+6i$ **35.** $-6$ **37.** $4$ **39.** $-\dfrac{1}{2}-\dfrac{3}{2}i$

# CHAPTER 9

## 9.1 Exercises

**1.** $\{-16,16\}$ **3.** $\left\{-\dfrac{5}{8},\dfrac{5}{8}\right\}$ **5.** $\{-11,11\}$ **7.** $\left\{-2\sqrt{7},2\sqrt{7}\right\}$ **9.** $\{-18,32\}$ **11.** $\left\{-3-3\sqrt{6},-3+3\sqrt{6}\right\}$

**13.** $\left\{-\sqrt{5},\sqrt{5}\right\}$ **15.** $\left\{-\dfrac{\sqrt{15}}{5},\dfrac{\sqrt{15}}{5}\right\}$ **17.** $\{-7,1\}$ **19.** $\left\{11-5\sqrt{7},11+5\sqrt{7}\right\}$ **21.** $\{-2i,2i\}$

**23.** $\left\{-2i\sqrt{2},2i\sqrt{2}\right\}$ **25.** $\left\{1-i\sqrt{5},1+i\sqrt{5}\right\}$ **27.** $\left\{2-4i\sqrt{2},2+4i\sqrt{2}\right\}$ **29.** $\{15-12i,15+12i\}$

**31.** $4\sqrt{13}\approx14.42$ **33.** The height is approximately 13.7 feet. **35.** The salmon's girth is approximately 15.8 inches. **37.** The stone will take 2.5 seconds to hit the water. **39.** The circle's radius is 15 feet.

## 9.2 Exercises

**1.** $x^2+8x+16=(x+4)^2$ **3.** $z^2+3z+\dfrac{9}{4}=\left(z+\dfrac{3}{2}\right)^2$ **5.** $\{-6,4\}$ **7.** $\left\{7-2\sqrt{10},7+2\sqrt{10}\right\}$

**9.** $\left\{3-\sqrt{6},3+\sqrt{6}\right\}$ **11.** $\left\{-\dfrac{7}{2}-\dfrac{\sqrt{7}}{2}i,-\dfrac{7}{2}+\dfrac{\sqrt{7}}{2}i\right\}$ **13.** $\left\{-3-\sqrt{11},-3+\sqrt{11}\right\}$ **15.** $\left\{-1-\sqrt{2},-1+\sqrt{2}\right\}$

**17.** $\left\{-2-\sqrt{3},-2+\sqrt{3}\right\}$ **19.** $\left\{\dfrac{3-\sqrt{65}}{4},\dfrac{3+\sqrt{65}}{4}\right\}$ **21.** $\left\{\dfrac{9-\sqrt{65}}{8},\dfrac{9+\sqrt{65}}{8}\right\}$ **23.** $\{6-2i,6+2i\}$

**25.** $\{2-3i,2+3i\}$ **27.** $\left\{-\dfrac{4}{5}-\dfrac{2}{5}i,-\dfrac{4}{5}+\dfrac{2}{5}i\right\}$

## 9.3 Exercises

**1.** $\{-9, 16\}$    **3.** $\left\{\dfrac{5}{2}, 8\right\}$    **5.** $\left\{\dfrac{-3-\sqrt{19}}{2}, \dfrac{-3+\sqrt{19}}{2}\right\}$    **7.** $\left\{\dfrac{-1-\sqrt{13}}{2}, \dfrac{-1+\sqrt{13}}{2}\right\}$    **9.** $\left\{-2-\sqrt{3}, -2+\sqrt{3}\right\}$

**11.** $\left\{\dfrac{1-\sqrt{10}}{3}, \dfrac{1+\sqrt{10}}{3}\right\}$    **13.** $\left\{\dfrac{3-\sqrt{15}}{3}, \dfrac{3+\sqrt{15}}{3}\right\}$    **15.** $\left\{-3, \dfrac{1}{4}\right\}$    **17.** $\left\{\dfrac{5-i\sqrt{55}}{2}, \dfrac{5+i\sqrt{55}}{2}\right\}$

**19.** $\left\{\dfrac{-3-i\sqrt{11}}{10}, \dfrac{-3+i\sqrt{11}}{10}\right\}$    **21.** $\{2-i, 2+i\}$    **23.** $D = 0$, so there is exactly one real solution.

**25.** $D = -256$, so there are two non-real solutions.    **27.** $D = 49$, so there are two real solutions.
**29.** The rock will take 2 seconds to hit the water.    **31.** The vertical rise is approximately 3.76 feet.

**33.** The rectangle is $3\dfrac{1}{3}$ centimeters wide and 10 centimeters long.

## 9.4 Exercises

**1. a.** $(2, 4)$    **b.** $x = 2$    **c.** $(0, 0)$    **d.** $(0, 0), (4, 0)$    **3. a.** $(2, 0)$    **b.** $x = 2$    **c.** $(0, 4)$    **d.** $(2, 0)$

**5.** $(-7, 0), (3, 0), (0, -21)$    **7.** $\left(-\dfrac{5}{3}, 0\right), \left(\dfrac{1}{3}, 0\right), (0, -5)$    **9.** $\left(-1-\sqrt{5}, 0\right), \left(-1+\sqrt{5}, 0\right), (0, -4)$

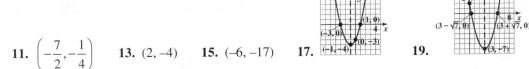

**11.** $\left(-\dfrac{7}{2}, -\dfrac{1}{4}\right)$    **13.** $(2, -4)$    **15.** $(-6, -17)$    **17.**    **19.**

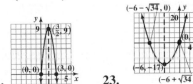

**21.**    **23.**    **25.** 160 jackets should be produced for a minimum daily cost of $2520.

# APPENDICES

## Appendix A Exercises

**1.** $x$ is independent variable; $y$ is dependent variable    **3.** $x$ is independent variable; $y$ is dependent variable
**5.** Domain: $\{0, 1, 4\}$; Range: $\{-2, -1, 0, 1, 2\}$    **7.** Domain: $\mathbb{R}$ or $(-\infty, \infty)$; Range: $[3, \infty)$    **9.** Function
**11.** Function    **13.** Function    **15.** Not a function    **17.** Function    **19.** Not a function    **21.** Function
**23.** Not a function    **25.** $f(x) = |2x - 5|$    **27.** $f(x) = -\sqrt{x} + 1$    **29.** $f(x) = -\dfrac{3}{4}x + 3$
**31.** $f(x) = 2x^2 - 5$    **33.** $f(x) = 2x^2 + 3$    **35.** 13    **37.** 1    **39.** 5    **41.** 54

## Appendix B Exercises

**1.** 24    **3.** 368    **5.** 40    **7.** 5.6    **9.** 214 mi    **11.** 4    **13.** 23    **15.** No mode
**17. a.** Mean: 78; Median: 78; Mode: 78    **b.** Mean: 74.25; Median: 78; Mode: 78; using the incorrect value decreased the mean but did not affect the other two measures.    **19. a.** Mean: 143.67; Median: 135; No mode
**b.** Mean: 130.3; Median: 117.5; No mode

# Index